120051

SCOTT, FORESMAN
PRECALCULUS

About the Cover
The picture below, and on the cover, shows the atrium of a hotel in Chicago.

SCOTT, FORESMAN
PRECALCULUS

Douglas E. Crabtree

Frank M. Eccles

Nathaniel B. Smith

Scott, Foresman and Company
Editorial Offices: Glenview, Illinois

Regional Offices: Sunnyvale, California •
Tucker, Georgia • Glenview, Illinois •
Oakland, New Jersey • Dallas, Texas

Authors

Douglas E. Crabtree
Dr. Crabtree is an instructor of mathematics and Chairman, Department of Mathematics, Phillips Academy, Andover, Massachusetts. He has had extensive teaching experience at both high school and college levels, and has had a number of articles published in mathematics journals.

Frank M. Eccles
Mr. Eccles is an instructor of mathematics, Phillips Academy, Andover, Massachusetts. He is Director of the Andover-Dartmouth Urban Teachers Institute, and has written two high school mathematics textbooks.

Nathaniel B. Smith
Mr. Smith is an instructor of mathematics, Phillips Academy, Andover, Massachusetts. He has had extensive teaching experience at high school and college levels both in the U.S.A. and overseas.

Mathematical Adviser

Edward D. Gaughan
Professor of Mathematical Sciences
New Mexico State University
Las Cruces, New Mexico

Reader/Consultant

Sidney Sharron
Supervisor in the Educational Communications and Media Branch
Los Angeles City Unified School District
Los Angeles, California

ISBN: 0-673-23434-7

Copyright © 1987, 1984,
Scott, Foresman and Company, Glenview, Illinois
All Rights Reserved.
Printed in the United States of America.

This publication is protected by Copyright and permission should be obtained from the publisher prior to any prohibited reproduction, storage in a retrieval system, or transmission in any form or by any means, electronic, mechanical, photocopying, recording, or otherwise. For information regarding permission, write to: Scott, Foresman and Company, 1900 East Lake Avenue, Glenview, Illinois 60025.

1 2 3 4 5 6 7 8 9 10—VHJ—92 91 90 89 88 87 86

Contents

1 Algebra Review — x

- 1-1 Numbers and Operations — 1
- 1-2 The Field Axioms — 5
- 1-3 Subtraction and Division — 9
- 1-4 Order — 15
- 1-5 Radicals — 22
- 1-6 Absolute Value and Exponents — 28
- 1-7 Quadratic Equations — 33
- *Special* The Binomial Theorem — 39
- 1-8 Complex Numbers — 40

Chapter Review Exercises 45   *Chapter Test* 47

2 Analytic Geometry—Lines and Circles — 48

- 2-1 Distance and Midpoint Formulas — 49
- 2-2 Slopes and Equations of Lines — 53
- 2-3 Parallel and Perpendicular Lines — 58
- 2-4 Circles — 63
- 2-5 Systems of Equations — 66
- 2-6 Distance—Point to Line — 72
- *Special* An Application — 76
- 2-7 Systems of Inequalities — 77
- 2-8 Linear Programming Theory — 81
- 2-9 Applications of Linear Programming — 86

Chapter Review Exercises 91   *Chapter Test* 93

3 Functions — 94

- 3-1 Relations — 95
- 3-2 Functions — 99
- 3-3 Zeros of a Function — 104
- 3-4 Some Special Functions — 107
- 3-5 Addition of Functions — 110
- 3-6 Multiplication of Functions — 114
- 3-7 Linear Functions and Variation — 119
- 3-8 Composition of Functions — 122
- *Special* int and the Computer — 126
- 3-9 Graphs of Composite Functions — 127

Chapter Review Exercises 133   *Chapter Test* 135

4 Quadratic Functions 136

4-1 Definition of a Quadratic Function 137
4-2 Vertex and Axis of Symmetry 141
4-3 Solving Quadratic Inequalities Algebraically 146
4-4 Solving Quadratic Inequalities Graphically 150
4-5 Maxima and Minima 153
Special An Application to Physics 158
4-6 The Method of Least Squares 159

Chapter Review Exercises 163 *Chapter Test* 165

5 Analytic Geometry—Conic Sections 166

5-1 Locus Problems 167
Special Analytic Geometry: The Conic Sections 170
5-2 The Parabola 171
5-3 Applications of the Parabola 177
5-4 Tangents to Parabolas 181
5-5 The Ellipse 186
5-6 The Hyperbola 191
5-7 Translation of Axes 197
5-8 Eccentricity and Conic Sections 202

Chapter Review Exercises 205 *Chapter Test* 207

6 Polynomial Functions 208

6-1 Polynomial Values 209
6-2 The Remainder and Factor Theorems 215
6-3 Rational Zeros 220
6-4 Narrowing the Search for Zeros 225
6-5 Approximating Irrational Zeros 228
6-6 Maximum and Minimum Values 230
6-7 Polynomial and Fractional Inequalities 234
6-8 Graphing Rational Functions 238
Special Interval Bisection 243

Chapter Review Exercises 244 *Chapter Test* 245

7

Inverse Functions   246

7-1 Finding Inverse Functions   247
7-2 One-to-One Functions   251
7-3 Increasing and Decreasing Functions   256
7-4 Graphing Inverses   261
7-5 Inverses of Power Functions   265
7-6 Inverses of Composite Functions   269
Special   Groups   272

Chapter Review Exercises   274   *Chapter Test*   275

8

Exponential and Logarithmic Functions   276

8-1 Exponential Functions   277
8-2 Exponential Functions on **R**   282
8-3 Logarithmic Functions   286
8-4 Properties of Logarithms   290
8-5 Exponential Equations   294
8-6 Applications of Exponential Functions   298
Special   Carbon Dating   306

Chapter Review Exercises   304   *Chapter Test*   307

9

Circular Functions   308

9-1 The Wrapping Function   309
9-2 Sine and Cosine Functions   313
9-3 Graphs of Sine and Cosine   316
9-4 Equations and Inequalities   320
9-5 Graphs Derived from Sine and Cosine   324
9-6 Secant and Cosecant   327
9-7 Tangent and Cotangent   330
9-8 The Graph of $g(x) = f(ax)$ 335
9-9 The Graph of $g(x) = f(ax + b)$ 340
Special   Hyperbolic Functions   343
9-10 Inverse Circular Functions   344

Chapter Review Exercises   349   *Chapter Test*   351

10 Trigonometric Functions — 352

- **10-1** Angles and Angular Velocity — 353
- **10-2** Trigonometric Functions — 356
- **10-3** Parametric Equations — 360
- **10-4** Right-Triangle Trigonometry — 364
- **10-5** The Law of Sines — 368
- **10-6** The Law of Cosines — 373
- **10-7** Area — 377
- **10-8** Polar Coordinates — 380
- **10-9** Changing Coordinate Systems — 383
- *Special* Conic Sections in Polar Coordinates — 386

Chapter Review Exercises 387   *Chapter Test* 389

11 Circular Functions—Identities — 390

- **11-1** Identities — 391
- **11-2** Equations on **R** — 394
- **11-3** Addition and Subtraction Identities — 398
- *Special* Wave Addition — 404
- **11-4** Double-Argument Identities — 405
- **11-5** Half-Argument Identities — 409
- **11-6** Further Equation Solving — 412
- **11-7** Complex Numbers in Polar Form — 416

Chapter Review Exercises 422   *Chapter Test* 423

12 Vectors — 424

- **12-1** Vectors and Arrows — 425
- **12-2** Vector Addition and Subtraction — 428
- **12-3** Scalar Multiplication — 432
- **12-4** Vectors in Physics — 437
- **12-5** Space — 439
- **12-6** Vectors and Lines in 3-Space — 442
- **12-7** The Dot Product — 447
- **12-8** Planes — 450
- *Special* Matrices — 454

Chapter Review Exercises 456   *Chapter Test* 457

13 Sequences and Series — 458

- **13-1** Arithmetic Sequences — 459
- **13-2** Geometric Sequences — 464
- **13-3** Sigma Notation — 468
- **13-4** Mathematical Induction — 473
- **13-5** De Moivre's Theorem — 479
- **13-6** Limits — 483
- *Special* The Number π — 489
- **13-7** Infinite Series — 490

Chapter Review Exercises 496 *Chapter Test* 497

14 Introduction to Calculus—Derivatives — 498

- **14-1** Tangents to Curves — 499
- *Special* The Limit of a Function — 503
- **14-2** The Derivative Function — 504
- **14-3** Theorems for Differentiation — 509
- **14-4** Curve Sketching Using Derivatives — 514
- **14-5** Maximum-Minimum Problems — 519
- **14-6** Velocity — 522

Chapter Review Exercises 527 *Chapter Test* 529

15 Introduction to Calculus—Integrals — 530

- **15-1** The Area Under a Curve — 531
- *Special* Computer Approximations — 536
- **15-2** The Definite Integral $\int_0^1 x^2\, dx$ — 538
- **15-3** The Definite Intregral $\int_a^b x^2\, dx$ — 541
- **15-4** Areas under Power Functions — 545
- **15-5** Area Functions — 551
- **15-6** The Fundamental Theorem of Calculus — 556

Chapter Review Exercises 560 *Chapter Test* 561

Acknowledgments 562   Tables 564
Symbols 571   Glossary 572
Selected Answers to Odd-Numbered Exercises 578
Index 593

Algebra Review

1

1–1 Numbers and Operations

If the Pittsburgh Steelers lose two yards on each of three consecutive downs, they lose a total of six yards. It seems reasonable to say that $3(-2) = -6$ and that the product of a positive and a negative number is a negative number.

Similarly, if someone loses two pounds of excess weight per week for five weeks, that person weighs 10 pounds less at the end of the five-week period. That is, $5(-2) = -10$. Working backward from this fact, you can conclude that the person weighed 10 pounds more at the beginning of the five-week period, that is, $(-5)(-2) = 10$. This is not a significant example, but it does suggest that the product of two negative numbers is a positive number.

More importantly, this rule and all the other familiar algebraic laws are logical consequences of a small number of axioms which establish the structure of the real-number system.

You learned in geometry that no proof can be constructed without first agreeing to use certain terms without definition and to accept certain statements without proof. This is just as true for the algebra of the real number system as it is for geometry.

It is possible to begin with *natural numbers* as a fundamental undefined notion and then, through a series of steps, to define the set of real numbers in terms of the natural numbers. This is such a long and sophisticated process that we take a shortcut and simply consider *real number* as an undefined term.

The set of real numbers includes several subsets.

1. The set of **natural numbers N** consists of the counting numbers.

    $N = \{1, 2, 3, 4, \ldots\}$

2. The set of **whole numbers W** is obtained by including zero with the set of natural numbers.

    $W = N \cup \{0\}$

3. The set of **integers J** is obtained by including the additive inverses of the natural numbers with the whole numbers.

    $J = W \cup \{-1, -2, -3, \ldots\}$

4. The set of **rational numbers Q** consists of those numbers that can be expressed as ratios of two integers.

    $Q = \left\{ x : \text{For some pair of integers } p \text{ and } q \neq 0, x = \frac{p}{q} \right\}$

    Thus -7, 0, and 6.23 are in Q, since $-7 = \frac{-7}{1}$, $0 = \frac{0}{1}$, and $6.23 = \frac{623}{100}$.

    Furthermore, it can be proved that any number which can be expressed as a repeating decimal is a rational number.

5. Numbers that cannot be expressed as ratios of integers are called **irrational numbers**. Examples are π, $\sqrt{2}$, and 0.01001000100001 . . . .

6. The set of **real numbers R** includes the set of rational numbers together with the set of irrational numbers.

The illustration shows how these subsets are related to each other. In symbols,

$$N \subset W \subset J \subset Q \subset R.$$

Recall that $A \subset B$, which is read *A is a subset* of *B*, means that every element of set *A* is an element of set *B*.

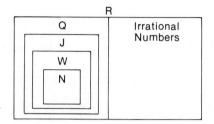

Example 1 Show that the repeating decimal 0.123123123 . . . is a rational number by expressing it as a ratio of two integers.

SOLUTION Let $x = 0.123123123 \ldots$
$1000x = 123.123123123 \ldots$ *Multiply both sides by 1000.*
$999x = 123.000000000 \ldots$ *Subtract the first equation from the second.*
$x = \dfrac{123}{999}$ *Solve for x.*

Example 1 uses the equality relation. Although equality is a familiar concept, its mathematical meaning rests on the following set of axioms.

The Equality Axioms

For all *a*, *b*, *c* in **R**:

E–1 The Reflexive Property $a = a$

E–2 The Symmetric Property   If $a = b$, then $b = a$.

E–3 The Transitive Property   If $a = b$ and $b = c$, then $a = c$.

E–4 The Substitution Property If $a = b$, then *a* can be replaced by *b* in any statement in which *a* occurs without changing the truth or falsity of the statement.

The substitution property, **E-4**, guarantees that if $a = b$, then *a* and *b* are simply different names for the same item. Thus the number 2 can be designated by a great many different symbols: $\sqrt{4} = 2$; $3 - 1 = 2$;

Chapter 1: Algebra Review

$8 \div 4 = 2$; and so on. The substitution property is used in the proof of the following theorem.

Theorem 1–1 If a, b, c, and d are real numbers such that $a = b$ and $c = d$, then $a + c = b + d$ and $ac = bd$.

PROOF
$a + c = a + c$ and $ac = ac$ Axiom E–1
$a = b$ and $c = d$ Given
$a + c = b + d$ and $ac = bd$ Axiom E–4

Theorem 1–1 involves addition and multiplication, two binary operations.

DEFINITION A **binary operation** on a set S is a rule which assigns to every ordered pair of elements in S exactly one element of S.

Thus addition is a binary operation on the set of natural numbers **N**, because if a and b are any two members of **N**, then $a + b \in$ **N**. Recall that $\in$ means *is an element of*. By operation we shall always mean a binary operation.

Example 2 Show that the process of averaging two numbers, denoted here by #, is not an operation on the set of even integers.

SOLUTION To show that # is an operation on the set of even integers, you must show that combining two even integers under # gives an even integer.

For the even integers 4 and 6: $4 \# 6 = \dfrac{4 + 6}{2} = 5$.

Since 5 is not an even integer, # is *not* an operation on the set of even integers. Notice that to show that # is not an operation on the set of even integers, all that was required was to exhibit one counterexample.

An equivalent way of saying that a rule ∗ is a binary operation on a set S is to state that S is *closed under* ∗. The set of odd integers is, for example, closed under multiplication.

Exercises

A Specify to which of sets N, W, J, W, and R each number belongs.

1. $\dfrac{0}{3}$
2. 0
3. -17
4. 9.7
5. $-\dfrac{1}{10}$
6. $\sqrt{3}$
7. 101
8. -0.043

9. $\dfrac{\pi}{2}$ 10. 12.83   11. 0.333...   12. 1.414

13. 3.14   14. $\sqrt{1.21}$ 15. $-\sqrt{2}$ 16. 0.1121231234...

Determine whether each set is closed under the specified operation. If it is not, give an appropriate counterexample.

17. The set of integers **J**; subtraction
18. The set of prime numbers; addition
19. The set of natural numbers **N**; $*$ defined by $a * b = (a + b)^2$
20. The set of odd integers; $*$ defined by $a * b = ab + 4$

B 21. $A = \{-1, 0, 1\}$; addition
22. $B = \{x : x \in \mathbf{R} \text{ and } 0 < x < 1\}$; multiplication
23. The set of rational numbers **Q**; division
24. $C = \{3n : n \in \mathbf{N}\}$; multiplication

Find the rational number represented by each repeating decimal.

25. 0.242424...   26. 0.999...   27. 3.267267...   28. 0.03151515...

29. If $A = \{1, 2, 5, c\}$ and $B = \{1, 3, b\}$ and $B \subset A$, what can be said about b and c?
30. Two sets A and B are equal if $A \subset B$ and $B \subset A$. If $A = \{3, 1, 7\}$ and $B = \{7, 1, 3, 1, 7\}$, does $A = B$?

C 31. If T is the set of odd integers, that is, if $T = \{2n + 1 : n \in \mathbf{J}\}$, prove that T is closed under multiplication. [HINT: Let $a = 2p + 1$ and $b = 2q + 1$, with $p, q \in \mathbf{J}$, be any two odd integers.]
32. If $S = \{7n : n \in \mathbf{N}\}$, prove that S is closed under addition.
33. Let $S = \{1, 2, 3, 4\}$ and let $W = \{A : A \subset S\}$.
    a. How many sets are in W?
    b. Is W closed under intersection, $\cap$? Under union, $\cup$? Give an example or counterexample in each case.
34. The operations # and $*$ are defined on a given plane M as follows: If A and B are any two distinct points of M, then
    i. $A \# B = P$ such that P is the midpoint of $\overline{AB}$ and $A \# A = A$, and
    ii. $A * B = Q$ such that B is the midpoint of $\overline{AQ}$ and $A * A = A$.
    a. Is # a commutative operation on M?
    b. Is $*$ an associative operation on M?
    c. Does # distribute over $*$?

1–2 The Field Axioms

The essential components of the **real-number system** are the set of real numbers **R**, the two operations addition and multiplication, and the axioms which support and shape the system. These axioms fall into three groups, the first of which consists of the field axioms.

The Field Axioms

F–1	Closure:	Addition and multiplication are operations on **R**.
F–2	Associative:	For all a, b, and c in **R**, $(a + b) + c = a + (b + c)$ and $(ab)c = a(bc)$.
F–3	Commutative:	For all a and b in **R**, $a + b = b + a$ and $ab = ba$.
F–4	Identity:	**a.** There exists a unique real number 0 such that for every a in **R**, $a + 0 = 0 + a = a$.
		b. There exists a unique real number 1 such that $1 \neq 0$ and for every a in **R**, $a \cdot 1 = 1 \cdot a = a$.
F–5	Inverse:	**a.** For every a in **R** there exists a unique real number denoted by $-a$ such that $a + (-a) = (-a) + a = 0$.
		b. For every a in **R** except 0 there exists a unique real number denoted by a^{-1} such that $a \cdot a^{-1} = a^{-1} \cdot a = 1$.
F–6	Distributive:	For all a, b, and c in **R**, $a(b + c) = ab + ac$.

You have been using these axioms ever since you started doing arithmetic. An awareness of them sometimes reduces the work in numerical computations, as the example below suggests.

Example 1 Simplify: **a.** $273 + (512 + 727)$ **b.** $17 \cdot 24 + 24 \cdot 13$

SOLUTION
a. $273 + (512 + 727) = 273 + (727 + 512)$
$= (273 + 727) + 512$
$= 1000 + 512 = 1512$

b. $17 \cdot 24 + 24 \cdot 13 = 24 \cdot 17 + 24 \cdot 13$
$= 24(17 + 13)$
$= 24 \cdot 30 = 720$

For any given nonzero real number a, the corresponding numbers $-a$ and a^{-1} are called the **additive** and **multiplicative inverses** of a, respectively. The existence of these inverses provides a basis for proving the cancellation properties.

Theorem 1–2 **Cancellation Properties**

For all $a, b, c \in \mathbf{R}$:

a. If $a + c = b + c$, then $a = b$.
b. If $ac = bc$ and $c \neq 0$, then $a = b$.

PROOF **b.**

$ac = bc,\ c \neq 0$	Given
c^{-1} exists.	Inverse Axiom
$(ac)c^{-1} = (bc)c^{-1}$	Theorem 1–1
$a(cc^{-1}) = b(cc^{-1})$	Associative Axiom
$a \cdot 1 = b \cdot 1$	Inverse Axiom
$a = b$	Identity Axiom

The proof of Theorem 1–2a is similar and is left as an exercise.

The inverse axiom states that for any given real number a the additive and multiplicative inverses are unique. Thus, if $a + 6 = 0$, by axiom **F–5a**, $a = -6$. Similarly if $3b = 1$, by axiom **F–5b**, $b = 3^{-1}$, or $\frac{1}{3}$. These results are used so often that we state them as a theorem.

Theorem 1–3 For all $a, x \in \mathbf{R}$:

a. If $a + x = 0$, then $x = -a$.
b. If $a \cdot x = 1$, then $x = a^{-1}$.

Since $-a + a = 0$ and $a^{-1} \cdot a = 1$, Corollary 1–3 follows immediately from Theorem 1–3.

Corollary 1–3 If $a \in \mathbf{R}$, then $-(-a) = a$, and if $a \neq 0$, then $(a^{-1})^{-1} = a$.

Theorem 1–3 provides a standard method of showing that one expression is the inverse of another, as in the following theorem.

Theorem 1–4 For all $a, b \in \mathbf{R}$:

a. $-(a + b) = -a + (-b)$.
b. $(ab)^{-1} = a^{-1} \cdot b^{-1}$, provided $a \neq 0$ and $b \neq 0$.

(cont. on p. 7)

PLAN FOR PROOF Show that $(a + b) + (-a + [-b]) = 0$.

PROOF **a.** $(a + b) + (-a + [-b]) = [(a + b) + (-a)] + (-b)$ Associative Axiom
$= [(-a) + (a + b)] + (-b)$ Commutative Axiom
$= [(-a + a) + b] + (-b)$ Associative Axiom
$= (0 + b) + (-b)$ Inverse Axiom
$= b + (-b)$ Identity Axiom
$= 0$ Inverse Axiom

Thus $-a + (-b) = -(a + b)$ Theorem 1–3

The proof of Theorem 1–4b parallels that of Theorem 1–4a and is left as an exercise. Notice that the rules for adding signed numbers depend on Theorem 1–4a. For example, $-3 + (-7) = -(3 + 7) = -10$.

The distributive axiom connects the two operations of addition and multiplication. Factoring, which is the process of converting an expression with two or more terms into a product, is basically an application of the distributive axiom.

Example 1 Factor: $c + d + 3(c + d)^2$.

SOLUTION $c + d + 3(c + d)^2 = ([c + d] \cdot 1) + (c + d)(3[c + d])$
$= (c + d)(1 + 3[c + d])$
$= (c + d)(1 + 3c + 3d)$

You should supply reasons for each step in the solution of example 1.

To show that for all $a, b \in \mathbf{R}$ the expression $a^2 + 2ab + b^2$ can be factored as $(a + b)^2$, the distributive axiom is applied, but this time starting with the product $(a + b)^2$.

$$(a + b)^2 = (a + b)(a + b)$$
$$= (a + b)a + (a + b)b$$
$$= a(a + b) + b(a + b)$$
$$= (aa + ab) + (ba + bb)$$
$$= a^2 + ab + ab + b^2$$
$$= a^2 + ab(1 + 1) + b^2$$
$$(a + b)^2 = a^2 + 2ab + b^2$$

Exercises

A Use field axioms to obtain the value of the given expression mentally.

1. $(287 + 468) + 113$
2. $83 \cdot 176 + 17 \cdot 176$
3. $999 \cdot 899 + 899$
4. $4(663 \cdot 25)$

Justify each step of the following simplifications or solutions by citing an appropriate definition, axiom, theorem, or corollary. Assume that sums and products of natural numbers are known.

5. $5 + (-3) = (2 + 3) + (-3)$
   $= 2 + (3 + [-3])$
   $= 2 + 0$
   $= 2$

6. $2 + (-7) = 2 + [-(2 + 5)]$
   $= 2 + [-2 + (-5)]$
   $= [2 + (-2)] + (-5)$
   $= 0 + (-5)$
   $= -5$

7. $3x = 15$
   3^{-1} exists.
   $3^{-1}(3x) = 3^{-1}(15)$
   $3^{-1}(3x) = 3^{-1}(3 \cdot 5)$
   $(3^{-1} \cdot 3)x = (3^{-1} \cdot 3)5$
   $1 \cdot x = 1 \cdot 5$
   $x = 5$

8. $x + 10 = -4$
   $(x + 10) + (-10) = -4 + (-10)$
   $x + (10 + [-10]) = -4 + (-10)$
   $x + 0 = -4 + (-10)$
   $x = -4 + (-10)$
   $x = -(4 + 10)$
   $x = -14$

Factor each expression.

9. $6x^3 + 9x$
10. $c + 3cd$
11. $y^2 + 12y + 36$
12. $49 + 121a^2 + 154a$

B 13. $(b + 2)^2 + 5(b + 2)$
14. $48x^2 + 27y^2 + 72xy$

15. Is division an associative operation? Is subtraction?
16. Prove that for all $a, b, c \in R$, $(a + b)c = ac + bc$.
17. Does multiplication distribute over subtraction?
18. Does division distribute over addition? Does addition distribute over division?

Complete the proof of each theorem by adding the necessary steps and giving a reason for each step.

19. **Theorem 1–2a:** For all $a, b, c \in R$, if $a + c = b + c$, then $a = b$.

    Proof: $a + c = b + c$
    $(a + c) + (-c) = (b + c) + (-c)$
    $a + [c + (-c)] = b + [c + (-c)]$

20. **Theorem 1–4b:** For all nonzero $a, b \in R$, $(ab)^{-1} = a^{-1}b^{-1}$.

    Proof: $(ab)(a^{-1}b^{-1}) = [(ab)a^{-1}]b^{-1}$
    $= [a^{-1}(ab)]b^{-1}$
    $= [(a^{-1}a)b]b^{-1}$

21. Prove Corollary 1–3: If $a \in R$, then $-(-a) = a$, and if $a \neq 0$, then $(a^{-1})^{-1} = a$.

C 22. Assume that the operations $\oplus$ and $\odot$ are defined for all $a, b \in \mathbf{R}$ by
$a \oplus b = \dfrac{a+b}{2}$ and $a \odot b = \dfrac{ab}{2}$.
   a. Are $\oplus$ and $\odot$ commutative operations?
   b. Are $\oplus$ and $\odot$ associative operations?
   c. Does $\odot$ distribute over $\oplus$?

23. An operation $*$ is defined on $\mathbf{N}$ as follows:

For all $a, b \in \mathbf{N}, a * b = \begin{cases} a \text{ if } a > b. \\ a \text{ if } a = b. \\ b \text{ if } a < b. \end{cases}$

   a. Evaluate $7 * 9$.
   b. Is $*$ commutative? Is it associative?
   c. Does $*$ distribute over $+$ (ordinary addition)? Does $+$ distribute over $*$?

24. Prove: For every $a \in \mathbf{R}, a \cdot 0 = 0$. [HINT: $a \cdot 0 = a(0 + 0)$ and $a \cdot 0 = a \cdot 0 + 0$]

1–3 Subtraction and Division

In this section we establish additional properties of real numbers, and provide rules for the subtraction and division of real numbers.

Theorem 1–5 For every $a \in \mathbf{R}, a \cdot 0 = 0$.

The proof of Theorem 1–5 was called for in Exercise 24, Section 1–2. Theorem 1–5 guarantees that 0 cannot have a multiplicative inverse. If 0^{-1} exists, then $0 \cdot 0^{-1} = 1$; but from Theorem 1–5, $0 \cdot 0^{-1} = 0$, which is a contradiction, since $0 \neq 1$.

Theorem 1–6 For all $a, b \in \mathbf{R}, (-a)b = -(ab) = a(-b)$.

PLAN FOR PROOF Show that $ab + (-a)b = 0$; then apply Theorem 1–3.

PROOF $ab + (-a)b = ba + b(-a)$
$= b[a + (-a)]$
$= b \cdot 0 = 0$
Since $ab + (-a)b = 0, (-a)b = -(ab)$. Now use this result to complete the proof.

$-(ab) = -(ba) = (-b)a = a(-b)$

Notice that in the proof of Theorem 1–6 only the statements are listed. It is left to you to supply an appropriate definition, axiom, theorem, or previously proven theorem to justify each step.

Corollary 1–6   **a.** For all $a, b \in R$, $(-a)(-b) = ab$.
b. For all $a \in R$, $(-1)a = -a$.
c. For all $a \in R$, $(-a)^2 = a^2$.

PROOF   **a.** $(-a)(-b) = -[a(-b)] = -[-(ab)] = ab$.

Corollary 1–6a shows that the product of two negative numbers is a positive number.

Subtraction is not a necessary operation for the real-number system, but it is very useful. The definition of subtraction is stated in terms of addition.

Definition   The operation **subtraction**, denoted by $-$, is defined for all $a, b \in R$ by
$$a - b = a + (-b).$$

The next theorem and its corollary are useful in simplifying expressions that involve differences. The proofs are left as exercises.

Theorem 1–7   For all $a, b, c \in R$, $a(b - c) = ab - ac$.

PLAN FOR PROOF   Convert subtraction to addition, that is, write $a(b - c)$ as $a[b + (-c)]$.

Corollary 1–7   For all $a, b \in R$, $-(a - b) = b - a$.

The important factorization $a^2 - b^2 = (a + b)(a - b)$ can now be derived by starting from the product.

$$(a + b)(a - b) = (a + b)(a + [-b])$$
$$= (a + b)a + (a + b)(-b)$$
$$= (aa + ba) + a(-b) + b(-b)$$
$$= aa + (ab + [-(ab)]) + [-(bb)]$$
$$= a^2 - b^2$$

Example 1 Factor $81 - x^2 - 4xy - 4y^2$

SOLUTION
$$81 - x^2 - 4xy - 4y^2 = 81 - (x^2 + 4xy + 4y^2)$$
$$= 9^2 - (x + 2y)^2$$
$$= [9 + (x + 2y)][9 - (x + 2y)]$$
$$= (9 + x + 2y)(9 - x - 2y)$$

Just as the definition of subtraction is stated in terms of addition, the definition of division is stated in terms of multiplication.

Definition The operation **division**, denoted by $\div$, is defined for all $a, b \in \mathbf{R}, b \neq 0$, by
$$a \div b = a \cdot b^{-1}.$$

The symbol $\frac{a}{b}$ is an alternative notation for $a \div b$ and is, in fact, more commonly used. Thus $a \div b = \frac{a}{b} = ab^{-1}$. This means that a fraction like $\frac{3}{4}$ can be regarded either as the real number 0.75 independent of the division operation or as the quotient obtained when 3 is divided by 4. The following familiar facts are immediate consequences of the definition of division.

$$\frac{0}{a} = 0 \qquad \frac{a}{a} = 1 \qquad \frac{1}{a} = a^{-1}$$

The rules for division of signed numbers are basically the same as for multiplication.

Theorem 1–8 For all $a, b \in \mathbf{R}, b \neq 0$, $-\left(\frac{a}{b}\right) = \frac{(-a)}{b} = \frac{a}{(-b)}$.

PROOF $-\left(\frac{a}{b}\right) = -(ab^{-1}) = (-a)b^{-1} = \frac{(-a)}{b}$

The proof that $-\left(\frac{a}{b}\right) = \frac{a}{(-b)}$ is left as an exercise.

Example 2 Simplify: $-\frac{a-7}{7-a}$.

SOLUTION
$$-\frac{a-7}{7-a} = \frac{a-7}{-(7-a)}$$
$$= \frac{a-7}{a-7} = 1$$

By converting fractions to products, the rules for simplifying or manipulating fractional expressions can be derived directly. These rules are summarized in the following theorem. We prove the addition rule and leave the others as exercises.

Theorem 1–9 **Operations on Fractions**

For all $a, b, c, d \in \mathbf{R}$, with $b \neq 0$ and $d \neq 0$:

a. Cancellation: $\dfrac{ac}{bc} = \dfrac{a}{b}$, $c \neq 0$ b. Addition: $\dfrac{a}{b} + \dfrac{c}{b} = \dfrac{a+c}{b}$

c. Multiplication: $\dfrac{a}{b} \cdot \dfrac{c}{d} = \dfrac{ac}{bd}$ d. Division: $\dfrac{a}{b} \div \dfrac{c}{d} = \dfrac{ad}{bc}$, $c \neq 0$

PROOF b. $\dfrac{a}{b} + \dfrac{c}{b} = ab^{-1} + cb^{-1}$

$\qquad\qquad = (a+c)b^{-1}$

$\qquad\qquad = \dfrac{a+c}{b}$

Example 3 Simplify: $\dfrac{2x-3}{5x^2+x-4} + \dfrac{3x+5}{10x^2+7x-12}$

SOLUTION $\dfrac{2x-3}{5x^2+x-4} + \dfrac{3x+5}{10x^2+7x-12} = \dfrac{2x-3}{(5x-4)(x+1)} + \dfrac{3x+5}{(5x-4)(2x+3)}$

$\qquad = \dfrac{(2x-3)(2x+3)}{(5x-4)(x+1)(2x+3)} + \dfrac{(3x+5)(x+1)}{(5x-4)(2x+3)(x+1)}$

$\qquad = \dfrac{(4x^2-9) + (3x^2+8x+5)}{(5x-4)(x+1)(2x+3)}$

$\qquad = \dfrac{7x^2+8x-4}{(5x-4)(x+1)(2x+3)}$

Example 4 Simplify: $\dfrac{a^3+2a^2-4a-8}{2a+1} \div \dfrac{4a^2-8a}{2a^2-5a-3}$

SOLUTION $\dfrac{a^3+2a^2-4a-8}{2a+1} \div \dfrac{4a^2-8a}{2a^2-5a-3} = \dfrac{a^2(a+2) - 4(a+2)}{2a+1} \cdot \dfrac{(2a+1)(a-3)}{4a(a-2)}$

$\qquad = \dfrac{(a+2)(a^2-4)(2a+1)(a-3)}{(2a+1)(4a)(a-2)}$

$\qquad = \dfrac{[(a+2)^2(a-3)][(2a+1)(a-2)]}{4a[(2a+1)(a-2)]}$

$\qquad = \dfrac{(a+2)^2(a-3)}{4a}$

As examples 3 and 4 indicate, manipulation of fractional expressions makes extensive use of factoring. To aid you in the factoring process, the following summary is provided.

1. Determine the greatest factor, other than 1, that is common to each term and factor it out.
2. Consider the following possibilities for expressions being factored.
   a. Expressions with two terms
      i. Difference of two squares: $a^2 - b^2 = (a + b)(a - b)$
      ii. Difference of two cubes: $a^3 - b^3 = (a - b)(a^2 + ab + b^2)$
      iii. Sum of two cubes: $a^3 + b^3 = (a + b)(a^2 - ab + b^2)$
   b. Expressions with three terms
      i. Trinomial perfect square: $a^2 \pm 2ab + b^2 = (a \pm b)^2$
      ii. Trinomial cross product: $ac + (bc + ad) + bd = (a + b)(c + d)$
   c. Expressions with four terms
      i. Grouping—three terms and one term leading to a difference of two squares: $(a^2 \pm 2ab + b^2) - c^2 = [(a \pm b) + c][(a \pm b) - c]$
      ii. Grouping—two pairs of terms leading to a common factor: $(ab + ac) \pm (db + dc) = a(b + c) \pm d(b + c) = (a \pm d)(b + c)$
3. After factoring once, check to see if any of the factors can be factored further.

Example 5 Factor: $54a^3 + 16b^6$.

SOLUTION
$$54a^3 + 16b^6 = 2(27a^3 + 8b^6)$$
$$= 2[(3a)^3 + (2b^2)^3]$$
$$= 2[3a + 2b^2][(3a)^2 - (3a)(2b^2) + (2b^2)^2]$$
$$= 2(3a + 2b^2)(9a^2 - 6ab^2 + 4b^4)$$

Exercises

A Remove parentheses and brackets and simplify.

1. $2x - [-3(x - 2y) + 7x]$
2. $-3(2rw - 5w[1 - 3w])$

Supply an appropriate reason to justify each step in the following.

3. $-(3x - 2) = 8$
   $-(3x) + 2 = 8$
   $-(3x) + 2 = 6 + 2$
   $-(3x) = 6$
   $(-3)x = 6$
   $(-3)x = (-3)(-2)$
   $x = -2$

4. $3x - 15 = -15$
   $3x + (-15) = -15$
   $3x + (-15) = 0 + (-15)$
   $3x = 0$
   $3^{-1}(3x) = 3^{-1} \cdot 0$
   $(3^{-1} \cdot 3)x = 3^{-1} \cdot 0$
   $1 \cdot x = 3^{-1} \cdot 0$
   $1 \cdot x = 0$
   $x = 0$

Factor each expression completely.

5. $8a^3 - 27b^3$ **6.** $64 + 125b^3$ **7.** $32x^4 - 162$

8. $3c^3 + 16c^2 + 5c$ **9.** $a^2 + 1 + 2a^2b + 2b$ **10.** $b^3 + 3b^2 - 3 - b$

B Complete the proof of each theorem by adding appropriate steps and giving a reason to justify each step. Do not use a theorem whose number follows that of the theorem being proved. Assume $a, b, c, d \in R$.

11. Theorem 1–7: $a(b - c) = ab - ac$

$a(b - c) = a(b + [-c])$
$\quad\quad\quad = ab + a(-c)$
$\quad\quad\quad = ab + (-[ac])$

12. Corollary 1–7: $-(a - b) = b - a$

$-(a - b) = -(a + [-b])$
$\quad\quad\quad = -a + (-[-b])$

13. Theorem 1–9a: $\dfrac{ac}{bc} = \dfrac{a}{b}$, b and $c \neq 0$

$\dfrac{ac}{bc} = (ac)(bc)^{-1}$
$\quad\;\; = (ac)(b^{-1}c^{-1})$

14. If $a, b \neq 0$, then $\left(\dfrac{a}{b}\right)^{-1} = \dfrac{b}{a}$.

$\left(\dfrac{a}{b}\right)^{-1} = (ab^{-1})^{-1}$
$\quad\quad\; = a^{-1}(b^{-1})^{-1}$

Simplify each expression. State any restrictions on the variables.

15. $\dfrac{x^2 - x - 2}{2y - 4} \cdot \dfrac{2 - y}{x - 2}$

16. $\dfrac{x - y}{4x} \div \dfrac{3x - 3y - x^2 + xy}{2x^2}$

17. $\dfrac{8a^3 + 27}{4a + 3} \div \dfrac{3 + 2a}{3 + a - 4a^2}$

18. $\dfrac{y - 5}{y^2 - 3y + 2} - \dfrac{y + 1}{y - 2}$

19. $\dfrac{2xy}{x^3 + y^3} - \dfrac{x}{x^2 - xy + y^2}$

20. $\left(2x + \dfrac{3}{x - 1}\right)\left(1 - \dfrac{3x - 1}{2x}\right)$

21. $\dfrac{y - 1 - \dfrac{5}{y + 3}}{3 - \dfrac{2y + 2}{y + 2}}$

22. $\dfrac{\dfrac{3x^2}{x^3 - 1} + \dfrac{1}{1 - x}}{1 - \dfrac{x^2 - x}{x^2 + x + 1}}$

23. Prove Theorem 1–9c: $\dfrac{a}{b} \cdot \dfrac{c}{d} = \dfrac{ac}{bd}$, b and $d \neq 0$.

24. Prove Theorem 1–9d: $\dfrac{a}{b} \div \dfrac{c}{d} = \dfrac{ad}{bc}$, b, c and $d \neq 0$. [HINT: Use Theorem 1–9c.]

25. Prove that if $b \neq 0$, then $\dfrac{a}{b} - \dfrac{c}{b} = \dfrac{a - c}{b}$.

26. Prove that if $b, d \neq 0$, then $\dfrac{a}{b} + \dfrac{c}{d} = \dfrac{ad + bc}{bd}$.

Factor each expression completely.

27. $a^3 - 1 + 3a - 3$ **28.** $t^2 + 4s^2 + 4st - 4t - 8s$

29. $x^3 - 6y - 3xy + 8$ **30.** $y^6 - 7y^3 - 8$

C **31.** Prove that for $b \neq 0$, $-(b^{-1}) = (-b)^{-1}$. Use this fact to prove

Theorem 1–8: $-\left(\dfrac{a}{b}\right) = \dfrac{a}{(-b)}$

32. The number $2^{48} - 1$ is divisible by two integers between 60 and 70. What are they?

1–4 Order

At an early age you acquired an intuitive sense of what it means to say that one number is greater than another. In terms of a horizontal number line, x is greater than y if and only if the position of x on the line is to the right of the position of y. But more than just an intuitive idea of ordering of numbers on a number line is needed to prove statements such as

$$\frac{a+b}{2} \text{ is greater than or equal to } \sqrt{ab}$$

for all positive numbers a and b.

We shall take the relation **greater than,** denoted by $>$, as an undefined term and give it meaning with the following set of axioms.

The Order Axioms

O–1   Trichotomy: For all $a, b \in \mathbf{R}$, exactly one of the following is true: $a > b$; $a = b$; $b > a$.

O–2   Transitive: For all $a, b, c \in \mathbf{R}$, if $a > b$ and $b > c$, then $a > c$.

O–3   Addition: For all $a, b, c \in \mathbf{R}$, if $a > b$, then $a + c > b + c$.

O–4   Multiplication: For all $a, b, c \in \mathbf{R}$ and $c > 0$, if $a > b$, then $ac > bc$.

The relation **less than,** denoted by $<$, is defined in terms of $>$.

Definition   For all $a, b \in \mathbf{R}$, a is less than b, denoted $a < b$, if and only if $b > a$.

By the trichotomy axiom, the real numbers can be partitioned into three nonoverlapping sets:

1. The positive numbers, $P = \{x \in \mathbf{R} : x > 0\}$,

2. $\{0\}$, and

3. The negative numbers, $M = \{x \in \mathbf{R} : x < 0\}$.

You can show that if a is positive, then $-a$ is negative. Extending this idea yields the following theorem.

Theorem 1–10   For all $a, b \in \mathbf{R}$, if $a > b$, then $-a < -b$.

PROOF
$$a > b$$
$$a + [-a + (-b)] > b + [-a + (-b)]$$
$$[a + (-a)] + (-b) > [b + (-b)] + (-a)$$
$$-b > -a$$
$$-a < -b$$

In using the multiplication axiom O–4, note carefully that the multiplier c must be positive. To see the need for this restriction, note that

$$5 > 2 \quad \text{but} \quad -3 \cdot 5 < -3 \cdot 2.$$

The following theorem implies that when multiplying both sides of an inequality by a negative number, it is necessary to reverse the inequality.

Theorem 1–11   For all $a, b, c \in \mathbf{R}$. If $a > b$ and $c < 0$, then $ac < bc$.

PROOF   Assume $a > b$ and $c < 0$. Then $0 > c$ and by Theorem 1–10, $0 < -c$, or $-c > 0$. Now apply the multiplication axiom.
$$a(-c) > b(-c)$$
$$-ac > -bc$$
Apply Theorem 1–10 again.
$$-(-ac) < -(-bc)$$
$$ac < bc$$

Corollary 1–11A   For all $a, b \in \mathbf{R}$:

a. If $a > 0$ and $b > 0$ or $a < 0$ and $b < 0$, then $ab > 0$.
b. If $a < 0$ and $b > 0$ or $a > 0$ and $b < 0$, then $ab < 0$.

Notice that the proof of Theorem 1–11 does not use the two-column format of statements and reasons commonly used in plane geometry. This form of proof, called a **paragraph proof**, is just as correct. From now on, we will often provide paragraph proofs, since they tend to be easier to read. In so doing, we obey the mathematician's standard of producing a convincing argument, with the understanding that, if requested, any assertion could be supported by an appropriate definition, axiom, or previously proven theorem. These references are included whenever the justification for a step might seem obscure.

With the results of Corollary 1–11A, you can now complete a table, as at the right, which indicates the sign of possible products when any two real numbers are multiplied together. From this table a second corollary follows.

	b		
a	+	0	−
+	+	0	−
0	0	0	0
−	−	0	+

Corollary 1–11B For all $a, b \in \mathbf{R}$:

a. If $ab > 0$, then $a > 0$ and $b > 0$, or $a < 0$ and $b < 0$.
b. If $ab < 0$, then $a < 0$ and $b > 0$, or $a > 0$ and $b < 0$.

You should realize that 1 and all the other natural numbers are positive. A significant result of the fact that 1 is positive is that the reciprocals of positive numbers are positive.

Theorem 1–12 If $a \in \mathbf{R}$ and $a > 0$, then $a^{-1} > 0$.

PLAN FOR PROOF By the trichotomy axiom, $a^{-1} > 0$ or $a^{-1} = 0$ or $0 > a^{-1}$. Show that $a^{-1} = 0$ and $0 > a^{-1}$ are both impossible.
The proof is required in exercise 22.

Since the natural numbers are positive, any number that can be expressed as a ratio of two natural numbers must be a positive number. Thus numbers like $\frac{3}{7}$ and 4.62 must be positive.

A valid counterpart of each axiom or theorem involving inequalities is created by interchanging $>$ and $<$ or by replacing $<$ with $\leq$ and $>$ with $\geq$. Thus from the transitive axiom, the following may be assumed.

If $a < b$ and $b < c$, then $a < c$.

As another example, you can use the statement "If $a \leq b$, then $-a \geq -b$" as a corollary of Theorem 1–10.

Inequalities are used in problem solving in much the same way that equations are used.

Example 1 Rachel has $40 and wishes to take three friends to a concert for which tickets are available at $3, $5, $8, and $12 each. If parking will cost $3, what price tickets can she afford?

(cont. on p. 18)

SOLUTION Let x represent the price paid per ticket, and solve the inequality $4x + 3 \le 40$.

$$(4x + 3) + (-3) \le 40 + (-3)$$
$$4x \le 37$$
$$\frac{1}{4}(4x) \le \frac{1}{4} \cdot 37$$
$$x \le \frac{37}{4}, \text{ or } x \le 9\frac{1}{4}$$

ANSWER She can buy $8, $5, or $3 tickets.

An inequality in one variable is very similar to an equation in one variable. We review a few terms by first recalling that a *variable* is just a symbol for any member of a given set called the *domain* of the variable. Like an equation, an inequality in one variable is an open sentence which involves one variable; it is neither true nor false. When the variable is replaced by a specific member of the domain, the open sentence then becomes a statement which is either true or false. The set consisting of all the members of the domain that yield true statements when substituted for the variable is called the *solution set* of the inequality. Any individual element that satisfies the inequality is called a *solution*. Thus in example 1 the domain of the inequality $4x + 3 \le 40$ is $\{3, 5, 8, 12\}$ and the solution set is $\{3, 5, 8\}$.

When an inequality is not associated with an applied problem, the domain of the inequality, that is, the domain of the variable, is not usually stated explicitly. In that case, the domain is assumed to be the largest appropriate subset of the real numbers. To determine this unstated domain, exclude those numbers that do not yield meaningful statements when substituted in the inequality. In particular, you must avoid division by zero and creating even roots of negative numbers.

Example 2 Determine the domain D of each inequality.

a. $x + \dfrac{1}{x - 3} > 5$ **b.** $x + \sqrt{x - 1} \le 2$

SOLUTION **a.** Since $\frac{1}{0}$ is undefined, 3 is not an acceptable replacement for x. Hence $D = \mathbf{R} - \{3\}$. (Recall that set subtraction is defined for all sets A and B by $A - B = \{x : x \in A \text{ and } x \notin B\}$.)

b. You cannot take the square root of a negative number, so $x - 1$ must be greater than or equal to 0. Hence, $D = \{x : x \ge 1\}$.

The reason square roots of negative numbers do not exist in $\mathbf{R}$ is that the square of a real number cannot be negative, a fact we state as a theorem.

Theorem 1–13 If $a \in \mathbf{R}$, then $a^2 \geq 0$.

An inequality of the form $ax + b > 0$, where a and b are constants, $a \neq 0$, is called a **linear inequality** in one variable. These and many other more complex inequalities can be solved by creating a sequence of equivalent inequalities, the last of which has an obvious solution. Two equations are equivalent if they have the same domain and the same solution set; equivalent inequalities are defined in the same manner.

Definition **Equivalent inequalities** are inequalities that have the same domain and the same solution set.

From the axioms and theorems it follows that two inequalities with the same domain are equivalent if one is obtained from the other by any of the following:

1. Adding to or subtracting from both sides of the inequality the same quantity.
2. Multiplying or dividing each term by the same positive quantity.
3. Multiplying or dividing each term by the same *negative* quantity and reversing the inequality.
4. Simplifying either side using axioms or theorems.
5. Substitution.

By using these procedures, you can see that $3x(x + 2) > 3x^2 - 5$ is equivalent to $6x + 5 > 0$ (add $5 - 3x^2$ to each side). Note, however, that $2 > \dfrac{1}{x - 1}$ is not equivalent to $2(x - 1) > 1$. The domains are different—1 is not in the domain of $2 > \dfrac{1}{x - 1}$, and the solution sets are different. For example, $\frac{1}{2}$ is a solution of $2 > \dfrac{1}{x - 1}$ but not of $2(x - 1) > 1$. The reason you don't get an equivalent inequality by multiplying each term of $2 > \dfrac{1}{x - 1}$ by $x - 1$ is that you may not assume that $x - 1$ is positive.

Example 3 Solve: $\dfrac{2x}{3} - 5 \leq \dfrac{3x + 7}{2}$.

SOLUTION The domain is unstated; assume it is **R**. Multiply both sides by 6.

$$4x - 30 \leq 9x + 21$$
$$-5x \leq 51$$
$$x \geq -\tfrac{51}{5} \qquad \text{Divide by } -5;\ \text{reverse the inequality.}$$

Since each inequality is equivalent to the preceding one, the solution set of the original inequality is $\{x \in \mathbf{R} : x \geq -\tfrac{51}{5}\}$.

Example 4 Solve: $(x^2 + 3)(x - 4) > 0$.

SOLUTION $(x^2 + 3)(x - 4) > 0$

By Theorem 1–13, $x^2 \geq 0$, so $x^2 + 3 > 0$. Dividing by the positive quantity $x^2 + 3$ yields an equivalent inequality, with the direction of the inequality unchanged.

$$x - 4 > 0$$
$$x > 4$$

The solution set is $\{x : x > 4\}$.

Example 5 Solve and graph the solution set on a number line: $x^2 + 5x < 0$.

SOLUTION $x^2 + 5x < 0$

$x(x + 5) < 0$

Now apply Corollary 1–11b.

$$[x > 0 \text{ and } x + 5 < 0] \text{ or } [x < 0 \text{ and } x + 5 > 0]$$
$$[x > 0 \text{ and } x < -5] \text{ or } [x < 0 \text{ and } x > -5]$$

ANSWER Since no number is both less than -5 and greater than 0, the solution set is $\{x : x > -5 \text{ and } x < 0\}$.

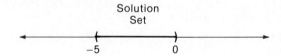

Exercises

A Solve each inequality. State any restrictions on the domain.

1. $5x + 3 > 2x + 7$
2. $x \geq -7x$
3. $\frac{3x}{2} - 5 < \frac{2}{3}(2x + 7)$
4. $3(x - 4) \leq 5x + 1$
5. $(x + 2)^2 < (x - 2)(x - 5)$
6. $3(x - 1) - 5x > 2(3 - x) + 4$
7. $\frac{x + 2}{-5} \geq 7$
8. $\frac{2x - 1}{3} - \frac{x}{2} < \frac{x + 2}{5}$
9. $(5x - 7)(3x^2) < 0$
10. $(2x + 3)(1 + x^2) \geq 0$

Determine the domain of each inequality.

11. $\dfrac{2x}{3-x} < x - 1$

12. $\sqrt{2x + 5} \geq 7$

13. $\sqrt{x^2 + 3} \geq 3x - 1$

14. $\dfrac{x}{x^2 - 3x - 4} < 5x^3$

B Solve each inequality.

15. $x(2x + 7) - 2(x + 2)(x - 1) < 3$
16. $x(x^2 - 3) + 4 \leq (x - 2)(x^2 + 2x + 4)$
17. $3cx + 5 > 6x + 1$; c is a constant and $c < 2$
18. $2k(x - k) < 2x + 3$; k is a constant and $k > 1$
19. $2x^3 - 6x^2 < 0$
20. $x^3 - 4x^2 + 2x - 8 \geq 0$

Prove each of the following statements.

21. Corollary 1–11A: For all $a, b \in R$ with $b > 0$, **(i)** if $a > 0$, then $ab > 0$ and **(ii)** if $a < 0$, then $ab < 0$.

22. Theorem 1–12: If $a \in R$ and $a > 0$, then $a^{-1} > 0$. [HINT: Use the plan for proof given in the lesson.]

23. Theorem 1–13: If $a \in R$, then $a^2 \geq 0$. [HINT: Consider three cases: $a > 0$, $a = 0$, and $a < 0$.]

24. For all $a, b \in R$, $a^2 + b^2 \geq 2ab$. [HINT: $(a - b)^2 \geq 0$.]

Solve each inequality and graph the solution set on a number line.

25. $3x(x + 7) < 0$
26. $x^2 - 4x > 0$
27. $(x - 2)(x + 5) > 0$
28. $(x + 3)(2x + 1) < 0$

C Prove each of the following statements.

29. For all $a, b, c, d \in R$, if $a > b$ and $c > d$, then $a + c > b + d$.

30. For all $a, b \in R$, if $a > b$ and $b > 0$, then $\dfrac{1}{a} < \dfrac{1}{b}$.

31. If a and b are positive, then $a^2 < b^2$ if and only if $a < b$.

1-5 Radicals

To indicate that a number k lies between two given numbers, say -3 and 5, you can combine the statements $k > -3$ and $k < 5$ and write the **compound inequality** $-3 < k < 5$. Note that there is no similar contraction available for $x < -3$ or $x > 5$.

Example 1 A biology teacher reports grades on a scale of 100 after rounding each grade to the nearest multiple of 5. If a student receives a grade of 85 on a test that contains 55 questions, how many questions did the student answer correctly?

SOLUTION From the teacher's grading system, it follows that the percentage of correct answers is between 82.5% and 87.5%. Thus if x represents the number of correct answers, then

$$0.825 \leq \frac{x}{55} < 0.875, \text{ or}$$

$$45.375 \leq x < 48.125.$$

In this case assume that the domain is $\{x \in \mathbf{N} : 0 \leq x \leq 55\}$. The solution set is $\{46, 47, 48\}$.

The set that consists of all real numbers between -3 and 5, that is, $\{x : -3 \leq x \leq 5\}$, is an example of an interval. Such sets can be described by using parentheses and brackets, and named as follows:

Closed interval
$[a, b] = \{x \in \mathbf{R} : a \leq x \leq b\}$

Open interval
$(a, b) = \{x \in \mathbf{R} : a < x < b\}$

Half-open interval
$(a, b] = \{x \in \mathbf{R} : a < x \leq b\}$

Half-open interval
$[a, b) = \{x \in \mathbf{R} : a \leq x < b\}$

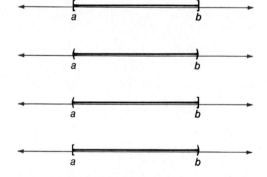

Notice how parentheses and brackets are used to indicate inclusion or exclusion of the endpoints of each interval.

To specify a ray, use $(-\infty, a]$ for $\{x \in \mathbf{R} : x \leq a\}$, and $[b, \infty)$ for $\{x \in \mathbf{R} : x \geq b\}$. To indicate numbers that lie outside the half-open interval $(-3, 5]$, write $(-\infty, -3] \cup (5, \infty)$.

Example 2 If $A = \{x \in R : x \geq 1\}$, $B = \{x \in R : 0 < x \leq 3\}$, and $C = \{x \in R : x < -2\}$, express $A \cup C$ and $A \cap B$ in interval notation.

SOLUTION Graph the sets on a number line. Then note that

$$A \cup C = [1, \infty) \cup (-\infty, -2), \text{ and}$$

$$A \cap B = [1, \infty) \cap (0, 3]$$

$$= [1, 3].$$

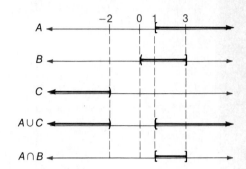

Now consider the question, "How long does it take a ball dropped from a height of 80 feet to hit the ground?" To answer the question, use the formula $s = 16t^2$, where s is the distance in feet that an object falls in t seconds. Solve the equation on the domain of positive numbers.

$$80 = 16t^2$$

$$t^2 = 5$$

$$t = \sqrt{5}$$

The symbol $\sqrt{5}$ denotes the positive number whose square is 5. Unlike $\sqrt{4}$, which is equal to 2, $\sqrt{5}$ must be evaluated by a trial-and-error process.

$2^2 = 4$	and	$3^2 = 9$,	so	$2 < \sqrt{5} < 3$.
$2.2^2 = 4.84$	and	$2.3^2 = 5.29$,	so	$2.2 < \sqrt{5} < 2.3$.
$2.23^2 = 4.9729$	and	$2.24^2 = 5.0176$,	so	$2.23 < \sqrt{5} < 2.24$.

This process is generating a sequence of numbers that are getting closer and closer to $\sqrt{5}$. Will this process really yield a number whose square is exactly 5?

It can be proved that there is no *rational* number whose square is 5. But if a ball is dropped, you expect that at some instant the ball must hit the ground. This is a good reason to think that there is a real number that satisfies the equation $t^2 = 5$. In purely mathematical terms, what is needed is an axiom that guarantees that there is a real number whose square is 5. Before stating this axiom, however, the concept of *boundedness* must be defined.

Definition A set of real numbers S is **bounded above** if there exists a number M such that for every $a \in S$, $a \leq M$.

Such a number M is called an **upper bound** of S.

Consider the following sets.

$$A = \{x \in \mathbf{Q} : 4 \leq x \leq 7\} \qquad D = \{-2, 3, 29\}$$
$$B = \{x \in \mathbf{N} : x < \tfrac{9}{2}\} \qquad E = \{x \in \mathbf{N} : x > 10\}$$
$$C = \{x \in \mathbf{R} : x < -2\} \qquad F = \{x \in \mathbf{Q} : x^2 > 4\}$$

Notice that 100 is an upper bound for sets A, B, C, and D. For each of these sets, every number greater than 100 is also an upper bound, as are 50 and 29, for that matter. But E is not bounded above, since no number is larger than every natural number. Set F is not bounded above either.

If there is a number m such that $m \leq a$ for every $a \in S$, then S is said to be **bounded below** and m is called a **lower bound** of S. For the sets listed above, -10 is a lower bound for A, D, and E, while B, C, and F are not bounded below. A **bounded set** is one that is bounded both above and below, such as sets A and D.

Any set which is bounded above has an infinite number of upper bounds. Note that sets A, B, C, and D have one upper bound which is smaller than any other upper bound. For A, the smallest of the upper bounds is 7; for B it is 4; for C it is -2; and for D it is 29.

Definition Let S be bounded above. A number b is the **least upper bound** of S if b is an upper bound of S and there is no upper bound smaller than b.

Set C illustrates the fact that the least upper bound of a set need not be a member of the given set.

Now we state the last of the axioms of the real-number system.

AXIOM **The Completeness Axiom**

Every nonempty set of real numbers that is bounded above has a least upper bound.

Now we can show that a real number t exists such that $t^2 = 5$. Consider the set $S = \{x \in \mathbf{Q} : x^2 < 5\}$. Set S includes rational numbers such as 0, 1, and $\tfrac{5}{3}$. Since $3^2 > 5$, every member of S is less than 3, so S is bounded above. By the completeness axiom, then, S has a *least* upper bound b. Now consider b^2. By the trichotomy axiom, there are three possibilities for b^2:

$$b^2 < 5 \quad \text{or} \quad b^2 = 5 \quad \text{or} \quad b^2 > 5.$$

It can be proved that the assumptions $b^2 < 5$ and $b^2 > 5$ both lead to contradictions. The conclusion is that $b^2 = 5$.

In the exercises you will be asked to show that b is the only positive number whose square is 5. We designate b as the **principal square root** of 5.

Definition Let $n \in \mathbb{N}$ and $c > 0$. The **principal nth root** of c, designated $\sqrt[n]{c}$, is that positive number b such that $b^n = c$.

By the argument outlined above with respect to $\sqrt[2]{5}$, the completeness axiom implies that every positive number c has a principal nth root and that $\sqrt[n]{c}$ is the least upper bound of the set S, where $S = \{x \in \mathbb{Q} : x^n < c\}$.

If b and c are numbers such that $b^n = c$, then b is called *an* nth root of c. Thus -2 and 2 are both fourth roots of 16, but 2 is the *principal* fourth root of 16 and $\sqrt[4]{16} = 2$, not -2.

If c is negative and n is odd, then there is exactly one negative number b such that $b^n = c$, and this nth root of c is designated $\sqrt[n]{c}$. Note that $\sqrt[n]{0} = 0$ and that when $n = 2$, the index is usually omitted, as in $\sqrt{9} = 3$. If $c < 0$ and n is even, then c has no real nth roots.

Examples $\sqrt{4} = 2$ $\sqrt[3]{27} = 3$ $\sqrt[5]{-32} = -2$

$-\sqrt{2.56} = -1.6$ $-\sqrt[3]{-8} = 2$ $\sqrt{-9}$ is undefined on $\mathbb{R}$.

The definition of the principal nth root implies that if $\sqrt[n]{a}$ is defined, then $(\sqrt[n]{a})^n = a$. This fact is used in proving the next theorem.

Theorem 1–14 If $a > 0$, $b > 0$, and $n \in \mathbb{N}$, then

a. $\sqrt[n]{a} \cdot \sqrt[n]{b} = \sqrt[n]{ab}$ and   **b.** $\dfrac{\sqrt[n]{a}}{\sqrt[n]{b}} = \sqrt[n]{\dfrac{a}{b}}$.

PROOF **a.** By the definition of the principal nth root, $\sqrt[n]{a} > 0$ and $\sqrt[n]{b} > 0$. Hence $\sqrt[n]{a} \cdot \sqrt[n]{b} > 0$. Also

$$(\sqrt[n]{a} \cdot \sqrt[n]{b})^n = (\sqrt[n]{a})^n (\sqrt[n]{b})^n = ab.$$

Thus $(\sqrt[n]{a})(\sqrt[n]{b})$ is that positive number p such that $p^n = ab$. Therefore, $\sqrt[n]{ab} = \sqrt[n]{a} \cdot \sqrt[n]{b}$, by the definition of the principal nth root.

The proof of **b** is similar and is left as an exercise.

Example 3 Simplify each expression.

a. $\sqrt{252}$ **b.** $\sqrt[3]{135}$ **c.** $\sqrt{\dfrac{5}{12}}$ **d.** $\sqrt[5]{\dfrac{3}{8}}$

(cont. on p. 26)

SOLUTION   a. $\sqrt{252} = \sqrt{36 \cdot 7} = \sqrt{36} \cdot \sqrt{7} = 6\sqrt{7}$
b. $\sqrt[3]{135} = \sqrt[3]{27 \cdot 5} = \sqrt[3]{27} \cdot \sqrt[3]{5} = 3\sqrt[3]{5}$
c. $\sqrt{\dfrac{5}{12}} = \dfrac{\sqrt{5 \cdot 3}}{\sqrt{12 \cdot 3}} = \dfrac{\sqrt{15}}{\sqrt{36}} = \dfrac{\sqrt{15}}{6}$
d. $\sqrt[5]{\dfrac{3}{8}} = \sqrt[5]{\dfrac{3 \cdot 4}{8 \cdot 4}} = \dfrac{\sqrt[5]{12}}{\sqrt[5]{32}} = \dfrac{\sqrt[5]{12}}{2}$

Example 4   Simplify: $2\sqrt{\dfrac{5}{3}} + \sqrt{60} - \dfrac{\sqrt{3}}{\sqrt{5}}$.

SOLUTION $2\sqrt{\dfrac{5}{3}} + \sqrt{60} - \dfrac{\sqrt{3}}{\sqrt{5}} = 2\sqrt{\dfrac{5 \cdot 3}{3 \cdot 3}} + \sqrt{4 \cdot 15} - \dfrac{\sqrt{3}\sqrt{5}}{\sqrt{5}\sqrt{5}}$

$= \dfrac{2\sqrt{15}}{3} + 2\sqrt{15} - \dfrac{\sqrt{15}}{5}$

$= \dfrac{37\sqrt{15}}{15}$

Example 5   Rationalize the denominator, that is, rewrite the expression without a radical in the denominator: $\dfrac{2\sqrt{3}}{5 - 2\sqrt{3}}$.

SOLUTION   Multiply the denominator and the numerator by $5 + 2\sqrt{3}$.

$\dfrac{2\sqrt{3}}{5 - 2\sqrt{3}} = \dfrac{2\sqrt{3}(5 + 2\sqrt{3})}{(5 - 2\sqrt{3})(5 + 2\sqrt{3})}$

$= \dfrac{10\sqrt{3} + 4 \cdot 3}{25 - 12}$

$= \dfrac{10\sqrt{3} + 12}{13}$

The numbers $5 + 2\sqrt{3}$ and $5 - 2\sqrt{3}$ are said to be **conjugates** of each other.

Looking back, you might ask, "Why outline such an elaborate procedure using the completeness axiom, to show that there exists a real number $\sqrt{5}$?" Why not simply replace that axiom by another axiom to the effect that for every $c > 0$ and $n \in \mathbf{N}$, there exists a unique real number b such that $b^n = c$? The answer is that with the completness axiom no additional axioms will ever be required for the real number system. Had we taken the shortcut, more axioms would have been needed later.

Exercises

A Express each set using interval notation.

1. $\{x: -2 \leq x \leq 3\}$
2. $\{x: x \geq 7\}$
3. $\{x: x < -4\}$
4. $\{x: -3 \leq x < 0\}$
5. $\{x: x < 2 \text{ or } x > 5\}$
6. $\{x: x \leq 1 \text{ or } x > 2\}$
7. $\{t: t > 1 \text{ and } t \geq 3\}$
8. $\{s: s < -1 \text{ and } s \leq 2\}$

Simplify each of the following.

9. $[-3, 5] \cap [2, 6]$
10. $(-3, \infty) \cap [1, 4)$
11. $(-2, \infty) \cup (3, \infty)$
12. $(-\infty, 4) \cup [4, 5)$
13. $(-\infty, 6) \cap [-1, \infty)$
14. $[-2, 3) \cap (-\infty, 1)$
15. $(-2, \infty) - (6, \infty)$
16. $[-4, 4] - (0, \infty)$

Simplify each radical expression.

17. $\sqrt{192}$
18. $\sqrt[3]{54}$
19. $\sqrt[4]{48}$
20. $\sqrt{0.008}$
21. $\sqrt{15} \cdot \sqrt{5}$
22. $\sqrt{3} \cdot \sqrt{24}$
23. $\sqrt[3]{-56}$
24. $\sqrt[5]{-64}$
25. $\dfrac{\sqrt{24}}{\sqrt{3}}$
26. $\dfrac{6}{\sqrt{3}}$
27. $\sqrt{\dfrac{10}{7}}$
28. $\dfrac{2\sqrt{7}}{\sqrt{6}}$

B 29. $\sqrt[3]{\dfrac{2}{9}}$
30. $\sqrt[5]{\dfrac{5}{4}}$
31. $\sqrt{79^2 + 79^2}$
32. $\sqrt{253^2 - 3^2}$
33. $\sqrt{2}(2\sqrt{2} - \sqrt{3})$
34. $(5 + 2\sqrt{3})^2$
35. $(2\sqrt{3} + 7)(2\sqrt{3} - 7)$
36. $\sqrt{18} - 3\sqrt{2} + \sqrt{32}$
37. $3\sqrt{12} - \sqrt{27} - 3\sqrt{18}$
38. $\sqrt{\dfrac{2}{3}} + 2\sqrt{\dfrac{8}{9}} - \dfrac{5\sqrt{6}}{27}$
39. $\sqrt{\dfrac{2}{7}} + 2\sqrt{56} - 9\sqrt{\dfrac{7}{2}}$

Simplify each expression by rationalizing the denominator.

40. $\dfrac{3\sqrt{2}}{\sqrt{5} + \sqrt{2}}$
41. $\dfrac{2\sqrt{5} - 1}{3\sqrt{5} + 3}$
42. $\dfrac{2\sqrt{7} + \sqrt{3}}{3\sqrt{7} - 5\sqrt{3}}$
43. $\dfrac{1}{(\sqrt{2} - 2\sqrt{3})^2}$

Specify the domain of each inequality.

44. $\sqrt{x + 7} < 1 + \sqrt{4 - 3x}$
45. $\dfrac{1}{x^2 - 9} > \sqrt{x + 2} - 7$

Determine whether the given set is bounded; if it is, specify the least upper bound.

46. $\{x: -2 \leq x < 7\}$
47. $\{x: x^2 > 2\}$
48. $\{x: x^2 + 1 < 3\}$
49. $\left\{\dfrac{n-1}{n}: n \in \mathbb{N}\right\}$
50. $\{2^{-n}: n \in \mathbb{N}\}$
51. $\{x \in \mathbb{N}: x^3 < 100\}$

52. Prove Theorem 1–14b: If $a > 0$ and $b > 0$ and $n \in \mathbb{N}$, then $\sqrt[n]{\dfrac{a}{b}} = \dfrac{\sqrt[n]{a}}{\sqrt[n]{b}}$.

C 53. Prove that if $a > 0$ and $b > 0$, then $\dfrac{a+b}{2} \geq \sqrt{ab}$. [HINT: $(\sqrt{a} - \sqrt{b})^2 \geq 0$]

Rationalize each denominator.

54. $\dfrac{1}{2 + \sqrt{3} - \sqrt{2}}$

55. $\dfrac{1}{4 - 2\sqrt[3]{3} + \sqrt[3]{9}}$

56. Prove that there is at most one positive number b such that $b^2 = 5$.
    [HINT: Assume $b > 0$ and $c > 0$ and that $b^2 = 5$ and $c^2 = 5$.]

57. Prove that if $a, b, c,$ and d are rational numbers and $a + b\sqrt{2} = c + d\sqrt{2}$,
    then $a = c$ and $b = d$.

1–6 Absolute Value and Exponents

It is important to recognize that if n is odd, then $\sqrt[n]{a^n} = a$. But if n is even, then this simplification may not be valid. For example, if $n = 4$ and $a = -2$, then

$$\sqrt[n]{a^n} = \sqrt[4]{(-2)^4} = \sqrt[4]{16} = 2 = -(-2) = -a$$

On the other hand, if $n = 4$ and $a = 2$, then

$$\sqrt[n]{a^n} = \sqrt[4]{2^4} = 2 = a.$$

This confusion can be resolved by using absolute value.

Definition For all $a \in \mathbf{R}$, the absolute value of a, denoted $|a|$, is given by

$$|a| = \begin{cases} a & \text{if } a \geq 0, \\ -a & \text{if } a < 0. \end{cases}$$

You can see that if n is even, then $\sqrt[n]{a^n} = |a|$; in particular $\sqrt{a^2} = |a|$.

Example 1 If $a < 0$, simplify $a - \sqrt{a^2 - 4a + 4}$.

SOLUTION $a - \sqrt{a^2 - 4a + 4} = a - \sqrt{(a-2)^2}$
$= a - |a - 2|.$

But since $a < 0$, $a - 2 < 0$. Hence $|a - 2| = -(a - 2)$.

$a - \sqrt{a^2 - 4a + 4} = a - [-(a - 2)]$
$= 2a - 2$

The definition implies that $|0| = 0$ and that for $a \neq 0$, $|a| > 0$. Thus $|a|$ is never negative.

Theorem 1–15   For all $a, b \in R$:

a. $|a| \geq 0$
b. $-|a| \leq a \leq |a|$
c. $|-a| = |a|$
d. $|a|^2 = a^2 = |a^2|$
e. $|ab| = |a||b|$
f. $\left|\dfrac{a}{b}\right| = \dfrac{|a|}{|b|},\ b \neq 0$

PROOF   b. Since $|a| \geq 0$ (from Part a), $-|a| \leq 0 \leq |a|$. Consider the two cases, $a \geq 0$ and $a \leq 0$.
If $a \geq 0$, then $|a| = a$. Hence $-|a| \leq 0 \leq a = |a|$.
If $a < 0$, then $-|a| = a$. Hence $-|a| = a < 0 \leq |a|$.

Since $a = |a|$ implies $a \leq |a|$ and $-|a| = a$ implies $-|a| \leq a$, it follows that for both cases, $-|a| \leq a \leq |a|$.

e. Recall that for $a \in R$, $\sqrt{a^2} = |a|$.
$|ab| = \sqrt{(ab)^2} = \sqrt{a^2 b^2} = \sqrt{a^2} \cdot \sqrt{b^2} = |a||b|$.

Parts **c**, **d**, and **f** of Theorem 1–15 are also proved by using the fact that $|a| = \sqrt{a^2}$.

Example 2   Solve: $|2x - 3| = -5$.

SOLUTION   By Theorem 1–15a, $|2x - 3| \geq 0$ for all $x \in R$. Hence the given equation has no solution.

Example 3   Solve: $|2x - 3| = 7$.

SOLUTION   From the definition, $|2x - 3| = 7$ if and only if
$$2x - 3 = 7 \text{ or } 2x - 3 = -7.$$
$$x = 5 \text{ or } \quad x = -2$$

ANSWER   The solution set is $\{-2, 5\}$.

The absolute value of a number is the measure of its distance from the origin on a number line. Thus $\{x : |x| < 3\}$ describes those numbers that correspond to points on a number line that lie within 3 units of the origin, that is, in the interval $(-3, 3)$. The numbers in $\{x : |x| \geq 2\}$ correspond to points on the number line that are at least two units from the origin.

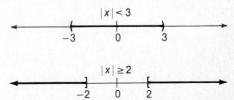

1–6: Absolute Value and Exponents

These examples suggest the next theorem, which is given without proof.

Theorem 1–16 For all $a \in R$ and $p > 0$:

    a. $|a| < p$ if and only if $-p < a < p$.

    b. $|a| > p$ if and only if $a < -p$ or $a > p$.

Example 4   Solve: $|3x - 2| > 7$.

SOLUTION   Use Theorem 1–16b: $3x - 2 > 7$ or $3x - 2 < -7$.

$$3x > 9 \text{ or } 3x < -5$$
$$x > 3 \text{ or } x < -\tfrac{5}{3}$$

ANSWER   The solution is $(-\infty, -\tfrac{5}{3}) \cup (3, \infty)$.

Next, the notion of exponentiation is extended to include rational powers of positive numbers. This extension is an outgrowth of the definition of the principal nth root.

Definition   If $a > 0$, $p \in J$, and $q \in N$, then $a^{p/q} = (\sqrt[q]{a})^p = \sqrt[q]{a^p}$.

Example 5   Simplify $81^{3/4}$ and $8^{-4/3}$.

SOLUTION $81^{3/4} = (\sqrt[4]{81})^3 = 3^3 = 27$

$8^{-4/3} = (\sqrt[3]{8})^{-4} = 2^{-4} = \dfrac{1}{2^4} = \dfrac{1}{16}$

You will recall the basic laws which hold for integer exponents:
For all $a, b \in R$ and $m, n \in J$,

$$a^m \cdot a^n = a^{m+n}; \quad (ab)^m = a^m b^m; \quad \text{and } (a^m)^n = a^{mn}.$$

Working from these laws and the definition of a rational exponent, it is possible to prove that the laws of exponents are valid for rational exponents. These laws are summarized in the next theorem, for which proof is omitted. Note carefully that the hypothesis specifies that the bases must be positive, since $a^{1/q}$ is undefined if $a < 0$ and q is even.

Theorem 1–17 Given that a and b are positive real numbers, and that r and s are rational numbers. Then:

a. $a^r \cdot a^s = a^{r+s}$
b. $(ab)^r = a^r b^r$
c. $(a^r)^s = a^{r \cdot s}$
d. $a^r \div a^s = a^{r-s}$
e. $\left(\dfrac{a}{b}\right)^r = \dfrac{a^r}{b^r}$
f. $a^{-r} = \dfrac{1}{a^r}$

Example 6 Simplify each expression. Assume $c > 0$ and $d > 0$.

a. $(4c)^{1/2} c^{-1/4}$
b. $\left(\dfrac{27c^3}{d^{-6}}\right)^{-2/3}$

SOLUTION
a. $(4c)^{1/2} c^{-1/4} = (4^{1/2} c^{1/2}) c^{-1/4}$
$= 4^{1/2}(c^{1/2} c^{-1/4})$
$= 4^{1/2} c^{1/2 - 1/4}$
$= 4^{1/2} c^{1/4}$
$= 2\sqrt[4]{c}$

b. $\left(\dfrac{27c^3}{d^{-6}}\right)^{-2/3} = \dfrac{(27c^3)^{-2/3}}{(d^{-6})^{-2/3}}$
$= \dfrac{27^{-2/3} c^{-2}}{d^4}$
$= \dfrac{1}{9c^2 d^4}$

Example 7 The time in days for a planet to complete its orbit about the sun is given by

$$T = (0.4076 \times 10^{-9}) R^{3/2},$$

where R is the average distance of the planet from the sun. If the average distance of Mars from the sun is 142 million miles, what is the approximate length of a Martian year?

SOLUTION $T = (0.4076 \times 10^{-9})(142 \times 10^6)^{3/2}$
$= (0.4076 \times 10^{-9})(142^{3/2} \times 10^9)$
$= 0.4076 \times 142^{3/2}$

ANSWER Using a calculator, you should find that $T \doteq 690$ days.

Exercises

A Simplify each expression.

1. $4^{3/2}$
2. $(-27)^{1/3}$
3. $-16^{1/2}$
4. $-216^{2/3}$
5. $\left(\dfrac{16}{81}\right)^{5/4}$
6. $\left(\dfrac{4}{100}\right)^{3/2}$
7. $\left(\dfrac{16}{9}\right)^{-3/2}$
8. $\left(\dfrac{-27}{125}\right)^{-2/3}$

9. $4a^0 + \left(\dfrac{4}{9}\right)^{-1/2} - 8^{2/3} + \left(\dfrac{1}{16}\right)^{-3/4}$

10. $\left(\dfrac{1}{27}\right)^{-2/3} - (2x)^0 - 27^{1/3} - 3^{-2}$

Solve for x.

11. $|x + 3| = 7$
12. $|7x + 1| = 0$
13. $|3x - 1| = -2$
14. $|4 - 2x| = 1$
15. $\sqrt{(x - 1)^2} = 2$
16. $\sqrt{(3 + x)^2} = 6$
17. $|x - 2| < 3$
18. $|2x + 1| \geq 1$
19. $|2x - 13| < -3$
20. $|4 - x| > 0$
21. $|5 - 2x| > 7$
22. $|3x + 4| \leq 4$

B Simplify each expression. The result should contain only positive exponents.

23. $p^{1/2} \cdot 2p^{4/3}$
24. $(4m^{1/3})^{-3/2}$
25. $3b^{2/3} \div 2b^3$
26. $\left[\dfrac{27a^3 b^{-2/3}}{c^{-6}}\right]^{1/3}$
27. $\dfrac{(16x)^{1/2}(x^{-2/3})^6}{x^{3/2}}$
28. $\left[\dfrac{-2c^{-1/3}}{b^{2/3} c^{1/2}}\right]^6$
29. $p^{2/3}(3p^{-2/3} + 2p^{1/3})$
30. $(a^{1/2} - 3)(a^{1/2} + 3)$
31. $(2m^{1/3} - 3)^2 - 9$
32. $(b^{1/3} + 2)(b^{2/3} - 2b^{1/3} + 4)$

Solve each inequality.

33. $\left|\dfrac{3x - 7}{2}\right| < 7$
34. $\left|\dfrac{5 - 2x}{3}\right| < 1$
35. $|x^2 + 1| > 5$
36. $\dfrac{1}{|x - 2|} > 3$
37. $\left|\dfrac{4}{2x + 1}\right| < 3$
38. $|3 + x^2| < 6$

39. Prove Theorem 1–15c and 1–15d: For all $a \in R$, **c.** $|-a| = |a|$ and **d.** $|a|^2 = a^2 = |a^2|$.

40. Prove Theorem 1–15f: For all $a, b \in R$, $b \neq 0$, $\left|\dfrac{a}{b}\right| = \dfrac{|a|}{|b|}$.

41. If the average distance of the planet Venus from the sun is 67.2 million miles, find the time in days required for Venus to complete one revolution about the sun.

42. Find Saturn's average distance from the sun if it requires 29.5 earth years to complete its orbit about the sun.

43. For each of the following, specify whether the equation is True or False and provide a counterexample for each False answer. Assume $a, b \in R$.

  a. $|b^3| = b^2|b|$
  b. $\sqrt[3]{a^3} = |a|$
  c. $|a + b| = |a| + |b|$
  d. $\sqrt{a^2 + b^2} = |a| + |b|$

C Solve for x.

44. $|x^2 - 4| < 5$

45. $|3 - x^2| > 2$

46. In calculus frequent use is made of the following theorem, which is called the triangle inequality: For all $a, b \in \mathbf{R}$, $|a + b| \leq |a| + |b|$.
  a. Find two values of a and of b such that $|a + b| < |a| + |b|$.
  b. Prove the theorem. [HINT: Use Theorem 1–15b.]

1–7 Quadratic Equations

In an earlier section you determined how long it would take a ball dropped from a height of 80 feet to reach the ground. Now suppose that instead of being dropped, the ball is thrown upward from that height with an initial vertical velocity of 64 ft./sec. Again the question is, "How long will it take for the ball to hit the ground?"

Under the given conditions, the height h of the ball is

$$h = -16t^2 + 64t + 80,$$

where t is the time in seconds after the ball is thrown. To answer the question, you must solve the quadratic equation.

$$-16t^2 + 64t + 80 = 0.$$

An equation of the form $ax^2 + bx + c = 0$, where a, b, and c are constants with $a \neq 0$, is called a **quadratic equation**. The solution of quadratic equations is based on the principle that if the product of two numbers is zero, then at least one of the factors is equal to zero.

Theorem 1–18 For all $a, b \in \mathbf{R}$, if $ab = 0$, then $a = 0$ or $b = 0$.

PROOF There are two possibilities for a; either $a = 0$ or $a \neq 0$.
If $a = 0$, the conclusion that $a = 0$ or $b = 0$ is satisfied.
If $a \neq 0$, then a^{-1} exists. By hypothesis, $ab = 0$. Hence $a^{-1}(ab) = a^{-1} \cdot 0$, and it follows that $b = 0$, and again the conclusion is satisfied.

Theorem 1–18 is the basis for using factoring to solve the quadratic equation which appeared above.

$$-16t^2 + 64t + 80 = 0$$
$$-16(t^2 - 4t - 5) = 0$$
$$(t - 5)(t + 1) = 0$$
$$t - 5 = 0 \quad \text{or} \quad t + 1 = 0 \qquad \text{Use Theorem 1–18.}$$
$$t = 5 \quad \text{or} \quad t = -1$$

The domain of the equation does not include negative numbers, so the conclusion is that the ball strikes the ground 5 seconds after it is thrown.

If the domain of the equation above is **R**, then the equation has two real roots. This is not true of every quadratic equation. Consider, for example, the equation $t^2 - 4t + 5 = 0$. By writing the equivalent equation

$$t^2 - 4t + 4 = -1, \quad \text{or}$$
$$(t - 2)^2 = -1,$$

you can see that the equation has no real solutions, because the square of any real number cannot be negative. On the other hand, the equation $t^2 - 4t + 4 = 0$ is equivalent to $(t - 2)^2 = 0$, and this equation has 2 as its only solution.

The question of whether an equation of the form $ax^2 + bx + c = 0$ has two, one, or no real solutions is determined by the number $D = b^2 - 4ac$, which is called the **discriminant**.

Theorem 1–19 Given the quadratic equation $ax^2 + bx + c = 0$ with $a \neq 0$.

a. If $b^2 - 4ac < 0$, the equation has no real solutions.

b. If $b^2 - 4ac = 0$, the equation has one solution: $x = -\dfrac{b}{2a}$.

c. If $b^2 - 4ac > 0$, the equation has two real solutions:
$$x = \frac{-b \pm \sqrt{b^2 - 4ac}}{2a}.$$

PROOF First note that the equation $x^2 = p$ has no real solutions if $p < 0$; has 0 as its only solution if $p = 0$; and has $\sqrt{p}$ and $-\sqrt{p}$ as its solutions if $p > 0$.

Now complete the square to transform the quadratic equation into an equivalent equation of the form $x^2 = p$.

$$ax^2 + bx + c = 0$$
$$x^2 + \frac{b}{a}x = -\frac{c}{a} \qquad \text{Divide each term by } a.$$

(cont. on p. 35)

$$x^2 + \frac{b}{a}x + \left(\frac{b}{2a}\right)^2 = -\frac{c}{a} + \left(\frac{b}{2a}\right)^2$$
Add the square of one-half the coefficient of x to both sides.

$$\left(x + \frac{b}{2a}\right)^2 = \frac{b^2 - 4ac}{4a^2}$$

Since $4a^2 > 0$, the sign of the right side depends only upon the sign of the numerator, or the discriminant, $D = b^2 - 4ac$.

a. If $D < 0$, then $\frac{D}{4a^2} < 0$ and the equation has no real solutions.

b. If $D = 0$, then $\frac{D}{4a^2} = 0$, which means the only solution is $x = -\frac{b}{2a}$.

c. If $D > 0$, then

$$x + \frac{b}{2a} = \pm\sqrt{\frac{b^2 - 4ac}{4a^2}}, \text{ or}$$

$$x = -\frac{b}{2a} \pm \frac{\sqrt{b^2 - 4ac}}{|2a|}.$$

By considering the two cases, $a > 0$ and $a < 0$, you will find that in either case

$$x = \frac{-b \pm \sqrt{b^2 - 4ac}}{2a}.$$

Example 1 Solve the equation $2x^2 - 3x + 7 = 0$.

SOLUTION Since $D = (-3)^2 - 4(2)(7) = -47$, the equation has no real solutions.

Example 2 Solve the equation $x^2 = 5 + \sqrt{2}x$.

SOLUTION First write the equation in standard quadratic form.

$$x^2 - \sqrt{2}x - 5 = 0$$

Since $D = (-\sqrt{2})^2 - 4(1)(-5) = 22$, the equation has two solutions:

$$x = \frac{-(-\sqrt{2}) \pm \sqrt{22}}{2(1)} = \frac{\sqrt{2} \pm \sqrt{22}}{2}$$

Example 3 What are the possible values of a constant k such that the equation $x(x + k) = 3(x - k) - 7$ has exactly one solution.

SOLUTION Write the equation in standard quadratic form.

$$x(x + k) = 3(x - k) - 7$$
$$x^2 + kx = 3x - 3k - 7$$

(cont. on p. 36)

1–7: Quadratic Equations

$$x^2 + (k - 3)x + (3k + 7) = 0$$

In this equation, $a = 1$, $b = k - 3$, and $c = 3k + 7$.

$$\begin{aligned} D &= (k - 3)^2 - 4(1)(3k + 7) \\ &= k^2 - 18k - 19 \end{aligned}$$

There will be exactly one solution when $D = 0$.

$$k^2 - 18k - 19 = 0$$
$$(k - 19)(k + 1) = 0$$

Thus 19 and -1 are the possible values for k. You should check these solutions in the original equation.

Quadratic equations may appear in the solution of equations with radical expressions. The usual strategy is to eliminate any radicals by squaring both sides of the equation, a process that may have to be used more than once. However, this technique sometimes leads to difficulties. Consider the equation

$$\sqrt{x + 2} = -3,$$

for which the domain is $[-2, \infty)$. Square both sides to get

$$x + 2 = 9.$$
$$x = 7$$

Even though 7 is in the domain, it does not satisfy the original equation, since $\sqrt{7 + 2} \neq -3$. What went wrong?

The problem is that squaring both sides of an equation is not one of those procedures which produces an equation that is equivalent to the given equation. For example, $x = 2$ and $x^2 = 4$ are not equivalent equations, since $\{-2, 2\}$ is the solution set for the second one. Although false solutions may be introduced by squaring, no solutions are lost by this process. Consequently it is permissible to use the squaring process in solving an equation, provided that all solutions are checked in the original equation.

Example 4 Solve the equation $1 + \sqrt{2 - x} = \sqrt{2x + 2}$.

SOLUTION

$$1 + \sqrt{2 - x} = \sqrt{2x + 2}$$
$$1 + 2\sqrt{2 - x} + (2 - x) = 2x + 2 \qquad \text{Square both sides.}$$
$$2\sqrt{2 - x} = 3x - 1$$
$$4(2 - x) = 9x^2 - 6x + 1 \qquad \text{Square both sides again.}$$
$$9x^2 - 2x - 7 = 0$$

(cont. on p. 37)

$$(9x + 7)(x - 1) = 0 \quad \text{Factor.}$$

$$x = -\frac{7}{9} \quad \text{or} \quad x = 1$$

These solutions must be checked in the original equation.

For $x = -\frac{7}{9}$: $1 + \sqrt{2 - x} = 1 + \sqrt{2 - (-\frac{7}{9})} = \frac{8}{3}$, and
$$\sqrt{2x + 2} = \sqrt{2(-\frac{7}{9}) + 2} = \frac{2}{3}.$$

So $-\frac{7}{9}$ is not a solution of the original equation.

For $x = 1$: $1 + \sqrt{2 - x} = 1 + \sqrt{2 - 1} = 2$, and
$$\sqrt{2x + 2} = \sqrt{2 \cdot 1 + 2} = 2.$$

Thus $x = 1$ is the only solution.

Exercises

A Evaluate the discriminant and determine the number of real solutions.

1. $3x^2 - 2x + 1 = 0$ **2.** $9x^2 - 2x = -4$ **3.** $25x^2 - 30x = -9$ **4.** $7x^2 = 4x + 10$

Solve by factoring.

5. $2x^2 + 5x = 3$ **6.** $18x^2 + 9x = 5$ **7.** $x(2x - 11) = 21$ **8.** $10x^2 - 61x = -33$

Solve by using the quadratic formula.

9. $x^2 + x - 3 = 0$ **10.** $x^2 - x - 1 = 0$

11. $(2x + 1)(3x - 1) = 1$ **12.** $x - \frac{2x - 1}{3} = x^2 - \frac{1}{2}$

B Solve for y, given that m is a nonzero constant.

13. $2y^2 - my = m^2$ **14.** $\frac{y^2 - m^2}{2} = m(y - m) + \frac{1}{2}$

15. $(y + 1)(y - 1) = m\left(y - \frac{m}{4}\right)$ **16.** $\frac{(y - 1)^2}{m} = 2y - m,\ m > 0$

Determine the constant k such that the given equation has only one solution.

17. $x^2 + kx + 1 = 0$ **18.** $2x(4x + 1) = k(2x - 1)$

19. For the equation $ax^2 + bx + c = 0$, with $a \neq 0$, suppose that the discriminant is positive and that the two solutions are r_1 and r_2. Show that

a. $r_1 + r_2 = -\frac{b}{a}$ and **b.** $r_1 r_2 = \frac{c}{a}$.

20. One solution of the quadratic equation $3x^2 - 12x + 2k = 0$ is $2 + \sqrt{3}$. Use the results of exercise 19 to find the other solution and the value of k.

Solve for x.

21. $\sqrt{x + 1} = x - 5$
22. $1 - \sqrt{2x + 1} = x$
23. $\sqrt{3x + 1} - \sqrt{9 - x} = 2$
24. $\sqrt{3x - 5} = 1 + \sqrt{2x - 5}$

Specify the domain and solve for x.

25. $\dfrac{1}{x - 1} - \dfrac{x - 2}{x + 1} = \dfrac{2}{x^2 - 1}$

26. $\dfrac{1}{x + 2} - \dfrac{2x - 3}{2x^2 - x - 1} = \dfrac{2x - 1}{x^2 + x - 2}$

27. A ball is thrown upward from a height of 100 ft. with an initial vertical velocity of 80 ft./sec. The height of the ball t seconds later is given by $h = -16t^2 + 80t + 100$.
    a. How long will it take for the ball to hit the ground?
    b. After how many seconds is the ball 150 feet above the ground?
    c. Does the ball ever reach a height of 250 ft.?

28. A rectangular pool in a water-purification plant requires a surface area of 1240 sq. ft. If the pool is situated in a room with dimensions 70 ft. × 28 ft. and the distance from the pool edge to the room wall is uniform, find the dimensions of the pool.

C Solve each equation.

29. $x^3 - 3x^2 - 4x + 12 = 0$
30. $x^3 = 2 + 2x - x^2$
31. $x^2 + \sqrt{3x^2 - 3} = 1$
32. $x^4 = 2x^2 + 2$

33. Consider the quadratic equation $x^2 + kx + 9 = 0$. For what values of k does this equation have exactly one solution? Two real solutions? No real solution?

34. If r and s are solutions of the quadratic equation $ax^2 + bx + c = 0$, show that

    a. $r^3 + s^3 = \dfrac{3abc - b^3}{a^3}$ and   b. $\dfrac{1}{r^3} + \dfrac{1}{s^3} = \dfrac{3abc - b^3}{c^3}$

    [HINT: Use the results of exercise 19.]

35. Find rational numbers a and b such that $\sqrt{17 + \sqrt{33}} = \sqrt{a} + \sqrt{b}$.

Special — The Binomial Theorem

You are familiar with the expansion

$(a + b)^2 = 1 \cdot a^2 + 2ab + 1 \cdot b^2$,

and you should recognize that

$(a + b)^0 = 1$ and $(a + b)^1 = 1 \cdot a + 1 \cdot b$.

It is easy to show that

$(a + b)^3 = 1 \cdot a^2 + 3a^2b + 3ab^2 + 1 \cdot b^3$ and

$(a + b)^4 = 1 \cdot a^4 + 4a^3b + 6a^2b^2 + 4a^3 + 1 \cdot a^4$

The coefficients in these expansions can be arranged in a triangle as follows:

							Row
$(a+b)^0$			1				0
$(a+b)^1$		1		1			1
$(a+b)^2$	1		2		1		2
$(a+b)^3$	1	3		3	1		3
$(a+b)^4$	1	4	6	4	1		4

Notice that all the numbers on the outside diagonals are 1's and that each interior number is the sum of the two numbers above it. If rows are supplied indefinitely, the resulting pattern is called **Pascal's Triangle**.

For convenience, label the rows of this triangle starting from 0, as shown above. Thus the next two rows of the triangle are the 5th and 6th rows.

1	5	10	10	5	1	Row 5
1	6	15	20	15	6	1	Row 6

To number terms in a given row, again start from 0. Thus in the 6th row the 2nd term is 15. The symbol $_nC_r$ is used to denote the rth term of the nth row. So, $_6C_2 = 15$. To evaluate $_7C_5$, write the 7th row of the triangle to obtain 21 as the required term.

It can be proved that for every natural number n, the numerical coefficients in the expansion of $(a + b)^n$ are precisely those numbers found in the nth row of Pascal's Triangle.

The Binomial Theorem

For all real numbers a and b and for all natural numbers n,
$(a + b)^n = {_nC_0}a^nb^0 + {_nC_1}a^{n-1}b^1 + {_nC_2}a^{n-2}b^2 + \cdots + {_nC_r}a^{n-r}b^r + \cdots + {_nC_{n-1}}a^1b^{n-1} + {_nC_n}a^0b^n$.

Notice that in each term of the expansion the sum of the exponents is n. The coefficients $_nC_r$ are called **binomial coefficients**.

Example 1 Expand $(a + b)^7$.

SOLUTION The 7th row of Pascal's Triangle is

1 7 21 35 35 21 7 1.

$(a + b)^7 = 1 \cdot a^7 + 7a^6b + 21a^5b^2 + 35a^4b^3 + 35a^3b^4 + 21a^2b^5 + 7ab^6 + 1 \cdot b^7$

Example 2 Estimate 0.997^{10} using 3 terms of the expansion of $(1 - 0.003)^{10}$.

SOLUTION From the Binomial Theorem,

$(1 - 0.003)^{10} = {_{10}C_0} \cdot 1^{10} + {_{10}C_1} \cdot 1^9(-0.003) + {_{10}C_2} \cdot 1^8(-0.003)^2 + $ (8 additional terms)

$_{10}C_0 = 1$, $_{10}C_1 = 10$, and $_{10}C_2 = 45$.

Since the 8 additional terms are small,

$(0.997)^{10} \doteq 1 \cdot 1 + 10(-.003) + 45(0.000009)$
$\doteq 1 - 0.03 + 0.000405 \doteq 0.9704$.

Exercises

1. Write the 8th row of Pascal's Triangle.

Expand each of the following.

2. $(x + 2)^7$

3. $(2x + 3)^5$

4. $\left(\dfrac{x}{2} - y\right)^6$

Use the first 3 terms of a binomial expansion to estimate each expression.

5. 0.98^9

6. 1.005^{12}

1-8 Complex Numbers

The equation $-4x + 5 = 0$ has no solution if the domain is the set of integers. This limitation can be overcome by creating a larger number system in which the numbers are ratios, or ordered pairs, of integers. Thus, $Q = \left\{\frac{a}{b} : a, b \in J \text{ and } b \neq 0\right\}$, together with suitable additive and multiplicative operations, constitute the system of rational numbers. In **Q** the number $\frac{5}{4}$ is a solution of the equation $-4x + 5 = 0$.

You know also that the equation $x^2 - 4x + 5 = 0$ has no solution in the set of real numbers **R**, nor, in fact, does any quadratic equation with a negative discriminant. In this section we create the set of complex numbers **C** from the set of real numbers **R** so that all quadratic equations will have solutions in **C**.

The elements of this new system are ordered pairs of real numbers, frequently written as $a + bi$. In this notation, the role of the symbol i may be seen as an indicator that b is the second element of an ordered pair. Later you will see why $i^2 = -1$. In this system, the complex number $2 + 1i$ is a solution of the equation $x^2 - 4x + 5 = 0$, as you will see in example 3.

Definition The **Complex-Number System** consists of the set **C** = $\{a + bi : a, b \in \mathbf{R}\}$ together with the operations of addition $\oplus$ and multiplication $\odot$ defined as follows:

a. $(a + bi) \oplus (c + di) = (a + c) + (b + d)i;$

b. $(a + bi) \odot (c + di) = (ac - bd) + (ad + bc)i.$

Note that in this definition the plus sign is used in three ways. In $a + bi$, it is a notational convenience for indicating that the complex number is the "sum" of two components. The symbol $\oplus$ indicates addition of complex numbers. In $a + c$ and $b + d$, on the right side of the definition of addition, the plus sign indicates real-number addition.

Example 1 Simplify each expression: **a.** $(3 + 2i) \oplus (-1 + 5i)$.
b. $(3 + 2i) \odot (-1 + 5i)$.

SOLUTION **a.** $(3 + 2i) \oplus (-1 + 5i) = (3 + [-1]) + (2 + 5)i$
$= 2 + 7i$

b. $(3 + 2i) \odot (-1 + 5i) = [3(-1) - 2(5)] + [3(5) + 2(-1)]i$
$= (-3 - 10) + [15 + (-2)]i$
$= -13 + 13i$

There is a one-to-one correspondence between complex numbers $a + 0i$ and real numbers a. So the set of real numbers is a subset of the set of complex numbers. It is customary to denote the complex number $a + 0i$ by a.

Complex numbers of the form $0 + bi$ are called **pure imaginary** numbers. The notation is usually simplified to bi, and $1i$ becomes just i. Every complex number $a + bi$ can be viewed as the "sum" of a real number and a pure imaginary number; a is called the **real part** and b the **imaginary part**.

Definition Complex numbers $a + bi$ and $c + di$ are **equal complex numbers** if $a = c$ and $b = d$.

Example 2 Solve for z on the domain **C**: $(3 + 2i) \odot z = 11 - 10i$.

SOLUTION Let $z = x + yi$, where $x, y \in \mathbf{R}$.

$$(3 + 2i) \odot (x + yi) = 11 - 10i$$
$$(3x - 2y) + (3y + 2x)i = 11 - 10i$$

Use the definition of equal complex numbers.

$$3x - 2y = 11 \quad \text{and} \quad 3y + 2x = -10$$

Solve these two equations simultaneously to get

$$x = 1 \text{ and } y = -4$$

ANSWER The answer is $z = 1 - 4i$.

By using the properties of the real number system, it can be proved that $\oplus$ and $\odot$ are commutative and associative operations on **C** and that $\odot$ distributes over $\oplus$.

The following demonstration that $\oplus$ is associative illustrates how to verify such statements.

Consider any three complex numbers $z_1 = a + bi$, $z_2 = c + di$, and $z_3 = e + fi$, where $a, b, c, d, e, f \in \mathbf{R}$, and show that

$$(z_1 \oplus z_2) \oplus z_3 = z_1 \oplus (z_2 \oplus z_3).$$

$$\begin{aligned}
(z_1 \oplus z_2) \oplus z_3 &= [(a + bi) \oplus (c + di)] \oplus (e + fi) \\
&= [(a + c) + (b + d)i] \oplus (e + fi) \\
&= [(a + c) + e] + [(b + d) + f]i \\
&= [a + (c + e)] + [b + (d + f)]i \\
&= (a + bi) \oplus [(c + e) + (d + f)i] \\
&= z_1 \oplus [(c + di) \oplus (e + f)i] \\
&= z_1 \oplus (z_2 \oplus z_3)
\end{aligned}$$

It is easy to see that $0 + 0i$ and $1 + 0i$ are the identities for $\oplus$ and $\odot$ respectively. You can also verify that $-a + (-b)i$ is the additive inverse of $a + bi$ and that $\left(\dfrac{a}{a^2 + b^2}\right) + \left(\dfrac{-b}{a^2 + b^2}\right)i$ is its multiplicative inverse, provided a and b are not both zero. (Recall that the additive identity in the real-number system has no multiplicative inverse.) These results suggest the following theorem.

Theorem 1–20 The field axioms are satisfied by the complex-number system.

Since the field axioms are satisfied by the complex-number system, all the theorems derived for the real-number system from the field axioms are valid in the complex-number system. For example, if z_1 and z_2 are complex numbers such that $z_1 \odot z_2 = 0 + 0i$, then $z_1 = 0 + 0i$ or $z_2 = 0 + 0i$. Note also that all of the techniques for producing equivalent equations are applicable to solving equations in **C**. All of this suggests that we dispense with the special symbols $\oplus$ and $\odot$.

It is now easy to verify that $i^2 = -1$.

$$i^2 = (0 + 1i) \cdot (0 + 1i) = (0 - 1) + (0 + 0)i = -1$$

The number i helps to establish the basic structural difference between the real-number system and the complex-number system. Although both systems incorporate the field axioms and theorems, the complex number system is not ordered; that is, the order axioms don't apply to complex numbers.

To demonstrate this fundamental difference, use indirect proof and proceed as follows. Suppose that the order axioms *do* apply. Then since i and 0 are both in **C**, the trichotomy axiom implies that

$$i > 0 \quad \text{or} \quad i = 0 \quad \text{or} \quad i < 0.$$

If $i > 0$, then $i \cdot i > 0$, or $-1 > 0$, which is a contradiction.

If $i = 0$, then $i \cdot i = 0 \cdot i$. But $-1 \neq 0$, so the assumption that $i = 0$ also leads to a contradiction.

If $i < 0$, then $i \cdot i > 0 \cdot i$, so that $-1 > 0$, which is also a contradiction.

Since all three cases are impossible, the conclusion is that the trichotomy axiom does not apply in the complex-number system. Therefore **C** is not ordered and there can be no order relationship on **C**. Statements like $10 + 2i > 1 + i$ simply are not meaningful.

We use the fact that $i^2 = -1$ to define the symbol $\sqrt{a}$ for $a < 0$.

Definition If p is a positive real number, then $\sqrt{-p} = i\sqrt{p}$.

From the definition, $\sqrt{-25} = i\sqrt{25} = 5i$, so that $5i$ is called a square root of -25.

A word of caution: Previous theorems dealing with $\sqrt{a}$ do not always apply when $a < 0$. For instance, if a and b are negative real numbers, then $\sqrt{ab} \neq \sqrt{a}\sqrt{b}$. For instance,

$$\sqrt{-2}\sqrt{-3} = (i\sqrt{2})(i\sqrt{3}) = i^2\sqrt{6} = -\sqrt{6},$$

but

$$\sqrt{(-2)(-3)} = \sqrt{6}.$$

Now you can see that the quadratic formula can be used to solve *any* quadratic equation with real-number coefficients.

Theorem 1–21 If $a, b, c \in R$ and $a \neq 0$, then the equation $ax^2 + bx + c = 0$

has solutions $\dfrac{-b \pm \sqrt{b^2 - 4ac}}{2a}$ which are

a. Real and unequal if $b^2 - 4ac > 0$.
b. Real and equal if $b^2 - 4ac = 0$.
c. Nonreal complex numbers if $b^2 - 4ac < 0$.

Example 3 Solve in C: $x^2 - 4x + 5 = 0$.

SOLUTION Use the quadratic formula.

$$x = \frac{-(-4) \pm \sqrt{(-4)^2 - 4(1)(5)}}{(2)(1)}$$

$$= \frac{4 \pm \sqrt{-4}}{2} = \frac{4 \pm 2i}{2}$$

ANSWER The solution set is $\{2 + i, 2 - i\}$.

Notice that if the discriminant of a quadratic equation is negative, then the quadratic formula generates solutions of the form $a + bi$ and $a - bi$, which are said to be **complex conjugates**.

Example 4 Solve in C: $x^3 - 7x^2 + 4x - 28 = 0$

SOLUTION
$$x^3 - 7x^2 + 4x - 28 = 0$$
$$x^2(x - 7) + 4(x - 7) = 0$$
$$(x - 7)(x^2 + 4) = 0$$
$$x - 7 = 0 \quad \text{or} \quad x^2 + 4 = 0$$
$$x = 7 \quad \text{or} \quad x^2 = -4$$
$$x = 7 \quad \text{or} \quad x = 2i \text{ or } x = -2i$$

ANSWER The solution set is $\{7, 2i, -2i\}$.

Exercises

A Simplify each expression.

1. $(2 - i) + (4 + 5i)$
2. $(3 + 2i) + (-4 - 6i)$
3. $(2 - i) \cdot (4 + 5i)$
4. $(3 + 2i) \cdot (-4 - 6i)$
5. $3i(4 - 7i)$
6. $(2i + 1)^2$
7. $4i^3$
8. $i^4 - 7$

Use the definition of equality for complex numbers to solve for real numbers x and y.

9. $x + yi = 3 + 4i$
10. $7 + yi = 2x - 3i$

Find the additive and multiplicative inverses of each complex number. Verify each answer by computation.

11. $2 + 3i$
12. $-5 + 2i$
13. $1 - 3i$
14. $2i$

Simplify each expression.

15. $\sqrt{-16}\sqrt{-9}$
16. $\sqrt{-10}\sqrt{-5}$
17. $\sqrt{-2}(\sqrt{3} - \sqrt{-2})$
18. $(\sqrt{-8} - \sqrt{3})^2$

B Solve for x on the domain C.

19. $x^2 + x + 1 = 0$
20. $x^2 - 4x + 6 = 0$
21. $-5x^2 + 2x = 2$
22. $x(3 - 4x) = 7$
23. $2x^3 - x^2 + 18x - 9 = 0$
24. $x^4 - 16 = 6x^2$

25. Prove that complex-number multiplication in C is commutative.
26. Prove that complex-number addition in C is associative.
27. Show that if $a + bi$ is not zero, then $\left(\dfrac{a}{a^2 + b^2}\right) + \left(\dfrac{-b}{a^2 + b^2}\right)i$ is the multiplicative inverse of $a + bi$.

Solve for z on the domain $\mathbf{C}$.

28. $(1 + 2i)z = 1$

29. $(3 - i)z = 2i$

30. Find real constants p and q so that the equation $x^2 + px + q = 0$ will have $2 - 5i$ as a solution.

31. Find real constants b and c so that the equation $x^2 + bx + c = 0$ will have $3 + 2i$ as a solution.

C **32.** Without reference to the quadratic formula, show that if the complex number $u + vi$ is a solution of the equation $ax^2 + bx + c = 0$, where $a, b, c \in \mathbf{R}$, then $u - vi$ is also a solution.

33. Show that if $u + vi$ is a solution of the cubic equation $ax^3 + bx^2 + cx + d = 0$, where $a, b, c, d \in \mathbf{R}$, then $u - vi$ is also a solution.

34. The complex number $a + bi$ can be graphed as the point (a, b). Graph the numbers u, v, w, ui, vi, and wi, where $u = 2 + 3i$, $v = 3i$, and $w = 2 - 2i$. What geometric effect does multiplication by i have?

35. Repeat question 34, this time multiplying u, v, and w by $1 + i$.

Chapter Review Exercises

1–1, page 1

1. Specify to which of the sets $\mathbf{N, W, J, Q}$, and $\mathbf{R}$ each of the following numbers belongs:

   a. $-\frac{51}{17}$ **b.** 1.414 **c.** $\frac{\sqrt{3}}{2}$ **d.** $-1.3151515\ldots$

2. Determine whether the given set is closed under the specified operation. If it is not, give an appropriate counterexample.

   a. $\{-1, -\frac{1}{2}, 0, \frac{1}{2}, 1\}$; multiplication

   b. $\{3m + 1 : m \in \mathbf{N}\}$; $\#$ defined by $a \# b = a + b + 2$

1–2, page 5

3. Does addition distribute over multiplication? Justify your answer.

4. Justify each step of the following simplification by citing an appropriate definition, axiom, or theorem (assume $3 + 4 = 7$).

$$-(-7 + 4) = -(-7) + (-4)$$
$$= 7 + (-4)$$
$$= (3 + 4) + (-4)$$
$$= 3 + [4 + (-4)]$$
$$= 3 + 0$$
$$= 3$$

1–3, page 9

5. Factor each of the following completely:

   a. $81x^3 - 3$

   b. $3a^3 + a - 6a^2b - 2b$

6. Simplify: $\dfrac{x + \dfrac{4y^2}{x - 2y}}{\dfrac{x}{x^2 - 4y^2}}$

1–4, page 15

7. Determine the domain of the inequality $\dfrac{1}{\sqrt{2x + 9}} \leq 4x + 2$.

8. Solve for x: $\dfrac{x + 2}{4} < 1 - \dfrac{1 - 2x}{6}$.

1–5, page 22

9. Simplify $A \cap B$ if $A = \{x : x < -5 \text{ or } x \geq 2\}$ and $B = [-3, 4)$.

10. Simplify: $\dfrac{\sqrt{6}}{2\sqrt{3} - 3\sqrt{2}}$.

1–6, page 28

11. Simplify each expression.

   a. $3a^{1/2}(2a^{-1/2} - 3a^{3/2})$

   b. $\left(\dfrac{-27x^2}{b^{-3/2}}\right)^{2/3}$

12. Solve for x.

   a. $|2x - 3| < 1$

   b. $\left|\dfrac{1 - 3x}{2}\right| \geq 4$

1–7, page 33

13. For what values of m does the equation $mx^2 + 8x = 4$ have no real solutions?

14. Solve for x: $x + \sqrt{2x + 3} = 6$.

1–8, page 40

15. Find the multiplicative inverse of $2 + 5i$, express it in the form $a + bi$, and simplify.

16. Solve the equation $3x^3 - 4x^2 + 3x = 0$ for x on the domain C.

Chapter Test

1. If $R = \{x : x \leq 1\}$ and $S = (0, 3]$, express $R \cap S$ using interval notation.

Factor each expression completely.

2. $125x^6 + y^3$

3. $4a^4 - a^2 - 1 + 2a$

4. If $\#$ is the operation defined for all real numbers a and b by $a \# b = \dfrac{ab}{2}$, prove or disprove that $\#$ distributes over addition.

5. Simplify: $\left(a + b - \dfrac{4ab - 5b^2}{a - b}\right) \div \left(\dfrac{a^2 - ab - 2b^2}{2a - 2b}\right)$.

6. Solve on domain C: $4x^2 + 1 = 3x$.

7. Simplify: $\sqrt{27} - \dfrac{\sqrt{3}}{2\sqrt{6} + 3}$.

8. Solve for x: $|3x - 4| < 17$.

9. Solve for x: $(x^2 + 2)(4 - 3x) \leq 0$.

10. Specify whether each statement is True or False.
    a. $\{3n - 1 : n \in J\}$ is closed under multiplication.
    b. $\sqrt{4x^2 - 12x + 9} = 2x - 3$
    c. $(2i)^5 = 32i$
    d. If $ab < 0$ and $a < 0$, then $a^2 b > 0$.
    e. $\sqrt[4]{16} = \pm 2$

11. What are the possible values of a constant p such that the equation
$$px^2 - 2\sqrt{2}x + p = 1$$
has exactly one solution?

12. Solve for x: $\sqrt{2x + 3} - 1 = \sqrt{3 - 4x}$.

Analytic Geometry
Lines and Circles

2

2-1 Distance and Midpoint Formulas

A manufacturer has orders for 50 Model PA-99 computers, 24 to be shipped to Boston, 16 to Trenton, and 10 to Rutland. These orders will be filled using 28 computers in a Hartford warehouse and 22 in a New York warehouse. Transportation costs per computer are as follows.

From \ To	Boston	Trenton	Rutland
Hartford	$11	$16	$12
New York	$20	$10	$24

For example, a decision to ship 24 computers from Hartford to Boston, 4 from Hartford to Trenton, 12 from New York to Trenton, and 10 from New York to Rutland would incur shipping costs of $(24)(11) + (4)(16) + (12)(10) + (10)(24)$, or $688. Is there a cheaper way to arrange the deliveries?

In this chapter you will learn how to solve this kind of problem by a method known as *linear programming*. First it is necessary to review some analytic geometry.

A triangle has vertices $A(2, 3)$, $B(3, -2)$, and $C(-5, -1)$. Is it an isosceles triangle? In particular, is it true that $AC = BC$? What is needed to answer this question is a formula that gives the distance between two points in the plane in terms of their coordinates.

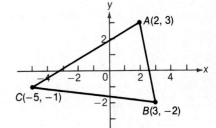

If two points are on a line parallel to a coordinate axis, the distance between them is found by using absolute value. In the figure,

$$QR = |y_2 - y_1| \text{ and } PR = |x_2 - x_1|.$$

Combining this with the Pythagorean theorem and the fact that $|x_2 - x_1|^2 = (x_2 - x_1)^2 = (x_1 - x_2)^2$, you can easily prove the following theorem concerning the distance between any two points in the plane. The proof is left as an exercise.

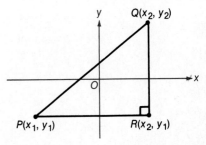

Theorem 2-1 Distance Formula

If $P(x_1, y_1)$ and $Q(x_2, y_2)$ are points in the coordinate plane, the distance between them is

$$PQ = \sqrt{(x_2 - x_1)^2 + (y_2 - y_1)^2}.$$

Now you can see that for $\triangle ABC$ shown on page 49.
$$BC = \sqrt{[3-(-5)]^2 + [-2-(-1)]^2} = \sqrt{65}, \text{ and}$$
$$AC = \sqrt{[2-(-5)]^2 + [3-(-1)]^2} = \sqrt{65},$$
and the triangle is isosceles.

Suppose that in the $\triangle ABC$ shown on page 49 you need to know the length of the median from vertex A to the midpoint of $\overline{BC}$. First you must find the coordinates of the midpoint of $\overline{BC}$.

Theorem 2–2 **Midpoint Formula**
For points $P(x_1, y_1)$ and $Q(x_2, y_2)$, the midpoint of $\overline{PQ}$ is
$$M\left(\frac{x_1 + x_2}{2}, \frac{y_1 + y_2}{2}\right).$$

PLAN FOR PROOF Establish that $PM = MQ$ and that $PM + MQ = PQ$.

PROOF Use the distance formula.
$$PM = \sqrt{\left[x_1 - \left(\frac{x_1 + x_2}{2}\right)\right]^2 + \left[y_1 - \left(\frac{y_1 + y_2}{2}\right)\right]^2}$$
$$= \sqrt{\left(\frac{x_1 - x_2}{2}\right)^2 + \left(\frac{y_1 - y_2}{2}\right)^2}$$
$$= \frac{1}{2}\sqrt{(x_1 - x_2)^2 + (y_1 - y_2)^2}$$
$$= \frac{1}{2}PQ$$

To complete the proof, first check that MQ is also equal to $\frac{1}{2}PQ$. (Why isn't it enough just to show that $PM = MQ$?)

Example 1 Find the midpoint M of segment PQ that has endpoints $P(-1, 3)$ and $Q(-2, -7)$.

SOLUTION Use the midpoint formula.
$$M = \left(\frac{-1 + (-2)}{2}, \frac{3 + (-7)}{2}\right)$$
$$= \left(-\frac{3}{2}, -2\right)$$

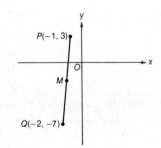

Example 2 One endpoint of $\overline{PQ}$ is $P(3, 1)$ and the midpoint is $M(1, 5)$. Find Q.

SOLUTION Let x_1 and y_1 denote the coordinates of endpoint Q. Then, by the midpoint formula

$$\frac{3 + x_1}{2} = 1 \text{ and } \frac{1 + y_1}{2} = 5, \text{ or}$$
$$x_1 = -1 \text{ and } y_1 = 9.$$

The endpoint is $Q(-1, 9)$.

The distance formula can be used to determine whether a point is inside, on, or outside a given circle.

Example 3 A circle C with radius 5 has its center at the origin O. Which of the points $R(-3, 4)$, $S(1, 4)$, and $T(3, 5)$ is in the interior and which is in the exterior of C?

SOLUTION Use the distance formula to find that

$$OR = 5, \ OS = \sqrt{17}, \text{ and } OT = \sqrt{34}.$$

So R is on C, S is in the interior of C, and T is in the exterior of C.

Exercises

A For points P and Q, calculate PQ and locate the midpoint of $\overline{PQ}$.

1. $P(-1, 0)$, $Q(4, 0)$
2. $P(6, -4)$, $Q(6, -8)$
3. $P(6, -4)$, $Q(-2, -4)$
4. $P(-1, 5)$, $Q(3, 5)$
5. $P(0, 0)$, $Q(6, -8)$
6. $P(-1, -6)$, $Q(2, -2)$
7. $P(14, 15)$, $Q(16, 17)$
8. $P(5, 12)$, $Q(0, 0)$
9. $P(0.01, -0.03)$, $Q(6.05, -7.97)$
10. $P(1.04, -6.22)$, $Q(13.17, -1.92)$

Determine whether the given points are vertices of an isosceles triangle.

11. $(0, 3)$, $(1, 0)$, $(2, 3)$
12. $(0, 3)$, $(-1, 1)$, $(1, -1)$
13. $(1, -4)$, $(-1, 2)$, $(2, 3)$
14. $(1, 0)$, $(1, 2)$, $(7, 1)$

In exercises 15–18, M is the midpoint of $\overline{PQ}$. Find the coordinates of Q.

15. $P(0, 3)$, $M(-1, 5)$
16. $P(1, -2)$, $M(5, -4)$
17. $P(10, -1)$, $M(0, 7)$
18. $P(4, k)$, $M(h, 3)$

Which of the following points are on a circle with radius 10 and center at the origin?

19. (−6, −8)   **20.** (3, 4)   **21.** (−4√3, 7)   **22.** (−3, √91)

Determine whether each point is in the interior or the exterior of the circle with radius √8 and center at (1, 2).

23. (√10, 4)   **24.** (−2, 3)   **25.** $\left(\sqrt{3}, -\dfrac{\sqrt{2}}{2}\right)$ **26.** (4, 5)

B 27. The vertices of △ABC are A(1, 10), B(−2, 0), and C(−3, 5).
   a. What is the perimeter of △ABC?
   b. What is the perimeter of the triangle whose vertices are the midpoints of the sides of △ABC?

For the given vertices, use the Pythagorean theorem to determine whether △ABC is a right triangle.

28. A(−2, 2), B(1, 1), C(0, 8)   **29.** A(−1, 1), B(−2, 6), C(6, 3)

30. The vertices of △RST are R(−2, 3), S(1, 1), and T(3, 4).
   a. Verify that △RST is a right triangle with right angle at S.
   b. Verify that the median drawn from vertex S is one-half as long as $\overline{RT}$.

31. A target for a dart game consists of rings formed by concentric circles of radii 1, 2, 3, 4, and 5 cm. A bullseye is worth 25 points. Darts landing in the remaining rings are scored 10, 5, 2, or 1 point, respectively. If a coordinate system is established with 1 cm as unit distance and the origin at the center of the bullseye, find the score for the player whose five darts land at the following points.

A(−4, 2), B(2, 2), C(−1, −3), D(0.9, 0.8), E(1, 1.3)

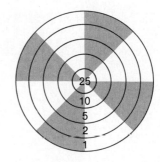

C 32. Prove Theorem 2–1.

33. Complete the proof of Theorem 2–2.

34. The distance between (k, k) and (0, 2) is 2√5. Find the value of k.

35. A point (k, 0) is on the x-axis, and its distance from (1, 2) is √13. Find the value of k.

36. Write a program to have a computer accept as input the coordinates of two points P and Q and print the distance PQ and the coordinates of the midpoint of $\overline{PQ}$.

2–2 Slopes and Equations of lines

In the problem stated at the beginning of section 2–1, let x be the number of computers to be shipped from Hartford to Boston and y the number of computers to be shipped from Hartford to Trenton. The Hartford warehouse contains 28 computers, so $x + y \leq 28$. In the extreme case that none are shipped from Hartford to Rutland, the equation $x + y = 28$ must be satisfied.

This equation is an example of a **linear equation**, so called because its graph in a coordinate plane is a straight line.

You can easily verify that the coordinate pairs in the table are points on the line $x + y = 28$, whose graph is shown.

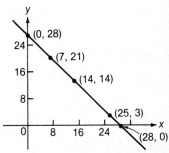

To distinguish among different lines that contain a given point, it is helpful to introduce a quantity called the slope of a line, which is a measure of the steepness of the line. First we define the slope of a segment.

Definition The **slope** m of a nonvertical segment $\overline{P_1 P_2}$ with endpoints $P_1(x_1, y_1)$ and $P_2(x_2, y_2)$ is the quotient $m = \dfrac{y_2 - y_1}{x_2 - x_1}$.

If a segment is vertical, then $x_1 = x_2$ and $x_2 - x_1 = 0$. Since division by zero is not defined, a vertical segment does not have a slope. If a segment is horizontal, then $y_1 = y_2$, $y_2 - y_1 = 0$, and the slope of the segment is 0.

Example 1 Find the slopes of segments $\overline{P_1 P_2}$ and $\overline{Q_1 Q_2}$ determined by $P_1(2, 2)$, $P_2(-3, -\frac{1}{2})$, $Q_1(0, -2)$, and $Q_2(-2, 2)$.

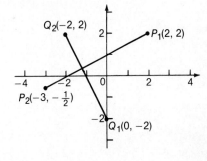

SOLUTION Slope m_1 of $\overline{P_1 P_2}$: $\dfrac{-\frac{1}{2} - 2}{-3 - 2} = \dfrac{1}{2}$

Slope m_2 of $\overline{Q_1 Q_2}$: $\dfrac{2 - (-2)}{-2 - 0} = -2$

If the coordinates of P_2 are subtracted from the coordinates of P_1, the resulting number $\dfrac{y_1 - y_2}{x_1 - x_2}$ is the same as $\dfrac{y_2 - y_1}{x_2 - x_1}$. Be careful to use the same order in both numerator and denominator. Otherwise an incorrect value for the slope will result.

In the expression $\frac{y_2 - y_1}{x_2 - x_1}$ for the slope of $\overline{P_1P_2}$, the numerator is a measure of the vertical change in moving from P_1 to P_2; it is sometimes called the **rise**. The corresponding horizontal change $x_2 - x_1$, is called the **run**. Thus the slope of the segment is $\frac{\text{rise}}{\text{run}}$. If a segment rises to the right, its slope is positive. If the segment falls to the right, its slope is negative.

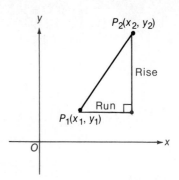

Theorem 2–3 If $P_1(x_1, y_1)$, $P_2(x_2, y_2)$, $P_3(x_3, y_3)$, and $P_4(x_4, y_4)$ are any points on a nonvertical line, then the segments $\overline{P_1P_2}$ and $\overline{P_3P_4}$ have the same slope.

The proof of this theorem is left as an exercise.

The fact that all the segments of any nonvertical line have the same slope is used to define the slope of that line.

Definition The **slope of a nonvertical line** is the slope of the segment determined by any two points on the line.

Thus to compute the slope of a line, select any two of its points and calculate the slope of the segment determined by those two points. Keep in mind that the slope of a vertical line is not defined.

In the next theorem, an equation of a nonvertical line is found from the slope and the coordinates of one of its points. The equation is said to be in **point-slope form**.

Theorem 2–4 If a line ℓ has slope m and contains the point $P(h, k)$, then ℓ is the graph of the equation

$$y - k = m(x - h).$$

PLAN FOR PROOF You must show that
a. Every point on ℓ has coordinates that satisfy the equation, and that
b. Every point whose coordinates satisfy the equation is on ℓ.

(cont. on p. 55)

PROOF **a.** If $Q(x, y)$ is on ℓ, and $Q \neq P$, then $\overline{PQ}$ has slope m (Theorem 2–3). Thus $\dfrac{y - k}{x - h} = m$. It follows that for all points of ℓ, including P, the coordinates (x, y) satisfy the equation $y - k = m(x - h)$.

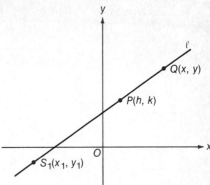

b. Suppose that $S_1(x_1, y_1)$ is a point whose coordinates satisfy the equation. Then $y_1 = k + m(x_1 - h)$. Since ℓ is not vertical, some point of ℓ will have abscissa x_1, say $S_2(x_1, y_2)$. Then the ordinate of S_2 must be $y_2 = k + m(x_1 - h)$. (Why?) Hence $y_1 = y_2$ and $S_1 = S_2$; so S_1 is a point on ℓ.

Example 2 Write an equation of the line ℓ with slope 7 that contains $P(-2, 5)$.

SOLUTION By Theorem 2–4, line ℓ is the graph of the equation $y - 5 = 7(x + 2)$.

Notice that we sometimes omit the word *graph* and speak of *an equation of line ℓ*, or say that a line *has equation* $\cdots$.

If two points are given, an equation of the line that contains them can be written by first computing the slope and then using Theorem 2–4.

Example 3 Write an equation of the line that contains $P(1, 6)$ and $Q(-1, 2)$.

SOLUTION The slope of $\overline{PQ}$, and therefore the slope of ℓ, is $\dfrac{6 - 2}{1 - (-1)}$, or 2. Now use Theorem 2–4, with $P(1, 6)$ as the known point. An equation of ℓ is $y - 6 = 2(x - 1)$.

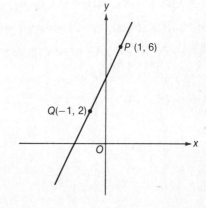

Had we used $Q(-1, 2)$ as the known point, the equation would have been $y - 2 = 2(x + 1)$. Each of these equations is equivalent to the simpler equation $y = 2x + 4$, and any of these three equations is a suitable answer. For this reason, we ask for *an equation* of a line rather than *the equation*.

Example 4 Given points $A(7, -3)$, $B(-3, 2)$, and $C(5, -6)$, write an equation of the line containing the median of $\triangle ABC$ drawn from vertex A.

SOLUTION The midpoint M of side $\overline{BC}$ is on the required median. By the midpoint formula, the coordinates of M are $x = 1$ and $y = -2$. The slope of $\overline{AM}$ is $\dfrac{-3 - (-2)}{7 - 1}$, or $-\dfrac{1}{6}$. By Theorem 2–4, an equation of the line containing the median is

$$y + 3 = -\tfrac{1}{6}(x - 7), \text{ or}$$
$$y = -\tfrac{1}{6}x - \tfrac{11}{6}.$$

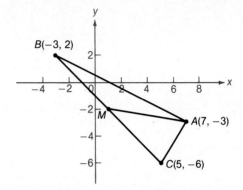

The simplified equation for example 4, $y = -\tfrac{1}{6}x - \tfrac{11}{6}$, has the advantage that it it is easy to see from it that if $x = 0$, then $y = -\tfrac{11}{6}$.

Theorem 2–5 If ℓ is a line with slope m that intersects the y-axis at $P(0, b)$, then ℓ is the graph of the equation $y = mx + b$.

The proof of Theorem 2–5 is left as an exercise.

The number b is usually called the **y-intercept** of ℓ because $(0, b)$ is the point of intersection of ℓ with the y-axis. The equation in Theorem 2–5 is said to be in **slope-intercept form**. Notice that both the slope and the y-intercept can be read directly from an equation of a line in slope-intercept form.

The x-coordinate of the point where ℓ intersects the x-axis is the **x-intercept**.

Example 5 A line ℓ has y-intercept -2 and x-intercept 6. Write an equation of ℓ in slope-intercept form.

SOLUTION Since $P_1(0, -2)$ and $P_2(6, 0)$ are points on ℓ, the line ℓ has slope $\tfrac{1}{3}$, and the desired equation is $y = \tfrac{1}{3}x - 2$.

Since a line parallel to the x-axis has slope 0, every horizontal line has as an equation $y = 0 \cdot x + b$, or $y = b$. Since slope is undefined for vertical lines, Theorems 2–4 and 2–5 do not apply to such lines. It is easy to verify that the equation of the vertical line that contains $P(h, k)$ is simply $x = h$. A horizontal line that is parallel to the x-axis has no x-intercept; a vertical line that is parallel to the y-axis has no y-intercept.

Exercises

A Compute the slope of the segment with the given endpoints.

1. (1, 4), (5, 6)
2. (−3, 5), (4, −1)
3. (7, 11), (8, 11)
4. (3, −9), (3, 5)

Write an equation of the line that contains the given point and has the given slope. Write the equation in point-slope form.

5. (2, 5), $m = 3$
6. (13, 4), $m = -2$
7. (2, −7), $m = \frac{1}{2}$
8. (−2, −5), $m = 0$
9. (−12, 17), $m = -\frac{5}{17}$
10. (1, 7), $m = 1$

Write an equation of the line that contains the given point and has the given slope. Write the equation in slope-intercept form.

11. (0, 9), $m = 2$
12. (0, −4), $m = 1$
13. (5, 0), $m = -2$
14. (1, 1), $m = -1$
15. (5, 10), $m = 3$
16. (−4, −2), $m = 6$

Write an equation of the line that contains the given points.

17. (2, 5), (3, 8)
18. (1, 0), (0, 1)
19. (−1, 6), (7, 6)
20. (3, 8), (3, 7)
21. (−2, −5), (1, −4)
22. (7, 0), (−7, 3)

B 23. Write an equation of the line with x-intercept 6 and y-intercept 3.

24. Write an equation of the line with x-intercept −2 and y-intercept 75.

25. Show that the line with x-intercept a and y-intercept b has an equation $ay + bx = ab$, provided $ab \neq 0$.

Rewrite each equation in slope-intercept form and determine the slope and the y-intercept.

26. $3x + 4y = 5$
27. $y - x = 2$
28. $2y + 7x + 6 = 11$
29. $y = 1 - x$
30. $5y = 4x + 9$
31. $4y + 3x = -1$

32. **a.** Write equations for the lines that contain the sides of the triangle with vertices $A(-1, 0)$, $B(10, 1)$, and $C(9, -2)$.
    b. Write an equation of the line containing the median of $\triangle ABC$ drawn from vertex B.

C 33. Determine k so that (−2, −3), (−1, 7), and (3, k) are collinear.

34. Write equations of the lines that contain the diagonals of the quadrilateral ABCD with vertices $A(-2, -1)$, $B(1, -3)$, $C(5, 2)$, and $D(-4, 7)$.

35. Prove Theorem 2–3.

36. Complete the proof of Theorem 2–4 by answering the "Why?" in part b.

37. Prove Theorem 2–5.

2–3 Parallel and Perpendicular lines

Consider the graphs of the lines

$\ell_1 : y = 2x + 4$, and

$\ell_2 : y = 2x - 1$.

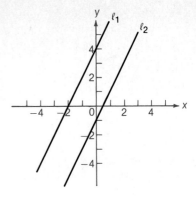

The fact that ℓ_1 and ℓ_2 have the same slope implies that they are parallel lines. To show this, suppose that the lines intersect and that $P(h, k)$ is a point common to both lines. Then $k = 2h + 4$ and $k = 2h - 1$. Since this implies $4 = -1$, which is a contradiction, the lines cannot intersect.

Any two horizontal lines are clearly parallel and have the same slope, namely 0. In fact, any two parallel, nonvertical lines must have the same slope. The next theorem states this connection between parallel lines and their slopes. The proof is omitted, but see exercise 29.

Theorem 2–6 Two nonvertical lines are parallel if and only if they have the same slope.

Example 1 Given points $A(-1, 2)$, $B(3, 4)$, $C(4, -1)$, and $D(0, -3)$. Show that $\square ABCD$ is a parallelogram.

SOLUTION Slope of $\overline{AB}$: $m = \dfrac{4 - 2}{3 - (-1)} = \dfrac{1}{2}$

Slope of $\overline{BC}$: $m = \dfrac{4 - (-1)}{3 - 4} = -5$

Slope of $\overline{CD}$: $m = \dfrac{-3 - (-1)}{0 - 4} = \dfrac{1}{2}$

Slope of $\overline{AD}$: $m = \dfrac{-3 - 2}{0 - (-1)} = -5$

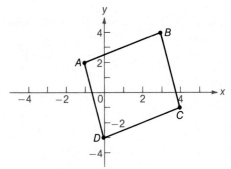

Since $\overline{AB}$ and $\overline{CD}$ both have slope $\frac{1}{2}$ and $\overline{BC}$ and $\overline{DA}$ both have slope -5, the corresponding pairs of lines must be parallel. Thus the opposite sides of $\square ABCD$ are parallel and the quadrilateral is a parallelogram.

Example 2 Write an equation of the line ℓ_1 that contains $(-1, 5)$ and is parallel to $\ell_2 : y = 4x - 6$.

SOLUTION Since ℓ_2 has slope 4, ℓ_1 must also have slope 4. By Theorem 2–4, ℓ_1 has an equation $y - 5 = 4(x + 1)$, or $y = 4x + 9$.

Example 3 The graph of $2x - 3y + 5 = 0$ is a line. Find its slope and write an equation of the line parallel to this given line and containing the origin.

SOLUTION First rewrite the given equation in slope-intercept form.

$$2x - 3y + 5 = 0$$
$$-3y = -2x - 5$$
$$y = \frac{2}{3}x + \frac{5}{3}.$$

Thus both lines must have slope $\frac{2}{3}$. Since the required line contains the origin, its y-intercept is 0, and its equation is $y = \frac{2}{3}x$.

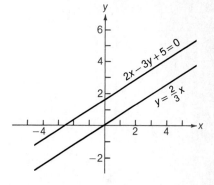

Example 3 illustrates that any equation of the form

$$ax + by + c = 0, b \neq 0,$$

can be rewritten in slope-intercept form. Hence the graph of any such equation is a line with slope $m = -\frac{a}{b}$. If $b = 0$ but $a \neq 0$, the equation $ax + by + c = 0$ simplifies to $x = -\frac{c}{a}$, which is the equation of a vertical line.

Property If a, b, and c are any real numbers with a and b not both zero, then the graph of the equation $ax + by + c = 0$ is a straight line.

The lines $\ell_1 : y = 3x - 1$ and $\ell_2 : y = -\frac{1}{3}x + 2$ are perpendicular, a fact which can be predicted merely by noting the relationship between their slopes.

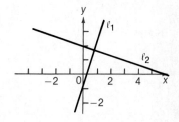

Theorem 2–7 Two nonvertical lines are perpendicular if and only if the product of their slopes is -1.

The proof of this theorem is in exercises 30 and 31.

Example 4 Write an equation of the line that contains (1, 2) and is perpendicular to the line $x + 2y = 3$.

SOLUTION Rewrite the equation in slope-intercept form.
$$y = -\tfrac{1}{2}x + \tfrac{3}{2}.$$
This line has slope $-\tfrac{1}{2}$. So the required line must have slope 2. Now use the point-slope form to write the equation
$$y - 2 = 2(x - 1), \text{ or } y = 2x.$$

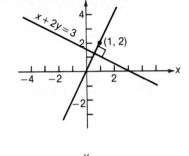

Example 5 If the vertices are $A(-2, 2)$, $B(1, 1)$, and $C(0, 8)$, is $\triangle ABC$ a right triangle?

SOLUTION You should verify that $\overline{AC}$ has slope 3 and that $\overline{AB}$ has slope $-\tfrac{1}{3}$. Thus $\angle BAC$ must be a right angle and $\triangle ABC$ is a right triangle.

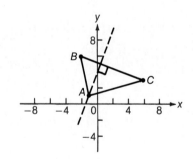

Example 6 Given a triangle with vertices $A(-1, 1)$, $B(-2, 6)$, and $C(6, 3)$. Write an equation of the line containing the altitude drawn from vertex A to side $\overline{BC}$.

SOLUTION Since $\overline{BC}$ has slope $-\tfrac{3}{8}$, the altitude must have slope $\tfrac{8}{3}$. The line that contains the altitude contains $A(-1, 1)$, so an equation is
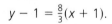
$$y - 1 = \tfrac{8}{3}(x + 1).$$

Exercises

A For the given vertices, show that $\square ABCD$ is a parallelogram.

1. $A(1, -1)$, $B(9, 1)$, $C(8, 5)$, $D(0, 3)$ **2.** $A(-6, 2)$, $B(-8, 11)$, $C(1, 5)$, $D(3, -4)$

Write an equation of the line that contains the given point and is parallel to the given line.

3. $(0, 1)$, $y = 5x - 4$ **4.** $(3, 0)$, $y = 2x + 1$

5. $(3, -7)$, $2x - 5y = 11$ **6.** $(-4, 3)$, $3y + 2x = 15$

Write an equation of the line that contains the given point and is perpendicular to the given line.

7. $(0, 4)$, $y = -2x + 1$
8. $(-1, 0)$, $y = 3x$
9. $(3, -2)$, $5x - 6y = 7$
10. $(2, -7)$, $2x - 5y = 11$

Determine whether the triangle with the given vertices is a right triangle.

11. $(8, 3)$, $(4, -1)$, $(0, 3)$
12. $(-4, 0)$, $(8, -2)$, $(3, 5)$

13. Write an equation of the line that contains the altitude to side $\overline{AB}$ of the triangle whose vertices are $A(-1, -4)$, $B(2, 5)$, and $C(-3, 12)$.

14. Write an equation of the line that contains the altitude to side $\overline{BC}$ of the triangle whose vertices are $A(0, 1)$, $B(5, -2)$, and $C(7, 10)$.

B 15. Find c so that the line that contains $(1, 4)$ and $(-2, c)$ is parallel to the line that contains $(5, -1)$ and $(9, 2)$.

16. Find c so that the line that contains $(-2, 3)$ and $(-1, c)$ is parallel to the line that contains $(-4, 1)$ and $(2, c)$.

17. What value of c will make the line that contains $(c, 3)$ and $(-2, 1)$ perpendicular to the line that contains $(5, c)$ and $(1, 0)$?

18. Find c so that the line that contains $(c, 3)$ and $(4, -9)$ will be perpendicular to the line that contains $(c, 2)$ and $(5, 3)$.

19. Determine k for points $A(5, -6)$, $B(1, 2)$, and $C(k, k)$ so that $\angle BAC$ is a right angle.

20. Determine k for points $R(-1, 2)$, $S(k, 2k)$, and $T(0, 0)$ so that $\angle RST$ is a right angle.

Write an equation of the perpendicular bisector of the segment with the given endpoints.

21. $(5, -2)$, $(7, 10)$
22. $(1, -1)$, $(9, 1)$

23. Show that the quadrilateral with vertices $(8, 5)$, $(0, 3)$, $(1, -1)$, and $(9, 1)$ is a rectangle.

24. Show that the quadrilateral with vertices $(-2, 2)$, $(2, -2)$, $(4, 2)$, and $(2, 4)$ is a trapezoid whose diagonals are perpendicular.

25. Find the y-intercept of a line that contains $(-2, 6)$ and is parallel to the line that contains $(1, -4)$ and $(6, 6)$.

26. Points $A(-1, 5)$, $B(6, 7)$, and $C(2, -1)$ are the vertices of a triangle. Write an equation of the line that joins the midpoints of sides $\overline{AB}$ and $\overline{AC}$ and show that this line is parallel to $\overline{BC}$.

27. Three vertices of a parallelogram are $(0, 0)$, $(5, 2)$, and $(1, 4)$. The fourth vertex can be in three different positions. What are they?

28. Show that for the quadrilateral with vertices (0, 0), (5, 0), (6, 4), and (−1, 8) the midlines form a parallelogram.

C 29. Prove that parallel lines have the same slope as follows:
   a. Treat the case where ℓ_1 and ℓ_2 are horizontal as a special case.
   b. If $\ell_1 \parallel \ell_2$, ℓ_1 intersects the x-axis at P, and ℓ_2 intersects the x-axis at Q, choose point R on the x-axis so that PQ = QR. Construct lines at Q and R perpendicular to the x-axis that intersect ℓ_1 and ℓ_2 at M and N, respectively. Now prove that $\triangle PQM \cong \triangle QRN$ and use this fact to prove that ℓ_1 and ℓ_2 have the same slope.

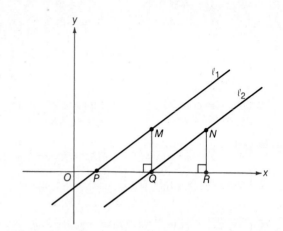

30. Prove that if $\ell_1 \perp \ell_2$ and neither line is vertical, then the product of their slopes is −1. Proceed as follows.
Suppose that the given lines have slopes m_1 and m_2, respectively. If $\ell_1 \perp \ell_2$, name their point of intersection P(h, k).
   a. Show that the vertical line $x = 1 + h$ intersects ℓ_1 at $Q(1 + h, m_1 + k)$ and intersects ℓ_2 at $R(1 + h, m_2 + k)$.
   b. Use the Pythagorean theorem to show that

   $$PQ^2 + PR^2 = QR^2$$

   so that

   $$(1 + m_1^2) + (1 + m_2^2) = (m_1 - m_2)^2.$$

   c. Simplify the result in **b** to show that $m_1 m_2 = -1$.

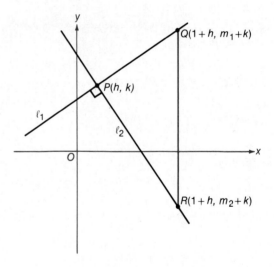

31. Prove that if the product of the slopes of two lines is −1, then the lines are perpendicular. [HINT: Reverse the steps in exercise 30 to show that $\triangle PQR$ is a right triangle with right angle at P.]

32. Write a program to have a computer accept as input the coefficients a, b, and c of two linear equations of the form $ax + by + c = 0$ and print "parallel," "perpendicular," or "neither parallel nor perpendicular," as appropriate.

2-4 Circles

You know that concentric circles with radii 1 and $2\sqrt{2}$, centered at the origin, have equations

$$x^2 + y^2 = 1 \text{ and } x^2 + y^2 = 8.$$

By using the distance formula you can write an equation of a circle with center at any point. Use the fact that the distance between (x, y) and (h, k) is

$$\sqrt{(x - h)^2 + (y - k)^2}$$

to prove the following theorem.

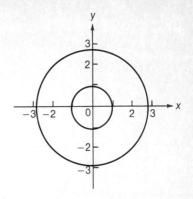

Theorem 2-8 An equation of a circle with center (h, k) and radius r is

$$(x - h)^2 + (y - k)^2 = r^2$$

For example, the circle with center $(1, 0)$ and radius 2 has an equation

$$(x - 1)^2 + y^2 = 4.$$

The circle with center $(4, 3)$ and radius 5 has an equation

$$(x - 4)^2 + (y - 3)^2 = 25.$$

Example 1 Find the center and the radius of the circle whose equation is
$(x + 2)^2 + (y - 3)^2 = 6.$

SOLUTION Write $x + 2$ as $x - (-2)$ and compare with the equation in Theorem 2-8 to see that the circle has radius $\sqrt{6}$ and center $(-2, 3)$.

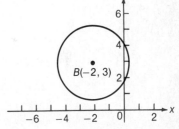

If an equation contains x^2- and y^2-terms, as well as additional terms, it may be an equation of a circle in disguise. To see if it is, try to rewrite the unfamiliar equation in the form in Theorem 2-8.

Example 2 Find the center and the radius of the circle $x^2 + y^2 + 10x - 2y = 74.$

SOLUTION Collect terms and complete the square as shown.

$$(x^2 + 10x) + (y^2 - 2y) = 74$$
$$(x^2 + 10x + 25) + (y^2 - 2y + 1) = 74 + 25 + 1$$
$$(x + 5)^2 + (y - 1)^2 = 100$$

Thus the center is $(-5, 1)$ and the radius is 10.

Deciding what constant to add to each side of the equation in completing the square on the x-terms (or the y-terms) depends upon the fact that $(x - h)^2 = x^2 - 2hx + h^2$, so that the required constant term will always be the square of one-half the coefficient of the x-term. If the coefficient of x^2 and y^2 is not 1, dividing both sides of the original equation by that coefficient produces an equivalent equation of the type just considered.

Example 3 Find the center and the radius of the circle $2x^2 + 8x + 2y^2 - 12y = 24$.

SOLUTION First divide both sides of the equation by 2.

$$x^2 + 4x + y^2 - 6y = 12$$

Now complete the squares on the x- and the y-terms.

$$(x^2 + 4x + 4) + (y^2 - 6y + 9) = 12 + 4 + 9$$

$$(x + 2)^2 + (y - 3)^2 = 25$$

Thus the center is $(-2, 3)$ and the radius is 5.

Example 4 Describe the graph of the equation $2x^2 + 8x + 2y^2 - 12y = -27$.

SOLUTION Use the procedure of example 3 to write the equivalent equation

$$(x + 2)^2 + (y - 3)^2 = -\tfrac{1}{2}.$$

But a sum of squares cannot be a negative number, so the graph of the equation is the empty set.

Example 4 illustrates that not every equation of the form

$$x^2 + y^2 + Dx + Ey + F = 0$$

describes a circle. The graph may, in fact, contain no points, or it may contain only one point (see exercises 21–24).

Example 5 Write an equation of the line tangent to the circle $x^2 + y^2 = 100$ at the point $(-6, 8)$.

SOLUTION The slope of the radius to the point $(-6, 8)$ is $-\tfrac{4}{3}$, so the tangent line, which is perpendicular to the radius at the point of tangency, must have slope $\tfrac{3}{4}$. Its equation is therefore

$$y - 8 = \tfrac{3}{4}(x + 6).$$

Exercises

A Write an equation of the circle whose center and radius are given.

1. $(-1, 5)$, $r = 2$
2. $(0, 7)$, $r = 1$
3. $(3, 4)$, $r = 5$
4. $(-5, 12)$, $r = 13$
5. $(-1, -2)$, $r = \sqrt{2}$
6. $(1, -3)$, $r = \sqrt{7}$

Find the center and the radius of each circle.

7. $(x - 1)^2 + (y - 3)^2 = 25$
8. $(x + 1)^2 + (y - 7)^2 = 400$
9. $(x + 3)^2 + (y + 2)^2 = 5$
10. $x^2 + (y + 11)^2 = 13$

Find the center and the radius of each circle by completing the square on the x terms and on the y terms.

11. $x^2 + 2x + y^2 = 8$
12. $x^2 + y^2 - 4y = 9$
13. $x^2 + 6x + y^2 - 10y = 2$
14. $x^2 + y^2 + 2y = 4x - 1$

B Find the center and the radius of each circle.

15. $y^2 = 4 - x^2$
16. $x^2 - 8x = 9 - y^2$
17. $3x^2 + 24x + 12y + 3y^2 = 15$
18. $2x^2 - 12x + 2y^2 + 4y = -18$
19. $2x + 6y = 6 + x^2 + y^2$
20. $5x^2 - 10x + 5y^2 - 30y + 50 = 45$

Describe the graph of each equation.

21. $(x - 3)^2 + (y + 2)^2 = 0$
22. $x^2 - 14x + y^2 + 49 = 0$
23. $x^2 - 14x + y^2 + 81 = 0$
24. $2x^2 - 12x + 2y^2 + 4y + 100 = 0$

Write an equation of the line tangent to $x^2 + y^2 = 25$ at the given point.

25. $(0, 5)$
26. $(-5, 0)$
27. $(-3, 4)$
28. $(3, -4)$

29. Prove Theorem 2–8.

Write an equation of the circle for which the endpoints of a diameter are given.

30. $(6, -8)$, $(-2, 4)$
31. $(-3, -4)$, $(1, -1)$

C 32. The equation $x^2 - 2x + y^2 - 16y = p$ represents a circle with radius 2. Find p.

33. If there are no points on the graph of $x^2 - 2x + y^2 - 16y = p$, describe p.

34. A circle has radius 4 and its equation is $x^2 - kx + y^2 + 2y = 10$. Determine k and find the coordinates of the center.

35. Write a program to have a computer accept as input the coefficients D, E, and F of an equation of the form $x^2 + y^2 + Dx + Ey + F = 0$ and print the radius and the coordinates of the center if the equation describes a circle.

2–5 Systems of Equations

Consider lines $\ell_1 : y = 2x + 5$ and $\ell_2 : y = 4x - 6$. Since the lines are not parallel, they must intersect in a single point (h, k). The coordinates h and k must satisfy both equations, so

$$k = 2h + 5, \text{ and}$$
$$k = 4h - 6.$$

By substitution,

$$2h + 5 = 4h - 6, \text{ or } h = \frac{11}{2}.$$

Now k can be found from either equation.

$$k = 2\left(\frac{11}{2}\right) + 5, \text{ or } k = 16.$$

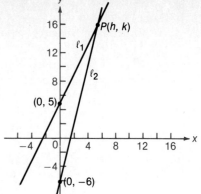

Thus the point of intersection of the two lines is $\left(\frac{11}{2}, 16\right)$.

This process is known as **solving a system** of equations. The coordinates of the point of intersection must satisfy both equations *simultaneously*. It is not necessary to introduce the letters h and k, provided that you maintain the point of view that a particular pair of numbers x and y is sought.

Example 1 Solve the following system of equations:

$$y = -2x + 6$$
$$y = 4x + 1.$$

SOLUTION By substitution,

$$-2x + 6 = 4x + 1, \text{ or } x = \frac{5}{6}.$$

Substituting $x = \frac{5}{6}$ in either of the equations

$$y = -2\left(\frac{5}{6}\right) + 6 \text{ or } y = 4\left(\frac{5}{6}\right) + 1$$

yields the value $y = \frac{13}{3}$.

The point of intersection of the graphs of the two equations is $\left(\frac{5}{6}, \frac{13}{3}\right)$.

If the given equations are not in slope-intercept form, they may, of course, always be rewritten in that form. However, it is also possible to approach such equations in another way.

Recall that two equations in a single variable with the same domain are equivalent if one is obtained from the other by

1. Adding or subtracting the same quantity from each side, or
2. Multiplying or dividing both sides by the same nonzero quantity, or
3. Simplifying either side by using the field axioms or theorems.

In working with equations in two variables, such as equations of lines, the same operations are possible. Thus the equation

$$\tfrac{3}{4}y - 9x = \tfrac{1}{4}y + 1$$

can be replaced by the equivalent equation

$$y - 18x = 2.$$

Moreover, in looking for the solution of a system of two equations in two variables, you may also add or subtract the two equations. The resulting equation will not have the same graph as either of the original equations, but any two numbers h and k which satisfy both of the original equations will also satisfy the new equation. This may lead to discovery of the solution.

Example 2 Solve the system of equations $8x - 4y = 5$
$-2x + 2y = 9.$

SOLUTION Multiply the second equation by 2.

$$8x - 4y = 5$$
$$-4x + 4y = 18$$

Now add these two equations.

$$4x = 23$$
$$x = \tfrac{23}{4}$$

The solution for y can be found by substituting $\tfrac{23}{4}$ for x in either of the original equations. The result is

$$y = \tfrac{41}{4}.$$

Having solved the system, you can conclude that the two lines intersect in the point $(\tfrac{23}{4}, \tfrac{41}{4})$.

To find the point or points of intersection, if any, of a line and a circle, you need to solve a system of two equations one of which is quadratic in x and y.

Example 3 Where does the line $y = -x + 1$ intersect the circle $x^2 + y^2 = 25$?

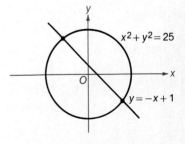

SOLUTION Solve the system $x^2 + y^2 = 25$
$y = -x + 1.$

Substitute the values of y from the linear equation in the quadratic equation.

(cont. on p. 68)

$$x^2 + (-x + 1)^2 = 25$$
$$2x^2 - 2x - 24 = 0$$
$$2(x - 4)(x + 3) = 0$$
$$x = 4 \quad \text{or} \quad x = -3$$

Since $y = -x + 1$, $y = -4 + 1 = -3$ when $x = 4$, and
$y = -(-3) + 1 = 4$ when $x = -3$.

Thus the points of intersection are $(-3, 4)$ and $(4, -3)$.

In general, when a linear equation is to be solved simultaneously with a quadratic equation, the method to use is that of substitution. As indicated in the last example, the steps to follow are these:

① Solve the linear equation for one variable in terms of the other.

② Substitute the value of the one variable into the quadratic to obtain a quadratic equation which involves only the second variable.

③ Solve this quadratic equation.

④ Go back to the first step to find corresponding values of the first variable.

Of course, it may happen that the given line and circle do not intersect, or intersect just once (when the line is tangent to the circle). In following the procedure just given, the discriminant of the quadratic equation in step ③ will tell you how many intersections to expect.

Example 4 Show that the line $y = -x + 8$ does not intersect the circle $x^2 + y^2 = 25$.

SOLUTION Substitute the expression for y from the linear equation in the quadratic equation.

$$x^2 + (-x + 8)^2 = 25$$
$$2x^2 - 16x + 39 = 0$$

Since the discriminant of this quadratic equation is $(-16)^2 - 4(2)(39)$, or -56, there are no solutions. Therefore there is no point of intersection.

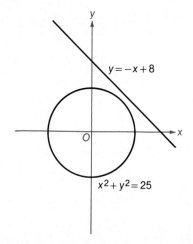

Example 5 Determine the y-intercept b of the line $y = -x + b$ that is tangent to the circle $x^2 + y^2 = 25$.

SOLUTION Using the four-step procedure outlined above yields, in step ③, the quadratic equation

$$2x^2 - 2bx + b^2 - 25 = 0.$$

When a line is tangent to a circle there is only one point of intersection between the line and the circle, and the quadratic equation must have only one solution.

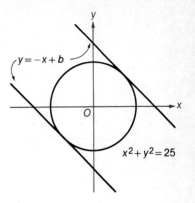

To find b, set the discriminant of the quadratic equation equal to zero.

$$D = (-2b)^2 - 4(2)(b^2 - 25) = 0$$
$$-4b^2 + 200 = 0$$
$$b^2 = 50$$
$$b = \pm 5\sqrt{2}$$

Thus there are two solutions, as shown in the figure; the lines $y = -x - 5\sqrt{2}$ and $y = -x + 5\sqrt{2}$ are each tangent to the circle.

There are three possibilities for points of intersection of two circles. The circles may intersect in two points, one point, or none at all, as shown in the figures below.

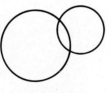 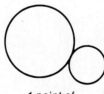

2 points of intersection | 1 point of intersection | No points of intersection

To find the intersection of two circles you must modify the substitution method slightly. Since any two circles can be defined by equations such as

$$(x - h)^2 + (y - k)^2 = r^2 \text{ and}$$
$$(x - p)^2 + (y - q)^2 = s^2,$$

it should be apparent that by subtracting one equation from the other a linear equation in two variables results. Solving that equation simultaneously with the equation of either of the circles will yield the points of intersection of the two circles.

Example 6 Find the points of intersection of the circles
$$x^2 + y^2 = 25 \text{ and}$$
$$x^2 + y^2 - 12x + 6y + 35 = 0.$$

SOLUTION Subtract the second equation from the first to get

$$12x - 6y - 60 = 0, \text{ or}$$
$$y = 2x - 10.$$

The line $y = 2x - 10$ contains the *common chord* of the two circles, as shown in the figure.

To complete the problem, solve the system
$$x^2 + y^2 = 25$$
$$y = 2x - 10.$$

You should verify that the points of intersection are $(3, -4)$ and $(5, 0)$.

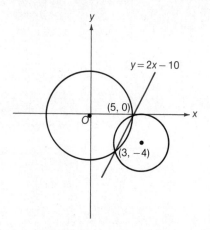

If two circles have either no points of intersection or only one such point, the quadratic equation which arises in the solution of the system will have a discriminant that is negative or equal to zero.

Exercises

A Solve each system to find the point of intersection of the lines.

1. $y = 7x - 2$
   $y = 3x + 5$

2. $y = 5x + 1$
   $y = 3x - 10$

3. $x + 2y = 1$
   $3x + y = 2$

4. $2x - 3y = 7$
   $4x + y = -5$

5. $x - y + 1 = 0$
   $x + y - 1 = 0$

6. $x + y + 3 = 0$
   $2x + 3y + 4 = 0$

Find the point or points of intersection, if any, of the given line and circle.

7. $x^2 + y^2 = 5$
   $y = -x + 3$

8. $x^2 + y^2 = 4$
   $y = -x + 4$

9. $x^2 + y^2 = 100$
   $x = -6$

10. $x^2 + y^2 = 1$
    $y = \dfrac{1}{\sqrt{2}}$

11. $x^2 + y^2 - 4x - 6y + 9 = 0$
    $2x + y = 5$

12. $x^2 + y^2 - 6x - 6y + 9 = 0$
    $x + y - 3 = 0$

Find the point or points of intersection, if any, of the given circles.

13. $x^2 + y^2 = 25$
    $x^2 + y^2 - 20y + 75 = 0$

14. $x^2 + y^2 - 8x = 0$
    $x^2 + y^2 - 16y = 0$

15. $x^2 + y^2 = 5$
    $x^2 + y^2 + 12x - 12y + 31 = 0$

16. $x^2 + y^2 = 1$
    $x^2 + y^2 - 12x + 2y = -33$

17. $x^2 + y^2 = 25$
    $x^2 - 12x + y^2 + 35 = 0$

18. $x^2 + y^2 = 100$
    $x^2 + y^2 - 18x - 24y = -200$

B 19. If the line $y = -x + b$ is tangent to the circle $x^2 + y^2 = 1$, what is the value of b?

20. If the line $y = m(x - 4)$ is tangent to the circle $x^2 + y^2 = 1$, what is the value of m?

21. For what numbers m will the line $y = mx$ intersect the circle $(x - 4)^2 + y^2 = 1$?

22. Find the points of intersection of the line that contains (1, 4) and (3, 6) with the circle of radius $2\sqrt{13}$ and center $(-3, -2)$.

The sides of a triangle lie on lines whose equations are given. Find the coordinates of each vertex of the triangle.

23. $y = 2x$; $4y = x$; $3x + 2y = 14$

24. $y = 2x + 3$; $y = -1$; $2y + x = 1$

A quadrilateral has as its sides segments on the lines given. Find the vertices and show that the quadrilateral is a rhombus.

25. $y = 3$; $4y - 3x = -3$; $y = 0$; $4y = 3x - 18$

26. $4y - x = 0$; $4y + x = 0$; $4y - x = 8$; $4y + x = 8$

C 27. Points $(-2, 2)$ and $(-3, 1)$ are on the circle $x^2 - 2x + y^2 + 4y = 20$. Tangents to the circle are drawn at the given points. Find the point of intersection of the two tangent lines.

28. Show that the perpendicular bisectors of the sides of the triangle whose vertices are $(-3, 0)$, $(3, 0)$, and $(1, 2)$ are concurrent.

29. Find the length of the perpendicular segment from the origin to the line $x + 2y = 4$.

30. For the circles $(x - h)^2 + (y - k)^2 = r^2$ and $(x - p)^2 + (y - q)^2 = s^2$ the line that contains the centers (h, k) and (p, q) is called the *line of centers* of the two circles. Show that if the circles intersect in two points, then their common chord is perpendicular to the line of centers.

31. Write a program to have a computer solve systems of equations for two nonparallel lines.

2–6 Distance—Point to Line

Consider the problem of finding the area of the triangle with vertices $A(-1, 1)$, $B(-2, 6)$, and $C(6, 3)$. If the length of the altitude $\overline{AD}$ were known, you could compute the area of $\triangle ABC$ as $\frac{1}{2}$ (base) · (height), or $\frac{1}{2} BC \cdot AD$. You can use the distance formula to find BC. One way of finding the perpendicular distance from A to $\overline{BC}$, which is $\overline{AD}$ in the figure, is to use the point-to-line distance formula.

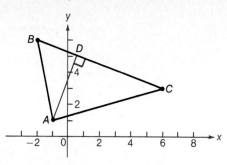

Theorem 2–9 **Point-to-Line Distance Formula**

The distance from the point (h, k) to the line $ax + by + c = 0$ is

$$d = \frac{|ah + bk + c|}{\sqrt{a^2 + b^2}}$$

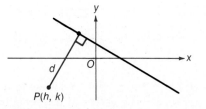

The proof of Theorem 2–9 is left as an exercise.

Example 1 Find the distance from $(-3, 5)$ to the line $\ell: y = 4x - 9$.

SOLUTION To use Theorem 2–9, first write the equation of ℓ as $4x - y - 9 = 0$, so that $a = 4$, $b = -1$, and $c = -9$. The required distance is

$$d = \frac{|(4)(-3) + (-1)(5) + (-9)|}{\sqrt{16 + 1}}$$

$$= \frac{26}{\sqrt{17}}.$$

Now you can complete the problem at the beginning of this section on finding the area of $\triangle ABC$. First find BC.

$$BC = \sqrt{64 + 9} = \sqrt{73}$$

The slope m of $\overline{BC}$ is $-\frac{3}{8}$, so an equation of the line that contains $\overline{BC}$ is

$$y - 3 = -\frac{3}{8}(x - 6).$$

Rewrite this equation in the form used in Theorem 2–9 as

$$3x + 8y - 42 = 0.$$

Then the length of the altitude from A to $\overline{BC}$ is

$$AD = \frac{|(3)(-1) + (8)(1) + (-42)|}{\sqrt{9 + 64}} = \frac{37}{\sqrt{73}}.$$

So the area of $\triangle ABC$ is $\frac{1}{2}(\sqrt{73})\left(\frac{37}{\sqrt{73}}\right)$, or $\frac{37}{2}$.

Theorem 2-9 can be used to find the distance between any two parallel lines ℓ_1 and ℓ_2. To do this, select any point on ℓ_1 and find the distance from this point to ℓ_2.

Example 2 Find the distance between the parallel lines $\ell_1: y = 3x - 7$ and $\ell_2: y = 3x + 5$.

SOLUTION Select $(0, -7)$ on ℓ_1 and use the formula in Theorem 2-9 to find its distance from ℓ_2.

$$d = \frac{|(3)(0) + (-1)(-7) + 5|}{\sqrt{9 + 1}}$$

$$= \frac{12}{\sqrt{10}} \doteq 3.79$$

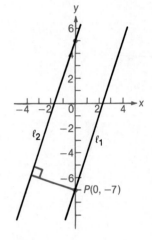

Example 3 Two lines ℓ_1 and ℓ_2 are parallel and the distance between them is $2\sqrt{5}$. An equation of ℓ_1 is $y = 2x + 3$. Write an equation of ℓ_2.

SOLUTION Since $\ell_1 \parallel \ell_2$, the slope of ℓ_2 must be 2. You also need to know the y-intercept b of ℓ_2. Use Theorem 2-9 to find that the distance from $(0, b)$ to ℓ_1 is $\frac{|3 - b|}{\sqrt{5}}$ which must equal $2\sqrt{5}$.

$$\frac{|3 - b|}{\sqrt{5}} = 2\sqrt{5}$$

$$|3 - b| = 10$$

$$b = -7 \text{ or } b = 13$$

So there are two lines that satisfy the description of ℓ_2,

$$y = 2x - 7 \text{ and } y = 2x + 13.$$

Another use of Theorem 2-9 is to find equations of the two tangents that can be drawn to a circle from a given point in the exterior of the circle.

2-6: Distance—Point to Line

Example 4 From the point (7, 1), two tangents can be drawn to the circle $x^2 + y^2 = 25$. Write an equation of each tangent line.

SOLUTION Any line that contains (7, 1) has an equation of the form

$$y - 1 = m(x - 7), \text{ or}$$
$$mx - y + (1 - 7m) = 0.$$

So $a = m$, $b = -1$, and $c = 1 - 7m$. Since the distance from the line to the origin is 5,

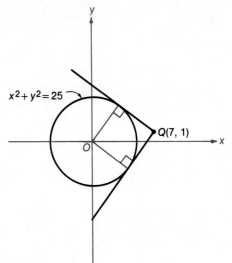

$$5 = \frac{|ah + bk + c|}{\sqrt{a^2 + b^2}}, h = k = 0.$$

$$= \frac{|(m)(0) + (-1)(0) + 1 - 7m|}{\sqrt{m^2 + 1}}$$

$$= \frac{|1 - 7m|}{\sqrt{m^2 + 1}}$$

It follows that

$$25(m^2 + 1) = (1 - 7m)^2, \text{ or}$$
$$24m^2 - 14m - 24 = 0.$$

$$m = -\frac{3}{4} \text{ or } m = \frac{4}{3}$$

Equations in point-slope form of the tangents from (7, 1) are

$$y - 1 = -\frac{3}{4}(x - 7) \text{ and } y - 1 = \frac{4}{3}(x - 7).$$

Exercises

A Find the distance from the given point to the given line.

1. $(-1, 3)$; $2x + 3y + 1 = 0$
2. $(-1, -5)$; $-2x - 3y + 4 = 0$
3. $(0, 6)$; $-3x + y + 5 = 0$
4. $(0, 0)$; $x + y - 1 = 0$
5. $(1, 2)$; $y = 2x + 5$
6. $(-1, 3)$; $y = -x + 1$
7. $(-1, -5)$; $2x + y = 12$
8. $(0, 1)$; $3y - 5x + 7 = 9$

B Find the area of the triangle whose vertices are given.

9. $(0, 0), (4, 1), (3, 8)$
10. $(1, 2), (-1, -3), (7, -2)$

Find the distance between the given parallel lines.

11. $y = 4x - 5$; $y = 4x + 3$
12. $2y + 8x + 3 = 0$; $y = -4x + 1$

13. From the point (5, 0) two tangents to the circle with equation $x^2 + y^2 = 1$ can be drawn. Find an equation of each tangent line.

14. From the point (0, −5) two tangents can be drawn to the circle $x^2 + y^2 = 9$. Find an equation of each tangent line.

15. Write equations for the two lines that are 2 units from $y = -3x + 2$.

16. Two lines $\ell_1: 2x - 4y = 5$ and ℓ_2 are parallel and the distance between them is 3 units. Write an equation for ℓ_2 if ℓ_2 lies below ℓ_1 in the coordinate plane.

C 17. A circle with radius 7 has its center in the second quadrant on the line $y = -x$. Write an equation of this circle if it is tangent to the line $y = \frac{3}{4}x$.

18. Find all points on the line $y = 3x - 2$ that are $\sqrt{5}$ units from the line $y = -2x + 3$.

19. The circle C has equation $x^2 + y^2 = 2x - 4y - 4$.
   a. Sketch a graph of C.
   b. Write equations for both tangent lines to C from (0, 0).
   c. Why does the algebraic solution give only one of the tangent lines?

20. An angle is formed by the positive x-axis and a ray along the line $y = 2x$. Find the slope of line ℓ that contains the bisector of the given angle. [HINT: Let $m > 0$ be the slope of ℓ, and consider the point $P(1, m)$ on ℓ. Since P is equidistant from the sides of the angle, you can find m by setting the distance from P to the line $y = 2x$ equal to m.]

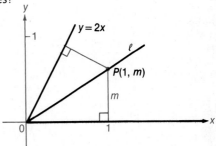

21. Given the points $A(0, 0)$, $B(5, 0)$, and $C(1, 4)$, find the slope of the line that contains the bisector of $\angle BAC$.

22. Write a program to have the computer accept as input the coordinates of a point P and the coefficients a, b, and c of an equation of a line $\ell: ax + by + c = 0$ and print the distance from P to ℓ.

23. Prove Theorem 2–9 as follows.
   a. Find the coordinates of points Q and R on line ℓ such that $\overleftrightarrow{PQ}$ is vertical and $\overleftrightarrow{PR}$ is horizontal.
   b. Find the area of $\triangle PQR$, which is equal to $\frac{1}{2}(PQ)(PR)$.
   c. Since the area of $\triangle PQR$ also equals $\frac{1}{2}(QR)(d)$, equate the two expressions for area and solve for d.
   d. Simplify algebraically to show that $d = \dfrac{|ah + bk + c|}{\sqrt{a^2 + b^2}}$.

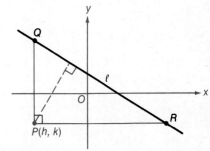

Special   An Application

A contractor plans to build two buildings at locations A and B, as shown, that are 100 m and 200 m, respectively, from the electric power line ℓ. One transformer is to be

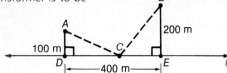

located at C between D and E on ℓ to serve both buildings. If the distance DE is 400 m, where should the transformer be placed so the total length of wire, AC + CB from the transformer to the buildings will be a minimum?

This could be called a "point-to-line-to-point" distance problem. As you might expect, the solution depends upon the relative lengths AD, BE, and DE.

Geometry is useful in finding the best location C. First find the reflection of B in ℓ, that is, extend $\overline{BE}$, perpendicular to ℓ, to B' so that B'E = BE. The best location for the transformer is at C, the point where $\overline{AB'}$ intersects ℓ.

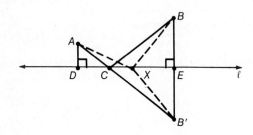

To verify this, let X be any other location along $\overleftrightarrow{DE}$. Using congruent triangles or the perpendicular bisector theorem, you can show that CB = CB' and XB = XB'. So AX + XB = AX + XB', while AC + CB = AC + CB' = AB'. In $\triangle AXB'$, you see that AX + XB' > AB', by the triangle inequality. Thus by transitivity, AX + XB > AC + CB, and C is indeed the best location for the transformer.

You can use ideas studied in this chapter to find a formula for the minimum distance AC + CB. In general, if A is a units from ℓ, and B is b units from

ℓ, you can introduce a coordinate system as shown with ℓ the x-axis and A on the positive y-axis. If d denotes the distance from D to E, then points A, B, D, and E will have the coordinates indicated.

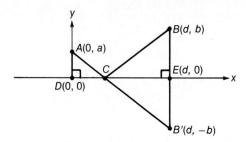

Then B' will have coordinates $(d, -b)$ and $\overleftrightarrow{AB'}$ will have slope $\dfrac{a+b}{-d}$ and equation $y = \left(\dfrac{a+b}{-d}\right)x + a$.
The required point C is the x-intercept of $\overleftrightarrow{AB'}$, so C has coordinates $\left(\dfrac{ad}{a+b}, 0\right)$. In the transformer problem, $a = 100$ m, $b = 200$ m, and $d = 400$ m, so the transformer should be placed

$$\dfrac{ad}{a+b} = \dfrac{(100)(400)}{(300)} = 133\tfrac{1}{3} \text{ m from } D \text{ towards } E.$$

Finally, the minimum distance, AC + CB, or AB', is given by the distance formula as

$$AC + CB = \sqrt{(a+b)^2 + d^2}$$
$$= \sqrt{(100+200)^2 + 400^2} = 500 \text{ m}.$$

Exercises

For A and B as given, find the minimum total distance from A to the x-axis to B

1. $A(0, 200)$, $B(600, 600)$

2. $A(0, 24)$, $B(25, 7)$

3. One pump at a river is to supply water to two cattle-feeding pens, one 100 ft. and the other 400 ft. from the river. How much pipe will be needed and where should the pump be placed if the distance along the river between the pens is 1200 ft. and the length of pipe used is to be a minimum?

2-7 Systems of Inequalities

Graphing inequalities is an extension of the process of graphing equations. The graph of an inequality in two variables is the set of all points with coordinates that satisfy the given inequality. For example, $(2, -7)$ lies on the graph of $2x - 3y^2 < -3$ because $2(2) - 3(-7)^2 < -3$ is a true statement.

Example 1 Sketch the graph of $y < 2x + 1$.

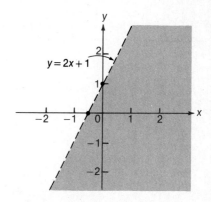

SOLUTION First sketch the graph of $y = 2x + 1$, which is the line with slope 2, y-intercept 1, and x-intercept $-\frac{1}{2}$.

It should be clear that points in the plane that are beneath this line have y-coordinates which satisfy the inequality $y < 2x + 1$.

Indicate the graph of $y < 2x + 1$ by shading the region below the line. The line should be drawn with a dashed line to indicate that it is a boundary of but not a part of the graph of the inequality. (For the inequality $y \leq 2x + 1$, the line would be drawn as a solid line.)

Example 1 is generalized in the following property.

Property The graph of $y < mx + b$ is the half-plane that lies below the line $y = mx + b$, and the graph of $y > mx + b$ is the half-plane that lies above the line $y = mx + b$.

Example 2 Sketch the graph of $4x - 10y \leq 15$.

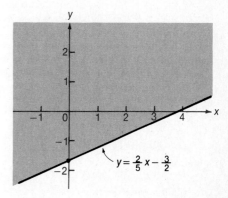

SOLUTION First rewrite the inequality as

$$y \geq \tfrac{2}{5}x - \tfrac{3}{2}.$$

Graph the equation $y = \tfrac{2}{5}x - \tfrac{3}{2}$; represent it with a solid line, since the symbol $\geq$ indicates that the boundary is included in the graph. Finally, shade the region above the boundary line.

The solution of a linear inequality will be either all points above the boundary line or all points below this line. To find which it is, all you need to do is check one point not on the line. For instance, you could determine whether the coordinates of the origin satisfy the inequality.

If they do, then the graph includes all the points on the same side of the boundary as the origin; otherwise the graph consists of the points on the opposite side from the origin. If the origin is a boundary point, the same test can be applied using any point not on the boundary. In example 2, note that, for the origin, $4(0) - 10(0) \leq 15$, so the required graph is the upper half-plane.

The same procedure applies to the graphing of some nonlinear inequalities.

Example 3 Sketch the graph of $(x - 1)^2 + (y + 3)^2 < 16$.

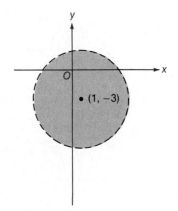

SOLUTION First, consider the graph of the corresponding equation,

$$(x - 1)^2 + (y + 3)^2 = 16,$$

which is a circle of radius 4 with its center at $(1, -3)$. The coordinates of the center satisfy the given inequality, so the interior of the circle should be shaded to represent the solution set. The boundary is drawn with a dashed line; it is not included in the graph.

You know how to solve a system of two linear equations and thus to find the ordered pair of numbers that satisfies both equations. (Under what condition will a system of two linear equations in two variables have no solution?) To find algebraically the set of all ordered pairs that satisfy a system of two linear inequalities is complicated, to say the least. The difficulty is compounded when the problem is to solve simultaneously a system of three or more linear inequalities, a situation which arises in many practical applications.

Fortunately, there is a simple graphical method for determining the solution set of a system of inequalities. First, graph each of the given inequalities. The solution set is the intersection of the regions which constitute the graphs of the inequalities.

Example 4 Graph the solution set of the following system of inequalities.

$$5x + y \leq 10$$
$$2x - y \geq -3$$
$$y \geq 0$$

SOLUTION Solve the corresponding systems of two linear equations to verify that the boundaries intersect at $A(-\frac{3}{2}, 0)$ $B(2, 0)$, and $C(1, 5)$.

(Cont. on p. 79)

The graphs of the separate inequalities are the points which lie below $\overleftrightarrow{AC}$, below $\overleftrightarrow{BC}$, and above $\overleftrightarrow{AB}$, respectively. These graphs indicate that the solution set of the system is the union of $\triangle ABC$ and its interior.

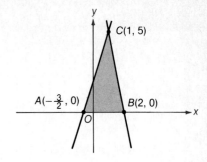

Systems of linear inequalities are important in applications. Consider the farmer who wishes to raise corn and alfalfa under the following conditions.

a. To plant, cultivate, and harvest the crops requires 20 hours of labor per acre for corn and 8 hours per acre for alfalfa.
b. There are at most 100 acres available for planting (but not all has to be planted).
c. There are 1600 hours of labor available.

If x is the number of acres of corn to be planted and y the number of acres of alfalfa, then the options for apportioning the acreage between the two crops must satisfy the following system of inequalities.

$$x + y \leq 100$$
$$20x + 8y \leq 1600$$
$$x \geq 0 \quad \text{and} \quad y \geq 0$$

The last two conditions signify that negative acreage is not possible.

The solution set is easily found by graphing the four inequalities and finding the intersection of the graphs.

The points that satisfy the first inequality are those which lie on or below the line $x + y = 100$. The points that satisfy the second inequality are those which lie on or below the line $20x + 8y = 1600$. Since points that satisfy $x \geq 0$ and $y \geq 0$ are those that are to the right of the y-axis and above the x-axis, the intersection of the four graphs is the polygonal region $ABCO$, including the boundary points.

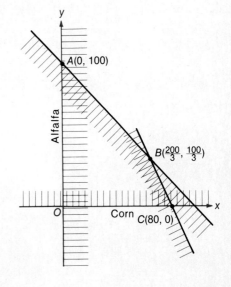

2-7: Systems of Inequalities

Thus to satisfy all four of the stated conditions that restrict the farmer's options, the farmer may plant c acres of corn and a acres of alfalfa if and only if (c, a) is in polygonal region $ABCO$. For example, the farmer could elect to plant 60 acres of corn and 30 acres of alfalfa; but 80 acres of corn and 10 acres of alfalfa do not correspond to a point in the polygon.

In summary, the polygonal region $ABCO$ reveals the set of options open to the farmer in deciding how to allocate acreage between corn and alfalfa with the stated restrictions. This set is called the **feasible region**. In the next section, you will learn how to select the best option from a feasible region.

Exercises

A Sketch a graph of each inequality.

1. $y < 2x + 7$
2. $y \leq 3x - 1$
3. $y \geq x - 4$
4. $y > 2x + 1$
5. $2x + 3y \geq 0$
6. $12x - 4y \leq 28$
7. $5y + 3x + 2 < 0$
8. $5x \leq 2y + 10$
9. $(x - 2)^2 + y^2 \leq 4$
10. $(x + 1)^2 + (y + 2)^2 \geq 5$
11. $x^2 + y^2 + 8x + 4y > 5$
12. $x^2 + y^2 + 2x - 6y + 6 \leq 0$

B In exercises 13–18, graph the solution set of the given system of inequalities. Label coordinates of any boundary corner points.

13. $y \leq 5x + 3$
    $y \geq 3x + 1$

14. $2x + 6y \leq 10$
    $3x - 2y \leq 8$

15. $-5x + 8y > 16$
    $x - y < 4$
    $x > 0$
    $y > 0$

16. $2x - y \geq -4$
    $4x + 5y \leq 6$
    $x \leq 1$
    $y \geq 0$

17. $x - y < 3$
    $5x - 9y > 3$
    $x - 5y > -7$

18. $2x - 3y \geq -7$
    $3x + 5y \leq 18$
    $3x + 5y \geq -1$

Write an inequality whose graph is the shaded region.

19.

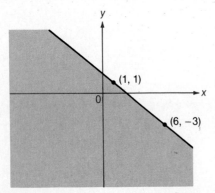

20.

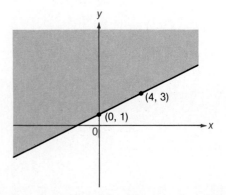

21.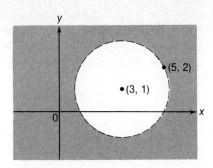

22.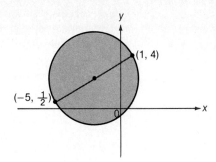

23. A car dealer can stock up to 300 cars on a storage lot. The dealer's franchise requires that at least 60 sedans and no more than 175 station wagons be stored. A deposit of $4000 for each sedan and $6000 for each station wagon must be made with the manufacturer. The total inventory may not exceed a deposit sum of $1,500,000.
    a. Using x for sedans and y for station wagons, specify the system of inequalities which restricts the way the lot is stocked.
    b. Sketch the feasible region and determine the coordinates of the vertices.

C In exercises 24–27, graph the solution set of the given system of inequalities and determine coordinates of any corner points of the boundary.

24. $x^2 + y^2 - 4y > 21$
    $x - 7y \le 11$

25. $x^2 + y^2 + 2x - 6y + 9 \le 0$
    $x + y \le 2$

26. $x^2 + y^2 + 8x + 15 \le 0$
    $x^2 + y^2 \le 10$

27. $x^2 + y^2 - 12x + 6y \le -35$
    $x^2 + y^2 \le 25$

2–8 Linear Programming Theory

The process of decision making is usually based on an estimate of what course of action will yield the greatest benefit. The benefit to be attained may be expressed in a variety of ways depending on the particular type of decision to be made. You might be interested in

a. Manufacturing a soft-drink can that requires as little aluminum as possible; or

b. Minimizing the time required to travel from New York to Washington; or

c. Maximizing the size of a television audience.

In every decision, the options are limited by certain constraints—perhaps the soft-drink can must hold at least 12 ounces, or you cannot exceed 55 miles per hour when driving on the New Jersey Turnpike, or the production costs for the TV program cannot exceed $300,000 per hour. For any given decision, the list of constraints may be endless. To make the problem manageable requires narrowing the field of limitations to a small finite number of conditions which are of primary importance.

It often turns out that these constraints can be expressed as a set of inequalities. The decision must be reached so as to satisfy the system of inequalities. When the system of constraints consists entirely of linear inequalities, then the decision-making process is called **linear programming**.

Suppose that a variable V depends upon two other variables x and y in accordance with the equation

$$V = 3x + 5y - 4,$$

and that the variables x and y must satisfy the following system.

$$0 \le y \le 5$$
$$0 \le x \le 6$$
$$y \le \frac{-3x}{2} + 11$$

If you graph the solution set of this system of conditions imposed on x and y, you obtain the feasible region for V.

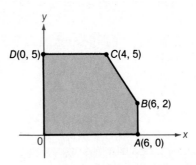

The graph above shows that polygonal region $OABCD$ is the feasible region. To determine, for example, whether 44 is a possible value of V, substitute 44 for V in the equation $V = 3x + 5y - 4$.

$$44 = 3x + 5y - 4$$

Rewrite this equation conveniently as

$$3x + 5y = 48$$

and graph this equation to obtain the line which is designated V_{44}. From the figure on page 83, you can see that the line V_{44} does not intersect the region *OABCD*. You can conclude that 44 is not a possible value of *V*.

To test whether *V* can equal 26, repeat the procedure, this time letting *V* equal 26 to obtain the equation

$$3x + 5y = 30.$$

Graph this equation and label it V_{26}. This line intersects the feasible region in segment $\overline{PQ}$. For every point on this segment *V* = 26.

By the same process, you can show that letting *V* equal 8 yields the line $\overleftrightarrow{RS}$.

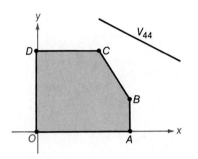 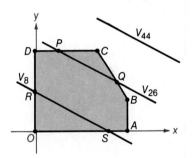

The parallel lines V_{44}, V_{26}, and V_8 are **lines of constant value** for the variable *V*. Notice that, in this case, as smaller values of *V* are chosen, the lines of constant value are lowered in the plane. The lowest line of constant value that intersects the feasible region will clearly be the one that contains the corner point *O*(0, 0). This point yields −4 for *V*.

To determine the greatest value *V* can assume, look for the highest of the parallel lines of constant value that intersect the feasible region. This is the line of constant value that passes through the vertex, or corner point, *C*(4, 5). Thus, the greatest value *V* can have is

$$V = 3(4) + 5(5) - 4 = 33.$$

Notice that maximum and minimum values of *V* occurred at vertices of the feasible region. It can be proved, though we shall not do so, that this will always be the case and we state this general property as follows.

Theorem 2–10 If *R* is a quantity related to two variables *x* and *y* by an equation of the form

$$R = ax + by + c,$$

where *a*, *b*, and *c* are real numbers, and if the feasible region for the choices of *x* and *y* is a convex polygonal region derived from a system of linear inequalities, then the maximum and minimum values of *R* will each occur at a vertex of the feasible region.

Thus to find the maximum and minimum values of a variable R as described above, you need to carry out the following procedure.

a. Determine the coordinates of the vertices of the feasible region.

b. Substitute each of these pairs of coordinates in the equation $R = ax + by + c$ to obtain the corresponding values of R.

c. Then choose from that set of values the greatest and least members.

Example 1 A variable V is determined by the equation
$$V = -4x + 6y - 1,$$
with x and y restricted by the following system of inequalities.
$$y \leq x + 3$$
$$y \leq -2x + 9$$
$$y \geq 0$$
$$1 \leq x \leq 4$$

Find the maximum and minimum values of V.

SOLUTION First, graph the system of inequalities. By solving systems of equations that correspond to pairs of adjacent sides, find the coordinates of the vertices of the feasible region. For instance, to find the coordinates of D, find the intersection of the lines
$$y = x + 3 \quad \text{and} \quad y = -2x + 9.$$
The result is $D = (2, 5)$. At D,
$$V = -4(2) + 6(5) - 1 = 21.$$

Repeat this process with the remaining vertices to complete the following table of values.

Vertex	A(1, 0)	B(4, 0)	C(4, 1)	D(2, 5)	E(1, 4)
V	−5	−17	−11	21	19

The maximum value of V is 21 and occurs when $x = 2$ and $y = 5$.

The minimum value of V is -17 and occurs when $x = 4$ and $y = 0$.

Exercises

A For the feasible regions shown in each figure, determine both the maximum and minimum values of the given variable V.

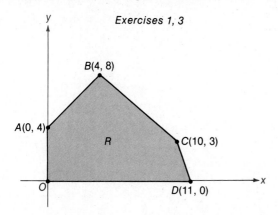
Exercises 1, 3

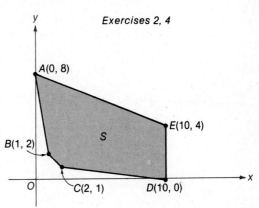
Exercises 2, 4

1. $V = 4x + 3y + 7$
2. $V = 2x - y + 10$
3. $V = 2x - 3y - 3$
4. $V = x + 4y - 5$

B Assume that P is a quantity related to x and y by the given equation and that the given system of inequalities defines the feasible region for P. Determine the maximum and minimum values of P.

5. $P = x + 2y + 10$
   $0 \le x \le 40$
   $y \ge 0$
   $x + y \le 60$

6. $P = x - 2y + 6$
   $x \ge 0$
   $0 \le y \le 20$
   $3x - y + 5 \ge 0$
   $4x + y \le 80$

7. $P = 3x + y - 2$
   $0 \le x \le 30$
   $y \ge 0$
   $x + 2y \ge 10$
   $x + 2y \le 40$
   $x - 4y \le 20$

8. $P = 11x + 19y - 15$
   $y \ge 0$
   $6x + 5y \ge 42$
   $3x - y \ge 0$
   $3x - 7y + 90 \ge 0$
   $9x + 4y \le 180$

C 9. The feasible region for a variable V is defined by the system

$$3y - x \le 15$$
$$2x + y \le 10$$
$$y \ge 0$$
$$0 \le x \le 5.$$

If $V = 4x + ky + 10$, for what value of the constant k will there be more than one point in the feasible region that will give the greatest possible value of V?

2–9 Applications of Linear Programming

Now you are ready to see how linear programming can be applied in a practical decision-making situation. Consider again the farmer and crops discussed in section 2–7.

Recall that the options in planting corn and alfalfa were subject to limitations imposed by the availability of land and labor, expressed in terms of the following system of inequalities,

$$x + y \leq 100$$
$$20x + 8y \leq 1600$$
$$x \geq 0$$
$$y \geq 0,$$

where x and y are the number of acres to be planted in corn and alfalfa, respectively. The farmer can elect to plant c acres of corn and a acres of alfalfa, provided (c, a) is in the feasible region $ABCO$.

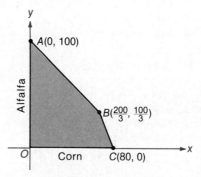

However, no conclusion has been reached about what acreage selection would be most advantageous for the farmer. More information is needed to reach such a conclusion.

Suppose that the following conditions apply.

a. The corn gives a gross profit of $340 per acre, the alfalfa $210 per acre.

b. The cost of labor is $4 per hour.

c. The costs of seed, fertilizer, and tractor fuel are $50 per acre for the corn and $28 per acre for the alfalfa.

With these conditions, together with the previous assumption that 20 hours of labor must be invested in each acre of corn and 8 hours for each acre of alfalfa, you can now determine what acreages will yield the maximum return.

The net return R, expressed in terms of x and y, is

R = income from corn + income from alfalfa −

labor costs − seed and other costs

$$= 340x + 210y - [4(20x + 8y) + (50x + 28y)].$$

86 Chapter 2: Analytic Geometry—Lines and Circles

This last equation simplifies to

$$R = 210x + 150y.$$

Since the maximum value of R will be reached at one of the vertices of the feasible region, compute R for each of vertices A, B, C, and O.

Vertex		Net return R
$A(0, 100)$	$R = 210(0) + 150(100)$	\$15,000
$B(\frac{200}{3}, \frac{100}{3})$	$R = 210(\frac{200}{3}) + 150(\frac{100}{3})$	\$19,000
$C(80, 0)$	$R = 210(80) + 150(0)$	\$16,800
$O(0, 0)$	$R = 210(0) + 150(0)$	0

Thus by planting $66\frac{2}{3}$ acres of corn and $33\frac{1}{3}$ acres of alfalfa, the farmer can realize a maximum net return of \$19,000.

In actual practice, problems may involve more than two variables and may include far more than three or four restricting inequalities for determining the feasible region. Such problems are so complex that computer solutions become a practical necessity. The process of solution is nevertheless basically the same as for the simpler problems presented in this chapter.

Example 1 provides a solution to the problem stated at the beginning of this chapter.

Example 1 A computer manufacturer has orders for 50 Model PA-99 computers, 24 to be shipped to Boston, 16 to Trenton, and 10 to Rutland. These orders will be filled using 28 computers in a Hartford warehouse and 22 in a New York warehouse. The transportation cost per computer is shown in the table below.

From \ To	Boston	Trenton	Rutland
Hartford	\$11	\$16	\$12
New York	\$20	\$10	\$24

How should the deliveries be managed to minimize shipping costs?

SOLUTION Let x be the number of computers shipped from Hartford to Boston. Let y be the number of computers shipped from Hartford to Trenton. The other shipments will then have to be made as shown in the table below.

From \ To	Boston	Trenton	Rutland
Hartford	x	y	$28 - (x + y)$
New York	$24 - x$	$16 - y$	$x + y - 18$

(Cont. on p. 88)

The last entry in the table, $x + y - 18$, is a simplification of the expression $22 - [(24 - x) + (16 - y)]$.

The following system of inequalities must be satisfied. Remember, no negative shipments can be made.

$$28 - (x + y) \geq 0 \qquad x + y - 18 \geq 0$$
$$y \geq 0 \qquad 16 - y \geq 0$$
$$x \geq 0 \qquad 24 - x \geq 0$$

The figure shows that the feasible region is the polygonal region *ABCDE*.

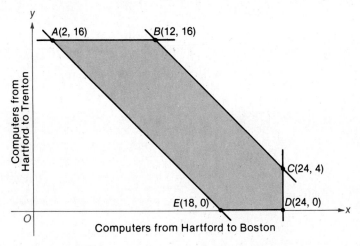

The total shipping cost S is

$$S = 11x + 16y + 12(28 - x - y) + 20(24 - x) + 10(16 - y) + 24(x + y - 18),$$

$$S = 3x + 18y + 544.$$

Now evaluate S at each of the five vertices of the feasible region.

Point	A	B	C	D	E
Coordinates	(2, 16)	(12, 16)	(24, 4)	(24, 0)	(18, 0)
Shipping Cost	$838	$868	$688	$616	$598

The minimum shipping cost occurs at *E*, so the following shipping schedule should be used.

From \ To	Boston	Trenton	Rutland
Hartford	18	0	10
New York	6	16	0

Exercises

A Suppose in the example about the farmer who plants corn and alfalfa, that the gross profits for corn and alfalfa change to the amounts specified. a. Determine the expression for net return R in terms of x and y. b. Evaluate R at each vertex of the feasible region. c. How should the farmer allocate acreage so as to have the greatest net return R?

1. Corn $390 per acre; alfalfa $160 per acre
2. Corn $300 per acre; alfalfa $290 per acre

Suppose the computer manufacturer in example 1 is faced with a new set of shipping costs as shown in the table below. How should the deliveries be managed so as to minimize the total shipping costs?

3.

From \ To	Boston	Trenton	Rutland
Hartford	$15	$12	$13
New York	$18	$10	$20

4.

From \ To	Boston	Trenton	Rutland
Hartford	$14	$14	$20
New York	$25	$13	$21

B 5. A certain machine can manufacture items either of type A or type B but not both at the same time. Suppose that daily production is desired that will satisfy the following conditions.

> The machine is to be used not less than 2 nor more than 8 hours per day.
>
> From 1 to 6 hours must be devoted to making type-A items.
>
> At most, 5 hours per day can be devoted to making items of type B.

Let x be the number of hours devoted to type-A items and
y be the number of hours devoted to type-B items.

Determine all the vertices of the feasible region.

6. A firm makes two kinds of articles, "glicks" and "widgets." Each day the firm can make at most 4 articles in all. It takes 2 hours to make each glick and 1 hour to make each widget, and the firm is open only 6 hours each day.
   a. Specify the inequalities that determine the feasible region.
   b. Sketch the graph of the feasible region and determine the coordinates of its vertices.
   c. If the profit is $50 on each glick and $40 on each widget, how many of each should be made in order that the profit be a maximum?

7. Suppose two warehouses supply three stores. Warehouse A has 12 items and warehouse B has 8 items. Store X needs 8 items, Y needs 6, and Z needs 6. Shipping costs per item are as follows.

From \ To	X	Y	Z
A	$7	$3	$2
B	$3	$1	$2

How many items should be shipped from each warehouse to each store to minimize the cost?

8. A store manager has room on her shelves for no more than a total of 100 items of brand X and brand Y. From experience, she knows she will need at least 30 items of brand X and not more than 50 items of brand Y. Her profit is 20 cents for each item of brand X and 25 cents for each item of brand Y. If she can sell her entire stock, how many of each brand should there be to yield maximum profit?

9. Two school systems wish to purchase large quantities of metal wastebaskets. Northampton must have at least 200 and will buy as many as 600 at $1.00 each. Easthampton requires at least 300 and will take as many as 500 at $1.10 each. As a seller of wastebaskets, you wish to meet the minimum needs of both school systems and to maximize your profit. Assume that you have 1000 wastebaskets available.
   a. How should you distribute your sales between the two school systems if you must pay delivery charges of 4 cents per basket on sales to Northampton and 8 cents on sales to Easthampton?
   b. If Northampton offers to pay $1.07 per wastebasket, will that affect your decision?

10. A manufacturing firm produces two different articles, toy cars, and toy trucks. To produce each car it takes 0.30 hours for assembly, 0.20 hours for inspection, and 0.06 hours for packing. To produce each truck takes 0.50 hours for assembly, 0.08 hours for inspection, and 0.20 hours for packing. The firm has available 1800 hours per week for assembly, 800 hours per week for inspection, and 600 hours per week for packing. The firm makes a profit of 10 cents for each car and 25 cents for each truck. How many cars and trucks should the firm produce each week to have maximum profit?

C 11. Write a program to have a computer find the maximum value of the net return $V = dx + ey + f$ on the feasible region determined by the inequalities $0 \leq x \leq a$, $0 \leq y \leq x + b$, and $x + y \leq c$. Input the constants a, b, c, d, e, and f.

Chapter Review Exercises

2–1, page 49

For the given points P and Q, calculate PQ and locate the midpoint of $\overline{PQ}$.

1. $P(-3, 1)$, $Q(2, 0)$
2. $P(1, -4)$, $Q(4, 9)$

3. Determine whether the points $(0, 2)$, $(6, 10)$, and $(-8, -4)$ are vertices of an isosceles triangle.

4. Use the Pythagorean theorem to determine whether the triangle in problem 3 is a right triangle.

2–2, page 53

Write an equation of the line that contains the given points.

5. $(-2, -4)$, $(-1, 6)$
6. $(3, -7)$, $(3, 10)$

7. Determine whether the points $(2, 5)$, $(-1, -2)$, and $(7, 17)$ are collinear.

8. For the triangle ABC with vertices $A(0, 1)$, $B(-1, 2)$, and $C(5, 7)$, write an equation of the line that contains the median drawn from vertex A.

2–3, page 58

Write an equation of the line that contains the given point and is parallel to the given line.

9. $(-1, 3)$; $y = 7x + 4$
10. $(0, 4)$; $3x + 2y = 10$

Write an equation of the line that contains the given point and is perpendicular to the given line.

11. $(3, -1)$; $y = 2x - 4$
12. $(-1, -2)$; $2y + 5x = 1$

13. Write an equation of the perpendicular bisector of the segment with endpoints $(2, -5)$ and $(-1, -7)$.

2–4, page 63

14. Write an equation of the circle with center $(-1, 2)$ and radius 3.
15. Find the center and radius of the circle $x^2 + 6x + y^2 - 12y = 4$.
16. Write an equation of the line tangent to the circle $x^2 + y^2 = 100$ at the point $(-6, 8)$.

2–5, page 66

17. Find the point of intersection of the lines $y = 3x - 2$ and $y = 4x + 5$.

18. Find the point or points of intersection, if any, of the line $y = 2x - 10$ and the circle $x^2 + y^2 = 25$.

19. Find the point or points of intersection, if any, of the circles $x^2 + y^2 = 25$ and $(x - 2)^2 + (y + 1)^2 = 10$.

20. If the line $y = m(x - 4)$ is tangent to the circle $x^2 + y^2 = 1$, what is the value of m?

2–6, page 72

Find the distance from the given point to the given line.

21. $(1, -3)$; $2x + 3y + 1 = 0$ **22.** $(-1, -2)$; $y = 7x + 2$

23. Find the area of the triangle with vertices $(1, -3)$, $(-1, -2)$, and $(0, 4)$.

24. Find the distance between the parallel lines $y = 4x + 1$ and $y = 4x - 2$.

2–7, page 77

Sketch a graph of each inequality.

25. $y \leq 2 - 4x$ **26.** $3x + 2y \leq 5$

27. Graph the solution set of the given system of inequalities.

$$x - y \leq 3$$
$$x \geq 0$$
$$y \geq 0$$
$$2x + 5y \leq 20$$

2–8, page 81

28. For the feasible region shown, determine both the maximum and the minimum values of the variable $V = 3x + 4y - 1$.

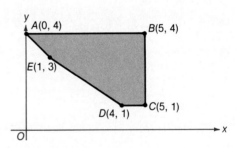

2–9, page 86

29. The Acme Syrup Co. produces two grades of pancake syrup, Grade A that is $\frac{3}{4}$ Vermont maple syrup and $\frac{1}{4}$ corn syrup, and Grade B that is $\frac{1}{8}$ Vermont maple and $\frac{7}{8}$ corn syrup. In a given season, Acme can buy up to 10,000 gallons of Vermont syrup and up to 13,750 gallons of corn syrup, but processing capacity is limited to 20,000 gallons per season. Acme makes a profit of $5 per gallon on Grade A syrup and $3 per gallon on Grade B. Assuming that Acme can sell its entire output, how many gallons of each grade should the company produce in order to maximize profit?

Chapter Test

1. For points $P(-1, -3)$ and $Q(1, -7)$, (a) calculate PQ, (b) find the midpoint of $\overline{PQ}$, and (c) write an equation of line PQ.

2. Find the constant c so that the line containing $(1, 3)$ and $(-2, c)$ is parallel to the line through $(2, -1)$ and $(4, 3)$.

3. Find the distance from the point $(-2, 3)$ to the line $3x - 4y = 11$.

4. For the circle with center $(-4, 2)$ and radius 5, (a) write an equation of the circle, and (b) write an equation of the line tangent to the circle at $(-1, 6)$.

5. Find the distance between the parallel lines $2x + 3y = 9$ and $2x + 3y = -1$.

6. Given line $\ell_1 : 3x + 4y + 7 = 0$, find the constant k so that the line $\ell_2 : kx + 5y - 3 = 0$ (a) is parallel to ℓ_1; (b) is perpendicular to ℓ_1; (c) intersects ℓ_1 at $(-1, -1)$ only.

7. For the circle $C: x^2 - 6x + y^2 + 2y = 15$, (a) find the center and radius of C; (b) find the points of intersection, if any, of C with the line $x + 7y = 21$.

8. Given $A(-2, 1)$, $B(3, 2)$, and $C(4, 0)$. (a) Write an equation of the altitude through B for $\triangle ABC$. (b) Find the area of $\triangle ABC$.

9. Graph the solution set of the following system of inequalities:

    $2x + y \leq 6$
    $x + y \leq 4$
    $x \geq 0$
    $y \geq 0$

10. The system of inequalities in problem 9 defines the feasible region for a quantity $V = 7x + 4y + 20$. Determine the maximum value of V.

11. Region $ABCDE$ is the feasible region for a variable P.

    a. Specify the system of inequalities that defines this region.

    b. If $P = 3x + 4y - 8$, what choices of x and y will maximize P?

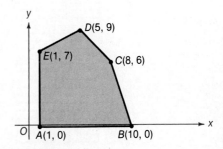

12. In order to comply with fire department regulations, a service station manager can store at most 10,000 gallons of gasoline. He sells two types, unleaded and regular, and from experience he knows that he will need at least 3000 gallons of unleaded and at most 6000 gallons of regular. His profit is 5¢ per gallon on unleaded gasoline and 7¢ per gallon on regular gasoline. Assuming that he will sell all of the fuel, how many gallons of each type should he obtain in order to maximize his profit?

Functions

3

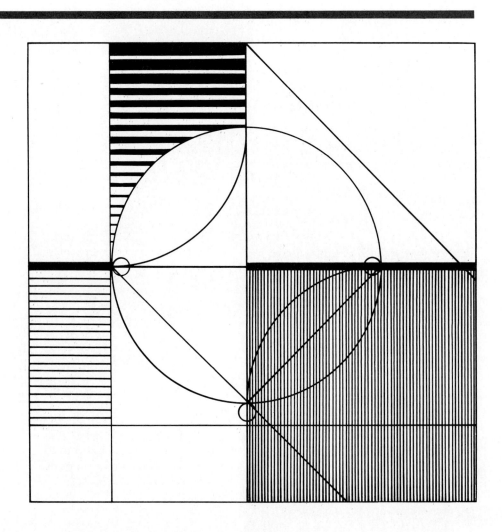

3-1 Relations

When studying equations in two variables, you became familiar with the idea of an *ordered pair* (a, b) of objects a and b, which were usually real numbers. In this chapter you will make more use of ordered pairs and sets of ordered pairs, and some additional language is needed.

Definition A **relation** is a set of ordered pairs.

For an ordered pair (a, b), we call a the **argument**, or first coordinate, and b the **value**, or second coordinate.

For a relation R, the set of arguments is called the **domain** of R and written Dom R. The set of values is called the **range** of R and written Rng R. A relation assigns values to arguments.

Example 1 Find the domain and range of each relation.

$S = \{(1, -1), (2, 5), (3, 0)\}$

$T = \{(0, 1), (0, 5)\}$

$W = \{(x, y) : y \in \mathbf{R} \text{ and } x = y^2\}$

SOLUTION For S, Dom $S = \{1, 2, 3\}$ and Rng $S = \{-1, 5, 0\}$.

For T, Dom $T = \{0\}$ and Rng $T = \{1, 5\}$.

For W, Dom $W = \{x \in \mathbf{R} : x \geq 0\}$ and Rng $W = \mathbf{R}$.

Example 2 Describe a relation whose domain is $A = \{0, 1\}$ and whose range is $B = \{-6, 0, 6\}$.

SOLUTION Several sets of ordered pairs have the given domain and range. Two of them are

$\{(0, -6), (1, 0), (1, 6)\}$ and $\{(0, 6), (0, 0), (1, -6)\}$.

In example 2, the relation with the greatest number of ordered pairs is the set K of all possible ordered pairs with first element in A and second element in B.

$K = \{(0, -6), (0, 0), (0, 6), (1, -6), (1, 0), (1, 6)\}$

This set is called the **Cartesian product** of A with B and written $A \times B$.

If the domain and range of a relation S are both subsets of **R**, you can graph S in the same way you graph the solutions of equations in two variables.

Each ordered pair (a, b) in S corresponds to exactly one point in the coordinate plane with x-coordinate a and y-coordinate b.

Example 3 Graph relations S, T, and W from example 1.

SOLUTION $S = \{(1, -1), (2, 5), (3, 0)\}$ $T = \{(0, 1), (0, 5)\}$

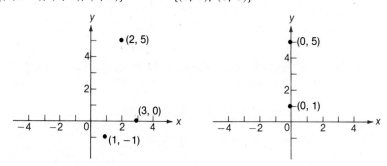

For S and T, the graph consists of only a few points. For $W = \{(x, y) : y \in \mathbf{R}$ and $x = y^2\}$, the graph consists of infinitely many points. Plot enough of them to indicate the shape of the graph.

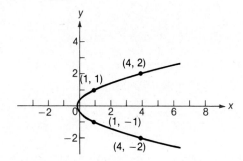

The graphs of some relations are regions in the coordinate plane, as shown in example 4.

Example 4 Graph the relation $C = \{(x, y) : x \geq 3$ and $|y| \leq 2\}$.

SOLUTION The graph is a region bounded by the lines $x = 3$, $y = 2$, and $y = -2$.

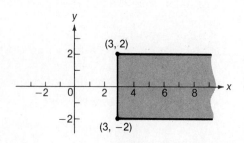

Note that boundary lines or curves which are included in the graph are drawn as solid lines. To indicate that a boundary line or curve is not included in the graph of a relation, use a dashed line. To indicate that any single point is omitted, use an open dot.

Example 5 Graph the relation $T = \{(x, y) : y > 2x + 3 \text{ and } y < 5\}$.

SOLUTION The graph is a region in the plane bounded by $y = 2x + 3$ and $y = 5$.

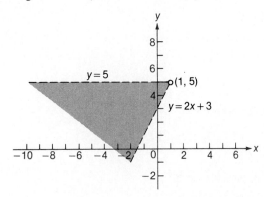

So far, the relations considered have consisted of ordered pairs of numbers. The ordered pairs may, however, contain elements other than numbers, as shown in the following examples.

$A = \{(\text{red}, 3), (\text{blue}, 4), (\text{green}, 5), (\text{violet}, 6)\}$

$B = \{(\text{Scott, Bob}), (\text{Scott, Steve}), (\text{Miguel, Bob})\}$

$C = \{(x, y) : x \text{ is a triangle and } y \text{ is its area}\}$

$D = \{(x, y) : x \text{ is a U.S. citizen and } y \text{ is a grandfather of } x\}$

Exercises

A Give the domain and range of each relation.

1. $\{(5, 2), (173, 3), (-1, 2)\}$
2. $\{(2, 1), (3, 1), (4, 0), (5, 1), (6, 0), (7, 1)\}$
3. $\{(\text{Saco, Maine}), (\text{Saco, New Hampshire}), (\text{Charles, Massachusetts})\}$
4. $\{(\text{one}, 3), (\text{two}, 3), (\text{three}, 5)\}$
5. $\{(7, -7), (7, 7), (6, -6), (0, 0)\}$
6. $\{(1, 0), (2, 0), (3, 0), (4, 0), (5, 0)\}$
7. $\{(\text{Kalamazoo}, 49008), (\text{Wheaton}, 60187), (\text{Champaign}, 61820)\}$
8. $\{(404, \text{Atlanta}), (512, \text{Austin}), (512, \text{San Antonio})\}$

Sketch the graph of each relation.

9. $\{(0, 5), (0, 4), (3, -1), (3, 2), (2, -1)\}$
10. $\{(-1, 2), (5, 3), (2, -1), (2, -2)\}$
11. $\{(x, y) : x + y = 4\}$
12. $\{(x, y) : x > 2\}$

B Give the domain and range of each relation.

13. $\{(x, y) : x \in J, y \in J \text{ and } y > x\}$
14. $\{(x, y) : x \geq y \text{ and } -1 \leq y \leq 1\}$
15. $\{(x, y) : x \geq 4 \text{ and } |y| \leq 2\}$
16. $\{(x, y) : x < 3 \text{ and } y < 2\}$

Sketch the graph of each relation.

17. $\{(x, y) : x \text{ is an even integer and } y = \frac{1}{2}x\}$
18. $\{(x, y) : 1 \leq x \leq 3 \text{ and } y > 1\}$
19. $\{(x, y) : x \text{ and } y \text{ are both in the set } \{0, 1, 2\}\}$
20. $\{(x, y) : y = 2x + 1 \text{ and } -1 \leq x \leq 3\}$

Sketch the graph of $A \times B$ for the given sets A and B.

21. $A = \{0, 1, 2\}$ and $B = \{1, 3\}$
22. The closed interval $A = [1, 3]$ and the half-open interval $B = (-1, 2]$

C 23. If $F = \{(x, y) : y^2 = 2x + 15\}$, list all values with argument -3.
24. If $G = \{(x, y) : y \leq 2x - 5\}$, determine all arguments that have the value $\frac{3}{2}$.
25. If $H = \{(x, y) : y \geq 3x^2 - 2x - 1\}$ and the ordered pair $(-2, k)$ is in H, what are the possible values of k?

Describe algebraically the relation whose graph is shown.

26.
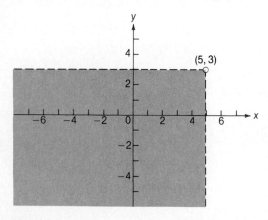

27.
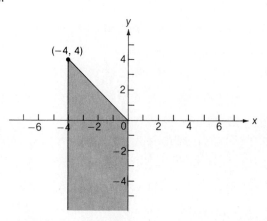

3–2 Functions

The relation in exercise 7 in section 3–1 may be called a "zip code relation." Since there is more than one zip code for many towns and cities, knowing the town does not always tell you the zip code. So the relation's arguments do not always determine its values. For some relations, however, arguments do determine unique values.

For example, in $\{(x,y) : x \in \mathbf{R} \text{ and } y = x^2\}$, values are uniquely determined by arguments; the argument 4 determines the one value 16, -7 determines 49, and so on. Relations such as this one are very important in mathematics and are called functions.

Definition A **function** is a relation that assigns exactly one value to each argument.

Stated symbolically, a relation S is a function if and only if whenever (a, b) and (a, c) are in S, then $b = c$. So a function is a set of ordered pairs such that no two ordered pairs have the same argument.

Example 1 Determine whether or not each relation listed in the chart is a function.

SOLUTION

Relation	Function	Explanation
$S = \{(1, -1), (2, 3), (4, 9)\}$	Yes	One value for each argument
$T = \{(1, 0), (2, 0), (3, 1)\}$	Yes	One value for each argument
$W = \{(0, 1), (0, -1), (5, 6), (-1, 2)\}$	No	Both 1 and -1 assigned to 0
$V = \{(x, y) : y \in \mathbf{R} \text{ and } x = y^2\}$	No	Pairs like $(4, 2)$ and $(4, -2)$ are both in V.
$P = \{(x, y) : x \in \mathbf{R} \text{ and } y = x^2\}$	Yes	If (a, b) and (a, c) are in P, then $b = a^2$, $c = a^2$, and $b = c$. P is a function.

Since every function is a relation, a function has a domain and range as defined on page 95.

Example 2 Specify the domain and the range of functions S, T, and P in example 1.

SOLUTION
Dom $S = \{1, 2, 4\}$ Rng $S = \{-1, 3, 9\}$
Dom $T = \{1, 2, 3\}$ Rng $T = \{0, 1\}$
Dom $P = \mathbf{R}$ Rng $P = \{y \in \mathbf{R} : y \geq 0\}$

It is easy to tell from the graph of a relation whether it is a function. If any vertical line intersects the graph in two or more points, then the relation is not a function. This is known as the **vertical-line test**. It is valid because any two points on a vertical line have the same x-coordinate. This means that if a vertical line intersects the graph of a relation in two or more points, then the relation has at least one argument with more than one value.

Example 3 Use the vertical-line test to decide which of the following are graphs of functions.

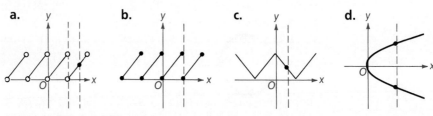

SOLUTION The illustrations show that **a** and **c** both pass the vertical-line test and are graphs of functions and that **b** and **d** fail the test.

To designate a given function, we usually use a single letter such as f, and to identify the value which a function f assigns to a given argument a we write f(a). If (a, b) is an ordered pair in f, then the value is b = f(a).

Example 4 For the function $f = \{(-2, 2), (0, 3), (1, 4)\}$, find $f(0)$, $f(1)$, and $f(-2)$.

SOLUTION $f(0) = 3$, $f(1) = 4$, and $f(-2) = 2$.

It is important to distinguish between the function f, which is a set of ordered pairs, and f(x), which is the value that f assigns to the argument x. We will usually describe a function f by specifying the domain and a rule—an equation or a formula—for assigning values f(x) to arguments x. In fact, whenever a function contains an infinite number of ordered pairs, as in the next example, it must be defined by a rule.

Example 5 A function g has domain **R** and its values are given by

$$g(x) = x^2 + 1.$$

Find $g(-2)$, $g(0)$, and $g(1)$.

SOLUTION
$g(-2) = (-2)^2 + 1 = 5$
$g(0) = (0)^2 + 1 = 1$
$g(1) = (1)^2 + 1 = 2$

Another way of displaying the rule of a function is to use the notation
$$f: x \to f(x).$$
For instance, the rule $f(x) = x^2 + 1$ may be written as
$$f: x \to x^2 + 1,$$
which is read "f sends x into $x^2 + 1$" or "f maps x onto $x^2 + 1$."

Example 6 Find the three ordered pairs of the function
$$g: x \to x^2 - 2x - 3$$
for the arguments 1, -3, and $1 + \sqrt{3}$.

SOLUTION
$$g(1) = 1^2 - 2(1) - 3 = -4$$
$$g(-3) = (-3)^2 - 2(-3) - 3 = 12$$
$$g(1 + \sqrt{3}) = (1 + \sqrt{3})^2 - 2(1 + \sqrt{3}) - 3 = -1$$
The three ordered pairs are $(1, -4)$, $(-3, 12)$, and $(1 + \sqrt{3}, -1)$.

Any function with domain **N**, the set of natural numbers, is called a **sequence**.

Example 7 A function f is a sequence and $f: x \to \dfrac{1}{x}$. Sketch a graph of f.

SOLUTION The graph consists of infinitely many isolated points $(1, 1)$, $\left(2, \frac{1}{2}\right)$, $\left(3, \frac{1}{3}\right)$, $\left(4, \frac{1}{4}\right)$, ..., $\left(n, \dfrac{1}{n}\right)$, ... Only the first six points are shown in the graph below.

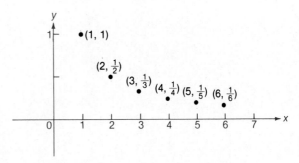

This function can also be written as $f = \left\{ \left(n, \dfrac{1}{n}\right) : n \in \mathbf{N} \right\}$.

The values $f(1)$, $f(2)$, $f(3)$, · · · of a sequence are called the **terms** of the sequence and are often written $a_1, a_2, a_3, \ldots$. The numbers 1, 2, 3, . . .

used in the symbols $a_1, a_2, a_3, \ldots$ are called **subscripts**. When working with a list of numbers, such as the terms of a sequence, subscripts help keep track of the position of each number in the list.

Exercises

A Determine whether each relation is a function.

1. $\{(0, 1), (0, 2), (0, 3)\}$

2. $\{(1, 0), (2, 0), (3, 0)\}$

3. $\{(6, 13), (2, 5), (-3, -5)\}$

4. $\{(-2, 4), (-1, 1), (0, 0), (1, 1), (2, 4)\}$

5. $\{(7, 1), (8, -3), (11, 9), (-5, 6), (4, 6)\}$

6. $\{(1, 0), (2, 0), (3, 3), (4, 4)\}$

7. $\{(x, y) : x \in \mathbf{R} \text{ and } y = 3x + 5\}$

8. $\{(x, y) : x \in \mathbf{N} \text{ and } 2x = y + 1\}$

9. $\{(x, y) : x \in \mathbf{R}, y \in \mathbf{R}, \text{ and } x^2 + y^2 = 100\}$

10. $\{(x, y) : x \in \mathbf{R}, y \in \mathbf{R}, \text{ and } x = |y|\}$

Use the vertical-line test to determine whether each graph represents a function.

11.
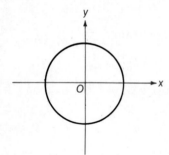

12.
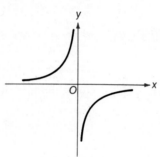

13.
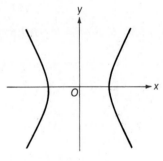

14.
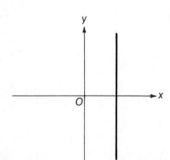

15.
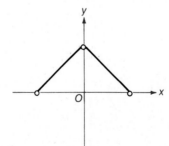

16.
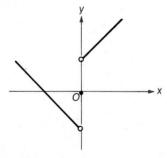

B 17. If $g(x) = 3x + (x - 1)^2$, find $g(0)$, $g(1)$, and $g(2)$.

18. If $f(x) = \dfrac{3x}{x - 1}$, find (if possible) $f(0)$, $f(1)$, and $f(2)$.

19. If $f: x \to x^2 - 1$, find $f(-3)$, $f(\frac{1}{3})$, and $f(1 + \sqrt{2})$.

20. If $g: x \to \frac{x-1}{x+2}$, find $g(0)$, $g(-x)$, and $g(2x)$.

List the first four terms of the given sequence.

21. $f: n \to \frac{n}{n+1}$

22. $f: n \to \frac{n(n-1)}{2}$

Graph the sequence with the given formula.

23. $f(n) = \frac{2}{n}$ **24.** $f(n) = n^2$ **25.** $a_n = 2n + 1$ **26.** $a_n = 2^n$

27. Evaluate the constant c if $(2, 3) \in f$ and $f: x \to x^2 - 3x + c$.

28. Evaluate the constant k if $(-1, 2) \in g$ and $g(x) = \frac{kx}{k+4}$.

For each function f, and $h \neq 0$, simplify the quotient
$$\frac{f(x+h) - f(x)}{h}.$$

29. $f(x) = 3x + 1$ **30.** $f(x) = x^2$ **31.** $f: x \to \frac{1}{x}$ **32.** $f: x \to \frac{x+1}{x}$

C 33. If $f(x) = \frac{1}{x+2}$ and $g(x) = 2x + 3$, find the arguments c for which $f(c) = g(c)$.

34. If $f(x) = \frac{1}{x+c}$ and $(3, c) \in f$, find c.

35. The graphs of two functions f and g are shown. Use the graphs to solve the following:

a. $f(x) = g(x)$
b. $f(x) \leq g(x)$
c. $f(x) = g(2)$

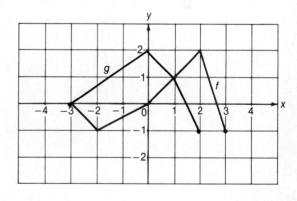

36. Write and run a program to have the computer print every tenth term (a_{10}, a_{20}, a_{30}, ...) of the sequence given by $a_n = f(n) = \left(1 + \frac{1}{n}\right)^n$, stopping with the term a_{200}.

3-3 Zeros of a Function

Often the rule for a function f is given without specifying the domain. When this happens, you are to assume that the domain is the largest possible set of real numbers for which $f(x)$ is defined and is a real number.

Example 1 If h assigns values by the rule $h(x) = \sqrt{2x - 6}$, what is the domain of the function h?

SOLUTION Since the square root of a negative number is not a real number,

$$2x - 6 \geq 0, \text{ or } x \geq 3.$$

$$\text{Dom } h = \{x \in \mathbf{R} : x \geq 3\} = [3, \infty)$$

Example 2 If $g(x) = \dfrac{1}{x - 1} + \dfrac{2x}{x + 7}$, what is the domain of g?

SOLUTION To avoid division by zero in the rule, $x - 1 \neq 0$ and $x + 7 \neq 0$.

$$\text{Dom } g = \{x \in \mathbf{R} : x \neq 1 \text{ and } x \neq -7\} = \mathbf{R} - \{1, -7\}$$

As in example 2, it is common mathematical practice to use informal language such as "the function $f(x) = x - 2x^2$" when more rigorously we would speak of "the function f whose values are given by the rule $f(x) = x - 2x^2$."

As the examples above indicate, an unstated domain of a function f is often found from the formula for $f(x)$ by eliminating those numbers which would require division by zero or taking an even root of a negative number.

In the last two examples, the domain was to be found when given a rule that implies the function. The problem of identifying the range of a function from its formula may be more involved than finding the domain, but you can determine whether a given number c is in the range by seeing whether the equation $f(x) = c$ has a solution.

Example 3 Determine whether 10 and 0 are in the range of the function $g(x) = 4 - x^2$ and find the range of g.

SOLUTION Solve $g(x) = 10$.

$$4 - x^2 = 10$$

$$x^2 = -6$$

This equation has no solution, so 10 is not in the range of g.
The solutions of $g(x) = 0$ are $x = \pm 2$. So g maps 2 and -2 onto 0; 0 is in the range of g.

(cont. on p. 105)

To find the range of g, use the fact that for any argument x,

$$x^2 \geq 0,$$
$$-x^2 \leq 0, \text{ and}$$
$$4 - x^2 \leq 4.$$

Thus $g(x) \leq 4$ for all x and $\text{Rng } g = \{y \in \mathbf{R} : y \leq 4\} = (-\infty, 4]$.

In example 3, it was shown that $g(2) = 0$ and that $g(-2) = 0$. The arguments 2 and -2 are called zeros of g.

Definition The number a is called a **zero of a function** f if and only if $f(a) = 0$.

Example 4 Find all the zeros of the function f defined on the domain $\mathbf{R}$ by
$$f : x \rightarrow x^2 - 2x - 3.$$

SOLUTION Find arguments for which the values $f(x)$ are zero.
$$x^2 - 2x - 3 = 0$$
$$(x + 1)(x - 3) = 0$$
$$x + 1 = 0 \quad \text{or} \quad x - 3 = 0$$
$$x = -1 \quad \text{or} \quad x = 3$$

The zeros of f are -1 and 3.

Exercises

A Find the domain of each function.

1. $f(x) = 3x - 11$
2. $g(x) = 7 - x^2$
3. $h : x \rightarrow \dfrac{3}{x} + 5x$
4. $g : x \rightarrow 11 - \sqrt{x - 5}$
5. $h(x) = \dfrac{-3}{\sqrt{x + 5}}$
6. $f(x) = \dfrac{1}{x} - \dfrac{x - 5}{x + 1}$

Find the zeros of each function.

7. $f : x \rightarrow x^2 - 1$
8. $g : x \rightarrow \dfrac{x - 1}{x + 2}$
9. $h(x) = 3x^2 + 2x - 1$
10. $f(x) = 9 - x^2$
11. If $f(x) = 2x^2 - 3$, is 29 in the range of f?
12. If $g : x \rightarrow \dfrac{x + 3}{x - 1}$, is 1 in the range of g?

Give the domain and range of each function.

13. $\{(x, y) : x \in R \text{ and } y = 2x + 3\}$ **14.** $\{(x, y) : x \in R \text{ and } y = |x|\}$

15. $f : x \to \dfrac{x}{2} + 3$ **16.** $g : x \to 3x^2 - 2$

17. $f(x) = \sqrt{x - 4}$ **18.** $g(x) = \dfrac{2}{x}$

19. $h(x) = 2 + (x - 2)^2$ **20.** $f(x) = 3 - 2x^2$

B Find the domain and the zeros of f.

21. $f(x) = x^3 + 8$ **22.** $f(x) = \dfrac{x^2 + 5x + 6}{x^2 - 9}$

23. $f : x \to 1 + \sqrt{4 - x}$ **24.** $f : x \to \dfrac{1}{x} - \dfrac{1}{x + 1}$

25. The graph of a function f is shown at the right.

   a. Find Dom f and Rng f.

   b. Find the zeros of f.

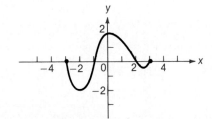

26. The graph of a function g is shown.

   a. Find Dom g and Rng g.

   b. Find the zeros of g.

   c. Solve for x: $g(x) = g(2)$.

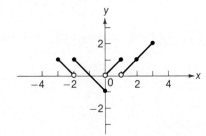

27. If $(2, 3) \in f$ and $(3, 7) \in f$, find $f(f(2))$.

28. For the function $g : x \to 2x^2 + 1$, find $g(g(0))$, $g(g(1))$, and $g(g(2))$.

C Find the domain and range of each function.

29. $f(x) = 2 + \dfrac{1}{x + 1}$ **30.** $g(x) = \dfrac{-1}{\sqrt{2x - 1}}$

31. Find the domain of $h(x) = \dfrac{1}{\sqrt{3x - 2}} + \sqrt{4 - x}$.

32. Write and run a program to have the computer print the values $f(1)$, $f(1.1)$, $f(1.2)$, $\ldots$, $f(2.9)$, $f(3)$ of the function $f(x) = x^4 - x^3 - x^2 - 2x - 6$. Use the results to estimate a zero of f.

3–4 Some Special Functions

Some functions occur so often that they are given special descriptive names. If you have worked with computers, you may already be familiar with some of these names.

THE IDENTITY FUNCTION

The function with domain **R** and values $f(x) = x$ is the **identity function** and is denoted by I.

$$I : x \to x$$

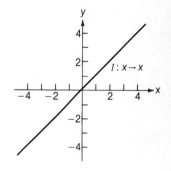

Thus $I(x) = x$, and all the ordered pairs in I are of the form (x, x). The graph of I is the line $y = x$ through the origin with slope 1.

THE CONSTANT FUNCTIONS

For any real number c there is a **constant function** described by

$$f(x) = c.$$

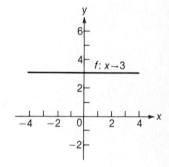

The domain of the function is **R**. For instance, if $c = 3$, then $f(5) = 3$, $f(-6) = 3$, $f(\pi) = 3$, and so on.

The graph of any constant function is a horizontal line.

THE RECIPROCAL FUNCTION

The function $f : x \to \dfrac{1}{x}$ is the **reciprocal function**. Its domain is the set of nonzero real numbers, that is, **R** − {0}. Notice that the graph of the reciprocal function never quite touches the x-axis, although it gets arbitrarily close to it. Such a line is called an **asymptote** of the graph. We also say that $f(x)$ has limit 0 as x increases without bound. Since $f(x)$ increases (or decreases) without bound as x tends to 0, the y-axis is also an asymptote of the graph.

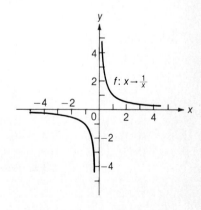

3–4: Some Special Functions **107**

The next three special functions are often used in computer programming, and so it is reasonable to adopt their names from the BASIC language.

THE ABSOLUTE-VALUE FUNCTION

The **absolute-value function abs** assigns the value $|x|$ to the argument x on the domain **R**. Thus,

$$\text{abs}(-7) = 7,$$
$$\text{abs}(0) = 0,$$
$$\text{abs}(5) = 5,$$

and for all $x \in \mathbf{R}$,

$$\text{abs}(x) = |x|.$$

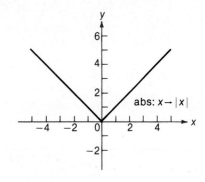

THE SIGNUM FUNCTION

The **signum function sgn** has domain **R** and assigns values as follows:

$$\text{sgn}(x) = \begin{cases} 1, & \text{if } x > 0 \\ 0, & \text{if } x = 0. \\ -1, & \text{if } x < 0 \end{cases}$$

This is a sign-sensing function; that is, for negative arguments it assigns the value -1 and for positive arguments it assigns the value $+1$.

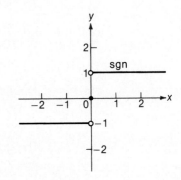

THE GREATEST-INTEGER FUNCTION

The **greatest-integer function int** is defined on the domain **R** by the rule

$$\text{int}(x) = \text{the greatest integer less than or equal to } x.$$

Here are some examples:

$$\text{int}(7) = 7 \qquad \text{int}(-\pi) = -4$$
$$\text{int}(-2.1) = -3 \qquad \text{int}(\sqrt{103}) = 10$$

If x is itself an integer, then $\text{int}(x) = x$. If x is between the integers k and $k + 1$, then $\text{int}(x) = k$. Sometimes the symbol $[x]$ is used for $\text{int}(x)$.

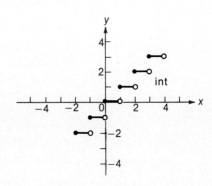

On a number line, int(x) is the integer just to the left of x, unless x is itself an integer.

The ability to graph functions is often helpful in solving equations and inequalities.

Example 1 Use graphs as an aid in solving the equation $|x| = x + 4$.

SOLUTION Sketch the graphs of abs and $f(x) = x + 4$ on the same coordinate axes and observe that abs and f intersect only at $(-2, 2)$. This means that only at $x = -2$ do abs and f have the same value. Hence -2 is the only solution of the equation.

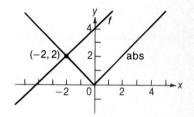

Example 2 Use graphs to solve the inequality $\frac{1}{x} - \text{sgn}(x) < 0$.

SOLUTION The given inequality is equivalent to

$$\frac{1}{x} < \text{sgn}(x).$$

Graph the reciprocal function and the signum function and note that they intersect at $(1, 1)$ and at $(-1, -1)$. Now observe where the graph of $f: x \to \frac{1}{x}$ lies below the graph of signum.

Solution set: $\{x : -1 < x < 0 \text{ or } x > 1\} = (-1, 0) \cup (1, \infty)$

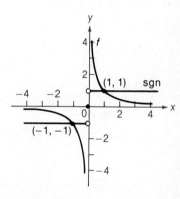

Exercises

A Evaluate each of the following.

1. $I(5)$
2. $I(x)$
3. $\text{sgn}(-4)$
4. $\text{sgn}(0)$
5. $\text{int}(2)$
6. $\text{int}(2.97)$
7. $\text{sgn}(\sqrt{2})$
8. $\text{sgn}(-0.1)$
9. $[-2.7]$
10. $[-0.1]$
11. $\text{sgn}(-7) + \text{abs}(-1)$
12. $I(-3) + \text{abs}(-2)$
13. $\text{int}(-2.1) + \text{int}(2.1)$
14. $[2.73] + I(2.73)$

B Use a calculator to evaluate each of the following.

15. $\text{int}(\sqrt{11.7 + \sqrt{12.5}})$
16. $\text{int}\left(\frac{(327.9)(\sqrt{0.0038})}{(68.7)(929.2)}\right)$

Sketch graphs of each pair of functions on the same coordinate axes. Use your graphs to determine the arguments, if any, for which the two functions have the same value.

17. $f(x) = x$ and $g(x) = |x|$
18. I and int
19. I and $f: x \to 1$
20. $f(x) = x$ and $g(x) = \text{sgn}(x)$
21. abs and sgn
22. $f(x) = [x]$ and $g(x) = 3$
23. $g: x \to \dfrac{1}{x}$ and $f: x \to 2$
24. $f(x) = \dfrac{1}{x}$ and $g(x) = \text{sgn}(x)$

Use graphs as an aid in solving each of the following.

25. $|x| = \dfrac{1}{x}$
26. $\text{sgn}(x) = |x|$
27. $\dfrac{1}{x} < 4$
28. $\text{int}(x) \geq \text{sgn}(x)$
29. $|x| - \text{int}(x) \leq 0$
30. $\dfrac{1}{x} - |x| > 0$

C **31.** Sketch the graph of $f(x) = \dfrac{|x|}{x}$ and compare it with the graph of sgn.

32. Write a program for the computer to print a list of the exact divisors of a given positive integer. $\left[\text{HINT: } a \text{ is divisible by } b \text{ if and only if } \left[\dfrac{a}{b} \right] = \dfrac{a}{b}. \right]$

3–5 Addition of Functions

One way of graphing $h: x \to |x| + 1$ is to begin with a graph of $f: x \to |x|$ and shift this graph upward 1 unit.

The function h is thought of as the sum of abs and the function with constant value 1. That is,

if $f(x) = \text{abs}(x)$,
and $g(x) = 1$,
then $h(x) = f(x) + g(x)$.

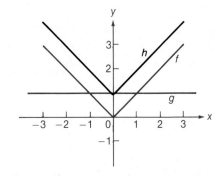

Definition Let f and g be real-valued functions, and let D be the intersection of Dom f and Dom g. Then $f + g$ is the function defined on D by

$$[f + g](x) = f(x) + g(x).$$

The illustration above shows that adding a constant function $g: x \to k$ to a given function f produces a new function $[f + g]: x \to f(x) + k$ with the same domain as f. To sketch the graph of $f + g$, shift points of f vertically $|k|$ units, upward if $k > 0$ and downward if $k < 0$.

Example 1 Sketch the graph of $f: x \to \dfrac{1}{x} - 2$.

SOLUTION First graph the reciprocal function $g(x) = \dfrac{1}{x}$. Then shift this graph downward 2 units. Notice that the asymptotes are now $y = -2$ and the y-axis.

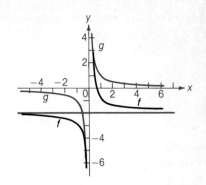

Example 2 Sketch a graph of $I + \text{sgn}$.

SOLUTION $[I + \text{sgn}](x) = I(x) + \text{sgn}(x)$
$= x + \text{sgn}(x)$

The function $I + \text{sgn}$ will differ from I by 1, 0, or -1, depending on the sign of x. For $x > 0$, add 1 to $I(x)$. For $x < 0$, subtract 1. Also,
$[I + \text{sgn}](0) = 0$.

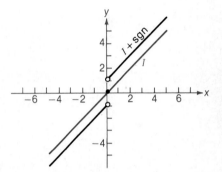

Changing the sign of every value of a given function f yields the function $-f$, which is called the **additive inverse** of f. Thus, $f(x) + [-f(x)] = 0$.

The graph of $-f$ is particularly easy to draw when the graph of f is given. If (a, b) is on the graph of f, then $(a, -b)$ is on the graph of $-f$. So the graph of $-f$ is obtained by flipping the graph of f over the x-axis. The two graphs below illustrate this procedure.

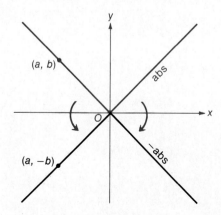 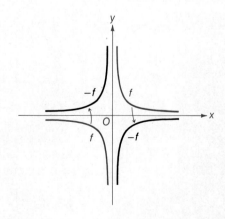

3–5: Addition of Functions

By combining function addition with the additive inverse, we define function subtraction.

Definition If f and g are real-valued functions, then $f - g$ is the function $f + (-g)$.

The domain D of $f - g$ is the intersection of the domains of f and g, and for all $x \in D$,

$$[f - g](x) = f(x) + [-g](x) = f(x) - g(x).$$

Example 3 Write a formula for and sketch the graph of $f - g$ if $f(x) = 2$ and $g(x) = \frac{1}{x}$.

SOLUTION The formula is $[f - g](x) = 2 - \frac{1}{x}$.

First graph g, then flip this graph over the x-axis and shift the result vertically upward 2 units.

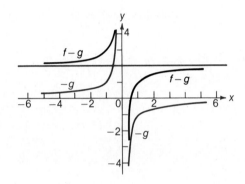

If neither of two functions f and g is a constant function, you can sketch the graph of $f + g$ by estimating the sum of the y-coordinates geometrically.

Example 4 Sketch the graph of $f: x \rightarrow \frac{1}{x} + |x|$.

SOLUTION First sketch the graphs of

$$g: x \rightarrow \frac{1}{x}$$

and

$$h: x \rightarrow |x|.$$

Now select an argument x, find the heights $g(x)$ and $h(x)$, and add these values geometrically. Do this for as many arguments x as you need to sketch the graph of f.

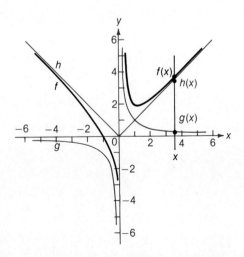

112 Chapter 3: Functions

Addition of functions is both commutative and associative, that is, $f + g = g + f$ and $(f + g) + h = f + (g + h)$. To prove this, as required in exercise 29, you must know what it means for two functions to be equal. Two functions f and g are equal if and only if they have the same domain D and for every $x \in D$, $f(x) = g(x)$. Note that just checking that $f(x) = g(x)$ does not guarantee that $f = g$. For example, if

$$f(x) = \frac{x^3}{x} \text{ and } g(x) = x^2,$$

then for every $x \in \text{Dom } f$, $f(x) = g(x)$. Yet $f \neq g$ because $g(0)$ is defined but $f(0)$ is not.

Exercises

A Write a formula for $[f + g](x)$, $-g(x)$, and $[f - g](x)$.

1. $f(x) = x^2$, $g(x) = 3x + 5$
2. $f(x) = \frac{1}{x}$, $g(x) = -x$
3. $f(x) = x^3$, $g(x) = x^2$
4. $f(x) = x^2 - 3x$, $g(x) = x - 4$

Calculate $(f + g)(3)$, $-f(3)$, and $(f - g)(3)$.

5. $f: u \to 2u + 5$, $g: t \to 3t - 6$
6. $f: u \to |u|$, $g: t \to t - 2$
7. $f: r \to -2r^2$, $g: s \to \frac{2}{3}s - 2$
8. $f: r \to 3r^3$, $g: s \to \frac{1}{2s}$

Sketch a graph of each function.

9. $f(x) = x + 1$
10. $f: x \to x - 3$
11. $f: x \to |x| + 3$
12. $f: x \to \frac{1}{x} + 2$
13. $f(x) = \text{sgn}(x) + 1$
14. $f(x) = |x| - 5$

B Sketch a graph of each function.

15. $f(x) = 3 - \frac{1}{x}$
16. $f(x) = 4 - |x|$
17. $f: x \to 2 - \text{sgn}(x)$
18. $f(x) = -2 - |x|$
19. $f(x) = \frac{1}{x} + \text{sgn}(x)$
20. $\text{sgn} - \text{abs}$

21. If $f: x \to \frac{x^2}{x}$, explain why $f \neq I$.

22. If $f(x) = \frac{x + 3}{x + 2}$ and $g(x) = \frac{x^2 + x - 6}{x^2 - 4}$, is $f = g$?

Name functions f and g so that the graph shown is approximately the graph of f + g.

23.
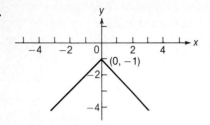

24.
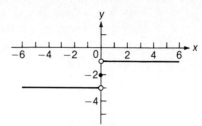

25.
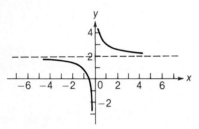

26.
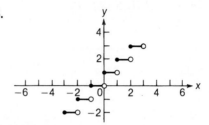

C Use a graph as an aid for solving each equation.

27. $|x| = \text{sgn}(x) + 1$ **28.** $2 - |x| = x$

29. Prove that function addition is commutative and associative; that is, prove that $f + g = g + f$ and $(f + g) + h = f + (g + h)$ for all real-valued functions f, g, and h.

Sketch a graph of each function.

30. $f(x) = x - |x|$ **31.** $f(x) = |x| - \dfrac{1}{x}$

32. Use a graph as an aid for solving the inequality $|x| - 2 \leq x$.

33. Sketch a graph of $I - \text{int}$.

3–6 Multiplication of Functions

Another way of combining functions to produce a new function is to use multiplication.

Definition Let f and g be real-valued functions, and let $D = \text{Dom } f \cap \text{Dom } g$. Then the product $f \cdot g$, or fg, is the function defined on D by

$$[f \cdot g](x) = f(x) \cdot g(x).$$

Suppose a given function f is multiplied by the constant function $g: x \to k$ to produce $[g \cdot f]: x \to kf(x)$. The result is to "stretch" the graph of f if $k > 1$ and to flatten the graph if $0 < k < 1$. We will often write kf to denote such a product.

Example 1 Graph abs, 2 abs, and $\frac{1}{2}$ abs on the same axes and compare the graphs.

SOLUTION All three functions are of the form $k \cdot$ abs. Note the stretching for $k = 2$ and the flattening for $k = \frac{1}{2}$.

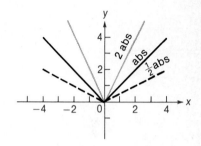

Example 2 Sketch the graph of $f: x \to 2(|x| - 1)$.

SOLUTION First sketch the graph of $g(x) = |x| - 1$, and then stretch this graph by doubling each ordinate to obtain the graph of f.

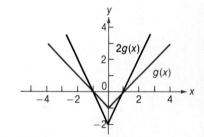

The graphs in examples 1 and 2 illustrate the fact that, for $k \neq 0$, the zeros of the functions f and kf are the same.

Example 3 Sketch the graph of $g: x \to -2|x|$.

SOLUTION Function g is the same as -2 abs. Sketch the graph of 2 abs and flip it over the x-axis.

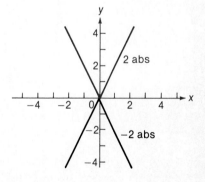

The product of the identity function with itself will prove to be very important.

3–6: Multiplication of Functions **115**

Example 4   Sketch the graph of $I \cdot I$.

SOLUTION   On the domain **R**,

$$[I \cdot I](x) = I(x) \cdot I(x)$$
$$= x \cdot x$$
$$= x^2.$$

Plot several ordered pairs (a, a^2) and connect them with a smooth curve.

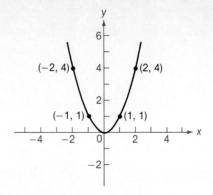

Since $I \cdot I$ is often used in this book, we adopt the shorthand used with multiplication of real numbers and denote the so-called **squaring function** by I^2. Then, $I^2(x) = x^2$. Similarly, for the product $I \cdot I^2$, we write I^3 and $I^3(x) = x^3$. For natural-number powers of I, the following definition applies.

Definition   If n is a natural number, then I^n denotes the function with domain **R** given by the rule

$$I^n(x) = x^n.$$

The graphs of I^3 and I^4 are shown below.

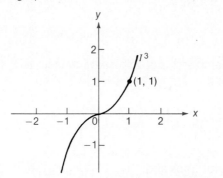
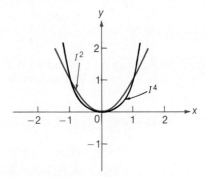

A comparison of these graphs suggests the following. If n is odd, $f(-x) = (-x)^n = -(x)^n = -f(x)$. So the value assigned to $-x$ is the negative of the value assigned to x. The graph of such a function is *symmetric about the origin*, which means that if the graph is rotated 180° around the origin, the final and original graphs will coincide. If n is even, the values assigned to x and $-x$ are the same, and the graph is symmetric about the y-axis.

These results suggest applying the words *odd* and *even* to certain functions, using the following definition.

Definition Let *f* be a function with domain *D*. Then

f is an **even function** if and only if $f(-x) = f(x)$ for all $x \in D$, and
f is an **odd function** if and only if $f(-x) = -f(x)$ for all $x \in D$.

Example 5 Graph abs and the reciprocal function and determine whether each function is odd or even.

SOLUTION Since (a, b) and $(-a, b)$ are both on the graph of abs for all $a \in$ Dom abs, it is an even function and its graph is symmetric about the *y*-axis.

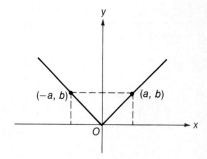

Since (a, b) and $(-a, -b)$ are both on the graph of the reciprocal function for all $a \in \mathbf{R}$, $a \neq 0$, it is an odd function and its graph is symmetric about the origin.

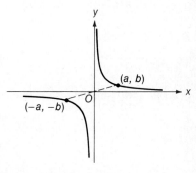

Most functions are neither odd nor even. An example is $f: x \rightarrow x + 2$.

Example 6 Determine algebraically whether the function defined by $f(x) = x^3 + \frac{1}{x}$ is an odd function, an even function, or neither.

SOLUTION Examine $f(-x)$:

$$f(-x) = (-x)^3 + \frac{1}{(-x)} = -(x)^3 - \frac{1}{x}$$

$$= -\left(x^3 + \frac{1}{x}\right)$$

$$= -f(x)$$

The function is an odd function.

Exercises

A Give a formula for $[f \cdot g](x)$ and calculate $[f \cdot g](5)$.

1. $f(x) = 2x;\ g(x) = 3$
2. $f(x) = \frac{1}{2}x;\ g(x) = 2x + 3$
3. $f(x) = x + 3;\ g(x) = x - 5$
4. $f(x) = x^2 - 1;\ g(x) = \dfrac{1}{2x + 1}$
5. $f: x \to -5,\ g = \text{abs}$
6. $f: x \to 2,\ g = I^2$

Graph each of the given functions on the same set of axes.

7. $f(x) = 2x;\ g(x) = \frac{1}{2}x;\ h(x) = -\frac{1}{2}x$
8. $f(x) = x^2,\ g(x) = 2x^2,\ h(x) = -\frac{1}{2}x^2$

Sketch and use a graph to decide whether the function is odd, even, or neither.

9. $f(x) = |x| + 1$
10. $f(x) = x^4 + 5$
11. $f(x) = x^2 + 1$
12. $f(x) = \dfrac{1}{x} + 1$
13. $f(x) = -2|x|$
14. $-\text{sgn}$

Use the definition to decide whether f is odd, even, or neither.

15. $f(x) = 2x^2 + 3$
16. $f(x) = x^3 + x^2 + 1$
17. $f(x) = \dfrac{x^3 + x}{2}$
18. $f(x) = x^4 + x^2 + 2$

B Use the definition to decide whether f is odd, even, or neither.

19. $f(x) = |x| \cdot x$
20. $f = \text{abs} \cdot \text{abs}$
21. $f = \text{sgn} \cdot \text{sgn}$
22. $f(x) = \dfrac{x^5 + x^3}{x}$

Sketch a graph of each function.

23. $f(x) = 2(|x| - 3)$
24. $f(x) = 2|x| - 3$
25. $g(x) = \dfrac{1}{x}|x|$
26. $h(x) = \dfrac{1}{2}\left(2 - \dfrac{1}{x}\right)$

C 27. Prove that function multiplication is both commutative and associative, that is, that $f \cdot g = g \cdot f$ and $(f \cdot g) \cdot h = f \cdot (g \cdot h)$ for all real-valued functions f, g, and h.

28. Prove: If f and g are odd functions, then $f + g$ is odd. [HINT: Evaluate $[f + g](-x)$ and use the definition.]

29. Prove: If f and g are odd functions, then $f \cdot g$ is even.

30. What can be said about the sum and product of two even functions?

31. If f is odd and g is even, what can be said of $f \cdot g$ and $f + g$?

32. True or false: If $f: x \to x^3 + 5x + 7$, then f is odd.

3-7 Linear Functions and Variation

Vertical lines are not the graphs of functions; horizontal lines are the graphs of constant functions. All other functions whose graphs are lines are called linear functions.

Definition A function f is a **linear function** if and only if $f(x) = mx + b$, where m and b are real numbers, $m \neq 0$, and the domain of f is **R**.

The graph of a linear function is the graph of an equation $y = mx + b$, which you should recognize as the line through $(0, b)$ with slope m. Thus a linear function $f(x) = mx + b$ can be described by specifying the *slope m* and the *y-intercept b* of its graph.

Example 1 Find and graph the linear function f whose graph contains $(1, -3)$, and $(-2, 1)$.

SOLUTION Find the particular numbers m and b for which $f(x) = mx + b$. Use the slope formula.

$$m = \frac{y_2 - y_1}{x_2 - x_1} = \frac{-3 - 1}{1 - (-2)} = -\frac{4}{3}$$

Thus $f(x) = -\frac{4}{3}x + b$.
Now find the y-intercept b.

Since $f(-2) = 1$ and $f(x) = -\frac{4}{3}x + b$,

$1 = \left(-\frac{4}{3}\right)(-2) + b$, and $b = -\frac{5}{3}$.

$f(x) = -\frac{4}{3}x - \frac{5}{3}$

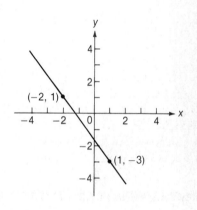

If the y-intercept b is zero, the linear function is described by an equation $f(x) = kx$ and is said to represent **direct variation**.

Definition The quantity y **varies directly** as x, or y is **directly proportional** to x, if and only if there is a nonzero constant k such that $y = kx$. The number k is the **constant of variation**.

Example 2 The labor cost for an automobile repair job is directly proportional to the number of hours a mechanic spends working on the car. If the labor cost for 2.5 hours is $85, what will be the labor cost for 4 hours of work?

SOLUTION Let c represent the cost and h the number of hours worked. By the definition, $c = kh$. From the data, $85 = k(2.5)$. So $k = 34$.
If $h = 4$, then $c = (34)(4) = 136$.

ANSWER The labor cost for 4 hours of work will be $136.

Some relationships between values and arguments that are *nonlinear* may also be described with the language of variation.

For instance, if $y = \dfrac{k}{x}$, where k is a nonzero constant, we say that y *varies inversely with* x. Notice that, in terms of absolute values, this means that as x increases, y decreases.

Similarly, if $y = kx^2$, we say that y is *directly proportional to the square* of x, or y varies directly as the square of x. If $y = kx^3$, we say that y is *directly proportional to the cube* of x. If $y = \dfrac{k}{x^2}$, then we say that y is *inversely proportional to the square* of x. Here are some examples.

The area of a circle varies directly as the square of the radius. ($A = \pi r^2$)

The volume of a sphere varies directly as the cube of the radius. ($V = \frac{4}{3}\pi r^3$)

The time required for a trip from Chicago to Seattle varies inversely with the average speed. $\left(t = \dfrac{k}{s}\right)$

Variation can be described with the usual function notation. For the examples above, we write $A(r) = \pi r^2$, $V(r) = \frac{4}{3}\pi r^3$, and $t(s) = \dfrac{k}{s}$.

Example 3 The distance d that a ball falls when dropped varies directly as the square of the time t. If a ball falls 144 feet in 3 seconds, how long will it take to fall a total of 400 feet?

SOLUTION The variation is described by $d = kt^2$. For $t = 3$, $d = 144$. So $144 = k(9)$, or $k = 16$.
Solve: $400 = 16t^2$
$t = 5$

ANSWER It takes 5 seconds for the ball to fall 400 feet.

Example 4 The force of gravitation g between two objects varies inversely as the square of the distance d between them. If two objects are approaching each other and are 10 km apart, at what distance will the force of gravitation between them be double its present value?

SOLUTION The variation is described by $g = \dfrac{k}{d^2}$. Let g_1 represent the present force of gravitation. From the data, $g_1 = \dfrac{k}{100}$. Now find d so that $g_2 = 2g_1 = \dfrac{k}{50}$.
$\dfrac{k}{d^2} = \dfrac{k}{50}$, or $d = \sqrt{50} = 5\sqrt{2}$ km.

Exercises

A Sketch graphs of the given functions on a single set of axes.

1. $f(x) = 3x$; $g(x) = 3x + 1$; $h(x) = 3x - 2$
2. $f(x) = \frac{1}{2}x + 2$; $g(x) = \frac{1}{2}x$; $h(x) = \frac{1}{2}x - 3$

Find the slope and the y-intercept for each linear function.

3. $f(x) = 5 - 3x$
4. $g(x) = \dfrac{2x + 3}{7}$
5. $f: x \to \dfrac{x + 1}{2}$
6. $g: x \to -4 - x$

Write an equation for the linear function that contains the given ordered pairs. The letters r and s denote constants.

7. $(2, -2), (-1, 4)$
8. $(0, 10), (1, 11)$
9. $(-2, -3), (0, 1)$
10. $(1, 2), (3, 4)$
11. $(1, r), (-1, 0)$
12. $(r, 0), (0, s)$

13. If y varies directly as x, and $y = 18$ when $x = 6$, find y when $x = 18$.
14. If p varies directly as t, and $p = 14$ when $t = 4$, find p when $t = 7$.

B Sketch a graph of each function.

15. $f: x \to \dfrac{2x - 7}{5}$
16. $g: x \to 3|x| - 2$
17. $f(x) = 2x^2 + 1$
18. $g(x) = 4\,\text{sgn}(x) + 2$

19. Which of the functions in exercises 15-18 are *not* linear functions?
20. If g is a linear function such that $g(2) = -2$ and $g(-1) = 4$, find $g(5)$.
21. If $f(1) = 9$ and $f(-1) = 1$ and f is a linear function, find the zeros of f.
22. The distance a plane flies is proportional to the weight of the fuel it carries. If a plane can fly 375 km on 700 kg of fuel, how far can it fly on 1000 kg of fuel?
23. The price of gold varies directly as the weight. If a nugget that weighs 3 ounces sells for $2127, what will an 8-ounce nugget cost?
24. If y varies inversely as x, and $y = 8.01$ when $x = 3.42$, find y when $x = 4.17$.
25. The price of a diamond varies directly as the square of its weight. How many times more expensive is a diamond weighing 1.293 grams than a diamond weighing 0.415 grams?

C 26. If f is a linear function such that $f(2) = 2f(1)$, what can you say about m and b in $f(x) = mx + b$?

27. True or False: If f is a linear function and u and v are real numbers, then $f(u + v) = f(u) + f(v)$ and $f(uv) = u \cdot f(v)$.

28. Under what conditions on m and b would the linear function $f(x) = mx + b$ satisfy the properties stated in exercise 27?

3-8 Composition of Functions

Celsius temperature C and Kelvin temperature K are related by the formula

$$C = K - 273.15.$$

Fahrenheit temperature F and Celsius temperature are related by the formula

$$F = \tfrac{9}{5}C + 32.$$

These formulas describe two functions:

$h: K \rightarrow K - 273.15$ h maps Kelvin onto Celsius temperature.

$g: C \rightarrow \tfrac{9}{5}C + 32$ g maps Celsius onto Fahrenheit temperature.

If values of h are used as arguments of g, a new function $g(h(K))$ is obtained that maps Kelvin temperature onto Fahrenheit temperature:

$$F(K) = g(h(K)) = \tfrac{9}{5}(K - 273.15) + 32.$$

Forming a new function in this way is an example of *composition of functions*, which is defined as follows.

Definition   Let f and g be functions, and let D denote $\{x : x$ is in the domain of g and $g(x)$ is in the domain of f$\}$. Then the **composite function** $f \circ g$ is defined on the domain D by

$$[f \circ g](x) = f(g(x)).$$

The notation makes clear why g is called the *inner function* and f is called the *outer function* of the composite $f \circ g$.

Example 1   List the elements of $[f \circ g]$ if

$f = \{(2, 7), (3, -1), (4, 9)\}$, and

$g = \{(-1, 4), (2, 3), (5, 7)\}$.

SOLUTION $[f \circ g](-1) = f(g(-1)) = f(4) = 9$

$[f \circ g](2) \ \ = f(g(2)) \ \ = f(3) = -1$

$[f \circ g](5) \ \ = f(g(5)) \ \ = f(7)$

Since 7 is not in the domain of f, $[f \circ g](5)$ is not defined; 5 is not in the domain of the composite function.

ANSWER $[f \circ g] = \{(-1, 9), (2, -1)\}.$

Example 1 illustrates that it is necessary to determine carefully the domain of the composite function.

Example 2 For the functions f and g in example 1, find $g \circ f$.

SOLUTION $[g \circ f](2) = g(f(2)) = g(7)$ Not defined

$[g \circ f](3) = g(f(3)) = g(-1) = 4$

$[g \circ f](4) = g(f(4)) = g(9)$ Not defined

ANSWER Thus, $g \circ f$ contains only a single ordered pair: $g \circ f = \{(3, 4)\}$.

Examples 1 and 2 show that composition of functions is not commutative, that is, that $f \circ g \neq g \circ f$.

Example 3 If $f(x) = 2x + 1$ and $g(x) = \dfrac{1}{x}$, find $[f \circ g](x)$ and $[g \circ f](x)$. Describe the domain of each composite function.

SOLUTION $[f \circ g](x) = f(g(x)) = f\left(\dfrac{1}{x}\right) = 2\left(\dfrac{1}{x}\right) + 1 = \dfrac{2}{x} + 1$

The domain of $f \circ g$ is the set of all real numbers except 0.

$[g \circ f](x) = g(f(x)) = g(2x + 1) = \dfrac{1}{2x + 1}$

The domain of $g \circ f$ is the set of all real numbers except $-\dfrac{1}{2}$, since the domain cannot contain any number that makes $f(x) = 0$.

Sometimes it is useful to be able to express a function f as a composite of two or more simpler functions.

Example 4 If $f(x) = x^2 + 4$, find functions g and h such that $f = g \circ h$.

SOLUTION Since f assigns values by first squaring the argument and then adding 4, select the squaring function as the inner function: $h = I^2$. The outer function g must add 4 to its arguments, so $g: x \rightarrow x + 4$.

As example 4 shows, to write a function f as a composite $g \circ h$, a good strategy is to ask what steps you would take to compute the value f assigns to an argument x. For instance, if $f(x) = |x + 1|$, then you must first add 1 and then take the absolute value. So $h(x) = x + 1$ and $g(x) = |x|$.

Composition of functions can be extended to three or more functions in forming composites such as $f \circ g \circ h$. You should convince yourself that composition of functions is associative, so that no parentheses are needed in writing a composite such as $f \circ g \circ h$.

Example 5 If $h(x) = \dfrac{1}{x}$, $g(x) = 3x - 1$, and $f(x) = 2x + 5$, find a formula for $[\text{abs} \circ g \circ h \circ f](x)$.

SOLUTION Apply the innermost function f first and then work from right to left until you reach the outermost function abs.

$$[\text{abs} \circ g \circ h \circ f](x) = [\text{abs} \circ g \circ h](2x + 5)$$
$$= [\text{abs} \circ g]\left(\dfrac{1}{2x + 5}\right)$$
$$= \text{abs}\left(3 \cdot \dfrac{1}{2x + 5} - 1\right)$$
$$= \left|\dfrac{3}{2x + 5} - 1\right|$$

In example 6 you are given the outer function and the composite function and are asked to determine the inner function.

Example 6 If $[f \circ g](x) = x$ and $f(x) = 3x + 5$, find a formula for $g(x)$.

SOLUTION
$f(x) = 3x + 5$
$f(g(x)) = 3(g(x)) + 5$
Thus $[f \circ g](x) = x$ is equivalent to
$3(g(x)) + 5 = x.$
$3(g(x)) = x - 5$
$g(x) = \dfrac{x - 5}{3}$

Exercises

A For each pair of functions compute $[f \circ g](3)$ and $[g \circ f](3)$.

1. $f = \text{sgn}$; $g = \text{abs}$

2. $f(x) = \dfrac{1}{x}$; $g(x) = x^2$

3. $f(x) = 3x + 1$; $g(x) = 2x - 1$

4. $f(x) = x^2 - 1$; $g(x) = \sqrt{x}$

Let $f = \{(0, 4), (1, 1), (2, 3), (4, 0)\}$ and $g = \{(-1, 3), (0, 2), (1, 2), (2, 4)\}$.
List the ordered pairs of each composite function.

5. $f \circ g$ **6.** $g \circ f$ **7.** $f \circ f$ **8.** $g \circ g$

Let $f(x) = 4x - 3$. Find a formula for each composite function.

9. $[I \circ f](x)$ **10.** $[f \circ I](x)$ **11.** $[-I \circ f](x)$ **12.** $[f \circ -I](x)$

Chapter 3: Functions

Find and simplify formulas for $[f \circ g](x)$ and $[g \circ f](x)$. Specify restrictions on the domain.

13. $f(x) = |x|;\ g(x) = x^2 - 4$
14. $f(x) = x + 3;\ g(x) = x^2$
15. $f(x) = x^3;\ g(x) = x - 1$
16. $f(x) = \dfrac{1}{x};\ g(x) = x^2 + 2$

Let $f(x) = 2x - 1$ and $g(x) = \dfrac{3x}{x - 1}$. Find the domain and a simplified formula for each composite function.

17. $[f \circ g](x)$
18. $[g \circ f](x)$
19. $[f \circ f](x)$

Let $f(x) = \dfrac{2}{x + 1}$ and $g(x) = 3x - 1$. Find the domain and a simplified formula for each composite function.

20. $[f \circ g](x)$
21. $[g \circ f](x)$
22. $[g \circ g](x)$

B For each function f identify functions g and h such that $f = g \circ h$. Neither g nor h may be I.

23. $f(x) = |x + 1|$
24. $f(x) = \dfrac{1}{|x|}$
25. $f(x) = 2(x - 1)^2$
26. $f(x) = \dfrac{1}{|x| + 1}$

Let $f(x) = \dfrac{1}{x} + 2$, $g(x) = 2|x| - 3$, and $h(x) = 1 - x$. Find a formula for each composite function.

27. $[f \circ g \circ h](x)$
28. $[h \circ g \circ f](x)$

Find functions f, g, and h, none of them I, so that $[f \circ g \circ h](x)$ will equal the given expression.

29. $2|x + 1| - 3$
30. $-3|x - 2| + 4$
31. $5(x + 6)^2 - 7$
32. $-9(x - 1)^3 + 13$

If $f(x) = x + 2$, find $g(x)$ from the following information.

33. $[g \circ f](x) = |x + 2|$
34. $[g \circ f](x) = x^2 + 4x + 4$

35. If $f(x) = |x| + 1$ and $g(x) = |x| - 3$, find the zeros, if any exist, of $g \circ f$.

36. If $f(x) = \dfrac{2x - 1}{3x + 1}$ and $g(x) = \dfrac{x - 2}{x - 1}$, find the zeros, if any exist, of $g \circ f$.

C 37. If $f(x) = \dfrac{1}{x}$ and $g(x) = 2x + 3$, find all arguments c such that $g(f(c)) = f(g(c))$.

38. If $f(x) = x - 2$ and $g(x) = \dfrac{1}{x} + 1$, find all arguments c such that $g(f(c)) = f(g(c))$.

39. Give an example of two functions f and g for which $g \circ f = f \circ g$.

40. Show that the composite $f \circ g$ is an odd function if f and g are both odd functions. [HINT: Use the definition of an odd function and examine $[f \circ g](-x)$.]

41. Show that if f is an even function and g is an odd function, then $f \circ g$ is an even function.

Special: The Function int and the Computer

In computer programming, the function int can be employed to round numbers to a desired number of decimal places. For example, to round x to the nearest hundredth, have the computer first multiply x by 100, then add 0.5, then apply int, and finally divide by 100. Try this yourself on a calculator with the numbers

$$x = 1.4142136 \text{ and}$$
$$x = 2.236068.$$

Another important use of int has to do with testing divisibility. In the following program, a search for the divisors of an integer K is carried out in order to see if K is a prime number. If $\frac{K}{J} = \text{int}\left(\frac{K}{J}\right)$, then $\frac{K}{J}$ is an integer, so that J is a divisor of K. In line 50, the variable D is increased by 1 each time a divisor of K is found. If only one divisor has been found by the time $J = \text{sqr}(K)$ [the square root of K], then K has no divisors other than 1 and K itself, so K is a prime number. In the following program, the input for N is 500.

LIST

```
10 PRINT "PRIMES FROM 2 to N; WHAT N";
20 INPUT N
30 FOR K = 2 TO N
40 FOR J = 1 TO SQR(K)
50 IF K/J = INT(K/J) THEN D = D + 1
60 NEXT J
70 IF D = 1 THEN PRINT K,
80 LET D = 0
90 NEXT K
100 END
```

Ready

RUN

PRIMES FROM 2 TO N; WHAT N? 500

2	3	5	7	11
13	17	19	23	29
31	37	41	43	47
53	59	61	67	71
73	79	83	89	97
101	103	107	109	113
127	131	137	139	149
151	157	163	167	173
179	181	191	193	197
199	211	223	227	229
233	239	241	251	257
263	269	271	277	281
283	293	307	311	313
317	331	337	347	349
353	359	367	373	379
383	389	397	401	409
419	421	431	433	439
443	449	457	461	463
467	479	487	491	499

Ready

Exercises

1. Modify the program on primes to have it ask the user for an interval $[a, b]$ and then print the primes in that interval.

2. Write a program to have the computer list all integers between 100 and 200 that are divisible by either 7 or 11.

3. Write a program to have the computer ask for a positive integer N (as input) and print either a list of the divisors of N or the message PRIME.

4. Modify the given program on primes so that instead of printing all primes from 2 to N, it prints only the *number* of primes in the given interval.

5. Write a program to have the computer print a table containing n, $\sqrt{n}$, and $\sqrt[3]{n}$ for $n = 1$ to 25. Print all values rounded to the nearest 0.001.

3-9 Graphs of Composite Functions

It is not always easy to sketch the graph of the composite of two given functions. There are, however, some standard composites that can be graphed without difficulty.

Since $I \circ f = f = f \circ I$ for any function f, the graph of either composite is the same as the graph of f.

If abs is the outer function in a composite $g = \text{abs} \circ f$, then

$$g(x) = |f(x)| = \begin{cases} f(x) \text{ if } f(x) \geq 0, \\ -f(x) \text{ if } f(x) < 0. \end{cases}$$

The graph of $\text{abs} \circ f$ coincides with the graph of f for $f(x) \geq 0$ and is the reflection in the x-axis for points such that $f(x) < 0$.

Example 1 Sketch the graph of $f(x) = |x^2 - 1|$.

SOLUTION Write f as the composite $\text{abs} \circ h$, where $h(x) = x^2 - 1$.
Since abs is the outer function, any portion of the graph of h that is below the x-axis is reflected in the x-axis. The graph is shown at the right.

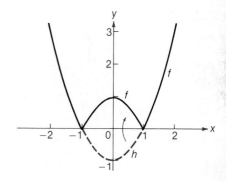

The composites $-I \circ f$ and $f \circ -I$ have different graphs. If $-I$ is the *outer* function, then $[-I \circ f](x) = -I(f(x)) = -f(x)$, and the graph of the composite is the *reflection of the graph of f in the x-axis*.

If $-I$ is the *inner* function, then $[f \circ -I](x) = f(-x)$. So, if (a, b) is on the graph of f, then $(-a, b)$ is on the graph of $f \circ -I$. The graph of the composite is the *reflection of the graph of f in the y-axis*.

Example 2 If $f(x) = x + 2$, write formulas for $-I \circ f$ and $f \circ -I$ and sketch graphs of the two composite functions.

SOLUTION $[-I \circ f](x) = -f(x) = -(x + 2) = -x - 2$

The graph of the composite function is the reflection in the x-axis of the graph of f.

$[f \circ -I](x) = f(-x) = -x + 2$

The graph of the composite function is the reflection in the y-axis of the graph of f.

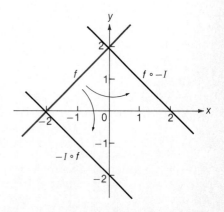

If $g(x) = x + k$ is the *outer* function in a composite $g \circ f$, then $[g \circ f](x) = f(x) + k$. The graph of the composite function is the graph of f shifted *vertically* $|k|$ units. The shift is *up* if $k > 0$ and *down* if $k < 0$.

Example 3 If $g(x) = x - 2$ and $f(x) = \dfrac{1}{x}$, graph the composite function $h = g \circ f$.

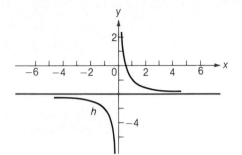

SOLUTION The graph of h is the graph of f shifted down 2 units since
$$g(f(x)) = \frac{1}{x} - 2.$$

If $g(x) = x + k$ is the *inner* function in a composite $f \circ g$, then $[f \circ g](x) = f(x + k)$. The graph of the composite is the graph of f shifted *horizontally* $|k|$ units.

Example 4 Sketch the function defined by $h(x) = (x + 2)^2$.

SOLUTION First rewrite h as the composite $f \circ g$, where $f(x) = x^2$ and $g(x) = x + 2$.

List some arguments and values of both f and h.

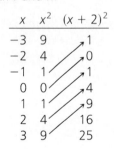

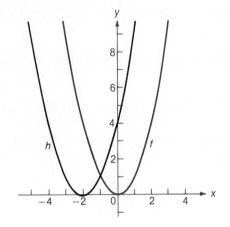

Values of h are the same as values of f, but they occur 2 units sooner (graphically, to the left). The graph of h is the same as the graph of f shifted 2 units to the *left*.

Example 4 shows that if $k > 0$ and if (a, b) is on the graph of $f(x) = x^2$, then $(a - k, b)$ is on the graph of $h(x) = (x + k)^2$, because $h(a - k) = [(a - k) + k]^2 = a^2$ and $f(a) = a^2$. So the graph of $h(x) = (x + k)^2$ is the same as the graph of $f(x) = x^2$ shifted horizontally $|k|$ units to the left.

Example 5 illustrates the fact that if $k < 0$, then the shift is $|k|$ units to the right.

Example 5 Sketch the graph of the function defined by $h(x) = \dfrac{1}{x-2}$.

SOLUTION Write h as the composite $f \circ g$ of $f(x) = \dfrac{1}{x}$ and $g(x) = x - 2$. The values of $\dfrac{1}{x-2}$ are the same as the values of $\dfrac{1}{x}$ but they occur 2 units later (graphically, to the *right*). The graph of h is the graph of f shifted 2 units to the *right*.

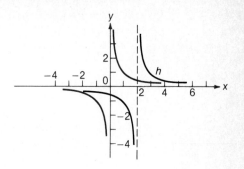

In general, if k is a positive number, then a composition that leads to values $f(x + k)$ results in a horizontal shift of the graph of f to the *left* k units. The graph of the function with values $f(x - k)$ is a horizontal shift of the graph of f to the *right* k units.

The following table summarizes results of this section. In this table, p represents a positive constant.

	To graph function g	do this to the graph of the function of f:
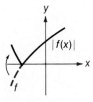	$g(x) = \lvert f(x) \rvert$	Reflect points below the x-axis in the x-axis.
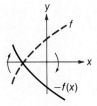	$g(x) = -f(x)$	Reflect the entire graph in the x-axis.
	$g(x) = f(-x)$	Reflect the entire graph in the y-axis.

3–9: Graphs of Composite Functions

	$g(x) = f(x) + p$	Shift vertically upward p units.
	$g(x) = f(x) - p$	Shift vertically downward p units.
	$g(x) = f(x + p)$	Shift horizontally to the left p units.
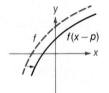	$g(x) = f(x - p)$	Shift horizontally to the right p units.

The key to graphing composite functions of the form $g(x) = f(x - h) + k$ is to identify first the shape of the graph of f and then shift this graph horizontally $|h|$ units and vertically $|k|$ units.

Example 6 Sketch the graph of the function $g: x \to |x - 2| + 1$.

SOLUTION Graph the basic function, abs, and then shift it to the right 2 units and up 1 unit.

Notice that, written as a composite, $g = s \circ \text{abs} \circ r$, where $s(x) = x + 1$ and $r(x) = x - 2$.

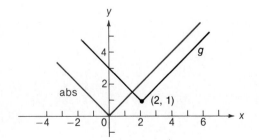

Example 7 Sketch the graph of the function defined by $g(x) = 3 - 2(x + 1)^2$.

SOLUTION The basic function is I^2 multiplied by -2. So first sketch I^2 stretched vertically and reflected in the x-axis. Then shift the resulting graph to the left 1 unit and upward 3 units.

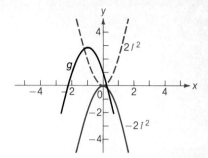

The results of this section can be used to solve inequalities, as shown in the following example.

Example 8 Solve the inequality $|x + 3| > |x - 3| - 2$.

SOLUTION If $f(x) = |x + 3|$ and $g(x) = |x - 3| - 2$, then the inequality to be solved can be written $f(x) > g(x)$. Arguments x for which the graph of f is above the graph of g are solutions of the inequality.
Graph f by shifting the graph of abs to the left 3 units. Graph g by shifting the graph of abs to the right 3 units and down 2 units.
Since $f(-1) = g(-1) = 2$, the point of intersection of the graphs is $(-1, 2)$. Everywhere to the right of this point the graph of f is above the graph of g. The solution set of the inequality is $\{x \in \mathbf{R}: x > -1\}$.

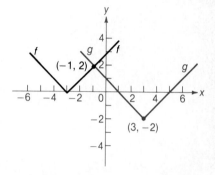

Exercises

A Sketch a graph of the given function.

1. $f: x \rightarrow |x - 1|$
2. $g(x) = |2 - x|$
3. $g(x) = \frac{1}{x} - 1$
4. $h(x) = \frac{1}{x - 1}$
5. $g: x \rightarrow (x - 1)^2$
6. $f: x \rightarrow x^2 - 1$

Sketch a graph of the composite function $f \circ g$.

7. $f = $ abs; $g(x) = x - 2$
8. $f = $ abs; $g(x) = \frac{1}{x}$
9. $f = -I$; $g = I^2$
10. $f = -I$; $g(x) = 1 + \frac{1}{x}$
11. $f = I^2$; $g = -I$
12. $f(x) = 1 + \frac{1}{x}$; $g = -I$
13. $f(x) = x - 4$; $g(x) = x^2$
14. $f(x) = x^3$; $g(x) = x - 4$

Find functions f and g such that the graph of f ∘ g will be as sketched.

15.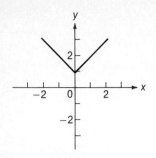

16.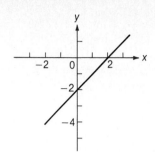

17.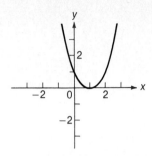

18.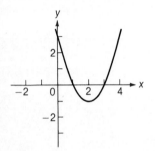

19.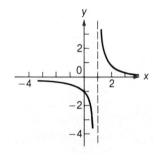

20.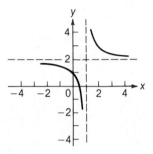

B For the given f and g, write and simplify a formula for $[f \circ g](x)$ and graph the composite f ∘ g.

21. $f(x) = x + 1$; $g(x) = x - 1$ **22.** $f(x) = 2x^2$; $g(x) = x + 1$

23. $f(x) = |x - 1|$; $g(x) = -x$ **24.** $f(x) = |x|$; $g(x) = |x| - 4$

Sketch a graph of each function.

25. $g(x) = -2(x - 1)^2$ **26.** $g(x) = 3(x + 2)^2$ **27.** $g(x) = (x + 1)^2 - 1$

28. $g(x) = -(x + 1)^2 + 2$ **29.** $g(x) = 4 + |x - 1|$ **30.** $g(x) = -3|x + 2| + 2$

31. $g(x) = (x + 1)^3 - 4$ **32.** $g(x) = 2(x - 1)^3 + 3$

C Sketch a graph of each function.

33. $g(x) = ||x + 3| - 1|$ **34.** $g(x) = ||x + 1| - 2|$

35. $g(x) = \left|\dfrac{1}{x + 3}\right|$ **36.** $g(x) = \dfrac{1}{x - 2} + 2$

Use a graph to solve each inequality.

37. $|x + 1| \leq |x - 1|$ **38.** $|x + 1| - 2 \leq |x - 1|$

39. $(x + 1)^2 + 1 > 2 - x$ **40.** $3 - 2(x + 1)^2 \geq |x + 1|$

Use a graph to determine the number of solutions of each equation.

41. $(x - 2)^2 + 3 = \dfrac{1}{x - 2}$ **42.** $|x - 8| = (x + 2)^3$

132 Chapter 3: Functions

Chapter Review Exercises

3–1, page 95

Give the domain and range of each relation.

1. $\{(-1, 3), (0, 3), (1, 0)\}$
2. $\{(x, y) : x = y^2 + 2 \text{ and } -1 \leq y \leq 1\}$

3–2, page 99

3. Which of the relations in exercises 1 and 2 above is a function?
4. If $f: x \to \dfrac{2x - 1}{x + 3}$, find $f(0)$, $f\left(\dfrac{1}{2}\right)$, and $f(-2)$.
5. Use the vertical-line test to determine whether the graph below represents a function.

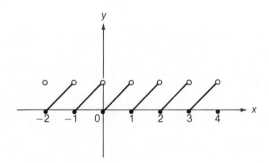

3–3, page 104

Find the domain and the zeros of each function.

6. $h(x) = \dfrac{3}{\sqrt{x - 1}}$
7. $f(x) = \dfrac{2 + x}{2 - x}$
8. $g: x \to \dfrac{x^2 + 3x + 2}{x^2 - 1}$

3–4, page 107

9. Evaluate: $\text{sgn}(-3) + I(2) + \text{int}(\sqrt{2})$.

Use graphs as an aid in solving the following.

10. $|x| = [x]$
11. $|x| > \dfrac{1}{x}$

3–5, page 110

Sketch a graph of each function.

12. $f(x) = 2 + |x|$
13. $g(x) = \dfrac{1}{x} + 1$
14. $h(x) = 3 - |x|$

Use graphs as an aid in solving the following.

15. $1 - |x| = x$
16. $\text{sgn}\, x = \dfrac{1}{x} + \dfrac{1}{2}$

3–6, page 114

17. If $f(x) = \dfrac{x}{1+x}$ and $g(x) = x - 2$, find $fg(0)$ and the zeros of fg.

18. Determine whether $f: x \to \dfrac{2x^3 + 3}{x}$ is odd, even, or neither.

3–7, page 119

19. If y varies inversely as w and $y = 3$ when $w = 5$, find y when $w = 10$.

20. Find the slope and the y-intercept of the linear function whose graph contains the points $(1, -3)$ and $(-2, 5)$.

3–8, page 122

21. If $f(x) = \dfrac{3-x}{x+1}$ and $g(x) = x - 1$, determine a simplified formula for $[f \circ g](x)$, and also find the domain of $f \circ g$.

22. Find functions f and g, neither of them I, so that $[f \circ g](x)$ will equal $3(x + 5)^2$.

3–9, page 127

Sketch a graph of each function.

23. $f: x \to |x - 3|$

24. $g: x \to 3 + (x - 1)^2$

25. $h: x \to 2 - (x + 1)^2$

26. Use graphs as an aid in solving the inequality $|x + 3| \geq 2 - 3x$.

Chapter Test

1. Give the domain and the range of the function $\{(x, y): y = x^2 + 1 \text{ and } -1 \leq x \leq 1\}$.

Tell whether the graph represents a function.

2.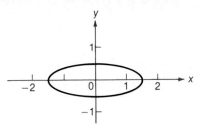

3.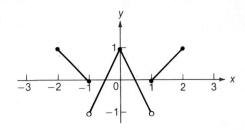

Let $f(x) = \dfrac{1 - 2x}{x}$ and $g(x) = x^2 - 4$.

4. Find the zeros and the domain of f.
5. Find the zeros and the domain of g.
6. Find the domain of $f \circ g$.
7. Find a simplified formula for $[f \circ g](x)$.

8. The number of grams of carbohydrate in a cereal varies directly as the number of grams of cereal. If a 28-gram serving of the cereal contains 19 grams of carbohydrate, how many grams of carbohydrate are in a box that contains 567 grams of cereal?

Sketch a graph of each function.

9. $f(x) = 2x + \text{sgn}(x)$
10. $g(x) = \dfrac{1}{x} - 1$
11. $h(x) = 2 - |x|$
12. $r(x) = -(x + 1)^2$
13. $s(x) = 1 - 2x^2$
14. $t(x) = 2 + 3|x - 1|$

Use graphs as an aid in solving the following.

15. $2 - \dfrac{1}{x} \geq 1$
16. $2x - 1 < |x - 5|$

Find functions f and g, neither of them I, so that $[f \circ g](x)$ will equal the given expression.

17. $\left|\dfrac{1}{x}\right|$
18. $\dfrac{1}{|x - 1|}$
19. $3(x - 1)^2$

Determine whether the given function is odd, even, or neither.

20. $f(x) = x^4 + 3x^2$
21. $g(x) = \text{sgn}(x) + 1$
22. $h(x) = 2(|x| - 1)$
23. $3I^2$

24. If $f(x) = \sqrt{x} + 3$ and $g(x) = \sqrt{4 - x}$, what are the domain and range of $g \circ f$?

25. If $f(x) = x + 1$ and $g(x) = \dfrac{1}{x} - 3$, find all arguments c (if any) for which $f(g(c)) = g(f(c))$.

Quadratic Functions

4

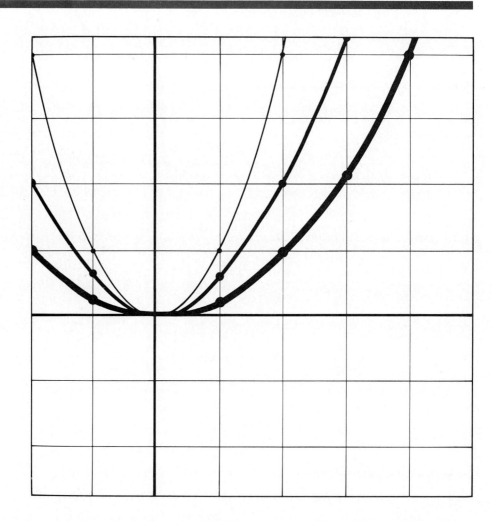

4–1 Definition of a Quadratic Function

Quadratic functions are common and useful in mathematics as well as in the physical and social sciences. In this chapter you will use your knowledge of quadratic equations, inequalities, the squaring function $f: x \to x^2$, and graphing techniques to investigate the quadratic function in one variable and some of its applications.

Definition A function which can be defined by $f(x) = ax^2 + bx + c$, where a, b, and c are real numbers and $a \neq 0$, is called a **quadratic function**.

Here are some examples of quadratic functions.

Rule	a	b	c	Comment
a. $f(x) = x^2 + 3x - 1$	1	3	-1	
b. $g: t \to t^2 + 3$	1	0	3	
c. $A: x \to x^2$	1	0	0	A is the area function for a square.
d. $h(k) = (2k + 1)^2 - 4(2k + 1) + 7$	4	-4	4	To determine a, b, and c, simplify the equation.

You know from the quadratic formula that the real zeros of the function $f: x \to ax^2 + bx + c$ are

$$x = \frac{-b \pm \sqrt{b^2 - 4ac}}{2a},$$

provided $b^2 - 4ac \geq 0$.

Example 1 Use the quadratic formula to find the zeros of the function defined by $f(x) = x^2 - 3x - 1$.

SOLUTION First identify the coefficients: $a = 1$, $b = -3$, $c = -1$. Substitute these values in the quadratic formula.

$$x = \frac{-(-3) \pm \sqrt{(-3)^2 - 4(1)(-1)}}{2(1)}$$

$$x = \frac{3 \pm \sqrt{13}}{2}$$

Hence there are two real zeros, $\frac{3 + \sqrt{13}}{2}$ and $\frac{3 - \sqrt{13}}{2}$.

For an application, as in the following example, it is sometimes convenient to use a decimal equivalent for a radical.

Example 2 From a point above the surface of Mars a stone is thrown upward with an initial velocity of 82 feet per second. Its height h in feet above the surface t seconds later is $h(t) = -5.6t^2 + 82t + 500$. From what height was the stone thrown and how long does it take to land?

SOLUTION Assume the stone is thrown when $t = 0$. Its initial height is $h(0) = 500$ ft. When the stone lands, the height is 0 feet. Set $h(t) = 0$ and solve the equation for t by using the quadratic formula.

$$t = \frac{-82 \pm \sqrt{(82)^2 - 4(-5.6)(500)}}{2(-5.6)}$$

$$= \frac{-82 \pm \sqrt{17{,}924}}{-11.2}$$

$$t \doteq 19.3 \text{ or } t \doteq -4.6$$

ANSWER In this situation, negative values of t are rejected; the stone lands after approximately 19.3 seconds.

For a function such as the one in example 2 you can approximate the greatest value of the function as follows. Use a table of values of $h(t)$ found by substituting for t in $h(t) = -5.6t^2 + 82t + 500$.

t	$h(t)$
0	$0 + 0 + 500 = 500$
2	$-5.6(4) + 82(2) + 500 \doteq 642$
4	$-5.6(16) + 82(4) + 500 \doteq 738$
6	$-5.6(36) + 82(6) + 500 \doteq 790$
8	$-5.6(64) + 82(8) + 500 \doteq 798$
10	$-5.6(100) + 82(10) + 500 \doteq 760$
12	$-5.6(144) + 82(12) + 500 \doteq 678$

Since $h(t)$ appears to be decreasing for $t > 8$, check values of $h(t)$ near $t = 8$.

$$h(7) = -5.6(49) + 82(7) + 500 \doteq 799.6$$
$$h(9) = -5.6(81) + 82(9) + 500 \doteq 784.4$$

It appears that the stone reaches a height of about 800 feet.

Example 3 Find a quadratic function whose graph contains the points $(-2, -9)$, $(2, 3)$, and $(3, -4)$.

SOLUTION Substitute coordinates of the points for x and $f(x)$ in the formula $f(x) = ax^2 + bx + c$.

(cont. on p. 139)

$$-9 = 4a - 2b + c$$
$$3 = 4a + 2b + c$$
$$-4 = 9a + 3b + c$$

To find the simultaneous solution of this system of three equations, subtract the second equation from the first. The result is

$$-12 = -4b, \text{ or } b = 3.$$

Substitute this value of b in the second and third equations and simplify.

$$-3 = 4a + c$$
$$-13 = 9a + c.$$

Subtract the second of these equations from the first. The result is

$$10 = -5a, \text{ or } a = -2.$$

Now substitute the values of a and b in any one of the equations to find $c = 5$.

ANSWER $f(x) = -2x^2 + 3x + 5$ is a quadratic function whose graph contains the three given points.

Example 4 What is the range of the quadratic function $f: x \to x^2 + 3$?

SOLUTION Since $x^2 \geq 0$ for all real numbers x, $x^2 + 3 \geq 3$. The range of the function is $[3, \infty)$.

Exercises

A Write each function in the form $f(x) = ax^2 + bx + c$ and determine the coefficients a, b, and c.

1. $f(x) = 3x^2 + 7x - 5$
2. $f(x) = 4x^2 - 3x + 9$
3. $f(x) = -3 + x^2$
4. $f(x) = -(-x - 2x^2)$
5. $f(x) = 1 - x^2 + 4x$
6. $f(x) = 2(x - 3)^2 + 3$
7. $f(x) = (\sqrt{3} - x)(\sqrt{3} + x + 1)$
8. $f(x) = 2(2 - 3x)^2 - 3x + 15$

Find the zeros of each quadratic function.

9. $f(x) = x^2 - 2x - 7$
10. $f(x) = 4x - (x^2 + 1)$
11. $f(t) = 4t - 3t^2 + 14$
12. $f(r) = -r^2 - 2r + 1$

B 13. Two concentric circles have radii x and $\frac{x}{2}$. Define a function that relates the area between the circles to the radius of the larger circle.

14. A semicircle is constructed on one side of a square so that its diameter is a side of the square. Let x represent the length of a side of the square. Define a function that relates the area of the entire figure to x.

15. A semicircle is drawn on each of the four sides of a square. Define a function that relates the area of the new figure to the length of a side of a square.

16. A gardener has 200 m of fence with which to enclose a rectangular plot. Define a function that relates the area of the plot to the length of one of the sides. What is the domain of the function?

17. To a 9 cm by 12 cm picture, a border of uniform width is added. Define a function that relates the area of the entire figure to the width of the border.

18. The height in feet of an object above ground at time t seconds is given by $h(t) = -2t^2 + 140t + 1600$.
    a. Find the height when $t = 0$.
    b. When does the object hit the ground?
    c. When is the height 1700 feet?

19. The speed of a moving particle is defined, in terms of the time elapsed t, by $s(t) = 2t^2 - 41t + 230$.
    a. What is the original speed of the particle?
    b. Does the particle ever come to rest?
    c. When is the speed 25 units?
    d. Approximately what is the slowest speed attained by the particle?

20. A real estate manager determines that the monthly profit P in dollars from a building with x stories is given by $P(x) = -3x^2 + 100x$. Determine a reasonable domain for the function and graph some ordered pairs of the function p in order to estimate the maximum profit.

For what values(s) of k is 3 a zero of the function f?

21. $f(x) = x^2 - kx + 1$ 22. $f(x) = kx^2 + 3$

Write a quadratic function whose graph contains the three given points.

23. $(-2, 1), (2, 7), (1, 2)$ 24. $(3, 1), (2, -1), (5, -7)$

Determine a constant k such that the quadratic function has only one zero.

25. $f(x) = x^2 + 6x + k$ 26. $f(x) = kx^2 - 4x + 1$

27. The difference of two numbers is k. Define a function which relates the product of the numbers to one of them.

C 28. Write and run a computer program to find the zeros of the function $f(x) = 10.2x^2 + 21.7x - k$. Have the program print the zeros or print that there are no zeros. Run your program for all integer values of k from 5 through 15.

4–2 Vertex and Axis of Symmetry

In Chapter 3 you learned how to graph the squaring function I^2, how to stretch or compress this graph to get the graph of aI^2, and how to shift it horizontally $|h|$ units and vertically $|k|$ units to obtain the graph of

$$f(x) = a(x - h)^2 + k.$$

All quadratic functions can be expressed in this form.

The function $q(x) = ax^2$ is an *even function*, since $a(-x)^2 = a(x)^2$ for all $x \in \mathbf{R}$. This means that the graph of q is symmetric about the y-axis. The y-axis is the **axis of symmetry** for this graph; its equation is $x = 0$. If the graph of $q(x) = ax^2$ is shifted to the right $|h|$ units, its axis of symmetry is also shifted $|h|$ units to the right. So for the function $g(x) = a(x - h)^2$, the axis of symmetry is the line $x = h$. Adding k to the value of $g(x)$ to get $f(x) = a(x - h)^2 + k$ shifts the graph of g vertically but does not move the axis of symmetry.

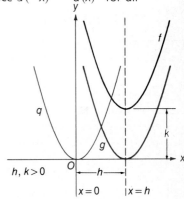

The axis of symmetry of the graph of a function acts like a "folding line," or **line of reflection**. When the graph to the left of the axis is reflected in the axis, it coincides with the graph on the right of the axis, and vice versa.

For example, the graph of the absolute-value function has the y-axis as its axis of symmetry.

The graph of the function $f(x) = 2(x - 3)^2$ has the line $x = 3$ as its axis of symmetry.

The graph of the area function for a square, $A: x \to x^2$, where x is the length of a side of the square, does not have an axis of symmetry, since x is always positive.

Example 1 What line is the axis of symmetry of the graph of $f(x) = 3(x - 2)^2 + 1$?

SOLUTION Since f is the function $3I^2$ shifted to the right 2 units and upward 1 unit, the axis of symmetry is the line $x = 2$.

Definition The **vertex** of the graph of a quadratic function is the point of intersection of the graph with its axis of symmetry.

The vertex of the graph of $f(x) = a(x - h)^2 + k$ is the point (h, k).

For the function $f(x) = a(x - h)^2 + k$, if $a > 0$ the graph of f opens upward, and if $a < 0$ it opens downward. So if you know the vertex of the graph of f and the sign of a you can determine the range of f.

Example 2 Graph the function $f(x) = -2(x + 1)^2 + 3$ and determine the vertex and the range.

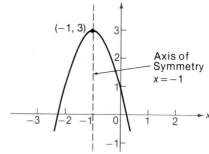

SOLUTION To graph f, reflect the graph of $2I^2$ in the x-axis and shift this graph to the left 1 unit and vertically upward 3 units.

The axis of symmetry is the line $x = -1$, so the vertex is the point $(-1, 3)$. Since the graph opens downward, the range is $(-\infty, 3]$.

Example 3 What is the vertex of the graph of the function $f(x) = -2x^2 - 4x + 1$?

SOLUTION By completing the square, rewrite $f(x)$ in the form $f(x) = a(x - h)^2 + k$.

$f(x) = -2x^2 - 4x + 1$

$= -2(x^2 + 2x \quad) + 1$

Complete the square by adding 1 inside the parentheses. Since this 1 is multiplied by the factor -2, add $+2$ outside the parentheses.

$f(x) = -2(x^2 + 2x + 1) + 1 + 2$

$= -2(x + 1)^2 + 3$

But this is the function from example 2. The vertex is the point $(-1, 3)$.

Theorem 4–1 Any quadratic function $f(x) = ax^2 + bx + c$, $a \neq 0$, can be written in the form $f(x) = a(x - h)^2 + k$, where

$$h = -\frac{b}{2a} \quad \text{and} \quad k = \frac{4ac - b^2}{4a}$$

PROOF $f(x) = ax^2 + bx + c$, $a \neq 0$

Factor a from the first two terms:

$f(x) = a[x^2 + \frac{b}{a}x + \quad] + c$

(cont. on p. 143)

Complete the square inside the brackets and proceed as follows.

$$f(x) = a\left[x^2 + \frac{b}{a}x + \left(\frac{b}{2a}\right)^2\right] + c - a\left(\frac{b}{2a}\right)^2$$

$$= a\left[x + \frac{b}{2a}\right]^2 + c - a\frac{b^2}{4a^2} \qquad \text{Simplify.}$$

$$= a\left[x + \frac{b}{2a}\right]^2 + \frac{4ac - b^2}{4a}$$

$$= a\left[x - \left(-\frac{b}{2a}\right)\right]^2 + \frac{4ac - b^2}{4a} \qquad \text{Rewrite } \frac{b}{2a} \text{ as } -\left(-\frac{b}{2a}\right).$$

This equation is in the form $f(x) = a(x - h)^2 + k$ where

$$h = -\frac{b}{2a} \quad \text{and} \quad k = \frac{4ac - b^2}{4a}.$$

Now you can see that the graph of f in the last example is the graph of aI^2 shifted horizontally $|h|$ units and vertically $|k|$ units.

Corollary 4–1 For the graph of $f(x) = ax^2 + bx + c$, the axis of symmetry is the line $x = h$ and the vertex is the point (h, k), where

$$h = -\frac{b}{2a} \quad \text{and} \quad k = \frac{4ac - b^2}{4a}, a \ne 0.$$

You may find it useful to memorize the first coordinate of the vertex as $h = -\frac{b}{2a}$ but then to find k, the second coordinate of the vertex, by substitution. Notice that $-\frac{b}{2a}$ is the first term of the quadratic formula. If a quadratic function has two real zeros, $-\frac{b}{2a} \pm \frac{\sqrt{b^2 - 4ac}}{2a}$, they are equidistant from the point on the x-axis whose x-coordinate is $-\frac{b}{2a}$. Graphically, this means that the axis of symmetry bisects the line segment that joins the x-intercepts of the graph of f.

Example 4 Sketch the graph of $f(x) = 4x^2 - 12x + 7$.

SOLUTION Since $-\frac{b}{2a} = -\frac{-12}{2(4)} = \frac{3}{2}$, the axis of symmetry is the line $x = \frac{3}{2}$. $f\left(\frac{3}{2}\right) = -2$; the vertex is $\left(\frac{3}{2}, -2\right)$. The graph opens upward, since $a > 0$. Use the point $(0, 7)$ as an aid in graphing.

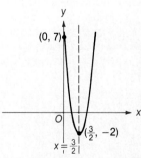

4–2: Vertex and Axis of Symmetry

Example 5 For the quadratic function $f(x) = 2x^2 - px + 11$, find p such that the vertex of the graph is $(3, -7)$.

SOLUTION $-\dfrac{b}{2a} = -\dfrac{-p}{2(2)} = \dfrac{p}{4}$

So $\dfrac{p}{4} = 3$, and $p = 12$. You should check that $(3, -7)$ is the vertex of the graph.

Exercises

A Determine the axis of symmetry of the graph of each function.

1. $f(x) = 2|x|$
2. $f(x) = x^2 + 3$
3. $f(x) = -x^2$
4. $f(x) = |x - 1|$
5. $f(x) = x^2 + 1$
6. $f(x) = x^2 - 5$

Tell whether or not each function is even.

7. $f(x) = \pi$
8. $f(x) = 7$
9. $f(x) = (x - 1)^2$
10. $f(x) = (x + 2)^2$
11. $f(x) = x^2 + 1$
12. $f(x) = x^2 - 3$

Determine the vertex and axis of symmetry of the graph of each function.

13. $f(x) = -2(x - 1)^2 - 3$
14. $f(x) = 3(x + 2)^2 - 7$
15. $f(x) = 3(x + 1)^2 + 1$
16. $f(x) = 4(x - 3)^2 + 2$

Find the axis of symmetry, the vertex, and the range. (Use Theorem 4–1 and its corollary.)

17. $f(x) = 2x^2 - 8x + 5$
18. $f(x) = 4x^2 - 8x - 1$
19. $f(x) = -x^2 + 2x$
20. $f(x) = -2x^2 - 4x - 5$

B 21. $f(x) = -(x^2 + 6x + 8)$
22. $f(x) = -\frac{1}{2}\left(x^2 - x - \frac{1}{2}\right)$
23. $f(x) = 1 + 2x(1 - x)$
24. $f(x) = \sqrt{3}x(x - 2) + 6(x - 1)$

25. A fireboat shoots water in a path described by the function $h(x) = -\frac{1}{10}x^2 + 2x + 20$, where x is the horizontal distance in meters from the nozzle and $h(x)$ is the height in meters. At what horizontal distance from the nozzle does the water reach a height of 30 meters?

144 Chapter 4: Quadratic Functions

26. Maria has a piece of poster board 24 inches wide and 36 inches long. She wants to reduce the size of the poster board by cutting the same amount from the length and the width. How much should she cut from each dimension so that the area of the new piece is 693 square inches?

Graph each function and label the vertex and one other point on the graph.

27. $f(s) = 2s^2 - 4s + 3$
28. $f(s) = 3s^2 - 24s + 38$
29. $f(t) = -3t^2 - 6t + 5$
30. $f(t) = -2t^2 + 4t - 2$
31. $f(x) = x(4 - x)$
32. $f(x) = 2x(2 + x)$

33. If $f(x) = 2x^2 - rx + 1$ and the axis of symmetry of the graph of f is the line $x = -3$, what is the value of r?

C 34. Write a quadratic function f such that $(4, 0) \in f$ and the point $(1, 3)$ is the vertex of its graph.

35. Write a quadratic function that contains $(0, 2)$ and $(4, 2)$ and has as its range the set of all nonnegative real numbers.

36. For every constant k, $f(x) = 2x^2 + 2x + k$ defines a quadratic function. If the domain of f is the set of positive real numbers, explain why none of the graphs of these functions has a vertex.

37. Find constants r and b such that $(-1, b)$ is the vertex of the graph of the function $f: x \rightarrow 3x^2 - 2rx + 1$.

38. The vertex of the graph of a quadratic function is $(-1, 2)$ and the curve contains the point $(0, 1)$. Find the zeros of the function.

39. Given the quadratic function defined by $f(x) = kx^2 + kx + 1$, where $k > 0$ and the range of f is the set of all nonnegative real numbers. Find k.

40. Find all the quadratic functions that have two positive zeros and the vertex of the graph at $(1, 2)$.

41. A batter hits a long fly ball along the curve $y = f(x) = -\dfrac{x^2}{600} + \dfrac{2x}{3} + 4$, where x is the horizontal distance in feet of the ball from home plate and y is the vertical height of the ball in feet. Find the coordinates of the vertex of the graph of f and determine if the ball will clear a ten-foot-high fence 390 feet from home plate.

4-3 Solving Quadratic Inequalities Algebraically

Quadratic inequalities of the form $ax^2 + bx + c > 0$ or $ax^2 + bx + c < 0$ may be solved by a variety of algebraic and graphical methods. One algebraic method depends upon the following theorem.

Theorem 4-2 For all $x \in R$ and $p \geq 0$,
a. $x^2 < p^2$ if and only if $-p < x < p$, and
b. $x^2 > p^2$ if and only if $x > p$ or $x < -p$.

We justify the theorem with the following graphical argument.

Let f be the function defined by $f(x) = x^2 - p^2$, $p \geq 0$. Since $x^2 < p^2$ if and only if $x^2 - p^2 < 0$, part **a** of the theorem corresponds to $f(x) < 0$ and part **b** to $f(x) > 0$. The graph of f is the graph of the function I^2 shifted downward p^2 units. Its x-intercepts are $-p$ and p. Observe from the graph that $f(x) < 0$ for all x such that $-p < x < p$. Similarly, $f(x) > 0$ for $x < -p$ or $x > p$.

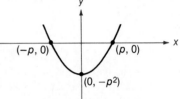

In Theorem 4-2 you may replace $<$ by $\leq$ and $>$ by $\geq$. We consider the theorem to include these variations.

The requirement that $p \geq 0$ is important in Theorem 4-2. For instance, if $x = 2$ and $p = -3$, then $2^2 < (-3)^2$ but $-(-3) < 2 < -3$ is false.

Corollary 4-2 For all nonnegative numbers a and b, $a^2 < b^2$ if and only if $a < b$.

Example 1 Solve for x: $x^2 < 16$.

SOLUTION
$x^2 < 16$
$x^2 < 4^2$
$-4 < x < 4$ Use theorem 4-2a.

Example 2 Solve for t: $(t - 2)^2 \geq 7$.

SOLUTION
$(t - 2)^2 \geq 7$
$(t - 2)^2 \geq (\sqrt{7})^2$
$t - 2 \geq \sqrt{7}$ or $t - 2 \leq -\sqrt{7}$ Use theorem 4-2b.
$t \geq 2 + \sqrt{7}$ or $t \leq 2 - \sqrt{7}$

The next example shows how Theorem 4–2 and the technique of completing the square can be used to solve quadratic inequalities.

Example 3 Solve for x: $x^2 - 6x + 7 \geq 0$.

SOLUTION
$x^2 - 6x + 7 \geq 0$
$x^2 - 6x \geq -7$
$x^2 - 6x + 9 \geq -7 + 9$ Add $\left(\frac{6}{2}\right)^2$ to both sides.
$(x - 3)^2 \geq 2$
$x - 3 \leq -\sqrt{2}$ or $x - 3 \geq \sqrt{2}$ Use theorem 4–2b.
$x \leq 3 - \sqrt{2}$ or $x \geq 3 + \sqrt{2}$

ANSWER The solution set is $(-\infty, 3 - \sqrt{2}] \cup [3 + \sqrt{2}, \infty)$.

For some quadratic inequalities, the solution is the set of all real numbers.

Example 4 Solve for t: $2t^2 - 16t + 37 > 0$.

SOLUTION
$2t^2 - 16t + 37 > 0$
$t^2 - 8t > -\frac{37}{2}$
$t^2 - 8t + 16 > -\frac{37}{2} + 16$ Complete the square; add 16 to both sides.
$(t - 4)^2 > -\frac{5}{2}$

But the square of any real number is nonnegative, so the last inequality is true for every real number t. Thus $(t - 4)^2 > -\frac{5}{2}$ is true for every $t \in \mathbf{R}$ and $2t^2 - 16t + 36 > 0$ is true for all real numbers t.

The solution set is **R**.

For some quadratic inequalities, the solution set may be the empty set ∅.

4–3: Solving Quadratic Inequalities Algebraically   **147**

Example 5 Solve for x: $x^2 - 6x + 13 < 0$.

SOLUTION
$$x^2 - 6x + 13 < 0$$
$$x^2 - 6x + 9 < -4$$
$$(x - 3)^2 < -4$$

Since the square of any real number is at least zero, $(x - 3)^2 \geq 0$ for all real numbers x. The solution set is $\emptyset$.

Exercises

A Graph the solution set on a number line.

1. $x^2 < 25$
2. $x^2 > 25$
3. $x^2 \geq 4$
4. $x^2 \leq 16$
5. $(x - 1)^2 < 1$
6. $(x + 1)^2 \geq 1$
7. $(x + 1)^2 \geq 4$
8. $(x - 2)^2 \leq 4$
9. $x^2 + 1 < 10$
10. $x^2 - 1 > 8$
11. $x^2 + 1 \geq 0$
12. $x^2 + 3 > 0$
13. $(x - 1)^2 + 2 < 0$
14. $1 + (x + 2)^2 \leq 0$

Complete the square and use Theorem 4–2 to solve each inequality.

15. $x^2 + 4x < 0$
16. $x^2 - 4x > 0$
17. $x^2 - 6x \geq 1$
18. $x^2 + 8x < -5$
19. $t^2 + 4 \leq 5t$
20. $s^2 \leq 7s - 13$

B 21. $2x^2 + 2x - 5 \leq 0$
22. $2x^2 + 3x > 0$
23. $4t - 3 > 5t^2$
24. $3x^2 - 7 \leq 2x$
25. $(x - 4)(2x + 5) \leq 1$
26. $(2x - 1)(x + 3) \leq x(4x - 3)$

Find the largest possible domain of each function if $f(x)$ is to be a real number.

27. $f(x) = \sqrt{x^2 - 3x}$
28. $f(x) = \sqrt{x^2 + x}$
29. $f(x) = \dfrac{1}{\sqrt{2x^2 - x}}$
30. $f(x) = \dfrac{1}{\sqrt{4x^2 + 2x}}$

31. For what values of k does the equation $2x^2 - kx + 3 = 0$ have no solutions in the set of real numbers?

32. For what values of x does the graph of $y = x^2$ lie below the graph of the line $3x - 2y + 2 = 0$?

Find values of a and b to show that each statement is *false*.

33. For all $a \in R$, $a^2 > a$.
34. For all $a, b \in R$, if $a^2 < b^2$ then $a < b$.

C **35.** For what values of p does the point $(0, p)$ lie inside the circle $x^2 + y^2 - 4x + 3y = \frac{11}{4}$?

36. Find the values of p for which the line $x + 3y = p$ does not intersect the circle of radius 2 with center at the origin.

Prove or find a counterexample which disproves each statement.

37. For all $b \in \mathbb{R}$, $b^2 - b + 1 > 0$.

38. For all $a > 0$, $\dfrac{1}{a^2 + 1} > \dfrac{1}{a^2 + 2a + 2}$.

39. a. Given $a, b > 0$, $\sqrt{ab}$ is the *geometric mean* of a and b, and $\dfrac{a+b}{2}$ is the *arithmetic mean* of a and b. For several pairs of positive numbers compare the geometric mean and the arithmetic mean and formulate a conjecture which relates them.

b. The following construction should verify your conjecture. For $a, b > 0$, mark off on a line segments AB and BC such that B is between A and C, $AB = a$ and $BC = b$. Draw a semicircle with $\overline{AC}$ as diameter. Construct segment BD perpendicular to $\overline{AC}$ with D on the semicircle. Then draw the radius to D. Explain how your conjecture is justified.

c. Give an algebraic proof of the inequality that relates $\sqrt{ab}$ and $\dfrac{a+b}{2}$.
[HINT: Look at $(\sqrt{a} - \sqrt{b})^2$.]

40. A manufacturing-plant manager determines that the monthly profit (or loss) from production using x employees is given by $p = -4x^2 + 100x$. How many employees must there be if the profit is to exceed $600?

41. Imagine a particle moving to the right along a horizontal line with velocity given by $v(t) = t^2 - 4t + 3$, where t is the time elapsed in seconds and velocity is measured in feet per second. Negative values of velocity indicate that the particle is moving to the left. Determine the time interval during which the particle is moving to the left.

42. Hightech Inc. manufactures and sells a particular model of calculator for $24 each. After subtracting costs of production the profit in dollars from sales of x calculators per week is $p(x) = 20.1x - 150 - 0.003x^2$. How many calculators should be produced and sold each week to ensure a weekly profit of at least $25,000?

43. If a ball is thrown upward with an initial velocity of 80 feet per second, its height s after t seconds is given by $s(t) = 80t - 16t^2$ in feet. During what time interval will the height of the ball be at least 96 feet?

4–4 Solving Quadratic Inequalities Graphically

Graphical methods are frequently used to solve quadratic inequalities. For example, to solve

$$ax^2 + bx + c > 0,$$

graph the function

$$f(x) = ax^2 + bx + c$$

and observe for what coordinates x the graph of $f(x)$ is above the x-axis. If the graph crosses the x-axis, the intercepts are the zeros of f. These zeros are found by factoring or by using the quadratic formula. The inequality

$$ax^2 + bx + c < 0$$

is solved in the same manner, as the first two examples illustrate.

Example 1 Solve for x: $2(x - 1)(x + 3) < 0$.

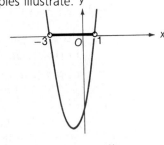

SOLUTION Use the zeros 1 and -3 of $f(x) = 2(x - 1)(x + 3)$ and the fact that the graph of f opens upward to sketch the graph of the function. Where the graph of f is below the x-axis, $f(x) < 0$, and that happens when x is in the interval $(-3, 1)$. The solutions are $-3 < x < 1$.

Example 2 Solve for x: $-x^2 - 3x + 1 < 0$.

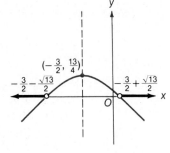

SOLUTION Use the quadratic formula to show that the zeros of the function $f(x) = -x^2 - 3x + 1$ are $-\frac{3}{2} \pm \frac{\sqrt{13}}{2}$. Since $a < 0$, the graph opens downward and $f(x) < 0$ when $x < -\frac{3}{2} - \frac{\sqrt{13}}{2}$ or $x > -\frac{3}{2} + \frac{\sqrt{13}}{2}$.

If the graph of $f(x) = ax^2 + bx + c$ has two distinct x-intercepts r_1 and r_2 such that $r_1 < r_2$, then the arguments x for which $f(x) > 0$ (or $f(x) < 0$) are either $\{x: r_1 < x < r_2\}$, as in example 1, or $\{x: x < r_1 \text{ or } x > r_2\}$, as in example 2.

If the graph of $f(x) = ax^2 + bx + c$ does not cross the x-axis, then the arguments x for which $f(x) > 0$ (or $f(x) < 0$) are either **R** or the empty set ∅. In such cases it helps to locate the vertex of the graph of f first.

Example 3 Solve for x: $x^2 - 2x + 3 > 0$.

SOLUTION For $f(x) = x^2 - 2x + 3$, $-\dfrac{b}{2a} = -\dfrac{-2}{2(1)} = 1$, $f(1) = 2$, and the vertex of the graph of f is $(1, 2)$. Since $a > 0$, the graph of f opens upward. A sketch of the graph shows that $f(x)$ is positive for all x in **R**. The solution set is **R**.

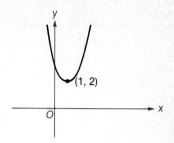

For quadratic functions f, the inequalities $f(x) \geq 0$ and $f(x) \leq 0$ are solved as in examples 1–3.

Example 4 Solve for x: $4(x - 1) \geq x^2 + 1$.

SOLUTION Collect all terms on the left-hand side: $-x^2 + 4x - 5 \geq 0$. For $f(x) = -x^2 + 4x - 5$, $-\dfrac{b}{2a} = -\dfrac{4}{2(-1)} = 2$, and $f(2) = -1$. The vertex is the point $(2, -1)$. Since $a = -1$, the graph opens downward. The graph shows that there are no arguments x such that $f(x) \geq 0$. The solution set is Ø.

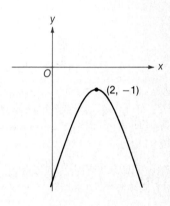

Exercises

A **1.** In example 2 verify that the zeros of f are $-\dfrac{3}{2} \pm \dfrac{\sqrt{13}}{2}$.

2. Rework the problem in example 2 with $<$ replaced by $\geq$.

Use a graphical method to solve each inequality.

3. $2(x - 1)(x + 3) > 0$ **4.** $3(x + 1)(2x - 1) \leq 0$

5. $6x^2 - x - 1 \leq 0$ **6.** $2x^2 - x - 3 \leq 0$

Each graph represents a quadratic function f. For each graph, find all x such that $f(x) \geq 0$.

7. 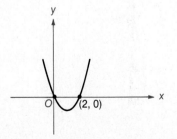 **8.** 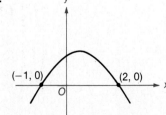 **9.**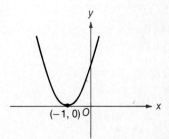

10.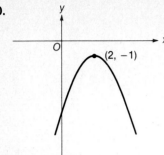
11.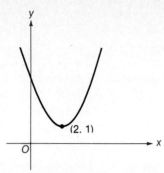
12.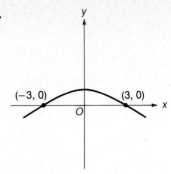

Find the vertex of the graph of each function and use the graph to solve the inequality.

13. $2x^2 - x + 1 < 0$
14. $-2x^2 + x - 1 < 0$
15. $-3x^2 + x - 1 \geq 0$
16. $x^2 + x + 1 \geq 0$

B Solve each inequality graphically.

17. $2x^2 - x + 1 < 0$
18. $(x - 1)^2 + 4 \geq 0$
19. $x^2 + 2x - 2 < 0$
20. $2x(x - 1) \geq 3$
21. $x^2 - x - 5 \leq 0$
22. $x^2 - 3x + 1 < 0$
23. $2x(x - 1) > 3(1 - x) - 4x$
24. $x(x - 1) \leq -x^2 + 3$

Find the domain of each function.

25. $f : x \to \dfrac{1}{\sqrt{(x + 1)(x + 3)}}$
26. $f(x) = \sqrt{(x - 1)(x + 2)}$
27. $g(t) = \dfrac{1}{\sqrt{t^2 - 9}}$
28. $g(t) = \dfrac{1}{\sqrt{t^2 - 4t + 3}}$

C For each pair of inequalities, sketch the solution set and label the points of intersection of the two graphs.

29. $y \leq x + 5$
    $y + 4x \geq x^2 + 5$
30. $y \geq -2(x + 1)$
    $y - 2 \leq -2x(x + 2)$

For each function determine the value(s) of k such that f has no real zeros.

31. $f(x) = x^2 - kx + k + 1$
32. $f(x) = 2kx^2 + x - k + 1$

For what arguments x is $f(x) \leq g(x)$?

33. $f(x) = 3x^2 + 2x + 1$
    $g(x) = x^2 - x + 3$
34. $f(x) = 2x^2 - 3x + 1$
    $g(x) = 4x^2 + x$

Find the domain of f.

35. $f(x) = \dfrac{1}{\sqrt{7 - x}} + \dfrac{2x}{\sqrt{x - 3}}$
36. $f(x) = \sqrt{x^2 - 4x + 3} + \dfrac{1}{\sqrt{x}}$

4–5 Maxima and Minima

You frequently hear people ask questions such as "Which is the *least* expensive?" or "Which is the *tallest* building?" When mathematical functions are used to describe real phenomena, such questions correspond to finding maximum or minimum values. In graphical terms, this means finding the highest or lowest points on the graph of a function.

From the graph of $f: x \to -x^2 + 3$, you can see that the maximum value of f is 3. But for the stone thrown from the surface of Mars, as in example 2 of section 4–1, it is not immediately evident what maximum height the stone reaches.

If a function is defined by $f(x) = 2x$ on the domain [1, 3), then a graph shows that 2 is the minimum value of f and that there is no maximum, since 6 is not in the range and no number less than 6 is a maximum.

A function f has a **maximum value** M if $M \in \text{Rng } f$ and $f(x) \leq M$ for all $x \in \text{Dom } f$. A function f has a **minimum value** m if $m \in \text{Rng } f$ and $m \leq f(x)$ for all $x \in \text{Dom } f$.

Some functions, such as $f: x \to \dfrac{1}{x}$, have neither a maximum nor a minimum.

Your knowledge of graphing quadratic functions from section 4–2 suggests the following theorem. A graphical justification of the theorem is given.

Theorem 4–3   For the quadratic function $f(x) = ax^2 + bx + c$ with domain **R**, $f\left(-\dfrac{b}{2a}\right)$ is the maximum value of f if $a < 0$ or the minimum value of f if $a > 0$.

Consider the two cases.

Case 1: $a < 0$

The graph of f opens downward and the vertex is the point $\left(-\dfrac{b}{2a}, f\left(-\dfrac{b}{2a}\right)\right)$. The graph shows that the y-coordinate of the vertex is the maximum value of the function.

Case 2: $a > 0$

The graph opens upward and the vertex is the point $\left(-\dfrac{b}{2a}, f\left(-\dfrac{b}{2a}\right)\right)$. The graph shows that the y-coordinate of the vertex is the minimum value of the function.

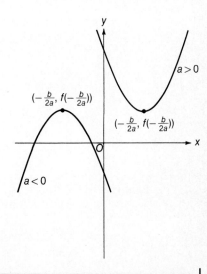

Example 1 What is the maximum value of the function $f: x \to -2x^2 + 3x - 1$?

SOLUTION Since $a = -2 < 0$, the function has a maximum value, $f\left(-\dfrac{b}{2a}\right)$.

$$-\frac{b}{2a} = -\frac{3}{2(-2)} = \frac{3}{4}$$

$$f(\tfrac{3}{4}) = -2\left(\tfrac{3}{4}\right)^2 + 3\left(\tfrac{3}{4}\right) - 1 = \tfrac{1}{8}$$

ANSWER The maximum value of f is $\tfrac{1}{8}$.

Example 2 Find the range of the function $g(x) = x^2 - 4x + 7$ defined on the domain $[3, \infty)$.

SOLUTION The x-coordinate of the vertex is

$-\dfrac{b}{2a} = -\dfrac{-4}{2(1)} = 2$. Because the domain of g is not **R**, $g(2)$ is not necessarily a minimum value. A graph of the function g shows that the minimum occurs at $x = 3$.

$$g(3) = (3)^2 - 4(3) + 7 = 4$$

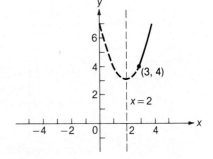

ANSWER The range is $[4, \infty)$.

Example 3 Find the positive number x such that the sum of its square and its additive inverse is as small as possible.

SOLUTION The number x which is sought is the argument at which the function $f(x) = x^2 + (-x)$ has its minimum value. The domain of f is $(0, \infty)$, since x must be positive. Graph the function and observe that $-\dfrac{b}{2a} = \dfrac{1}{2}$, which is the x-coordinate of the graph of the vertex. So $f(x)$ is a minimum at $x = \tfrac{1}{2}$.

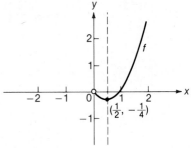

Example 4 Prove: Of all rectangles with a given perimeter, the square is the one which has the greatest area.

SOLUTION Let p be the given perimeter and x the width of the rectangle. Then the length of the rectangle is $\dfrac{p - 2x}{2}$. Create a function A which relates the area

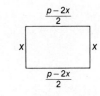

(cont. on p. 155)

to an argument, here the width x of the rectangle.

$$A: x \to \frac{x(p-2x)}{2} = -x^2 + \frac{p}{2}x.$$

The domain of this function is $0 < x < \frac{p}{2}$.
The maximum value of A occurs when

$$x = -\frac{b}{2a} = \frac{-\frac{p}{2}}{2(-1)} = \frac{p}{4},$$

which is in the domain of the function. The maximum area occurs when $x = \frac{p}{4}$, that is, when the width of the rectangle is one fourth the perimeter. Hence the rectangle with maximum area is a square.

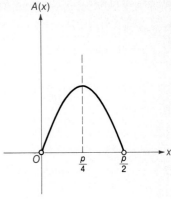

Example 5 A farmer has exactly 240 m of fencing with which to enclose a rectangular plot, one side of which runs along a river and does not need fencing. Find the dimensions and area of the plot which has maximum area.

SOLUTION You might be tempted from the previous example to use a square plot with sides each 80 m long, which encloses an area of 6400 m². But notice:

Length of sides	Area
75 m, 75 m, 90 m	6750 m²
70 m, 70 m, 100 m	7000 m²
55 m, 55 m, 130 m	7150 m²

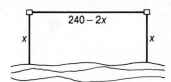

So the square is not the correct answer. Denote the width of the rectangle by x. Then the length of the rectangle must be $240 - 2x$. Thus the area function, which must be maximized, is

$$A(x) = x(240 - 2x) = -2x^2 + 240x.$$

The domain of this function is $0 < x < 120$. (Why?) Since $a < 0$, the graph of A opens downward and the maximum value occurs at

$$x = -\frac{b}{2a} = -\frac{240}{2(-2)} = 60.$$

Thus the maximum value of A occurs when the width of the rectangle is 60 m.

$$A(60) = -2(60)^2 + 240(60)$$
$$= 7200$$

ANSWER The largest area is 7200 m², which corresponds to a rectangle 60 m wide and 120 m long.

Exercises

A 1. Sketch the graph of the function in example 1.

2. Find the range of the function in example 1.

For each function find the maximum or minimum value and specify the range.

3. $f(x) = -x^2 + 4x$
4. $f(x) = x^2 + 2x + 5$
5. $f(x) = 3x^2 - 6x + 1$
6. $f(x) = x^2 + 2x + 1$
7. $f(t) = 3 - 2t^2 + 7t$
8. $g(s) = 2s^2 - 3s + 5$

9. In example 2, if the domain had been $(3, \infty)$ instead of $[3, \infty)$, show that g would have had neither a maximum nor a minimum.

10. For the function $f: x \rightarrow -x^2 + 7$, the number 10 is greater than all values of f. Why is 10 not the maximum value of f?

11. In example 2 of section 4–1, the maximum height attained by the stone whose height h at time t is $h(t) = -5.6t^2 + 82t + 500$ was estimated to be 800 feet. What is the error in this estimate?

12. Find a positive number such that twice its square added to its additive inverse is as small as possible.

B For each function g find the maximum or minimum value and specify the range.

13. $g(x) = |x - 1| + 2$
14. $g(r) = |r + 3| - 4$

15. A pencil manufacturer determines that the daily cost c in dollars of producing n dozen pencils is given by

$$c = \frac{n^2}{25{,}000} - \frac{2n}{5} + 4000.$$

   a. How many dozen pencils should be produced per day to minimize the cost?

   b. What is the minimum cost?

16. A gardener has 200 m of fencing with which to enclose a rectangular plot. What are the dimensions of the plot with greatest area?

17. Find two numbers whose sum is 65 and whose product is as large as possible.

18. A real estate manager determines that the monthly profit p in dollars from a building with x stories is given by $p = -3x^2 + 102x$. What is the maximum profit?

19. The height in feet of a particle above ground at time t seconds is given by $h(t) = -2t^2 + 140t + 1600$. What is the greatest height attained by the particle?

20. Find two numbers whose sum is 20 so that their product added to the larger number is as great as possible.

21. In a particular trapezoid the sum of the lengths of the two bases and the altitude is 12 feet.

   a. Define a function which relates the area of the trapezoid to its altitude.

   b. Find the largest possible area for the trapezoid.

22. A kennel operator has 400 m of fencing with which to enclose a rectangular area, one side of which is the wall of the main kennel building.

   a. Define a function which relates the area of the rectangle to one of its sides.

   b. Find the dimensions of the rectangle with maximum area.

C 23. If $y \geq x^2 - 10x + 24$ and $y + 4 \leq x$, what is the maximum value of y?

24. Prove Theorem 4–3 algebraically for $a > 0$; that is, show that $f(x) \geq f\left(-\dfrac{b}{2a}\right)$ for all $x \in \mathbf{R}$.

25. Write a computer program that inputs the coefficients of a quadratic function, prints out the maximum or minimum value of the function, and tells which it is printing.

26. The base of a scalene triangle is 12 feet long and its altitude is 4 feet long. If a rectangle is inscribed in the triangle with one side along the base, define a function which relates the area of the rectangle to its width and determine the maximum area of such a rectangle.

27. The graph of the quadratic function $f(x) = (k - 1)x^2 - 2(k^3 - 1)x + 1$ has a vertical axis of symmetry which depends upon the choice of the constant k. If $x = r$ is the axis of symmetry, what is the minimum value of r?

28. A wire of length L is cut into two pieces. One piece is bent to form a circle and the other a square. How should the wire be cut if the sum of the areas of the square and the circle is to be a minimum? A maximum?

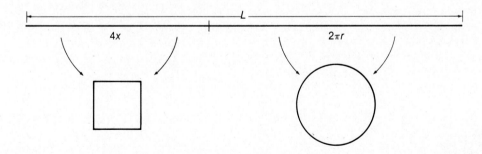

4–5: Maxima and Minima

Special   An Application to Physics

A particle P is traveling along the y-axis and its directed distance from (0, 0) at time $t \geq 0$ is given by

$$f(t) = 3 - t.$$

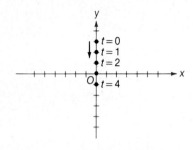

Another particle Q is traveling along the x-axis and its directed distance from (0, 0) at time $t \geq 0$ is given by

$$g(t) = 3t - 4.$$

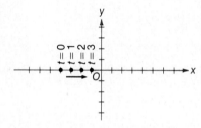

If both particles start at the same time, $t = 0$, what is the closest they ever get to one another?

Notice that the particles do not collide at the origin O. Particle P passes O when $t = 3$. Particle Q passes O when $t = \frac{3}{4}$. Particle Q starts further from the origin than P but travels faster.

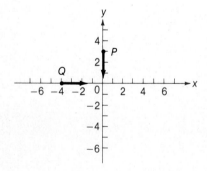

When $t = 0$, P is at (0, 3) and Q is at (−4, 0). So the distance D between them is $\sqrt{3^2 + 4^2} = 5$.

When $t = 1$, the distance D is $\sqrt{2^2 + (-1)^2} = \sqrt{5}$. To minimize this distance you first need to create a distance function in terms of some convenient variable such as t:

$$D(t) = \sqrt{(3t - 4)^2 + (3 - t)^2}.$$

This function $D(t)$ is not quadratic. The quantity under the radical is quadratic, and it must be made as small as possible for D to be a minimum. So let

$$h(t) = (3t - 4)^2 + (3 - t)^2,$$
$$= 10t^2 - 30t + 25.$$

So if $t = -\dfrac{b}{2a} = -\dfrac{-30}{2(10)} = 1.5$ seconds, $h(t)$ and $D(t)$ are minimum.

$$D(1.5) = \sqrt{0.5^2 + 1.5^2} = \dfrac{\sqrt{10}}{2}$$
$$\doteq 1.58$$

So the closest particles P and Q come to each other is approximately 1.58 units.

Exercises

Particle P is moving along the y-axis with position at time $t \geq 0$ given by $f(t)$, and particle Q is moving along the x-axis with position at time $t \geq 0$ given by $g(t)$. Find the minimum distance between the two particles.

1. $f(t) = 3 - 2t$
   $g(t) = 3t - 5$

2. $f(t) = 3t - 4$
   $g(t) = 1 - t$

4-6 The Method of Least Squares

A behavioral psychologist wishes to use data from an experiment on rats to obtain a function that relates the change in reaction time *y* to the amount *x* of a drug administered ten minutes earlier. She plots the data points (2, 1), (4, 2), and (5, 2) as shown.

The psychologist has theoretical reasons to believe that the graph of the function should be linear and should contain the origin. But since no line passes through the origin and through the three data points, she decides to look for a line through the origin which comes as close as possible to the data points.

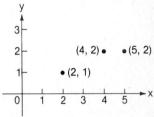

The question she must answer is, "What line best approximates the experimental data and how do I find it?"

A common approach to answering this question is to draw a number of such lines and pick the one which looks best. For the data given, the psychologist might draw a line like the one shown. What would you estimate the slope of this line to be?

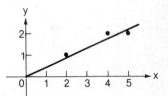

Mathematicians and scientists prefer to have a standard method for finding this line of "best fit." If, as is usually the case, there were many more points, it might be difficult to sketch even a good line, much less the best one. Furthermore, if a scientist is going to make a prediction based on the line of best fit, it is important that other scientists know the method used.

One such method involves minimizing the sum of the squares of the vertical distances of data points from the line. For the psychologist's data, the distances are PP', QQ', and RR', as shown in the diagram below.

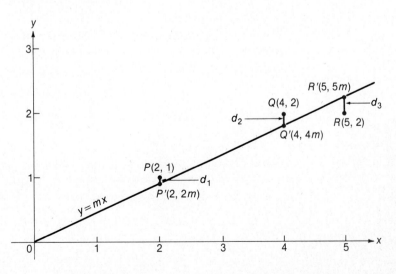

The **method of least squares** requires that the quantity
$$(PP')^2 + (QQ')^2 + (RR')^2$$
be a minimum.

Any line through the origin has an equation $y = f(x) = mx$. You may already have estimated m from the graph shown to be about 0.4. Now try to determine the value of m for the line of best fit. In order to find each vertical distance, you need both the observed value y_i of each data point (x_i, y_i) and the coordinate y_ℓ of the point on $y = mx$ that corresponds to each x_i.

Arrange the data in a table and include a column for the square of each distance d_i.

Point	x_i	y_i	y_ℓ	$(d_i)^2$
P	2	1	$2m$	$(1 - 2m)^2 = 4m^2 - 4m + 1$
Q	4	2	$4m$	$(2 - 4m)^2 = 16m^2 - 16m + 4$
R	5	2	$5m$	$(5m - 2)^2 = 25m^2 - 20m + 4$

Now create a function T which relates the sum of the d_i^2 to the argument m.

$$T(m) = (4m^2 - 4m + 1) + (16m^2 - 16m + 4) + (25m^2 - 20m + 4)$$
$$= 45m^2 - 40m + 9$$

To make $T(m)$ as small as possible, observe that T is a quadratic function and therefore, by theorem 4–3, $T(m)$ is a minimum for $m = -\dfrac{b}{2a}$.

$$m = -\frac{b}{2a} = -\frac{-40}{2(45)} = \frac{4}{9}$$

Since $a = 45 > 0$, the graph of T opens upward and $\frac{4}{9}$ is the value of m for which the sum of the squares of the distances d_i is a minimum. By the method of least squares, the line of best fit that contains the origin is

$$y = \tfrac{4}{9}x$$

Notice that the slope of this line, $\frac{4}{9} \doteq 0.444$, is close to the earlier estimate of 0.4.

For the behavioral psychologist's data, the line of best fit was forced to contain the origin. In other experimental situations, it could just as well contain some other fixed point, as in the next example.

Example 1 Find the line of best fit that contains (0, 4) for the data points (1, 3), (2, 1), and (3, 1).

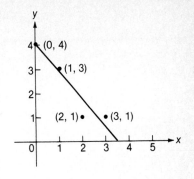

SOLUTION Let m be the slope of the line of best fit. Since that line passes through (0, 4), an equation of the line is $y = mx + 4$.

If $x_1 = 1$, then $y_\ell = m(1) + 4$.

If $x_2 = 2$, then $y_\ell = m(2) + 4$.

If $x_3 = 3$, then $y_\ell = m(3) + 4$

x_i	y_i	y_ℓ	$(d_i)^2$
1	3	$m + 4$	$(m + 1)^2 = m^2 + 2m + 1$
2	1	$2m + 4$	$(2m + 3)^2 = 4m^2 + 12m + 9$
3	1	$3m + 4$	$(3m + 3)^2 = 9m^2 + 18m + 9$

Now create a function T that relates the sum of the $(d_i)^2$ to the argument m. You should verify that

$$T(m) = 14m^2 + 32m + 19.$$

The function value $T(m)$ is a minimum for

$$m = -\frac{b}{2a}$$

$$= \frac{-32}{2(14)} = -\frac{8}{7},$$

and the line of best fit is therefore

$$y = -\frac{8}{7}x + 4.$$

Notice in the figure that a graph of this line does indeed appear to fit the data well.

In each case there has been a restriction that the line of best fit contain a specified point. Without such a restriction, it can be proved that the line of best fit must always pass through the point (x_a, y_a), where x_a is the average of the x_i's and y_a is the average of the y_i's.

Hence to find the line of best fit for the data points (1, 2), (2, 3), (5, 7), and (8, 8), you first compute

$$x_a = \frac{1 + 2 + 5 + 8}{4} = 4.$$

Also compute

$$y_a = \frac{2 + 3 + 7 + 8}{4} = 5.$$

Then let the line of best fit be $y - 5 = m(x - 4)$, or $y = m(x - 4) + 5$. For each x_i, find y_ℓ and $(d_i)^2$ and proceed to find m as in example 1.

Exercises

A Write an expression in terms of m for the vertical distance between the given point and the given line.

1. $(1, 2)$; $y = mx$
2. $(3, 5)$; $y = mx$
3. $(-1, 4)$; $y = mx$
4. $(-2, -5)$; $y = mx$
5. $(2, 3)$; $y = mx + 3$
6. $(3, 7)$; $y = mx + 2$

B 7. The function T for a line of best fit is given in terms of the slope m as $T(m) = 23m^2 + 36m + 53$. For what value of m is $T(m)$ a minimum?

8. Rework exercise 7 for $T(m) = 13m^2 + 17m + 28$.

For the given data find the line of best fit which contains the origin.

9. $(1, 2), (2, 3), (3, 5)$
10. $(2, 1), (3, 1), (4, 2), (6, 3), (7, 3)$

For the given data find the line of best fit that contains the fixed point $(0, 2)$.

11. $(1, 3), (2, 3), (3, 4), (4, 4)$
12. $(2, 2), (3, 1), (4, 1), (5, 1)$

C 13. For the data points $(2, 1), (4, 2),$ and $(5, 2)$, find x_a, y_a, and the line of best fit which contains the point (x_a, y_a).

14. A forester wishes to find the linear relationship between the circumference of a balsam pine and its age. Over a period of time, he collects the following data on 100 trees.

Age in years	4	6	8	10
Average circumference (inches)	$1\frac{1}{2}$	$2\frac{3}{4}$	$4\frac{1}{2}$	$7\frac{1}{2}$

By the method of least squares determine the line of best fit for the data and use it to predict the circumference of a tree that is 13 years old.

15. Finish the last example of this section by finding the line of best fit for the data points (1, 2), (2, 3), (5, 7), and (8, 8).

Chapter Review Exercises

4–1, page 137

Write each function that is quadratic in the form $f(x) = ax^2 + bx + c$.

1. $f(x) = (x - 2)^2 + 1$
2. $f(x) = |x + 1| - 2$
3. $f(x) = \sqrt{x} + x^2$
4. $f(x) = 1 - 2(x^2 + 3x)$

5. The height of an isosceles trapezoid is the same as the length of one of the bases and the other base is 2 units longer. Define a function that relates the area of the trapezoid to its height. What is the domain of the function?

6. Find the quadratic function whose graph contains the points $(1, -2)$, $(-1, 2)$, and $(2, -1)$.

7. The height in meters of a particle above ground at time t seconds is given by
$$h(t) = -2t^2 + 10t + 12.$$
When does the particle hit the ground, and when is the height of the particle 4 meters?

4–2, page 141

Determine the vertex and axis of symmetry of the graph of each function.

8. $f(x) = 2x^2 - 4x + 3$
9. $f(t) = -3t^2 + t$

Graph each function and determine its range.

10. $f(x) = 1 - 2x - x^2$
11. $f(a) = 2a^2 - a + 1$

4–3, page 146

Solve each inequality by completing the square.

12. $4x^2 - 12x + 5 \geq 0$
13. $9x^2 + 12x + 7 < 0$
14. $x^2 + 2x \leq 5$

4–4, page 150

Solve each inequality by graphing.

15. $3(x - 1)(x + 2) < 0$
16. $-2(x - 1)(2x + 5) \leq 0$
17. $x^2 - 2x \geq 1$
18. $2x(x + 2) < 1$

4–5, page 153

19. Find the maximum value of the function $f(x) = -2.1x^2 + 5.6x - 4.3$.

20. A manufacturer of ball bearings determines that the daily cost C in dollars of producing n ball bearings is given by

$$C(n) = \frac{n^2}{100{,}000} - \frac{n}{5} + 2200.$$

How many ball bearings should be manufactured each day so as to minimize the daily cost C?

4–6, page 159

21. An engineer measures the total waste in kilograms from a chemical process at three different times and plots the findings as shown below.

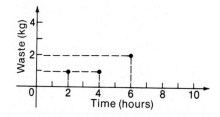

Assuming that there is a straight-line relationship between time in hours and waste in kilograms, find the line of best fit through the origin for these points and then use it to predict the amount of waste when $t = 9$ hours.

Chapter Test

1. Write the quadratic function $f(x) = 3 - 5(x - 2x^2)$ in the form $f(x) = ax^2 + bx + c$.

2. Determine the vertex and axis of symmetry of the graph of $f(x) = 2x^2 - 4x + 5$.

3. Solve the inequality $2(x - 1)(3x + 4) \leq 0$.

4. Graph the function $f(x) = -4x^2 - 12x - 1$. Label the coordinates of two points on the graph and the axis of symmetry.

5. What is the minimum value of the function $f(x) = 3x^2 + 12x + 11$?

6. Solve the inequality $4x - 1 < x^2$.

7. Write a quadratic function whose graph has the vertex $(3, -1)$ and contains the point $(2, 1)$.

8. Find the zeros, the axis of symmetry, and the range of the function $g(t) = 1 + 4t^2 - 12t$. Sketch a graph of g.

9. An economic analysis for Jane's Shop shows that the function

$$P(S) = \frac{S^2}{2500} - \frac{2}{5}S - 100$$

describes profit P in dollars for S units sold.

   a. Graph the function P.

   b. How many units must be sold in order for the profit to be positive?

   c. For what value of S is the profit P the least?

10. A farmer has 100 meters of fencing with which to enclose two adjacent and congruent rectangular pigpens. What are the dimensions of the largest region the farmer can fence?

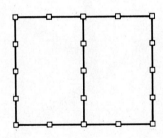

Analytic Geometry
Conic Sections

5

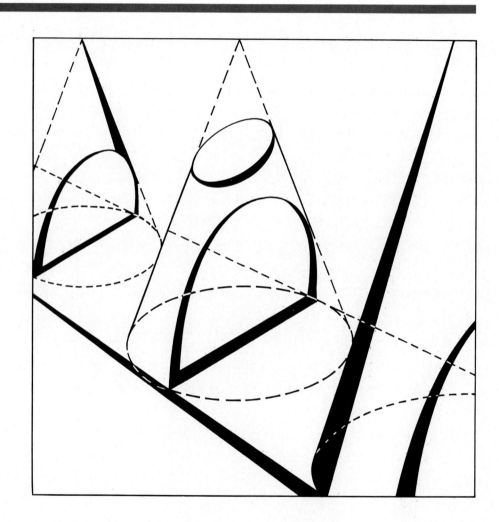

5-1 Locus Problems

You know how to graph sets described by certain equations. Now we reverse that procedure and show how to find the equation when a condition is given geometrically or in words.

For example, what is the set of points (x, y) each of which is a distance r from a fixed point (h, k)? From geometry you know that this is a circle, and you have already learned to translate the given condition into the equivalent algebraic statement

$$\sqrt{(x - h)^2 + (y - k)^2} = r.$$

The set S of all points P that satisfy some condition is called the **locus** of points P. For example, the locus of points $P(x, y)$ such that P is equidistant from $O(0, 0)$ and $A(4, 0)$ is the perpendicular bisector of segment OA. Algebraically, it is $S = \{(x, y): x = 2\}$. Any point on the line $x = 2$ is in the set S and any member of the set S is a point on the line $x = 2$.

Example 1 Given points $Q(-2, 0)$ and $R(4, 0)$, find the locus of points $P(x, y)$ such that $\overline{PQ} \perp \overline{PR}$.

SOLUTION Start with a sketch in the coordinate plane. The condition $\overline{PQ} \perp \overline{PR}$ means that the product of the slopes of the segments is -1. Calculate the slopes.

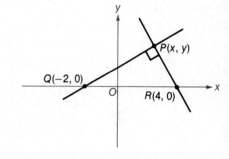

Slope of $\overline{PQ} = \dfrac{y - 0}{x - (-2)} = \dfrac{y}{x + 2}$

Slope of $\overline{PR} = \dfrac{y - 0}{x - 4} = \dfrac{y}{x - 4}$

The required set is $\left\{(x, y): \dfrac{y}{x + 2} \cdot \dfrac{y}{x - 4} = -1\right\}$. (Why are -2 and 4 not admissible values of x?) This equation does not look familiar, so simplify it.

$$y^2 = (-1)(x + 2)(x - 4)$$
$$y^2 = -x^2 + 2x + 8$$
$$y^2 + x^2 - 2x = 8$$
$$y^2 + (x^2 - 2x + 1) = 8 + 1$$
$$y^2 + (x - 1)^2 = (3)^2$$

You should recognize this as the equation of a circle with radius 3 and center $(1, 0)$, but remember, the endpoints of $\overline{QR}$, $(-2, 0)$ and $(4, 0)$, are deleted because -2 and 4 are not admissible values of x.

Example 2 Find the locus of points $P(x, y)$ such that P is twice as far from $A(2, 1)$ as it is from $B(6, -3)$.

SOLUTION Use the distance formula.

$$PA = \sqrt{(x-2)^2 + (y-1)^2} \quad \text{and} \quad PB = \sqrt{(x-6)^2 + (y+3)^2}$$

The condition to be met is $PA = 2(PB)$, or algebraically,

$$\sqrt{(x-2)^2 + (y-1)^2} = 2\sqrt{(x-6)^2 + (y+3)^2}.$$

Square and simplify this equation to remove the radicals.

$$x^2 - 4x + 4 + y^2 - 2y + 1 = 4(x^2 - 12x + 36 + y^2 + 6y + 9)$$

$$0 = 3x^2 - 44x + 3y^2 + 26y + 175$$

This is an equation of a circle.

Example 3 Given the point $F(0, 2)$ and the line $\ell: y = -2$. Find a radical-free equation for the locus of points $P(x, y)$ such that $PF = P\ell$, where $P\ell$ is the perpendicular distance from point P to line ℓ.

SOLUTION Make a sketch that shows F and ℓ. By the distance formula,

$$PF = \sqrt{(x-0)^2 + (y-2)^2}.$$

The coordinates of the point R on ℓ are $(x, -2)$. So the vertical distance PR is $|y - (-2)|$. The given condition is

$$\sqrt{x^2 + (y-2)^2} = |y - (-2)|.$$

Eliminate the radical by squaring.

$$x^2 + y^2 - 4y + 4 = y^2 + 4y + 4$$

$$x^2 = 8y$$

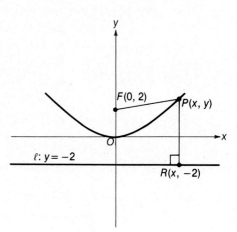

Exercises

A Find an expression for the distance between the given points in terms of x and y.

1. $(-3, 4), (x, y)$ **2.** $(x, y), (2, -5)$

Find an expression for the distance from (x, y) to the given line in terms of x and/or y.

3. $y = 4$ **4.** $y = -3$ **5.** $x = -1$ **6.** $x = 2$

7. Redo example 1, page 167, using $Q(2, 0)$ and $R(6, 0)$.

8. Redo example 2, page 168, using $A(1, 2)$ and $B(7, 0)$.

In exercises 9–23, write a radical-free equation for the locus of points $P(x, y)$ which satisfy the given conditions.

9. P is equidistant from points $(1, 2)$ and $(1, -6)$.
10. P is equidistant from points $(2, -1)$ and $(6, 5)$.
11. The distance from P to $(1, -4)$ always equals $\sqrt{2}$.
12. The distance from P to $(-3, 2)$ always equals 5.
13. The distance from P to $(-2, 1)$ is twice its distance to $(1, 5)$.
14. The distance from P to $(1, -3)$ is twice its distance to $(3, 2)$.
15. The sum of the distances from P to $(-4, 0)$ and to $(4, 0)$ is 10.
16. The sum of the distances from P to $(-3, 1)$ and to $(3, 1)$ is 12.

B 17. Segments PA and PB are perpendicular, given $A(-4, 1)$ and $B(6, -2)$.
18. The distance from P to the point $(4, -3)$ is twice its distance to the line $y = 5$.
19. The distance from P to the line $y = -2$ is equal to the slope of the segment joining P to the point $(-1, 1)$.
20. The slope of $\overline{PA}$ is twice the slope of $\overline{PB}$, given $A(-1, 2)$ and $B(3, 4)$.
21. P is equidistant from lines with equations $x = y$ and $y = -6$.
22. Given $A(2, -1)$ and $B(4, 3)$, segments PA and PB are perpendicular.
23. Given the point $F(4, 2)$, the line $\ell: y = -6$, and P, $PF = P\ell$.

C 24. Find an equation of the line ℓ with positive slope that bisects the angle formed by lines ℓ_1 and ℓ_2, where
$\ell_1: y = 3x + 2$ and $\ell_2: y = 2$.

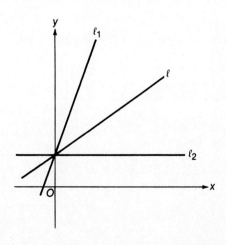

Special Analytic Geometry—The Conic Sections

Here is a construction in solid geometry. Begin with a fixed line called an **axis**, and a point called the **vertex** on that axis. Through the vertex pass another line called the **generator** that is at an angle α with the axis. Now rotate the generator line about the axis, keeping the angle α constant. The resulting figure is a *double cone* of infinite extent. Now cut this cone with a plane that makes an angle β with the axis, as in the figures. The intersection of this cutting plane with the cone is called a **conic section**. The section is called a **hyperbola** if $\beta \in [0, \alpha)$, a **parabola** if $\beta = \alpha$, and an **ellipse** if $\beta \in (\alpha, 90°)$.

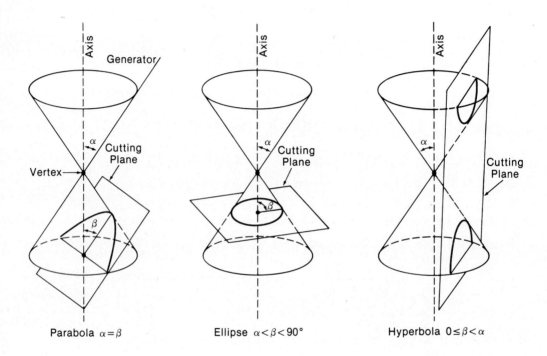

Parabola $\alpha = \beta$ Ellipse $\alpha < \beta < 90°$ Hyperbola $0 \leq \beta < \alpha$

These descriptions of the three classes of conic sections are the ones used by the great Greek mathematician Appolonius (c. 200 B.C.) in his extensive study of the conic sections and their properties. If $\beta = 90°$, the section is a circle, which is commonly considered a limiting case of an ellipse.

These descriptions are indeed graphic, as the figures show, but nowadays we find it more convenient to study each of these curves as a relation consisting of ordered pairs of real numbers. The study of plane curves described algebraically is called **analytic geometry**.

Exercises

Determine the intersection of the cutting plane and the double cone under the following conditions:

1. The cutting plane contains the vertex and $\beta = \alpha$.
2. The cutting plane does not contain the vertex and $\beta = 90°$.
3. The cutting plane does contain the vertex and $\beta = 90°$.
4. The cutting plane is parallel to the axis and contains the axis.

5–2 The Parabola

If you are given the point $F(0, c)$, $c \neq 0$, and the line $\ell: y = -c$, what is the locus of points $P(x, y)$ such that $PF = P\ell$, where $P\ell$ is the distance from P to line ℓ?

As before, introduce the point R on ℓ with coordinates $(x, -c)$. Then the condition $PF = P\ell$ is described algebraically by

$$\sqrt{x^2 + (y - c)^2} = |y - (-c)|.$$

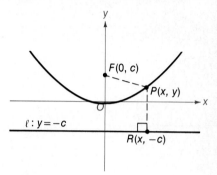

Square both sides of this equation:

$$x^2 + y^2 - 2cy + c^2 = y^2 + 2cy + c^2,$$

$$x^2 = 4cy, \text{ or } y = \frac{1}{4c}x^2.$$

Note that $x^2 = 4cy$ is the equation of a quadratic function of the form $y = ax^2$ with $a = \frac{1}{4c}$. When c is positive, so is $\frac{1}{4c}$, and the curve opens upward. When c is negative, the curve opens downward. In either case, the resulting curve is an example of a class of curves called conic sections. This curve is a parabola.

Definition   A **parabola** is the set of all points in the plane equidistant from a fixed line ℓ and a fixed point F which is not on ℓ.

The line ℓ is the **directrix** and the point F is the **focus** of the parabola. Any given line and any point not on the line determine a unique parabola. Two examples are sketched below.

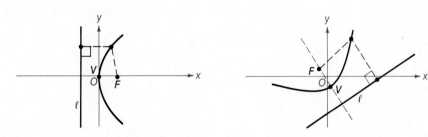

Every parabola opens from a vertex V, around a focus F, and away from the directrix ℓ. Every parabola has an axis of symmetry and a vertex. Only if the directrix is parallel to the x-axis is the parabola the graph of a function. In exercises 39 and 40 you will be asked to find the equations of parabolas with directrices that are not horizontal. It will be clear from the equations and from the graphs that such parabolas are not the graphs of functions. For now we consider only parabolas whose graphs represent functions. Each such parabola has a vertical axis of symmetry and will be called a *vertical parabola*.

The standard vertical parabola whose equation is $x^2 = 4cy$ has its vertex at (0, 0). Here is a theorem about the vertical parabola with vertex at (h, k). The focus F and the directrix ℓ are still $|c|$ units from the vertex.

Theorem 5–1 The equation of the parabola with vertex $V(h, k)$, focus $F(h, k + c)$, and directrix $\ell: y = k - c$ is

$$(x - h)^2 = 4c(y - k).$$

PROOF Make a sketch which shows points $P(x, y)$, $F(h, k + c)$, $V(h, k)$, and $R(x, k - c)$. Then the condition $PF = PR$ implies

$$\sqrt{(x - h)^2 + (y - [k + c])^2} = |y - (k - c)|.$$

Square both sides of this equation.

$$(x - h)^2 + ([y - k] - c)^2 = ([y - k] + c)^2$$

This equation simplifies to

$$(x - h)^2 = 4c(y - k).$$

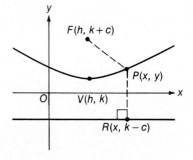

Observe that if the vertex (h, k) is $(0, 0)$, then this equation simplifies to $x^2 = 4cy$, as on page 171. The diagram for Theorem 5–1 is drawn for $c > 0$. What happens if $c < 0$?

By using Theorem 5–1, you can find an equation of a parabola without considering it as a locus problem, provided that values of constants h, k, and c are known.

Example 1 Write an equation of the parabola with vertex (3, 1) and focus (3, 4).

SOLUTION The vertex is (3, 1), so $h = 3$ and $k = 1$. The focus is 3 units above the vertex. Thus $c = 3$ and, by substitution, the equation is

$$(x - 3)^2 = 4(3)(y - 1), \text{ or}$$
$$(x - 3)^2 = 12(y - 1).$$

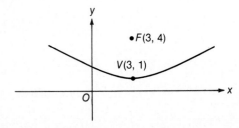

Note that shifting the parabola horizontally h units and vertically k units is algebraically equivalent to replacing x with $x - h$ and y with $y - k$ in the equation in standard form.

Example 2 Find the coordinates of the vertex and the focus of the parabola with equation $(x - 3)^2 = 20(y + 7)$.

SOLUTION Rewrite the equation in the form given in Theorem 5–1.

$$(x - 3)^2 = 4(5)(y - [-7])$$

So $h = 3$, $k = -7$, and $c = 5$.
The vertex is $(3, -7)$ and the focus is $(3, -2)$.

Sometimes values of h, k, and c that uniquely determine a parabola are not apparent and the equation must be rewritten, as in the next example.

Example 3 Sketch the parabola whose equation is $12y + x^2 + 2x = 23$. Name its focus and directrix.

SOLUTION Transform the equation into the form given in Theorem 5–1. This requires collecting the terms in x on the left side and completing the square.

$$x^2 + 2x = -12y + 23$$
$$x^2 + 2x + 1 = -12y + 23 + 1$$
$$(x + 1)^2 = -12y + 24$$
$$(x + 1)^2 = -12(y - 2)$$

So $h = -1$, $k = 2$, and $4c = -12$, or $c = -3$. The parabola opens downward because $c < 0$. The vertex is at $(-1, 2)$; the focus is 3 units below it at $(-1, -1)$; the directrix is 3 units above the vertex with equation $y = 5$. The parabola is easier to sketch if you first locate points such as $\left(3, \frac{2}{3}\right)$ and its symmetric point $\left(-5, \frac{2}{3}\right)$ [the axis of symmetry is $x = -1$].

Example 4 Write an equation of the parabola with focus $F(2, 1)$ and directrix $\ell: y = 7$.

SOLUTION In the sketch, the vertex V is midway between F and ℓ at $(2, 4)$. So $h = 2$ and $k = 4$. The distance between V and F is 3 units. The parabola opens downward, so $c = -3$. Substitute in the equation $(x - h)^2 = 4c(y - k)$ to get

$$(x - 2)^2 = 4(-3)(y - 4).$$

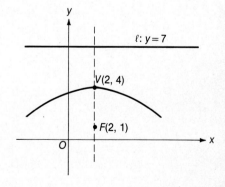

Example 5 Write an equation of the vertical parabola that contains $(-1, 2)$ with focus $(3, -1)$, and which opens upward.

SOLUTION By Theorem 5–1, the equation is of the form

$$(x - h)^2 = 4c(y - k).$$

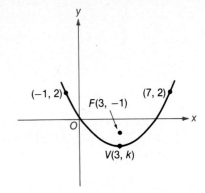

The vertex is directly below the focus, so $h = 3$, and $c = -1 - k$. Thus for all points (x, y) on the parabola,

$$(x - 3)^2 = 4(-1 - k)(y - k).$$

Since $(-1, 2)$ is on the parabola, these coordinates satisfy the equation above.

$$(-1 - 3)^2 = 4(-1 - k)(2 - k)$$
$$16 = -8 - 4k + 4k^2$$
$$0 = k^2 - k - 6$$
$$0 = (k - 3)(k + 2)$$

Thus $k = 3$ or $k = -2$. If $k = 3$, the vertex would be above the focus. So $k = -2$ and $c = -1 - k = -1 - (-2) = 1$.

ANSWER An equation of the parabola is $(x - 3)^2 = 4(y + 2)$.

You have seen that every vertical parabola is the graph of a quadratic function. Conversely, is the graph of every quadratic function a parabola? Theorem 5–2 answers this question.

Theorem 5–2 The graph of any quadratic function $f(x) = ax^2 + bx + c$, $a \neq 0$, is a parabola.

In Theorem 5–1 the directrix is horizontal and the parabola has a vertical axis of symmetry. The theorem has a counterpart in which the roles of x and y are reversed, thereby yielding a parabola with a horizontal axis of symmetry, as shown below.

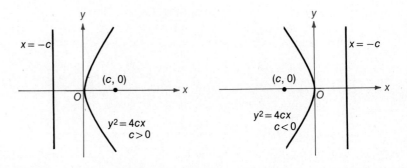

If the vertex of the parabola with a horizontal axis of symmetry is shifted from (0, 0) to (h, k), then x is replaced in the equation by x − h and y is replaced by y − k to yield the equation in the following theorem.

Theorem 5–3 An equation of the parabola with vertex $V(h, k)$, focus $F(h + c, k)$ and directrix $x = h - c$ is

$$(y - k)^2 = 4c(x - h).$$

The locus proof of Theorem 5–3 is left as an exercise.

Exercises

A In each diagram F is the focus and ℓ is the directrix of a parabola. Sketch the parabola.

1.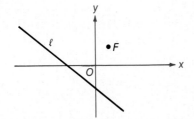

2.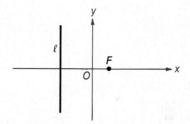

Determine the coordinates of the vertex for each parabola.

3. $x^2 = y$
4. $x^2 = y - 2$
5. $(x - 3)^2 = 8(y)$
6. $(x + 2)^2 = -8(y - 3)$

From the given information on vertex, focus, and directrix ℓ, determine an equation of the parabola.

7. $V(2, 3)$, $F(2, 7)$
8. $V(-3, -1)$, $F(-3, 0)$
9. $V(-1, 3)$, $\ell: y = 0$
10. $V(2, -5)$, $\ell: y = 2$

For each parabola determine the vertex and directrix and sketch its graph.

11. $(x - 3)^2 = -20(y + 1)$
12. $(x + 2)^2 = 4y - 4$
13. $x^2 = -4y - 8$
14. $x^2 = -2y + 1$
15. $x^2 - 2x = 4y - 1$
16. $x^2 - 6x = 4y - 5$
17. $(y - 1)^2 = 4(x + 2)$
18. $(y + 3)^2 = -8(x - 1)$
19. $y^2 + 2y = -8x + 15$
20. $y^2 + 4 = 4x$

The point H lies on the graph of each parabola. Find the value of b.

21. $x^2 + 2x = 2y - 5$; $H(1, b)$ **22.** $x^2 - 3x = -3y + 1$; $H(-2, b)$

B 23. $x^2 = 4y - 8$; $H(a, b)$ **24.** $x^2 - x = -3y + 3$; $H(a, b)$

Write an equation of the parabola obtained by shifting the given parabola to the right 2 units and downward 1 unit.

25. $x^2 - 6x + 4y + 17 = 0$ **26.** $x^2 - 8y + 17 + 2x = 0$

For what values of the constant r does the parabola open upward?

27. $rx^2 = y + 2x^2$ **28.** $(r - 1)x^2 = ry + r$

Sketch a graph of each parabola.

29. $4(x^2 + 6y) = 4x - 1$ **30.** $4x + 7 = 4(2y - x^2)$

31. $y^2 = 2(x + y)$ **32.** $y^2 + 3y + x + 2 = 0$

Find an equation for a vertical parabola such that:

33. Its focus is (0, 4), it opens upward, and contains (10, 4).

C 34. It contains the points (−6, 0), (0, 3), and (8, 0).

35. It has zeros at $x = 1$ and $x = 3$, and its maximum height is 5.

36. Its graph is symmetric with respect to the y-axis, it opens downward, its focus is on the x-axis, and it contains $\left(2, \frac{3}{2}\right)$.

37. It contains $A(0, 1)$ and $B(4, 1)$, opens upward, and its focus is on $\overline{AB}$.

38. Prove Theorem 5–2.

Use the definition of a parabola and the formula for the distance from a point to a line to find a radical-free equation for each parabola.

39. Focus (1, 2); directrix $2x + y + 3 = 0$ **40.** Focus (1, −1); directrix $y = x + 2$

41. Prove Theorem 5–3 from the definition of a parabola.

42. If a line is drawn through the focus of a parabola perpendicular to the axis of symmetry, the line intersects the parabola at two points. The line segment joining these points is called the *latus rectum* of the parabola. Prove that the length of the latus rectum depends only upon the value of c and is therefore a measure of the sharpness or flatness of the bend of the parabola.

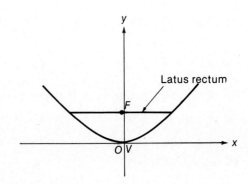

5–3 Applications of the Parabola

Many natural and man-made forms have parabolic shapes. In this section we look at three of them: a parabolic mirror, a bridge whose lower surface is parabolic, and the path of a projectile.

Example 1 A mirror has a parabolic cross section and dimensions as shown at the right. Find the distance from the vertex to the focus of the parabola.

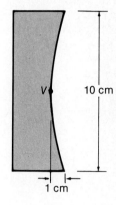

SOLUTION Place the parabola that represents the mirror on a coordinate system in a vertical position with vertex at (0, 0) and the focus at $F(0, c)$. From the dimensions you know that the coordinates of two points on the parabola are (5, 1) and (−5, 1).

The equation of the parabola is of the form $x^2 = 4cy$, and the point (5, 1) is on its graph. By substitution,

$$(5)^2 = 4c(1), \text{ or } c = \frac{25}{4} = 6\frac{1}{4}$$

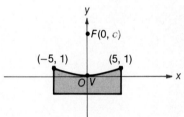

ANSWER The distance from the vertex to the focus of the mirror is $6\frac{1}{4}$ cm. (This is called the *focal length* of the mirror.)

Example 2 A rectangular barge 70 feet wide and projecting 20 feet above the water is steered down the center of a straight canal where there is a perfectly centered bridge with the following sign on it.

Maximum height	30 feet
Maximum width	120 feet

The bridge opening is in the shape of a parabola. Can the barge pass under the bridge?

SOLUTION To answer the question, the barge captain uses the coordinate system shown and labels three points determined from the given dimensions. The opening of the bridge is a parabola with vertex (0, 30). The captain knows that the equation of the parabola is

$$(x - 0)^2 = 4c(y - 30).$$

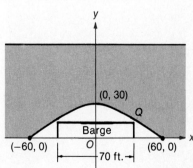

He sets $x = 60$ and $y = 0$ to find $c = -30$. The equation of the parabola is

$$x^2 = -120(y - 30).$$

(cont on p. 178)

If point Q is the upper right-hand corner of the barge, then from the barge's dimensions the captain knows that Q is the point (35, 20). Is the height y of the parabola at $x = 35$ greater than 20, as needed for safe passage?

For $x = 35$:

$$(35)^2 = -120(y - 30)$$
$$1225 = -120y + 3600$$

$$y = \frac{2375}{120} \doteq 19.79 \text{ feet}$$

ANSWER The answer is no. The barge cannot pass beneath the bridge.

In physics, the position $P(x, y)$ of a particle moving in a plane is often expressed by two functions of time t such that $x = f(t)$ and $y = g(t)$. The variable t is called a **parameter** and the equations for x and y are called **parametric equations**.

Example 3 A projectile is thrown into the air from point P_0 and its position after t seconds is $P(x, y)$. Coordinates x and y are given in terms of t by the following parametric equations

$$x = 1 + t$$
$$y = 6t - t^2.$$

Determine the coordinates of P_0, sketch the path of the projectile, and show that this path is part of a parabola.

SOLUTION Since P_0 corresponds to $t = 0$,

$$x = 1 + t = 1 + 0 = 1,$$
$$y = 6t - t^2 = 6(0) - 0^2 = 0.$$
$$P_0 = (1, 0)$$

To find an equation of the path in terms of x and y, eliminate the parameter t by solving $x = 1 + t$ for t and substituting in the equation $y = 6t - t^2$. The result is as follows.

$$y = 6(x - 1) - (x - 1)^2$$
$$y = -x^2 + 8x - 7$$

(cont. on p. 179)

Thus the path is parabolic, that is, part of a parabola. To facilitate sketching the curve, complete the square on the left side of the equation

$$x^2 - 8x = -y - 7.$$
$$x^2 - 8x + 16 = -y - 7 + 16$$
$$(x - 4)^2 = -1(y - 9)$$

The vertex is (4, 9); the parabola opens downward.

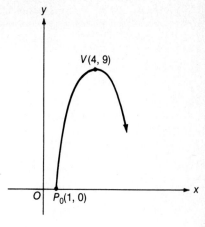

Exercises

A 1. If the barge in example 2 had been 68 feet wide and the same height, 20 feet, would it have passed under the bridge?

2. If the barge in example 2 had been 69 feet wide and the same height, 20 feet, would it have passed under the bridge?

For each parabolic mirror pictured, find the distance from the vertex of the mirror to the focus.

3. 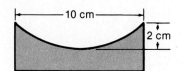   4.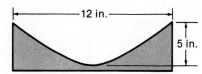

Graph the locus of points $P(x, y)$ where x and y are given by the following parametric equations on the restricted domain indicated.

5. $x = t, y = 2t^2; t \in [0, 4]$

6. $x = t^2, y = t + 1; t \geq 0$.

B 7. From a point above level ground a projectile is thrown into the air and its position $P(x, y)$ at any time t is given by the parametric equations $x = t + 3$ and $y = t^2 + 2$. Find x when $y = 10$.

8. The height h of a rocket above the ground is given as a function of time by $h(t) = 2t - t^2$, and its distance d down range is given by $d(t) = t + 5$. What is the maximum height attained by the rocket and how far down range is it at that time?

9. Given a parabolic mirror as in the illustration. If the distance from the vertex V to the focus F is 2 inches, what is the diameter TL of the mirror?

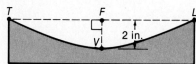

10. The position of a particle $P(x, y)$ in terms of time t is given by the parametric equations $x = t$ and $y = 1 - t^2$. Show that the particle travels along a parabolic

5–3: Applications of the Parabola

path and find the maximum height y that it attains. How long does it take from time $t = 0$ for the particle to hit the x-axis?

11. If the bridge opening in example 2 is circular instead of parabolic, show that the equation of the circle is $x^2 + (y + 45)^2 = (75)^2$ and use this equation to verify that if $x = 35$, then $y > 20$ and the barge can pass under the bridge.

12. A truck that is 10 feet wide has a rectangular cross section. It enters a tunnel with a parabolic cross section. If the tunnel is 20 feet high and 30 feet wide at its base, what is the maximum height of the truck if all it takes to pass through the tunnel is to let a little air out of the tires?

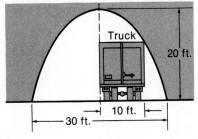

13. Let $P(x, y)$ be a point on the parabola $x^2 = y$ and let D^2 be the square of the distance from $(0, 1)$ to P. Define a function f that relates D^2 to y and find the value of y which minimizes f.

14. Show that the parametric equations $x = 1 + 3t$ and $y = -2 + t$ describe a line.

C 15. Suppose that the supporting cable of a suspension bridge hangs in the shape of a parabola, as in the figure. The supporting columns are 110 feet apart and their tops are 30 feet above the low point of the parabola. From the top of a column to the roadway below is 40 feet. Ten feet from a column is the first vertical cable that supports the roadway. How long is that cable?

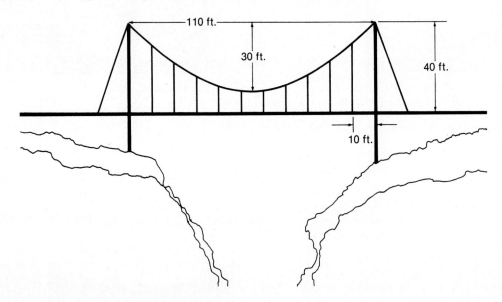

5–4 Tangents to Parabolas

In geometry you studied tangent lines to circles. We now extend the concept to include tangents to parabolas.

In example 3 of section 5–3, a projectile was thrown upward from (1, 0) along a parabolic path described by $(x - 4)^2 = -(y - 9)$. The line of sight along which the projectile was originally propelled was, in fact, a tangent to the curve at the point (1, 0).

In the case of circles, the tangent line and the radius to its point of contact are perpendicular. Unfortunately, no such simple idea exists for parabolas. The tangent line to a circle intersects the circle in exactly one point, and this idea can be used with parabolas. Examples 1 and 2 below investigate the intersection of a line with a parabola.

Example 1 Sketch the graphs of the parabola $R: x^2 = y - 1$ and the line $\ell: y = 2x$ and find the coordinates of their point, or points, of intersection.

SOLUTION Solve the two given equations simultaneously by substituting $y = 2x$ in $x^2 = y - 1$.

$$x^2 = (2x) - 1$$
$$x^2 - 2x + 1 = 0$$
$$(x - 1)^2 = 0$$

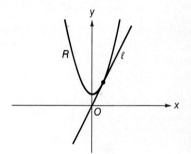

The only solution is $x = 1$. Since $y = 2$ when $x = 1$, the only point of intersection is (1, 2).

The result in example 1 suggests that ℓ is tangent to parabola R.

Example 2 For what values of m does the line $y = mx$ intersect the parabola $(x + 1)^2 = y$?

SOLUTION By substitution,

$$(x + 1)^2 = mx$$
$$x^2 + (2 - m)x + 1 = 0$$

Since the line is to intersect the parabola at either one or two points, this equation in x must have either one or two real solutions. Hence the discriminant D must be nonnegative.

$$D = b^2 - 4ac = (2 - m)^2 - 4(1)(1)$$
$$= 4 - 4m + m^2 - 4$$
$$= m^2 - 4m$$

(cont. on p. 182)

If D is nonnegative, then

$$m^2 - 4m \geq 0.$$
$$m(m - 4) \geq 0$$
$$m \leq 0 \quad \text{or} \quad m \geq 4$$

If $m > 4$ or $m < 0$, then $y = mx$ will intersect the parabola in two points. If $m = 0$ or if $m = 4$, the line $y = mx$ intersects the parabola in exactly one point. The lines $y = 0$ and $y = 4x$ are called tangent lines.

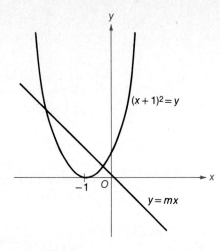

Definition   A **tangent to a parabola** is a line that

   a. intersects the parabola at exactly one point and
   b. is not parallel to the axis of symmetry of the parabola.

You can see now that in example 2 the lines given by $y = 0$ and $y = 4x$ are tangents to the parabola.

Here is a four-step method for finding a line through a given point Q that is tangent to a parabola P, expressed in terms of x and y.

① Write in slope-intercept form the equation of the line ℓ with slope m through Q.

② Substitute the expression for y in the equation for the parabola to get a quadratic equation in x.

③ Find the discriminant D of this quadratic equation.

④ Set $D = 0$ and solve for m.

Notice that in the equation $D = 0$,
   a. the variable is m, not x, and
   b. if there is more than one solution for m, then there is more than one tangent line.

This four-step method is used in the next two examples to find equations of lines tangent to a parabola.

Example 3 Write equations of lines through $Q(\frac{1}{2}, 1)$ that are tangent to the parabola $x^2 = -2y$.

SOLUTION

① The equation of any line ℓ through $(\frac{1}{2}, 1)$ with slope m is
$$y - 1 = m(x - \tfrac{1}{2}), \text{ or}$$
$$y = mx - \tfrac{1}{2}m + 1.$$

② Substitute this value of y in $x^2 = -2y$.
$$x^2 = -2(mx - \tfrac{1}{2}m + 1)$$
$$x^2 = -2mx + m - 2$$
$$x^2 + 2mx + (2 - m) = 0$$

③ The discriminant of this equation is
$$D = (2m)^2 - 4(1)(2 - m).$$
$$D = 4m^2 + 4m - 8$$

④ Set $D = 0$ and solve for m.
$$4m^2 + 4m - 8 = 0$$
$$m^2 + m - 2 = 0$$
$$(m + 2)(m - 1) = 0$$
$$m = -2 \quad \text{or} \quad m = 1$$

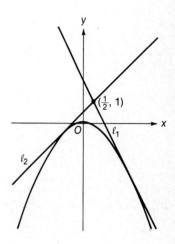

ANSWER The lines through $(\frac{1}{2}, 1)$ tangent to the parabola are $\ell_1: y - 1 = -2(x - \tfrac{1}{2})$ and $\ell_2: y - 1 = x - \tfrac{1}{2}$.

Example 4 Find the line or lines through $(2, 3)$ that are tangent to the parabola $y = 2x^2 - 4x + 3$.

SOLUTION

① $\ell: y - 3 = m(x - 2)$
$$y = mx - 2m + 3$$

② $$mx - 2m + 3 = 2x^2 - 4x + 3$$
$$-2x^2 + 4x + mx - 2m = 0$$
$$(-2)x^2 + (m + 4)x + (-2m) = 0$$

③ $D = (m + 4)^2 - 4(-2)(-2m)$
$$= m^2 - 8m + 16$$

④ $m^2 - 8m + 16 = (m - 4)^2 = 0$
$$m = 4$$

ANSWER There is only one tangent line, $\ell: y - 3 = 4(x - 2)$.

Tangent lines to parabolas have an interesting property, which is stated in the next theorem.

Theorem 5–4 If the tangent to a parabola at a point Q intersects the axis of the parabola at W, and ℓ is the line perpendicular to that axis at the vertex, then Q and W are the same distance from ℓ.

PROOF Start by introducing a coordinate system so that the origin is the vertex of the parabola and the y-axis is its axis of symmetry. An equation of the parabola is $x^2 = 4cy$. Let Q be the point (a, b). Since Q is on the parabola, $a^2 = 4cb$. Hence Q is the point $\left(a, \dfrac{a^2}{4c}\right)$. Now use the four-step method to find the equation of tangent line ℓ.

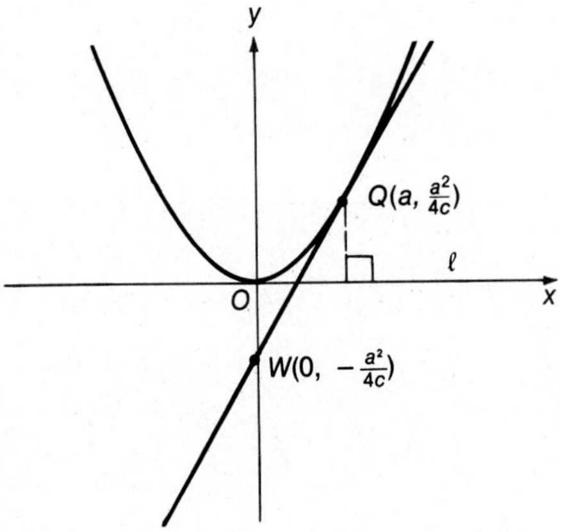

① $\ell: y - \dfrac{a^2}{4c} = m(x - a)$

$\qquad y = mx - ma + \dfrac{a^2}{4c}$

② $x^2 = 4c\left(mx - ma + \dfrac{a^2}{4c}\right)$

$\qquad x^2 - 4cmx + 4cma - a^2 = 0$

③ $D = (-4cm)^2 - 4(1)(4cma - a^2)$

$\qquad = 16c^2m^2 - 16cma + 4a^2$

$\qquad = (4cm - 2a)^2$

④ $4cm - 2a = 0$

$\qquad m = \dfrac{2a}{4c} = \dfrac{a}{2c}$

$\ell: y - \dfrac{a^2}{4c} = \dfrac{a}{2c}(x - a)$

$\qquad y = \dfrac{a}{2c}x - \dfrac{a^2}{4c}$

In the equation for ℓ, let $x = 0$ to find $y = -\dfrac{a^2}{4c}$, so that W is $\left(0, -\dfrac{a^2}{4c}\right)$. But Q is $\left(a, \dfrac{a^2}{4c}\right)$. So Q and W are equidistant from the line ℓ.

Exercises

A For what values of the constant k does the line $y = -x + k$ intersect each curve?

1. $x^2 + y^2 = 4$
2. $x^2 + y^2 = 9$
3. $x^2 = y$
4. $x^2 = -2y$

Find any line(s) that contain Q and are tangent to the given parabola.

5. $y = 2x^2$; $Q(1, 0)$
6. $y = -x^2$; $Q(0, 2)$
7. $x^2 + 2x - 4y = 3$; $Q(3, 3)$
8. $x^2 - 6x - 2y + 9 = 0$; $Q(4, -4)$
9. $(x - 2)^2 = 8(y + 1)$; $Q\left(-\frac{1}{2}, -\frac{1}{2}\right)$
10. $x^2 + 8y = 16$; $Q(3, 1)$
11. $y = x^2 + 1$; $Q(1, 3)$
12. $y = x^2 - 1$; $Q(1, 0)$

13. Given a parabola R. Where is point Q if there is no line that contains Q and is tangent to R?

14. Given a parabola R. For what points Q is there exactly one tangent to R that contains Q?

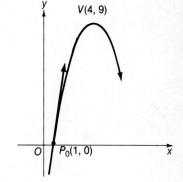

B 15. An object is projected into the air from point P_0 at time $t = 0$ and its position $P(x, y)$ later is given by $x = 1 + t$ and $y = 6t - t^2$. Find an equation of the line of sight, that is, of the tangent line through P_0.

16. Use the four-step method to find an equation of the line tangent to the graph of $y = \dfrac{1}{x - 1}$ at the point $(0, -1)$.

17. Use the four-step method to find an equation of the line(s) that contain $Q(1, -3)$ and are tangent to the graph of $f(x) = \dfrac{1}{x}$. Sketch the graph of f and of the tangent line(s).

18. Find equation(s) of the tangent line(s) from $(2, 0)$ to the parabola with vertex $(1, 0)$ and directrix $y = -\dfrac{1}{2}$.

19. Let Q be a point (r, s) with $s > r^2$. Show algebraically that there is no tangent to the parabola $y = x^2$ that contains Q.

C Two curves that intersect at a point P are said to be *orthogonal* if their tangents at the common point P are perpendicular. Sketch each pair of curves on the same axes, determine their points of intersection, and show algebraically that the curves are orthogonal at these points.

20. $x^2 + (y + 2)^2 = 17$
    $y = 2x^2$
21. $x^2 = 8y$
    $x^2 + (y + 2)^2 = 32$

22. Let the point $Q(a, b)$ be on the parabola $x^2 = 4(y - 1)$. Find, in terms of a, an equation of the tangent to the parabola that contains Q.

23. Find a radical-free equation for the locus of points $P(x, y)$ such that P is the center of a circle that is tangent both to the line $y = -3$ and to the circle $x^2 + y^2 = 4$.

24. Let $P(t, s)$ be any point on the graph of $f(x) = x^2$. Create a function g which relates the slope of the tangent at P to t and use this function g to find where on the graph the tangent has slope 5.

5-5 The Ellipse

The planet Earth travels in a plane around the sun, tracing out each year a path which is a conic section called an ellipse. We use a locus approach in the following definition of an ellipse.

Definition An **ellipse** is the set of all points in the plane, the sum of whose distances from two fixed points is a constant.

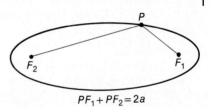

The mathematicians of ancient Greece worked out most of the properties of an ellipse, using Euclidean geometry. It is more convenient to use a coordinate approach and represent each ellipse algebraically. An ellipse has two axes of symmetry, the primary one being the line that contains the two fixed points, each of which is called a **focus**. We first write the equation of an ellipse when this line is horizontal.

Place one focus F_1 at $(c, 0)$, $c > 0$, and the other focus F_2 at $(-c, 0)$, and let $2a$ be a given constant such that $a > c$. Then the midpoint of $\overline{F_1 F_2}$, called the **center of the ellipse**, is at the origin. By definition, any point P such that $PF_1 + PF_2$ is $2a$ is on the ellipse.

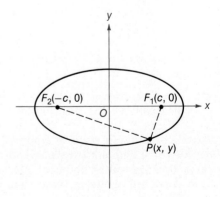

186 Chapter 5: Analytic Geometry—Conic Sections

Use the distance formula twice to find the locus of points P.

$$\sqrt{(x-c)^2 + y^2} + \sqrt{(x+c)^2 + y^2} = 2a$$
$$(x-c)^2 + y^2 = 4a^2 - 4a\sqrt{(x+c)^2 + y^2} + (x+c)^2 + y^2$$
$$a\sqrt{(x+c)^2 + y^2} = a^2 + cx$$
$$a^2(x^2 + 2cx + c^2 + y^2) = a^4 + 2a^2cx + c^2x^2$$
$$(a^2 - c^2)x^2 + a^2y^2 = a^4 - a^2c^2$$
$$= a^2(a^2 - c^2)$$
$$\frac{x^2}{a^2} + \frac{y^2}{a^2 - c^2} = 1.$$

Since $a^2 > c^2$, let $b^2 = a^2 - c^2$, with $b > 0$. Then the equation in standard form of an ellipse is

$$\frac{x^2}{a^2} + \frac{y^2}{b^2} = 1.$$

Since $b^2 = a^2 - c^2$, you know that $a > b$. When $y = 0$, $x^2 = a^2$ and the x-intercepts of the graph are a and $-a$. The points $V_1(a, 0)$ and $V_2(-a, 0)$ are the **vertices** of the ellipse and the line segment V_1V_2 is the **major axis**. When $x = 0$, $y^2 = b^2$ and the y-intercepts of the graph are $T_1(0, b)$ and $T_2(0, -b)$. The line segment T_1T_2 is the **minor axis**. Since $a > b$, the major axis is longer than the minor axis.

Theorem 5–5 An equation of the ellipse with center $(0, 0)$, foci $(\pm c, 0)$, and vertices $(\pm a, 0)$ is

$$\frac{x^2}{a^2} + \frac{y^2}{b^2} = 1,$$

where $b^2 = a^2 - c^2$.

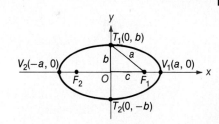

Example 1 Sketch the graph of $4x^2 + 9y^2 = 36$.

SOLUTION Divide both sides by 36 to write the equation in standard form.

$$\frac{x^2}{9} + \frac{y^2}{4} = 1$$

So $a^2 = 9$ and $b^2 = 4$ and the vertices are $(3, 0)$ and $(-3, 0)$; the endpoints of the minor axis are $(0, 2)$ and $(0, -2)$. Use $b^2 = a^2 - c^2$ to find that $4 = 9 - c^2$, or $c = \pm\sqrt{5}$. The foci are $(\sqrt{5}, 0)$ and $(-\sqrt{5}, 0)$.

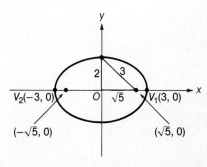

Example 2 Write an equation of the ellipse with vertices $(\pm 4, 0)$ and foci $(\pm 3, 0)$.

SOLUTION Substitute $a = 4$ and $c = 3$ in $b^2 = a^2 - c^2$ to find that $b^2 = 7$.

In standard form, the equation of the ellipse is

$$\frac{x^2}{16} + \frac{y^2}{7} = 1.$$

Sometimes it is convenient to place the major axis of the ellipse along the y-axis. Then the foci are also on the y-axis and have coordinates $(0, \pm c)$. The equation in standard form of this vertical ellipse is

$$\frac{x^2}{b^2} + \frac{y^2}{a^2} = 1,$$

where $a > b$ and $b^2 = a^2 - c^2$. The vertices are $V_1(0, a)$ and $V_2(0, -a)$. The endpoints of the minor axis are $(b, 0)$ and $(-b, 0)$.

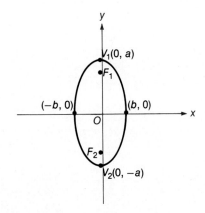

Example 3 Sketch the graph of $4x^2 + y^2 = 1$.

SOLUTION The given equation is equivalent to

$$\frac{x^2}{\frac{1}{4}} + \frac{y^2}{1} = 1.$$

This is the standard form of the equation of an ellipse with center at the origin and vertices on the y-axis. So $a^2 = 1$, $b^2 = \frac{1}{4}$, and $c = \frac{\sqrt{3}}{2}$.

The vertices are $(0, \pm 1)$; the foci are $\left(0, \pm \frac{\sqrt{3}}{2}\right)$, and the endpoints of the minor axis are $\left(\pm \frac{1}{2}, 0\right)$.

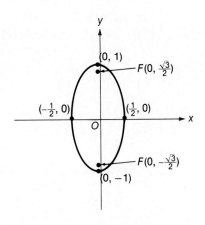

Example 4 Halley's comet orbits the sun in an elliptical orbit with the sun at one focus. It is about 90,000,000 km from the sun at its closest point (called *perihelion*). It is about 5,340,000,000 km from the sun at its furthest point (called *aphelion*). Calculate the lengths of the major and the minor axes of the orbit and the distance between the foci.

(cont. on p. 189)

SOLUTION The length of the major axis, $2a$, is the sum of the distances to perihelion and to aphelion.

$$2a = (5.34 \times 10^9) + (0.09 \times 10^9)$$
$$= 5.43 \times 10^9 \text{ km}$$
$$a \doteq 2.72 \times 10^9 \text{ km}$$

From the figure:
$$c = (2.72 \times 10^9) - (0.09 \times 10^9)$$
$$\doteq 2.63 \times 10^9$$

Now calculate b.
$$b^2 = a^2 - c^2$$
$$b^2 = [(2.72 \times 10^9)^2 - (2.63 \times 10^9)^2]$$
$$b = 0.694 \times 10^9 \text{ km}$$
$$2b \doteq 1.39 \times 10^9 \text{ km}$$

ANSWER The length of the major axis, $2a$, is 5.43×10^9 km.
The length of the minor axis, $2b$, is 1.39×10^9 km.
The distance between the foci, $2c$, is 5.26×10^9 km.

Exercises

A Write a radical-free equation for the locus of points (x, y) the sum of whose distances from the given points F_1 and F_2 is the constant $2a$.

1. $F_1 = (-1, 0), F_2 = (1, 0), 2a = 4$
2. $F_1 = (2, 0), F_2 = (-2, 0), 2a = 6$

Tell whether the major axis of the given ellipse is on the x-axis or the y-axis.

3. $\dfrac{x^2}{6.1} + \dfrac{y^2}{4.3} = 1$
4. $\dfrac{x^2}{3.7} + \dfrac{y^2}{3.5} = 1$
5. $4x^2 + 3y^2 = 1$
6. $5x^2 + 10y^2 = 1$

Sketch a graph of the ellipse and label the coordinates of foci and vertices.

7. $\dfrac{x^2}{25} + \dfrac{y^2}{9} = 1$
8. $\dfrac{x^2}{9} + \dfrac{y^2}{4} = 1$
9. $\dfrac{x^2}{4} + \dfrac{y^2}{9} = 1$
10. $\dfrac{x^2}{1} + \dfrac{y^2}{4} = 1$
11. $4x^2 + 3y^2 = 12$
12. $3x^2 + 4y^2 = 12$

Write an equation of the ellipse with the given characteristics.

13. Vertices $(\pm 3, 0)$; foci $(\pm \sqrt{5}, 0)$
14. Vertices $(\pm 5, 0)$; foci $(\pm 4, 0)$
15. Vertices $(0, \pm 4)$; minor axis length 2
16. Vertices $(0, \pm 6)$; foci $(0, \pm 4)$

B 17. Vertices $(\pm 4, 0)$; contains $(2, \pm \sqrt{3})$
18. Foci $(\pm 3, 0)$; contains $(0, \pm 2)$

Sketch the ellipse and label the coordinates of the foci.

19. $x^2 + 2y^2 = 1$ **20.** $4x^2 + 5y^2 = 10$

Find the coordinates of the two points of intersection of the ellipse $x^2 + 2y^2 = 2$ with

21. The line $y = x + 1$.   **22.** The parabola $y = x^2$.

23. Suppose that the bridge opening in example 2 of section 5–3 is elliptical, 120 feet wide, and 30 feet high. Can a barge 70 feet wide and 20 feet high pass under the bridge?

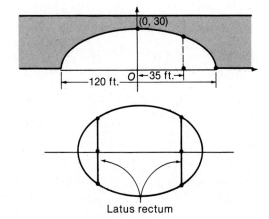

24. The *latus rectum* of an ellipse is a line segment through a focus perpendicular to the major axis that joins two points of the ellipse.

   a. Find the length of the latus rectum of the ellipse $4x^2 + 9y^2 = 36$.

   b. Find the length of the latus rectum of the ellipse $b^2x^2 + a^2y^2 = a^2b^2$, where $a > b$.

Latus rectum

C 25. Discuss the graph of the equation $mx^2 + y^2 = m$ in terms of the nonnegative constant m.

26. Find an equation of the ellipse with its center at the origin, its foci on the x-axis, containing (3, 3), and with the length of its major axis 3 times the length of its minor axis.

27. Write equations of lines that are parallel to $x - 3y = 25$ and tangent to the ellipse $x^2 + 3y^2 = 4$.

28. Write a radical-free equation of the locus of points $P(x, y)$ such that $PF_1 + PF_2 = 12$ where F_1 is $(4, -1)$ and F_2 is $(4, 7)$.

29. Let $P(x, y)$ be a point in the first quadrant of the ellipse $4x^2 + 9y^2 = 36$. If A is $(0, y)$, B is $(x, 0)$, and O is $(0, 0)$, define a function that relates the area of rectangle $AOBP$ to x. What is the domain of this function?

30. An arch for a bridge over a highway is to have the shape of half an ellipse as shown. There are to be four traffic lanes each 4 meters wide and a median strip 4 meters wide. If the arch opening has a maximum height of 7 meters and a minimum clearance of 4 meters at the sides of the highway, how wide must the arch be?

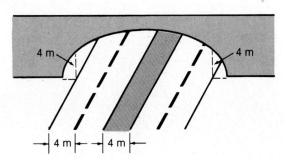

31. Prove Theorem 5–5.

32. Suppose line ℓ has equation $x = \dfrac{a^2}{c}$, $c > 0$. For the ellipse $\dfrac{x^2}{a^2} + \dfrac{y^2}{b^2} = 1$ with a focus F at $(c, 0)$, the vertical line ℓ is called a *directrix*. Let the distance from point P on the ellipse to the line ℓ be $P\ell$. Prove that the ratio of distances $\dfrac{PF}{P\ell}$ is equal to the constant $\dfrac{c}{a}$.

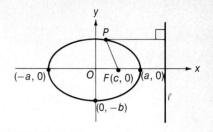

5–6 The Hyperbola

Gas in a container at a constant temperature obeys the gas law $PV = c$, where c is a constant. The graph of pressure P versus volume V is a hyperbola, the last type of conic section. Again, we use a locus approach in the following definition.

Definition A **hyperbola** is the set of all points in the plane the difference of whose distances from two fixed points is a constant.

Choose a coordinate system such that the two fixed points, the foci, are $F_1(c, 0)$ and $F_2(-c, 0)$, $c > 0$. Represent the constant in the definition by $2a$. Then a hyperbola is the locus of points $P(x, y)$ such that $|PF_1 - PF_2| = 2a$. Suppose that $PF_2 > PF_1$.

By the triangle inequality,

$PF_2 < PF_1 + F_1F_2$.

Therefore, $PF_2 - PF_1 < F_1F_2$,

$2a < 2c$, and

$a < c$.

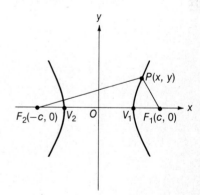

So $a < c$. If $PF_1 > PF_2$, you can verify that the result is the same. Now use the distance formula to write an equation of a hyperbola.

$|\sqrt{(x + c)^2 + y^2} - \sqrt{(x - c)^2 + y^2}| = 2a$.

If $PF_1 > PF_2$, this equation becomes

$\sqrt{(x - c)^2 + y^2} - \sqrt{(x + c)^2 + y^2} = 2a$.

We omit the algebra required to show that squaring and simplifying yields

$$\frac{x^2}{a^2} - \frac{y^2}{c^2 - a^2} = 1.$$

Since $c > a > 0$, $c^2 - a^2 > 0$. The substitution $b^2 = c^2 - a^2$ yields the **standard form** of the equation of a hyperbola.

$$\frac{x^2}{a^2} - \frac{y^2}{b^2} = 1$$

If $y = 0$, then $x^2 = a^2$ and the graph intersects the x-axis at a and $-a$. The points $V_1(a, 0)$ and $V_2(-a, 0)$ are the **vertices** of the hyperbola. The segment $V_1 V_2$ is called the **transverse axis**. The midpoint of $V_1 V_2$ is the **center** of the hyperbola. There are no y-intercepts since $x = 0$ implies $-\frac{y^2}{b^2} = 1$, which has no real solutions. Nevertheless, the line segment that joins $(0, b)$ and $(0, -b)$ is called the **conjugate axis**. This proves the following theorem.

Theorem 5–6 An equation of the hyperbola with center $(0, 0)$, foci $(\pm c, 0)$, and vertices $(\pm a, 0)$ is

$$\frac{x^2}{a^2} - \frac{y^2}{b^2} = 1,$$

where $a^2 + b^2 = c^2$.

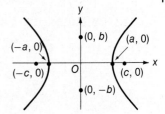

Graphing hyperbolas is made easier by noting symmetries in the equation in standard form. Since replacing x by $-x$ does not change the equation, the graph is symmetric about the y-axis. A similar argument on y-values shows that the graph is symmetric about the x-axis. Hence you need only consider points in quadrant I in order to sketch the entire graph.

Example 1 Sketch the graph of $\frac{x^2}{16} - \frac{y^2}{9} = 1$.

SOLUTION Solve the equation in standard form for y.

$$y = \tfrac{3}{4}\sqrt{x^2 - 16}$$

Clearly $x^2 - 16 \geq 0$ so that $x \geq 4$ in quadrant I.

x	4	5	6	7	8	9	10
y	0	2.25	3.35	4.31	5.20	6.05	6.87

Now sketch the graph using symmetries.

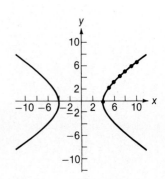

A useful property of hyperbolas for graphing is presented in the following theorem.

Theorem 5–7 As $|x|$ gets larger, the hyperbola $\dfrac{x^2}{a^2} - \dfrac{y^2}{b^2} = 1$ gets closer and closer to the lines $y = \pm \dfrac{bx}{a}$.

PROOF The equation in standard form is equivalent to

$$y = \pm \frac{bx}{a} \sqrt{1 - \frac{a^2}{x^2}}.$$

As $|x|$ gets large, $\dfrac{a^2}{x^2}$ gets very close to 0 and $1 - \dfrac{a^2}{x^2}$ gets close to 1. Thus y gets close to $\pm \dfrac{bx}{a}$, as was to be verified.

The lines $y = \pm \dfrac{bx}{a}$ are **asymptotes** of a hyperbola and are used in graphing as follows. The rectangle with vertices (a, b), $(a, -b)$, $(-a, -b)$, and $(-a, b)$ is called the **auxiliary rectangle**. Lines through its opposite vertices are the asymptotes of the hyperbola. Constructing the auxiliary rectangle helps locate the foci. Since $a^2 + b^2 = c^2$, half the length of a diagonal of the auxiliary rectangle is c. Therefore a circle with radius c and center at the origin intersects the x-axis at the foci, F_1 and F_2.

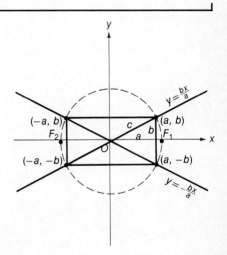

Example 2 Sketch the hyperbola $2x^2 - y^2 = 2$ and locate its foci.

SOLUTION Divide both sides of the given equation by 2 to write it in standard form.

$$\frac{x^2}{1} - \frac{y^2}{2} = 1$$

So $a^2 = 1$, $b^2 = 2$, $a = 1$, and $b = \sqrt{2}$. Draw the auxiliary rectangle, and the asymptotes. Sketch the hyperbola.

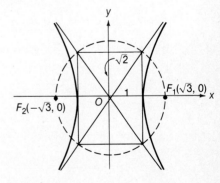

Notice that the hyperbola is tangent to the sides of the auxiliary rectangle at its vertices. Since $a^2 + b^2 = 3$, $c = \sqrt{3}$, and the foci are $(\pm\sqrt{3}, 0)$.

Example 3 Write an equation of the hyperbola with center at the origin, foci on the x-axis, asymptotes $y = \pm x$, and that contains (3, 1).

SOLUTION From the equation of the asymptotes, $a = b$, and therefore an equation of the hyperbola in standard form is

$$\frac{x^2}{a^2} - \frac{y^2}{a^2} = 1.$$

Since (3, 1) lies on the hyperbola,

$$\frac{9}{a^2} - \frac{1}{a^2} = 1,$$

and $a^2 = 8$. The equation of the hyperbola is then

$$\frac{x^2}{8} - \frac{y^2}{8} = 1.$$

Recall that for ellipses, $\frac{x^2}{a^2} + \frac{y^2}{b^2} = 1$, the relative sizes of the positive numbers a and b determine which is the major axis, as well as the shape of the ellipse. On the other hand, all hyperbolas with standard equation $\frac{x^2}{a^2} - \frac{y^2}{b^2} = 1$ have their foci on the x-axis regardless of the relative size of a and b. Reversing the roles of x and y in the equation yields

$$\frac{y^2}{a^2} - \frac{x^2}{b^2} = 1,$$

which is the equation in standard form of the original hyperbola rotated 90°. The transverse axis is now vertical and the graph opens upward and downward. This situation is summarized in the next theorem. The proof is left as an exercise.

Theorem 5–8 An equation of the hyperbola with center (0, 0), foci (0, $\pm c$), and vertices (0, $\pm a$) is

$$\frac{y^2}{a^2} - \frac{x^2}{b^2} = 1,$$

where $a^2 + b^2 = c^2$. The asymptotes of the hyperbola are the lines

$$y = \pm \frac{a}{b}x.$$

Example 4 Sketch the graph of the hyperbola $x^2 - 4y^2 + 2 = 0$. Label its asymptotes, vertices, and foci.

SOLUTION Rewrite the equation of the hyperbola as

$$2y^2 - \frac{x^2}{2} = 1,$$

and then as

$$\frac{y^2}{\frac{1}{2}} - \frac{x^2}{2} = 1.$$

Since the y^2-term has the positive coefficient, this hyperbola opens upward and downward and Theorem 5–8 applies. From $a^2 = \frac{1}{2}$ and $b^2 = 2$, compute $a = \frac{\sqrt{2}}{2}$, $b = \sqrt{2}$, and $c = \frac{\sqrt{10}}{2}$.
The sketch is shown at the right.

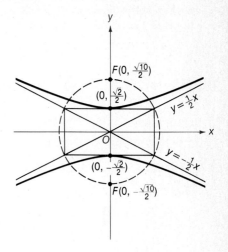

Example 5 Write a radical-free equation for the set of points (x, y) the difference of whose distances from $(0, 3)$ and $(0, -3)$ is always 4.

SOLUTION By definition, the set of points is a hyperbola. Its center is $(0, 0)$ and $c = 3$, since the fixed points are known to be the foci $(0, \pm c)$. Likewise you know that $2a = 4$, since the fixed distance was represented by $2a$ in the algebraic development of the equation of a hyperbola. So $a = 2$, and from $a^2 + b^2 = c^2$ you find $b^2 = 5$. Since the transverse axis is vertical, apply Theorem 5–8 to write an equation of the hyperbola as

$$\frac{y^2}{4} - \frac{x^2}{5} = 1.$$

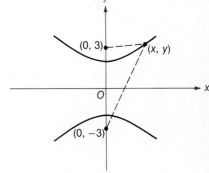

Exercises

A Write a radical-free equation for the locus of points (x, y) the difference of whose distances from the given points F_1 and F_2 is the constant $2a$.

1. $F_1 = (4, 0)$, $F_2 = (-4, 0)$, $2a = 2$ **2.** $F_1 = (-3, 0)$, $F_2 = (3, 0)$, $2a = 4$

Tell whether the transverse axis is on the x-axis or the y-axis.

3. $\dfrac{x^2}{4.1} - \dfrac{y^2}{9.2} = 1$ **4.** $\dfrac{x^2}{43} - \dfrac{y^2}{44} = 1$

5. $35x^2 - 36y^2 - 1 = 0$ **6.** $7x^2 - 6y^2 + 42 = 0$

Sketch a graph of the hyperbola by first drawing the auxiliary rectangle and the asymptotes. Label the coordinates of the foci and the vertices.

7. $\dfrac{x^2}{4} - \dfrac{y^2}{9} = 1$ **8.** $\dfrac{x^2}{9} - \dfrac{y^2}{4} = 1$ **9.** $\dfrac{y^2}{4} - \dfrac{x^2}{1} = 1$ **10.** $\dfrac{y^2}{1} - \dfrac{x^2}{4} = 1$

11. $4x^2 - 3y^2 = 12$ **12.** $4y^2 - 3x^2 = 12$ **13.** $2y^2 - x^2 = 1$ **14.** $x^2 - 2y^2 = 1$

Write an equation of the hyperbola with the given characteristics.

15. Foci $(\pm 4, 0)$, vertices $(\pm 3, 0)$ **16.** Foci $(0, \pm 4)$, vertices $(0, \pm 1)$

17. Vertices $(\pm 3, 0)$, contains $(5, 4)$ **18.** Vertices $(\pm 4, 0)$, contains $(8, 2)$

19. Asymptotes $y = \pm 2x$, foci on the y-axis, contains $(1, 4)$

20. Asymptotes $y = \pm \sqrt{3}x$, contains $(\sqrt{6}, 3)$, foci on the x-axis

B 21. Do the work to show that $\sqrt{(x-c)^2 + y^2} - \sqrt{(x+c)^2 + y^2} = 2a$ simplifies to $\dfrac{x^2}{a^2} - \dfrac{y^2}{c^2 - a^2} = 1$, as asserted at the beginning of this section.

Find the foci and asymptotes of each hyperbola and sketch its graph.

22. $\dfrac{y^2}{3} - \dfrac{x^2}{6} = 1$ **23.** $\dfrac{y^2}{6} - \dfrac{x^2}{3} = 1$ **24.** $4x^2 - y^2 = 2$ **25.** $2x^2 - 4y^2 = 1$

26. Find the coordinates of the two points of intersection of the hyperbola $2x^2 - y^2 = 8$ with the line $y = x - 1$.

27. For what values of m does the line $y = mx$ intersect the hyperbola $4x^2 - y^2 + 4 = 0$?

28. For what values of m does $mx^2 - y^2 = m$ describe a hyperbola?

C 29. The graphs of the equations

$$\dfrac{x^2}{a^2} - \dfrac{y^2}{b^2} = 1 \quad \text{and} \quad \dfrac{y^2}{b^2} - \dfrac{x^2}{a^2} = 1$$

are called *conjugate hyperbolas* since the transverse axis of one is the conjugate axis of the other and vice versa. Sketch graphs of the two equations when $a = 3$ and $b = 5$.

30. Prove Theorem 5–8 by following the steps for the proof of Theorem 5–6.

31. Prove that the product of the distances from any point on a hyperbola to the two asymptotes is a constant.

5-7 Translation of Axes

The equation of a circle with center at the origin is $x^2 + y^2 = r^2$. If its center is translated to the point (h, k), you need only replace x by $x - h$ and y by $y - k$ to obtain the equation in standard form,

$$(x - h)^2 + (y - k)^2 = r^2.$$

Similarly, translating the vertex of a parabola from $(0, 0)$ to (h, k) changes the standard form of the equation from $x^2 = 4cy$ to

$$(x - h)^2 = 4c(y - k).$$

The result of this translation for an ellipse is as follows.

Theorem 5-9

Equation	$\dfrac{(x-h)^2}{a^2} + \dfrac{(y-k)^2}{b^2} = 1$	$\dfrac{(x-h)^2}{b^2} + \dfrac{(y-k)^2}{a^2} = 1$	
Center	(h, k)	(h, k)	
Transverse Axis	Parallel to x-axis	Parallel to y-axis	
Foci	$(h \pm c, k)$	$(h, k \pm c)$	$[a^2 = b^2 + c^2]$

The equation $\dfrac{(x-1)^2}{9} + \dfrac{(y+2)^2}{4} = 1$

represents an ellipse with major axis horizontal and center $(1, -2)$. Since $c^2 = 9 - 4 = 5$, and thus $c = \sqrt{5}$, the foci are $\sqrt{5}$ units to the left and to the right of the center, as shown in the figure.

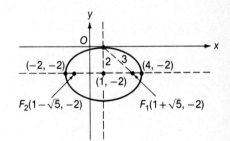

For hyperbolas, the translation has the following result.

Theorem 5-10

Equation	$\dfrac{(x-h)^2}{a^2} - \dfrac{(y-k)^2}{b^2} = 1$	$\dfrac{(y-k)^2}{a^2} - \dfrac{(x-h)^2}{b^2} = 1$	
Center	(h, k)	(h, k)	
Transverse Axis	Parallel to x-axis	Parallel to y-axis	
Foci	$(h \pm c, k)$	$(h, k \pm c)$	$[c^2 = a^2 + b^2]$
Asymptotes	$y - k = \pm \dfrac{b}{a}(x - h)$	$y - k = \pm \dfrac{a}{b}(x - h)$	

The equation $\dfrac{(x-1)^2}{9} - \dfrac{(y-2)^2}{16} = 1$ represents a hyperbola with a horizontal transverse axis and center (1, 2). To sketch this hyperbola it is easiest to work from dashed lines drawn parallel to the coordinate axes through the center (1, 2). Label these lines x'- and y'-axes. Draw the auxiliary rectangle relative to the x'- and y'-axes, sketch the asymptotes through its opposite vertices, and finally sketch the hyperbola. Since $c^2 = 9 + 16 = 25$, the foci are 5 units to the left and to the right of the center (1, 2). The equation in standard form of this hyperbola with respect to x'- and y'-axes is

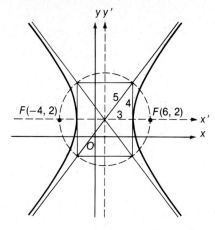

$$\dfrac{x'^2}{9} - \dfrac{y'^2}{16} = 1.$$

This equation and the original equation both describe the same hyperbola, but with respect to different axes.

Now consider the translated equations for the three classes of conic sections.

Parabola $(x - h)^2 = 4c(y - k)$ and $(y - k)^2 = 4c(x - h)$

Ellipse $\dfrac{(x-h)^2}{a^2} + \dfrac{(y-k)^2}{b^2} = 1$ and $\dfrac{(x-h)^2}{b^2} + \dfrac{(y-k)^2}{a^2} = 1$

Hyperbola $\dfrac{(x-h)^2}{a^2} - \dfrac{(y-k)^2}{b^2} = 1$ and $\dfrac{(y-k)^2}{a^2} - \dfrac{(x-h)^2}{b^2} = 1$

In every case, multiplying by a lowest common denominator and collecting terms on the left side yields an equation of the form

$$Ax^2 + Cy^2 + Dx + Ey + F = 0.$$

For parabolas, either A or C is zero but not both, so that $AC = 0$.
For ellipses, A and C have the same sign, so that $AC > 0$.
For hyperbolas, A and C have opposite signs, so that $AC < 0$.

Thus the equation represents each of the translated standard conic sections. But it also describes other so-called degenerate conic sections as well. For instance, $2x^2 + 3y^2 = 0$ describes the point (0, 0), which is not a conic section; $x^2 - 3y^2 = 0$ describes two lines which are the asymptotes of a hyperbola and not a conic section.

Careful analysis of all cases like these leads to the following theorem.

Theorem 5–11 If the equation $Ax^2 + Cy^2 + Dx + Ey + F = 0$ describes a curve other than a point, line, or pair of lines, then that curve is a conic section and its graph is

a. A parabola if $AC = 0$;

b. An ellipse if $AC > 0$ (a circle if $A = C$);

c. A hyperbola if $AC < 0$.

Example 1 Show that $5x^2 + 4y^2 - 16y - 4 = 0$ describes an ellipse. Find its center and its vertices.

SOLUTION Since $AC = 20 > 0$, the curve represents an ellipse if it is a conic section.

$$5x^2 + 4y^2 - 16y - 4 = 0$$
$$5x^2 + 4(y^2 - 4y \quad) - 4 = 0$$
$$5x^2 + 4(y - 2)^2 = 20$$
$$\frac{x^2}{4} + \frac{(y - 2)^2}{5} = 1$$

This is the equation in standard form of an ellipse with center $(0, 2)$. Since the y-term has the greater denominator, the major axis is vertical. Hence $a = \sqrt{5}$ and the vertices are $\sqrt{5}$ units above and below the center, as shown in the figure.

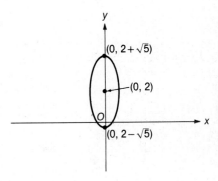

Example 2 Graph $\{(x, y): 4x^2 - 9y^2 + 8x + 36y + 4 = 0\}$.

SOLUTION Since $AC = -36 < 0$, the curve represents a hyperbola if it is a conic section.

$$4x^2 - 9y^2 + 8x + 36y + 4 = 0$$
$$4(x^2 + 2x \quad) - 9(y^2 - 4y \quad) = -4$$
$$4(x + 1)^2 - 9(y - 2)^2 = -4 + 4 - 36 = -36$$
$$\frac{(y - 2)^2}{4} - \frac{(x + 1)^2}{9} = 1$$

This is the equation in standard form of a hyperbola with center $(-1, 2)$ and vertical transverse axis. Use the fact that $a = 2$ and $b = 3$ to draw the auxiliary rectangle and the asymptotes as aids in sketching the hyperbola.

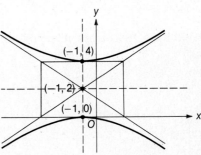

Example 3 The equation $x^2 - my^2 + 4y - 8 = 0$ represents a nondegenerate conic section. For what values of the constant m is it a parabola? An ellipse? A hyperbola?

SOLUTION Since the equation describes a conic section and $AC = -m$, use Theorem 5–11 to conclude that the curve is

a parabola if $m = 0$,
an ellipse if $m < 0$, and
a hyperbola if $m > 0$.

If the term Bxy appears in the general equation of a conic section, so that the equation is

$$Ax^2 + Bxy + Cy^2 + Dx + Ey + F = 0,$$

then the two foci lie on a line that is not parallel to the x-axis or the y-axis.

Exercises

A Write the given equation in the form $Ax^2 + Cy^2 + Dx + Ey + F = 0$, find the product AC, and identify the conic section.

1. $\dfrac{(x-1)^2}{4} - \dfrac{(y-3)^2}{5} = 1$

2. $\dfrac{(x+1)^2}{3} + \dfrac{(y-1)^2}{5} = 1$

3. $(x-1)^2 = -4(y+2)$

4. $(y+2)^2 = 8(x-1)$

Graph the conic section with the given equation. Specify the coordinates of the center, vertices, and foci.

5. $\dfrac{(x-1)^2}{9} + \dfrac{(y+2)^2}{4} = 1$

6. $\dfrac{(x+2)^2}{1} + \dfrac{(y-3)^2}{4} = 1$

7. $\dfrac{(x-1)^2}{9} - \dfrac{(y+2)^2}{4} = 1$

8. $\dfrac{(x+2)^2}{1} - \dfrac{(y-3)^2}{4} = 13$

9. $4(y+2)^2 - 9(x-1)^2 = 36$

10. $4(y-3)^2 - (x+2)^2 = 1$

11. $x^2 + 4y^2 + 4x - 8y + 4 = 0$

12. $4x^2 + y^2 + 16x - 2y + 13 = 0$

Find the vertex or vertices of the conic section described by the equation.

13. $x^2 - 2x + 4y - 3 = 0$

14. $y^2 + 4x + 6y + 17 = 0$

15. $4x^2 - y^2 + 8x + 2y + 1 = 0$

16. $4x + 2y^2 - 4x + 4y + 2 = 0$

B For what values of the constant m does the equation describe a parabola?

17. $3y^2 - mx + 6y = 0$

18. $2x^2 - mx + 2y - 1 = 0$

19. $mx^2 + (m + 1)y + 1 = 0$

20. $mx^2 + 2my^2 + 4x = 0$

Write an equation in standard form of the given conic section.

21. The sum of the distances from (x, y) to $(-2, 0)$ and $(4, 0)$ is 8.

22. The difference of the distances of (x, y) from $(-1, 2)$ and from $(5, 2)$ is 2.

23. The ellipse with vertices $(6, 0)$ and $(-4, 0)$ that contains $(5, \frac{9}{5})$.

24. The hyperbola with $y = x$ and $y = -x + 2$ as asymptotes that contains $(3, 1)$.

25. The hyperbola with vertices $(-2, 3)$ and $(-2, 5)$ and a focus at $(-2, 6)$.

26. The ellipse with vertices $(5, -1)$ and $(10, -1)$ and a focus at $(6, -1)$.

27. For what values of the constant m does the equation $2x^2 - 2y^2 - 4x - 4y - m = 0$ describe a hyperbola?

For what values of the constant m does the equation $mx^2 + y^2 - 4x = 0$ describe

28. A parabola?

29. An ellipse with major axis parallel to the x-axis?

30. An ellipse with major axis parallel to the y-axis?

31. A hyperbola?

C 32. Under what conditions on the constants A, C, and F does $Ax^2 + Cy^2 + F = 0$ describe an ellipse or a circle?

33. Under what conditions on the constants A, C, and F does $Ax^2 + Cy^2 + F = 0$ describe a hyperbola?

Analyze the equation and discuss its graph for all values of m.

34. $x^2 + my^2 - 2my = 0$

35. $x^2 + my^2 - 2my - 1 = 0$

36. Show that the line $y = mx + \sqrt{4m^2 - 9}$ intersects the hyperbola $9x^2 - 4y^2 = 36$ at exactly one point if $4m^2 > 9$.

37. For the ellipse $2x^2 + y^2 - 4x - 6 = 0$, find the line through $(0, \sqrt{6})$ that is tangent to the ellipse.

5–8 Eccentricity and Conic Sections

In a Special near the beginning of this chapter, the conic sections were described geometrically in terms of the cone. In the text we chose to define the conic sections in terms of the locus of points since this approach led more easily to algebraic equations. We now prove a theorem that shows that the conic sections could have been defined algebraically all at once. (Doing this at the outset would, however, have made writing the equations more difficult.)

Theorem 5–12 Let F be a fixed point, ℓ a fixed line not containing F, $P\ell$ the distance between the point P and the line ℓ, and $e > 0$ a constant.

Then $\left\{ P : \dfrac{PF}{P\ell} = e \right\}$ is **a.** a parabola if $e = 1$;
 b. an ellipse if $0 < e < 1$; or
 c. a hyperbola if $e > 1$.

PROOF For any given focus F and directrix ℓ, there is no loss in generality by drawing the x-axis through the focus perpendicular to the directrix. We are free to place the origin anywhere on the x-axis; so place it conveniently at the focus. Then the coordinates of the focus are $(0, 0)$. Determine the scale by calling the distance from $F(0, 0)$ to the directrix d units so that the equation of the directrix ℓ is then $x = d$. For points $P(x, y)$, $F(0, 0)$, and line ℓ,

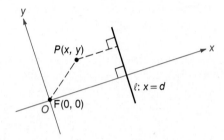

$$PF = \sqrt{x^2 + y^2} \quad \text{and} \quad P\ell = |x - d|.$$

So the locus statement $PF = e(P\ell)$ becomes

$$\sqrt{x^2 + y^2} = e|x - d|.$$
$$x^2 + y^2 = e^2 x^2 - 2de^2 x + e^2 d^2$$
$$(1 - e^2)x^2 + y^2 + 2de^2 x - e^2 d^2 = 0$$

This equation is in the form $Ax^2 + Cy^2 + Dx + Ey + F = 0$, and the product AC is $1 - e^2$. By Theorem 5–11, this equation represents

a. a parabola if $1 - e^2 = 0$, or $e = 1$.
b. an ellipse if $1 - e^2 > 0$, or $e < 1$;
c. a hyperbola if $1 - e^2 < 0$, or $e > 1$.

So you can distinguish between the conic sections by the value of e, which is called the **eccentricity**.

The value of e determines the "roundness" of an ellipse. Look at the first two terms of the last equation, namely, $(1 - e^2)x^2 + y^2$. As e approaches zero, their sum approaches $x^2 + y^2$ and the shape of the ellipse approaches the shape of a circle. As e approaches 1, the x^2-term becomes very small and the ellipse is "stretched out."

Example 1 Write in standard form the equation of the hyperbola with directrix $\ell: x = 1$, focus $F(2, 0)$, and eccentricity $e = \sqrt{2}$.

SOLUTION For points $P(x, y)$,
$$PF = \sqrt{(x-2)^2 + y^2} \text{ and}$$
$$P\ell = |x - 1|.$$

Substitute in $PF = e(P\ell)$ and simplify.
$$(x - 2)^2 + y^2 = 2(x - 1)^2$$
$$x^2 - 4x + 4 + y^2 = 2x^2 - 4x + 2$$
$$\frac{x^2}{2} - \frac{y^2}{2} = 1$$

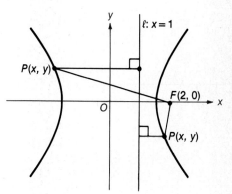

In exercise 13 you are asked to show that using the focus $(-2, 0)$ and the other directrix $\ell: x = -1$ yields the same equation.

If the directrix is not parallel to a coordinate axis, you must use the formula for the distance from a point to a line to find $P\ell$.

Example 2 Write a radical-free equation for the hyperbola with focus $(\sqrt{2}, \sqrt{2})$, directrix $\ell: y = -x + \sqrt{2}$, and eccentricity $\sqrt{2}$.

SOLUTION Locate the focus and the directrix, and estimate the location of a few points on the hyperbola before writing the equation. For $P(x, y)$,
$$PF = \sqrt{(x - \sqrt{2})^2 + (y - \sqrt{2})^2}.$$

Rewrite $y = -x + \sqrt{2}$ as $x + y - \sqrt{2} = 0$. Then use the formula for the distance from a point to a line to write
$$P\ell = \frac{|x + y - \sqrt{2}|}{\sqrt{2}}.$$

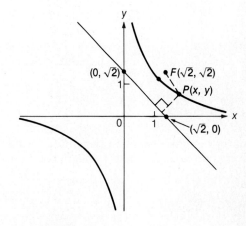

Substitute this expression in $PF = e(P\ell)$ and simplify.
$$x^2 - 2\sqrt{2}x + 2 + y^2 - 2\sqrt{2}y + 2 = 2\left(\frac{x + y - \sqrt{2}}{\sqrt{2}}\right)^2$$

which simplifies to $xy = 1$.

The hyperbola in example 2, $xy = 1$, does not have an equation in standard form since its transverse axis is not parallel to either coordinate axis. Nevertheless, since $e > 1$, the equation $xy = 1$ does, in fact, represent a hyperbola.

Recall that the equation $Ax^2 + Cy^2 + Dx + Ey + F = 0$ has no term with coefficient B. Now you can see that the term Bxy appears when the axes of the conic section are not parallel to the coordinate axes. In example 2, a standard hyperbola has been rotated 45° so that its transverse axis makes a 45° angle with the x-axis. Whenever such rotations are made, the term Bxy appears.

Hence the general equation for all conic sections is

$$Ax^2 + Bxy + Cy^2 + Dx + Ey + F = 0.$$

For the equation $xy = 1$, $B = 1$, $F = -1$, and all the other coefficients are zero. For this hyperbola, $(1, 1)$ is a vertex and the coordinate axes are the asymptotes. If you select this vertex, the focus $(\sqrt{2}, \sqrt{2})$, the directrix, the asymptotes, and the graph and rotate them all 45°, you will get the hyperbola in example 1 which has the equation

$$\frac{x^2}{2} - \frac{y^2}{2} = 1.$$

The following theorem is a natural extension of Theorem 5–11. The proof is omitted.

Theorem 5–13 Any conic section represented by the equation

$$Ax^2 + Bxy + Cy^2 + Dx + Ey + F = 0$$

is **a.** a parabola if $B^2 - 4AC = 0$;
b. an ellipse or a circle if $B^2 - 4AC < 0$; or
c. a hyperbola if $B^2 - 4AC > 0$.

Exercises

A For the given equation, compute $B^2 - 4AC$ and determine whether the conic section described by the equation is a parabola, an ellipse, or a hyperbola.

1. $x^2 - 4xy + 4y^2 + 4x + 2y - 1 = 0$
2. $-3x^2 - 8xy + 3y^2 + 8x + 4y - 2 = 0$
3. $3x^2 + 2xy + 3y^2 + 2x + 2y - 4 = 0$
4. $9x^2 - 4xy + 6y^2 + 2x + 4y - 1 = 0$

Sketch the curve and write an equation of the conic section with the given focus, directrix, and eccentricity.

5. $F(0, 0)$; $\ell: x = 1$; $e = 2$
6. $F(0, 0)$; $\ell: x = -2$; $e = \frac{1}{2}$
7. $F(1, 1)$; $\ell: y = 2$; $e = 1$
8. $F(1, -2)$; $\ell: y = -1$; $e = \sqrt{2}$

B 9. $F(1, 2)$; $\ell: x + y - 1 = 0$; $e = \frac{1}{2}$ 10. $F(2, -1)$; $\ell: 2x + y + 1 = 0$; $e = 1$
11. $F(1, -1)$; $\ell: 2x - y + 1 = 0$; $e = \sqrt{2}$ 12. $F(0, 1)$; $\ell: x + y + 2 = 0$; $e = 2$

13. Show that the equation of the conic section with focus $(-2, 0)$, directrix $x = -1$, and eccentricity $\sqrt{2}$ is the same as that in example 1.

14. For a specific conic section, a focus F is 2 units away from the directrix ℓ and the eccentricity is $\frac{1}{2}$. Place F and ℓ on convenient coordinate axes and write an equation for the conic section.

C 15. For a conic section with focus $F(0, 0)$, directrix $\ell: x = d$, and eccentricity e, you know that its equation is

$$(1 - e^2)x^2 + y^2 + 2de^2x - e^2d^2 = 0.$$

If this equation represents the hyperbola $\dfrac{(x - h)^2}{a^2} - \dfrac{y^2}{b^2} = 1$ where $c^2 = a^2 + b^2$, find c in terms of e and d and prove that $e = \dfrac{c}{a}$.

Chapter Review Exercises

5–1, page 167

1. Write a radical-free equation for the locus of points $P(x, y)$ where the distance from P to $(1, 2)$ is twice the distance from P to $(3, -6)$.

2. Given points $A(-1, 2)$ and $B(2, 3)$. Write a radical-free equation for the locus of points $P(x, y)$ such that segments PA and PB are perpendicular.

5–2, page 171

For each parabola determine the vertex, the focus, and the directrix. Sketch the graph of each equation.

3. $(x - 2)^2 = -8(y + 1)$ 4. $y^2 = 8x$
5. $x^2 + 4x - 2y + 5 = 0$ 6. $x^2 + 4y = 4$

7. Write an equation of the vertical parabola that opens upward, has its vertex at $(3, 0)$, and contains $(1, 1)$.

5–3, page 177

8. Given a parabolic mirror as pictured, find the distance from the vertex of the mirror to the focus.

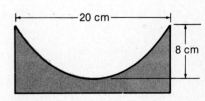

9. The position $P(x, y)$ of a particle at time t is given by the parametric equations $y = 4t^2 + 1$ and $x = t + 1$. Show that the particle is traveling along a parabola and determine its equation in x and y. Find y when $x = 9.4$ units.

10. A quonset hut has a parabolic cross section. If the width at ground level is 24 feet and the maximum height is 16 feet, what is the height of the hut 3 feet in from the base of an outside wall?

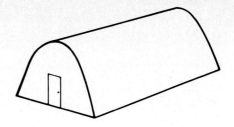

5–4, page 181

11. For what values of k does the line $y - 3x + k = 0$ fail to intersect the parabola $y = x^2 + 1$?

12. Write equations for the tangents from $(3, 5)$ to the parabola $y = x^2$.

5–5, page 186

13. Given points $A(-1, 0)$ and $B(3, 0)$. Write a radical-free equation for the locus of points $P(x, y)$ such that $PA + PB = 5$ and find x when $y = 2$.

14. Sketch a graph of the ellipse $9x^2 + y^2 = 9$ and label the coordinates of the foci and the vertices.

15. Write an equation of the ellipse with vertices $(\pm 3, 0)$ and foci $(\pm 2, 0)$.

5–6, page 191

16. Sketch a graph of the hyperbola $4x^2 - 9y^2 = 36$ and label the coordinates of its foci.

17. Write an equation of the hyperbola with foci $(0, \pm 4)$ and vertices $(0, \pm 3)$.

5–7, page 197

Find the vertex, or vertices, and focus, or foci, of the conic section described by each equation.

18. $x^2 + 4y^2 - 2x + 8y + 1 = 0$

19. $4x^2 - y^2 + 8x + 4y - 4 = 0$

20. $9y^2 - 4x^2 = 36y$

21. $1 - x^2 = 2(x + y)$

22. For what values of m does the equation $x^2 - 2x - m^2 = (m - 1)y^2$ describe an ellipse?

5–8, page 202

Compute $B^2 - 4AC$ for each equation and determine what conic section is described.

23. $x^2 - 4xy + 12y^2 - 1 = 0$

24. $x^2 + 2xy + y^2 - 1 = 0$

For the conic section with the given focus, directrix, and eccentricity, write its equation in the general form.

25. $F(2, 1);\ \ell: x = 4;\ e = \dfrac{1}{\sqrt{2}}$

26. $F(1, -1);\ \ell: y = 2x;\ e = 2$

Chapter Test

1. Write a radical-free equation for the locus of points $P(x, y)$ such that for $A(1, 3)$, $PA = 5$.

2. What is the focus of the parabola $x^2 - 2x = -4y - 5$?

3. Does the equation $3x^2 - 2xy + 3y^2 + 2x + 2y = 0$ describe an ellipse, a parabola, or a hyperbola?

4. Write a radical-free equation for the parabola with focus $F(1, 2)$ and directrix $\ell: y = 2x - 1$.

5. Write an equation of the tangent to the parabola $x^2 - 2x = y$ at the point $(-1, 3)$.

Sketch the given conic section and label the coordinates of its foci and vertices.

6. $4y^2 - x^2 = 4$
7. $3x^2 + 2y^2 - 6x + 8y + 5 = 0$
8. $2x^2 - y^2 - 4x + 4y - 4 = 0$

9. Write an equation of the hyperbola with asymptotes $y = \pm 3(x - 1)$ and foci $(1 \pm \sqrt{10}, 0)$.

10. For what values of m does $mx^2 + 4y^2 - 8y - 1 = 0$ describe

    a. an ellipse?
    b. a hyperbola?
    c. a parabola?
    d. a circle?

11. Find the vertices of the hyperbola with a focus at $(0, 0)$, a directrix $\ell: x = 1$, and eccentricity $e = \sqrt{2}$.

12. Write equations for the asymptotes of the hyperbola $x^2 - 4y^2 - 2x - 4y - 2 = 0$.

Polynomial Functions

6

6–1 Polynomial Values

Quadratic functions (Chapter 4) and linear and constant functions (Chapter 3) are members of a large class of functions known as polynomial functions, which we investigate in this chapter.

We want the definition of polynomial function to be such that $f(x) = 3x^2 + 2x$, $g(x) = -\sqrt{2}x^5 + 13x^2 + 2$, and $h(x) = 3x + 7$ are formulas for polynomials, but $p(x) = 2x + \dfrac{3}{x}$, $q(x) = 3|x| + 7$, and $r(x) = \sqrt{x^4 + 7x^2}$ are not.

Definition A **polynomial function** is a function f whose values can be given by
$$f(x) = a_n x^n + a_{n-1} x^{n-1} + \cdots + a_2 x^2 + a_1 x + a_0,$$
where n is a nonnegative integer.

The numbers $a_0, a_1, \ldots, a_n$ are called **coefficients** of the polynomial function. The coefficient a_n of the highest power of x is called the **leading coefficient** of the polynomial function. The positive integer n is the **degree** of f. The number a_0 is called the **constant term**. If $f(x) = a_0$, a constant function with $a_0 \ne 0$, then the degree of f is zero. By convention, the polynomial function $f: x \to 0$ is said not to have a degree.

Thus for the polynomial function $f(x) = -3x^4 + \sqrt{6}x^2 + x - 7$, the coefficients are $a_4 = -3$, $a_3 = 0$, $a_2 = \sqrt{6}$, $a_1 = 1$, and $a_0 = -7$. The leading coefficient is -3, the degree of f is 4, and the constant term is -7.

Example 1 If $f: x \to x^2 - 5x^3$, name the leading coefficient, the degree, and the constant term.

SOLUTION Arrange the powers of x in descending order.
$$f(x) = -5x^3 + x^2 + 0 \cdot x + 0$$
The leading coefficient is -5, the degree is 3, and the constant term is 0.

The sum, difference, or product of two polynomial functions is also a polynomial function.

Example 2 If $f(x) = 3x^2 + 2x + 4$ and $g(x) = -x + 2$, find formulas for the polynomial functions $f + g$, $f - g$, and $f \cdot g$.

SOLUTION
$[f + g](x) = (3x^2 + 2x + 4) + (-x + 2) = 3x^2 + x + 6$
$[f - g](x) = (3x^2 + 2x + 4) - (-x + 2) = 3x^2 + 3x + 2$
$[f \cdot g](x) = (3x^2 + 2x + 4) \cdot (-x + 2) = -3x^3 + 4x^2 + 8$

In example 2, the degree of $f + g$ and of $f - g$ is the same as the degree of f, namely, the larger of the two original degrees. The degree of $f \cdot g$ is the sum of the degrees of f and g. If f and g have the same degree, either $f + g$ or $f - g$ may have degree less than the original degree. For instance, if $f(x) = 4x^3 - 2x + 1$ and $g(x) = -4x^3 + x^2 + x$, then the degree of $f + g$ is 2.

An expression of the form

$$a_n x^n + a_{n-1} x^{n-1} + \cdots + a_1 x + a_0$$

is called a polynomial in x, while the function defined by the rule

$$f(x) = a_n x^n + a_{n-1} x^{n-1} + \cdots + a_1 x + a_0$$

is called a **polynomial function**. Sometimes we will relax the terminology and refer simply to "the polynomial f . . ." when it would be more precise to say "the polynomial function f . . . ." The meaning should be clear from the context.

In order to graph a polynomial function, and for other purposes as well, it is important to be able to find values of the function. Using the $\boxed{Y^x}$ key on a calculator, you can evaluate $f(2)$ for

$$f(x) = 5x^4 + 2x^3 + x^2 - 3x - 7$$

with the following sequence of keystrokes:

5 $\boxed{\times}$ 2 $\boxed{Y^x}$ 4 $\boxed{+}$ 2 $\boxed{\times}$ 2 $\boxed{Y^x}$ 3 $\boxed{+}$ 2 $\boxed{X^2}$ $\boxed{-}$ 3 $\boxed{\times}$ 2 $\boxed{-}$ 7 $\boxed{=}$.

(Try this with your calculator; the correct value is $f(2) = 87$.) If you try this sequence of keystrokes to compute $f(-2)$, you soon get the displayed message error. The key $\boxed{Y^x}$ cannot be used to find a power of a negative number. There is, however, a way of rewriting $f(x)$ so that the key $\boxed{Y^x}$ is not needed.

$$f(x) = 5x^4 + 2x^3 + x^2 - 3x - 7$$
$$= (5x^3 + 2x^2 + x - 3)x - 7$$
$$= [(5x^2 + 2x + 1)x - 3]x - 7$$
$$= [[((5x + 2)x + 1)x] - 3]x - 7$$

On a calculator $f(2)$ is now found with the following sequence of keystrokes:

5 $\boxed{\times}$ 2 $\boxed{+}$ 2 $\boxed{=}$ $\boxed{\times}$ 2 $\boxed{+}$ 1 $\boxed{=}$ $\boxed{\times}$ 2 $\boxed{-}$ 3 $\boxed{=}$ $\boxed{\times}$ 2 $\boxed{-}$ 7 $\boxed{=}$.

Example 3 For the polynomial function f given by the rule

$$f(x) = [((4x + 2)x + 1)x - 3]x - 7,$$

find $f(5)$, $f(3.1418)$, and $f(-3.1418)$.

(cont. on p. 211)

SOLUTION $f(5) = [((4 \cdot 5 + 2)5 + 1)5 - 3]5 - 7$
$= ((22 \cdot 5 + 1)5 - 3)5 - 7$
$= (111 \cdot 5 - 3)5 - 7$
$= 552 \cdot 5 - 7 = 2753$

For the argument 3.1418, use a calculator. Begin by storing 3.1418 in memory. The key [RCL], or its equivalent, will recall this argument each time it is needed.

4 [×] [RCL] [+] 2 [=] [×] [RCL] [+] 1 [=] [×] [RCL] [−] 3 [=] [×] [RCL] [−] 7 [=]

The final display is 445.20958.

If you follow the same sequence of steps for the argument −3.1418, the final display will be 340.01071. You should check this.

Individual calculators vary in their array of keys and their rules of operation. If the calculator you are using does not lead to the correct result for the steps shown in example 3, study the manufacturer's instruction booklet carefully.

Now look more closely at the procedure used in example 3. To help keep track of the sequence of multiplications and additions, arrange the coefficients of the polynomial as follows.

$$\begin{array}{|ccccc} 4 & 2 & 1 & -3 & -7 \\ \end{array}$$

To compute $f(5)$ place the argument at the left of this array. Then add on the descending arrows and multiply that sum by the argument on the ascending arrows.

$$\begin{array}{r|rrrrr} & 4 & 2 & 1 & -3 & -7 \\ 5 & \downarrow & 20\downarrow & 110\downarrow & 555\downarrow & 2760\downarrow \\ \hline & 4 & 22 & 111 & 552 & 2753 \end{array}$$

So $f(5) = 2753$, as found in example 3. This process is called **synthetic substitution**.

Example 4 Use synthetic substitution to find $f(-4)$ and $f(5)$ if $f(x) = 2x^3 - 7x^2 + x + 2$.

SOLUTION

$$\begin{array}{r|rrrr} & 2 & -7 & 1 & 2 \\ -4 & & -8 & 60 & -244 \\ \hline & 2 & -15 & 61 & -242 \end{array} \quad f(-4) = -242$$

$$\begin{array}{r|rrrr} & 2 & -7 & 1 & 2 \\ 5 & & 10 & 15 & 80 \\ \hline & 2 & 3 & 16 & 82 \end{array} \quad f(5) = 82$$

In using synthetic substitution, if any power of x is missing, insert 0 as a coefficient for that term in the array.

Example 5 If $f(x) = 3x^4 - 9x^2 + 5x - 7$, evaluate $f(-3)$.

SOLUTION The coefficient of x^3 is 0, which must be inserted in the array.

$$
\begin{array}{r|rrrrr}
 & 3 & 0 & -9 & 5 & -7 \\
-3 & & -9 & 27 & -54 & 147 \\
\hline
 & 3 & -9 & 18 & -49 & 140
\end{array}
\qquad f(-3) = 140
$$

If the zero is not inserted, the coefficients in the array represent $g(x) = 3x^3 - 9x^2 + 5x - 7$ and you would be evaluating $g(-3)$ rather than $f(-3)$.

By plotting several points a graph of a polynomial can be sketched.

Example 6 For the polynomial function P defined by $P(x) = x^3 - 4x + 1$, compute $P(-3)$, $P(-2)$, $P(-1)$, $P(0)$, $P(1)$, $P(2)$, and $P(3)$. Use the results to sketch a graph of P.

SOLUTION Use synthetic substitution except for cases such as $P(1)$ and $P(0)$, which can be quickly calculated mentally.

$$
\begin{array}{r|rrrr}
 & 1 & 0 & -4 & 1 \\
-3 & & -3 & 9 & -15 \\
\hline
 & 1 & -3 & 5 & -14
\end{array}
\qquad P(-3) = -14
$$

Calculation of the other values in the table below is left as an exercise.

x	-3	-2	-1	0	1	2	3
$P(x)$	-14	1	4	1	-2	1	16

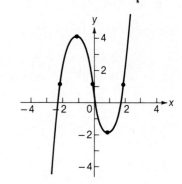

The graph is shown at the right.

The graph above shows that the polynomial function $P(x) = x^3 - 4x + 1$ has three real zeros. If the graph is shifted upward 5 units, it will cross the x-axis only once, so that the corresponding function, $Q(x) = x^3 - 4x + 6$, has only one zero. The following theorem asserts that every polynomial equation, such as $P(x) = 0$ or $Q(x) = 0$, has at least one root in the domain **C** of complex numbers.

Theorem **Fundamental Theorem of Algebra**

Every polynomial equation $a_n x^n + a_{n-1} x^{n-1} + \cdots + a_1 x + a_0 = 0$ of degree $n \geq 1$ with real or complex coefficients has at least one solution in the domain **C**.

In fact, it turns out that, provided repeated zeros are counted once for each occurrence, such a polynomial equation has exactly n solutions in **C**. For example, on the domain **C**, the polynomial $f(x) = x^3 - 2x^2 + x - 2 = (x - 2)(x^2 + 1)$ has the three zeros, 2, i, and $-i$.

In all the work in this chapter we restrict the coefficients of polynomials to be real numbers and use **R** as domain. Thus you should keep in mind that *a polynomial function of degree n has at most n zeros.*

Exercises

A Which of the following are polynomial functions?

1. $f(x) = x + \dfrac{3}{x}$
2. $f: x \to x^2 + 3$
3. $f(x) = 2x^4$
4. $s(x) = |x|$
5. $h: x \to 0$
6. $h(x) = \sqrt[3]{x}$
7. $f(x) = ax^2 + bx + c$
8. $g(x) = mx + b$
9. $p(x) = x^3 + \sqrt{3}x - \pi$

Evaluate $P(0)$, $P(1)$, and $P(-1)$ by substitution.

10. $P(x) = x^3 - 2x^2 + 3x - 7$
11. $P(x) = 9x^5 + 4x^6 - x$
12. $P(x) = 3x - x^2 + 2$
13. $P: x \to 0$
14. $P(x) = mx + b$
15. $P(x) = \sqrt{3}x^3 + \sqrt{2}x^2 + x$

16. For the polynomial functions in exercises 10, 12, and 14, find the leading coefficient, the degree, and the constant term.

17. For the polynomial functions in exercises 11, 13, and 15, find the leading coefficient, the degree, and the constant term.

18. If f has degree m and g has degree n, what can you say about the degree of $f \cdot g$? What can you say about the degree of the composite $f \circ g$?

Use synthetic substitution to find $f(2)$ and $f(-2)$ for each polynomial function.

19. $f(x) = 2x^3 - x^2 + 10x + 1$
20. $f(x) = x^4 - 2x^3 + x^2 + 3x - 7$
21. $f(x) = 2x^3 - x + 1$
22. $f(x) = x^4 - 2x^2 + x + 3$
23. $f(x) = 5x^5 + 3x^2 + x + 1$
24. $f(x) = 3x^3 + 2x^2 - 5x$

B Rewrite each polynomial as on page 210. Then find the specified values.

25. $f(x) = 2x^3 + 3x^2 - x - 4$; $f(1)$, $f(-1)$, $f(3)$, $f(-3)$

26. $g(x) = 2x^4 - x^3 - x^2 + 3x + 4$; $g(2)$, $g(-2)$, $g(\frac{1}{2})$, $g(-\frac{1}{2})$

With $f(x)$ and $g(x)$ as given in exercises 25 and 26, use a calculator to find each value.

27. $f(1.207)$; $f(-2.23)$; $f(0.01)$

28. $g(\sqrt{2})$; $g\left(\frac{231}{467}\right)$; $g\left(\frac{1 - \sqrt{5}}{2}\right)$

Use a calculator and synthetic substitution to find $f(3.15)$, $f(-1.27)$, and $f(\sqrt{7})$ for each polynomial function.

29. $f(x) = 2x^3 - 14x^2 + 3x - 9$

30. $f(x) = -12x^2 + 15x + 29$

31. $f(x) = 8x^3 - 19x + 7$

32. $f(x) = 11x^4 - 123x^2 + 42x - 9$

For each function, find the values $f(-3)$, $f(-2)$, $f(-1)$, $f(0)$, $f(1)$, $f(2)$, and $f(3)$, plot the corresponding points, and sketch a graph of f.

33. $f(x) = x^3 - 2x$

34. $f(x) = 2x^4 + 4x^2 - 8$

35. $f(x) = x^5 - 2x^4 + 3x^3$

36. $f(x) = -2x^3 + 3x^2 - 1$

37. A factory that employs x-hundred workers can produce $P(x) = x^3 - 11x^2 + 30x$ electric automobiles per day. Plot some points on the graph of P, and decide on a reasonable domain for this polynomial function.

38. Four congruent squares are cut from a sheet of copper that measures 8 inches by 6 inches. The copper sheet is then folded to make an open box. Determine the polynomial function $V(x)$ that gives the volume of the resulting box. Sketch a graph of V, and determine a reasonable domain for this polynomial function.

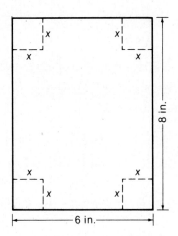

C Determine the constant k.

39. $f(x) = x^3 - 2x^2 - x + k$ and 2 is a zero of f.

40. $f(x) = x^3 - 3x^2 + kx + 1$ and $f(-3) = -50$.

41. $f(x) = 2x^3 + kx^2 - x - 2$ and $f(-2) = 0$.

42. Show that if r is a zero of $f(x) = ax^3 + bx^2 + cx + d$, then $2r$ is a zero of $g(x) = ax^3 + 2bx^2 + 4cx + 8d$.

43. Show that if r is a zero of $f(x) = x^4 + x^3 - 1$, then r^2 is a zero of $g(x) = x^4 - x^3 - 2x^2 + 1$.

6–2 The Remainder and Factor Theorems

The polynomial $g(x) = x^4 - 2x^3 + 9x - 8$ has $x - 1$ as a factor; that is,

$$g(x) = x^4 - 2x^3 + 9x - 8 = (x - 1)(x^3 - x^2 - x + 8).$$

From the factored form you can see that $g(1) = 0$. Thus factors of polynomials make it easy to find zeros of the polynomial function.

One way to try to factor a polynomial is to use long division with linear divisors of the form $x - a$ and search for divisors that will have 0 as a remainder. For instance, to see whether $h(x) = x^3 + x^2 - 4x - 4$ has $x - 3$ as a factor divide as follows.

$$\begin{array}{r}
x^2 + 4x + 8 \quad \leftarrow \text{Quotient} \\
x - 3 \overline{\smash{\big)}\, x^3 + x^2 - 4x - 4} \\
\underline{x^3 - 3x^2} \\
4x^2 - 4x \\
\underline{4x^2 - 12x} \\
8x - 4 \\
\underline{8x - 24} \\
20 \quad \leftarrow \text{Remainder}
\end{array}$$

The quotient is $Q(x) = x^2 + 4x + 8$ and the remainder is 20. So $x - 3$ is not a factor of $h(x)$. The results of the division can be displayed in two ways.

$$\frac{h(x)}{x - 3} = Q(x) + \frac{20}{x - 3} \quad \text{or} \quad h(x) = (x - 3)Q(x) + 20$$

If a linear divisor of the form $x - a$ is used, the remainder will always be a constant, r, which may be zero.

In general, for a given polynomial function f and any real number a, $f(x)$ can be written in the form

$$f(x) = (x - a)Q(x) + r.$$

The degree of the quotient function $Q(x)$ is one less than the degree of f, and r is a constant. Substituting the argument $x = a$ in both sides of the above equation leads to the following theorem.

Theorem 6–1 **The Remainder Theorem**

For any polynomial function f, $f(a)$ is equal to the remainder when $f(x)$ is divided by $x - a$.

PROOF Substitute $x = a$ in the identity

$$f(x) = (x - a)Q(x) + r.$$

$$\begin{aligned}
f(a) &= (a - a)Q(a) + r \\
&= 0 \cdot Q(a) + r
\end{aligned}$$

Thus $f(a) = r$, as was to be proved.

Example 1 Use synthetic substitution to determine the remainder when $h(x) = x^3 + x^2 - 4x - 4$ is divided by (a) $x - 3$ and by (b) $x + 1$.

SOLUTION Use the remainder theorem to find $h(3)$ and $h(-1)$.

(a)
$$
\begin{array}{r|rrrr}
 & 1 & 1 & -4 & -4 \\
3 & & 3 & 12 & 24 \\
\hline
 & 1 & 4 & 8 & 20
\end{array} \quad h(3) = 20
$$

The remainder is 20, as found by long division on the previous page.

(b)
$$
\begin{array}{r|rrrr}
 & 1 & 1 & -4 & -4 \\
-1 & & -1 & 0 & 4 \\
\hline
 & 1 & 0 & -4 & 0
\end{array} \quad h(-1) = 0
$$

Since $h(-1) = 0$, division of $h(x)$ by $x + 1$ has a remainder of zero. Since the remainder is 0, $x + 1$ is a factor of $h(x)$.

From the identity $f(x) = (x - a)Q(x) + r$, the remainder r is zero if and only if $x - a$ is a factor of $f(x)$. Combining this with the remainder theorem yields the following theorem.

Theorem 6–2 **The Factor Theorem**

The number a is a zero of the polynomial function f if and only if $x - a$ is a factor of $f(x)$.

Example 2 Show that $x + 4$ is a factor of $f(x) = x^3 - 3x^2 - 6x + 88$.

SOLUTION Using the factor theorem, you can show that $x - (-4)$ is a factor of $f(x)$ by verifying that $f(-4) = 0$.

$$
\begin{array}{r|rrrr}
 & 1 & -3 & -6 & 88 \\
-4 & & -4 & 28 & -88 \\
\hline
 & 1 & -7 & 22 & 0
\end{array} \quad f(-4) = 0
$$

Example 3 Find the polynomial function h of degree 3 having zeros at 2, 4, and -1 with the property that $h(1) = 18$.

SOLUTION By the factor theorem, $x - 2$, $x - 4$, and $x + 1$ must be factors of h. Since multiplication by a nonzero constant k does not change the zeros of a function,

$$h(x) = k(x - 2)(x - 4)(x + 1) \text{ and } h(1) = k(-1)(-3)(2) = 6k.$$

But $h(1) = 18$, so k must be 3. Thus

$$h(x) = 3(x - 2)(x - 4)(x + 1) = 3x^3 - 15x^2 + 6x + 24.$$

If you compare the results of dividing $x^3 + x^2 - 4x - 4$ by $x - 3$ with the array used to determine $h(3)$, you will see that the coefficients of the quotient, $Q(x) = x^2 + 4x + 8$, appear in the bottom row of the array. Thus synthetic substitution leads not only to the remainder but also to the quotient. Because of this connection with long division, synthetic substitution is often called **synthetic division.**

Example 4 Use synthetic division to find the quotient and remainder when $g(x) = 2x^3 + x^2 - 5x - 9$ is divided by $x + 2$.

SOLUTION The synthetic substitution array shows that $g(-2) = -11$ is the remainder and the coefficients of the quotient are the numbers 2, -3, and 1 from the bottom row of the array.

$$\begin{array}{r|rrrr} & 2 & 1 & -5 & -9 \\ -2 & & -4 & 6 & -2 \\ \hline & 2 & -3 & 1 & -11 \end{array}$$

Thus $g(x) = (x + 2)(2x^2 - 3x + 1) - 11$.

Example 5 Find all the linear factors of f if $f(x) = x^3 - 7x + 6$.

SOLUTION It is easy to see by inspection that $f(1) = 0$, so that $x - 1$ is a factor of $f(x)$. Use synthetic substitution to verify this result and find the quotient.

$$\begin{array}{r|rrrr} & 1 & 0 & -7 & 6 \\ 1 & & 1 & 1 & -6 \\ \hline & 1 & 1 & -6 & 0 \end{array} \quad f(1) = 0$$

The quotient is $Q(x) = x^2 + x - 6$.

Thus $f(x) = (x - 1)(x^2 + x - 6)$. Now use simple factoring methods to find that

$$x^2 + x - 6 = (x + 3)(x - 2).$$
$$f(x) = (x - 1)(x + 3)(x - 2)$$

After one factor of a polynomial function has been found, the quotient is called the **reduced polynomial.** Any further factors of f can be found by working with the reduced polynomial. Each time a new linear factor is found the degree of the reduced polynomial is reduced by one. Once the reduced polynomial is a quadratic, simple factoring methods or the quadratic formula can be used to find any remaining factors.

Example 6 If 1 and -2 are zeros of the polynomial function defined by $P(x) = 3x^4 - 2x^3 - 13x^2 + 8x + 4$, find all the linear factors of $P(x)$.

SOLUTION Use synthetic division to find the quotient when $P(x)$ is divided by $x - 1$.

$$\begin{array}{r|rrrrr} & 3 & -2 & -13 & 8 & 4 \\ 1 & & 3 & 1 & -12 & -4 \\ \hline & 3 & 1 & -12 & -4 & 0 \end{array}$$

The quotient has coefficients 3, 1, -12, and -4.

$$P(x) = (x - 1)(3x^3 + x^2 - 12x - 4)$$

The reduced polynomial must have -2 as a zero. Use synthetic division again to divide the reduced polynomial by $x + 2$.

$$\begin{array}{r|rrrr} & 3 & 1 & -12 & -4 \\ -2 & & -6 & 10 & 4 \\ \hline & 3 & -5 & -2 & 0 \end{array}$$

$$P(x) = (x - 1)(x + 2)(3x^2 - 5x - 2)$$

Finally, factor the quadratic reduced polynomial.

$$P(x) = (x - 1)(x + 2)(3x + 1)(x - 2)$$

Exercises

A Use the remainder theorem to find the remainder when $f(x)$ is divided by $x - 1$ and by $x + 1$.

1. $f(x) = x^3 + 2x - 1$
2. $f(x) = -2x^5 + 10$
3. $f(x) = x^3 + 1$
4. $f(x) = x^4 + x^3 + x^2 + x$

Find the remainder when $f(x)$ is divided by $x - 2$ and by $x + 2$.

5. $f(x) = \frac{1}{2}x^4 + 4x^3 + x^2 + 4$
6. $f(x) = x^4 + 2x^2 + 5$
7. $f(x) = 2x^5 + 8x$
8. $f(x) = -3x^3 - x^2 + x + 8$

Use the factor theorem to show that $x - 2$ is a factor of each polynomial.

9. $f(x) = x^2 - 4$
10. $f(x) = x^3 - 8$
11. $f(x) = x^2 - 5x + 6$
12. $f(x) = x^3 + 3x^2 - 4x - 12$
13. $f(x) = x^5 - 32$
14. $f(x) = x^4 - 9x^2 + 4x + 12$

Show that $x + \frac{1}{3}$ is a factor of each polynomial and determine the other factor.

15. $f(x) = 9x^2 - 1$
16. $f(x) = 6x^2 + 5x + 1$

Find a polynomial of smallest possible degree that has the given numbers as zeros.

17. 1, 2, −3

18. $\frac{1}{2}, -\frac{1}{3}, 3$

19. −1, 1, 0, 2, −2

Use synthetic division to find the quotient and the remainder when $f(x)$ is divided by $x - 1$ and by $x + 2$.

20. $f(x) = x^4 + 3x^2 - 4$

21. $f(x) = x^3 + 8$

22. $f(x) = -3x^3 + 7x^2 - 4x + 1$

23. $f(x) = x^2 - x - 2$

B Give an example of a polynomial function that has the given description.

24. f has −1 and 2 as zeros, and $f(1) = 5$.

25. f has −1 and 2 as zeros, and $f(0) = -3$.

26. f has −1 and 2 as zeros, and f has degree 3.

27. f has 2 and 4 as zeros, and $f(1) = 18$.

28. Divide $2x^4 + 3x^3 - 4x^2 - 8x + 1$ by $x - 2$ by using long division and by using synthetic division. Verify that the coefficients of the quotient and the remainder are the entries in the bottom row of the array for synthetic division.

If $x + 4$ is a factor of the given polynomial, what are the remaining factors?

29. $f(x) = x^3 + 5x^2 + 2x - 8$

30. $f(x) = x^3 - x^2 - 16x + 16$

31. $f(x) = x^3 + 5x^2 - 2x - 24$

32. $f(x) = x^3 - 3x^2 - 6x + 88$

33. The results of a synthetic-division problem are shown in the array at the right. What was the original division problem, and what are the results of the division?

$$\begin{array}{r|rrrrr} & 2 & 0 & -1 & 3 & 0 \\ -1 & & -2 & 2 & -1 & -2 \\ \hline & 2 & -2 & 1 & 2 & -2 \end{array}$$

34. The volume of a box constructed (as in exercise 38 in section 6–1) by cutting four congruent squares from the corners of a piece of copper measuring 5 in. by 6 in. is given by the polynomial function $f(x) = 4x^3 - 22x^2 + 30x$. Use the factor theorem to show that both $x - 3$ and $x - \frac{5}{2}$ are factors of $f(x)$.

If 2 and −3 are zeros of the polynomial functions in the following exercises, find all the linear factors of $P(x)$.

35. $P(x) = x^3 - 6x^2 - 13x + 42$

36. $P(x) = x^4 + x^3 - 7x^2 - x + 6$

37. $P(x) = x^4 + 8x^3 + 7x^2 - 36x - 36$

38. $P(x) = x^4 + 3x^3 - 7x^2 - 15x + 18$

C 39. If f is a polynomial function of degree 3 with 0, −1, and 2 as zeros, and $f(1) = 4$, find $f(-3)$.

Determine the value of k so that

40. $x - 1$ is a factor of $x^4 + 2x^2 - 4x + k$.

41. $x - 2$ is a factor of $x^3 - 3x^2 + 6x + k$.

42. $x + 3$ is a factor of $x^3 - 4x^2 + kx - 6$.

43. Determine values of A and B such that $(x - 3)^2$ is a factor of $x^4 - 2x^3 - 3x^2 - Ax - B$.

44. Show that $\dfrac{x^n - 1}{x - 1} = 1 + x + x^2 + \cdots + x^{n-1}$ for $x \neq 1$ and $n = 2, 3, 4,$ and 5.

6–3 Rational Zeros

You know how to find the zeros of any quadratic function by using the quadratic formula. For polynomials of higher degree, no such simple formula is available. For cubic (degree 3) and quartic (degree 4) polynomials, complicated methods of finding the zeros were devised over 400 years ago and are known as Cardan's and Ferrari's formulas, respectively. For polynomials of degree $n \geq 5$, it has been proved that, in general, no algebraic formulas exist which will express the zeros in terms of sums, products, and rational powers of the coefficients.

It is not difficult, however, to find the *rational* zeros of many polynomials of whatever degree.

Begin by considering four cubic polynomials written in factored form.

$$P_1(x) = (x - 2)(x + 1)(x + 3)$$
$$P_2(x) = (x - 2)(x^2 + 1)$$
$$P_3(x) = (x - 2)^2(x + 1)$$
$$P_4(x) = (x - 2)^3$$

By the factor theorem, P_1 has zeros $2, -1,$ and -3.

Since the quadratic equation $x^2 + 1 = 0$ has no real solutions, the only real zero of P_2 is 2.

For P_3 there are two distinct zeros, -1 and 2. Since the factor $x - 2$ is squared, we say that 2 is a zero of **multiplicity** two for P_3.

For P_4 the only real zero is 2 and its multiplicity is 3.

The graphs of these four functions are on the next page. Note that the graph of P_3 does not cross the x-axis at the zero 2. Instead, the axis is tangent to the graph there. This occurs whenever a polynomial function has a zero whose multiplicity is an even number.

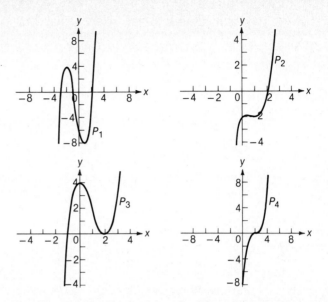

If the zeros of polynomials such as P_1, P_2, P_3, and P_4 are not given, you can plot many points and sketch a graph from which you can gain information about the number of real zeros. Next we consider this problem of finding the zeros of a polynomial given in unfactored form.

The first hope will be to discover some zeros by trial and error. But what numbers are worth considering?

Suppose that $f(x) = ax^3 + bx^2 + cx + d$ is a cubic polynomial with coefficients a, b, c, and d that are integers. It turns out that if the nonzero rational number $\dfrac{p}{q}$ in lowest terms is a zero of f, then p divides the constant term d and q divides the leading coefficient a. To see why this is true proceed as follows.

Since $f\left(\dfrac{p}{q}\right) = 0$, you know that

$$a\left(\frac{p}{q}\right)^3 + b\left(\frac{p}{q}\right)^2 + c\left(\frac{p}{q}\right) + d = 0.$$

$$ap^3 + bp^2q + cpq^2 + dq^3 = 0 \qquad (1)$$

$$ap^3 = -(bp^2q + cpq^2 + dq^3)$$

$$\frac{ap^3}{q} = -(bp^2 + cpq + dq^2)$$

The number on the right side of this equation is an integer, so $\dfrac{ap^3}{q}$ must also be an integer. By hypothesis, p and q have no common factors. It follows that

6–3: *Rational Zeros* 221

q is a divisor of a. Now rewrite equation (1) as

$$-(ap^3 + bp^2q + cpq^2) = dq^3.$$

$$-(ap^2 + bpq + cq^2) = \frac{dq^3}{p}, p \neq 0$$

Since the number on the left side of the equation is an integer, $\frac{dq^3}{p}$ is an integer, and it follows that p is a divisor of d.

The method of proof used for the cubic polynomial function can be extended to polynomial functions of higher degree, thus proving the following theorem.

Theorem 6–3 **The Rational-Zero Theorem**

If the nonzero rational number $\frac{p}{q}$, a fraction in lowest terms, is a zero of the polynomial function of degree n defined by

$$f(x) = a_n x^n + a_{n-1} x^{n-1} + \cdots + a_2 x^2 + a_1 x + a_0,$$

where the coefficients $a_0, a_1, \ldots, a_n$ are integers, then p is a divisor of a_0 and q is a divisor of a_n.

Example 1 Use the rational-zero theorem to make a list of the possible rational zeros of $f(x) = 2x^3 + x^2 - 3x + 3$.

SOLUTION The numerator p of any rational zero must be a divisor of 3, and the denominator q must be a divisor of 2.

So $p = 1, 3, -1,$ or -3, and $q = 1, 2, -1,$ or -2. The only possible rational zeros of f are

$$1, 3, \tfrac{1}{2}, \tfrac{3}{2}, -1, -3, -\tfrac{1}{2}, \text{ or } -\tfrac{3}{2}.$$

There is no assurance that any of the possible rational zeros will actually be zeros of the polynomial function in example 1. You would have to test the candidates to see whether any of them are zeros.

Example 2 Find all the rational zeros of the polynomial function defined by $f(x) = 3x^3 + 4x^2 - 5x - 2$.

SOLUTION If $r = \frac{p}{q}$ is a rational zero in lowest terms, p must be a divisor of -2 and q must be a divisor of 3. Hence $p = \pm 1, \pm 2$ and $q = \pm 1, \pm 3$ are the possible values. So any rational zero of f must be among the numbers $1, 2, \tfrac{1}{3}, \tfrac{2}{3}, -1, -2, -\tfrac{1}{3},$ and $-\tfrac{2}{3}$.

(cont. on p. 223)

Begin testing these numbers to see which, if any, are actually zeros of f.

$$\begin{array}{r|rrrr} & 3 & 4 & -5 & -2 \\ 1 & & 3 & 7 & 2 \\ \hline & 3 & 7 & 2 & 0 \end{array}$$

By synthetic substitution, $f(1) = 0$. It is easiest now to work with the reduced polynomial,

$$Q(x) = 3x^2 + 7x + 2,$$

which factors as

$$Q(x) = (3x + 1)(x + 2).$$
$$f(x) = (x - 1)(3x + 1)(x + 2)$$

The zeros of f are 1, $-\frac{1}{3}$, and -2.

The rational-zero theorem can be used to prove that certain numbers are irrational.

Example 3 Prove that $\sqrt{2}$ is an irrational number.

SOLUTION You know that $\sqrt{2}$ is a zero of $f(x) = x^2 - 2$. But the only possible rational zeros of f are 1, -1, 2, and -2. None of these numbers is a zero of f, so $\sqrt{2}$ cannot be a rational number.

The rational-zero theorem requires working with a polynomial function whose coefficients are all integers. If any of the coefficients are irrational numbers, as in $x^3 - \pi x^2 + \sqrt{2}x - 7$, the theorem is of no use. If all the coefficients are rational numbers, you can use the theorem after first multiplying by the lowest common denominator of the coefficients.

Example 4 List the possible rational zeros of the polynomial function

$$f(x) = x^3 - 2x^2 - \tfrac{1}{2}x - \tfrac{1}{6}.$$

SOLUTION The function g given by

$$g(x) = 6f(x) = 6x^3 - 12x^2 - 3x - 1$$

has the same zeros as f, and its coefficients are all integers. So the possible rational zeros are

$$\pm 1, \pm \tfrac{1}{2}, \pm \tfrac{1}{3}, \text{ and } \pm \tfrac{1}{6}.$$

Exercises

A Use the rational-zero theorem to make a list of the possible rational zeros of each polynomial function.

1. $f(x) = x^3 - 7x + 2$
2. $f(x) = x^3 + 3x^2 - 5$
3. $f(x) = 3x^3 - 4x^2 + 5x - 1$
4. $f(x) = 3x^3 - 7x + 2$
5. $f(x) = -4x^3 + x^2 + x - 2$
6. $f(x) = 6x^3 - 6x^2 + 6x - 1$
7. $f(x) = 4x^3 + 3x^2 - 5x + 3$
8. $f(x) = 2x^3 - 5x^2 + 3x + 4$
9. $f(x) = -6x^5 + 4x^2 - 5$
10. $f(x) = -9x^4 + 6x^2 + 5x - 3$

If the rational-zero theorem can be applied, as in example 4 in this section, make a list of the possible rational zeros of f.

11. $f(x) = x^3 + \frac{3}{2}x^2 - \frac{1}{3}x + 1$
12. $f(x) = 2x^3 - 3x^2 + \frac{1}{4}x + \frac{1}{2}$
13. $f(x) = -\sqrt{3}x^3 + \sqrt{2}x^2 + x + 1$
14. $f(x) = x^4 - \pi x^3 + 3x^2 + 2$
15. $f(x) = \frac{2}{3}x^4 + \frac{3}{2}x^2 + \frac{1}{5}$
16. $f(x) = \frac{1}{2}x^3 - \frac{3}{4}x^2 + \frac{1}{2}$

B 17. $f(x) = x^5 + x^3 + x$
[HINT: Factor $f(x)$ first.]

18. $f(x) = 3x^5 - \frac{7}{2}x^2 + \frac{1}{2}x$

Find all the rational zeros of each polynomial.

19. $f(x) = x^3 - 6x^2 + 11x - 6$
20. $f(x) = -x^3 - x^2 + 24x - 36$
21. $f(x) = 2x^3 - 3x^2 - 11x + 6$
22. $f(x) = 8x^3 - 12x^2 + 6x - 1$
23. $f(x) = 2x^3 + 5x^2 - x - 4$
24. $f(x) = 2x^3 - x^2 + 9x + 5$

25. Prove that $\sqrt{3}$ is an irrational number.

C 26. Prove that if p is a prime number, then $\sqrt{p}$ is an irrational number.

27. Write a program to have the computer print a list of the divisors of any given positive integer. Use this program as a subroutine to quickly find all rational roots of a quartic equation $a_4x^4 + a_3x^3 + a_2x^2 + a_1x + a_0 = 0$, where $a_4, a_3, a_2, a_1,$ and a_0 are integers.

28. What does the synthetic-substitution array shown below reveal about the rational zeros of the polynomial function $f(x) = 3x^3 - 8x^2 + 3x + 2$?

	3	−8	3	2
1		3	−5	−2
	3	−5	−2	0

29. What does the synthetic-substitution array shown below reveal about the rational zeros of the polynomial function $g(x) = x^4 + 4x^3 - 6x^2 - 7x - 10$?

	1	4	−6	−7	−10
2		2	12	12	10
	1	6	6	5	0

6-4 Narrowing the Search for Zeros

Once you have used the rational-zero theorem to list the possible rational zeros of a polynomial function, you can rule out some of the numbers in the list on the basis of their sign alone.

Theorem 6-4 If all the coefficients of a polynomial function

$$f(x) = a_n x^n + a_{n-1} x^{n-1} + \cdots + a_2 x^2 + a_1 x + a_0, a_n \neq 0,$$

are nonnegative, then no positive number can be a zero of f.

PROOF If c is any positive number, then by direct substitution $f(c) > 0$. Thus c is not a zero of f.

Example 1 What rational numbers can be zeros of $f(x) = 3x^3 + 10x^2 + 9x + 2$?

SOLUTION By the rational-zero theorem, the possible rational zeros are

$$1, 2, \tfrac{1}{3}, \tfrac{2}{3}, -1, -2, -\tfrac{1}{3}, -\tfrac{2}{3}.$$

Since every coefficient of $f(x)$ is positive, no positive numbers can be zeros of f (Theorem 6-4). Thus the list of possible rational zeros is reduced to

$$-1, -2, -\tfrac{1}{3}, -\tfrac{2}{3}.$$

Notice that a polynomial whose coefficients are all nonpositive, like $g(x) = -7x^5 - x - 4$, has the same zeros as one whose coefficients are all nonnegative, namely, $f(x) = -g(x) = 7x^5 + x + 4$. So no positive number can be a zero of g.

Theorem 6-4 provides a way of ruling out positive arguments in searching for zeros. Theorem 6-5 provides a way of ruling out negative arguments.

Theorem 6-5 If the coefficients of a polynomial function

$$f(x) = a_n x^n + a_{n-1} x^{n-1} + \cdots + a_2 x^2 + a_1 x + a_0$$

alternate in sign, then no negative number can be a zero of f.

PROOF Suppose that all terms with even powers of x have negative coefficients, while all terms with odd powers of x have positive coefficients. An even power of a negative number will be positive, while an odd power of a negative number will be negative. Thus if $r < 0$, then $f(r) < 0$. Thus $f(r) \neq 0$. (The other case, in which the odd terms have negative coefficients, is proved in a similar way.)

If some, but not all, of the coefficients are zero, the method of proof of Theorem 6–5 is still valid. Any missing terms can be written with 0, or −0, as coefficients. For example, $g(x) = x^3 + x - 2$ can be written as $g(x) = x^3 - 0x^2 + x - 2$, so no negative number can be a zero of g.

Example 2 What rational numbers could be zeros of the polynomial $f(x) = x^3 - 7x^2 + 5x - 9$?

SOLUTION Use the rational-zero theorem and Theorem 6–5 to see that $f(x)$ has as possible rational zeros only the numbers 1, 3, and 9. No negative numbers can be zeros of f.

In contrast to the rational-zero theorem, nothing in Theorems 6–4 or 6–5 requires that the coefficients of the polynomial be integers. The coefficients of f may, in fact, be irrational numbers. For instance, no negative number can be a zero of the polynomial

$$f(x) = \sqrt{3}x^3 - \pi x^2 + 7x - \sqrt{2}.$$

Example 3 Find all the real zeros of $f(x) = 4x^5 + 4x^4 + x^3 - 4x^2 - 4x - 1$.

SOLUTION First list the possible rational zeros: $\pm 1, \pm \frac{1}{2}, \pm \frac{1}{4}$.

Try 1 by synthetic substitution.

$$\begin{array}{r|rrrrrr} & 4 & 4 & 1 & -4 & -4 & -1 \\ 1 & & 4 & 8 & 9 & 5 & 1 \\ \hline & 4 & 8 & 9 & 5 & 1 & 0 \end{array}$$

Thus $f(x)$ can be factored as $f(x) = (x - 1)(4x^4 + 8x^3 + 9x^2 + 5x + 1)$.

Since all the coefficients of the reduced polynomial are positive, only negative numbers from the list of possible rational zeros need be tried. The polynomial can have no additional positive zeros, rational or irrational, by Theorem 6–4.

You can check that -1 is not a zero.

Now try $-\frac{1}{2}$.

$$\begin{array}{r|rrrrr} & 4 & 8 & 9 & 5 & 1 \\ -\frac{1}{2} & & -2 & -3 & -3 & -1 \\ \hline & 4 & 6 & 6 & 2 & 0 \end{array}$$

So $-\frac{1}{2}$ is a zero and the polynomial can now be factored as

$$f(x) = (x - 1)(x + \tfrac{1}{2})(4x^3 + 6x^2 + 6x + 2), \text{ or}$$

$$f(x) = 2(x - 1)(x + \tfrac{1}{2})(2x^3 + 3x^2 + 3x + 1).$$

(cont. on p. 227)

The advantage of removing the factor 2 from the last quotient is that it reduces the list of possible rational zeros. Any further zeros of f must be zeros of the doubly-reduced polynomial $2x^3 + 3x^2 + 3x + 1$. The possibilities are -1 and $-\frac{1}{2}$. The number $-\frac{1}{2}$ could be a zero of multiplicity 2, so try it.

$$
\begin{array}{r|rrrr}
 & 2 & 3 & 3 & 1 \\
-\frac{1}{2} & & -1 & -1 & -1 \\
\hline
 & 2 & 2 & 2 & 0
\end{array}
$$

Now you can write the polynomial in factored form as

$$f(x) = 2(x-1)(x+\tfrac{1}{2})(x+\tfrac{1}{2})(2x^2 + 2x + 2), \text{ or}$$

$$f(x) = 4(x-1)(x+\tfrac{1}{2})^2(x^2 + x + 1).$$

The discriminant of $x^2 + x + 1$ is negative so there are no additional real zeros of f.

ANSWER The real zeros of f are $-\frac{1}{2}, -\frac{1}{2},$ and 1.

Exercises

A Use Theorems 6–4 and 6–5 to decide which of the following polynomials have no positive zeros and which have no negative zeros.

1. $f(x) = 4x^3 + x^2 + 2x + 3$
2. $f(x) = 6x^3 - 6x^2 + 6x - 1$
3. $f(x) = 3x^3 - 4x^2 + 5x - 1$
4. $f(x) = 9x^3 + 4x^2 + 3x + 4$
5. $f(x) = -5x^2 - 2x - 3$
6. $f(x) = x^3 + 3x^2 - 5$
7. $f(x) = -4x^3 + x^2 + x - 2$
8. $f(x) = -x^3 - 3x^2 - 5$
9. $f(x) = 3x^3 - 7x + 2$
10. $f(x) = x^3 - 17x + 12$
11. $f(x) = 4x^3 + 2x - 1$
12. $f(x) = x^4 - 2x^3 - 3x + 6$

B Find all the real zeros of each polynomial.

13. $f(x) = 2x^4 + x^3 - 2x^2 - 4x - 3$
14. $f(x) = 2x^4 + 5x^3 - 6x^2 - 2x + 1$
15. $f(x) = 4x^5 + 12x^4 + x^3 - 10x^2 - 6x - 1$
16. $f(x) = 3x^5 + 13x^4 + 26x^3 + 30x^2 + 19x + 5$

17. Use the rational-zero theorem to list the possible rational zeros of

$$f(x) = x^4 + 4x^3 - 6x^2 - 6x - 12.$$

Then use the information in the given synthetic substitution array to revise your list, reducing it as much as possible.

$$
\begin{array}{r|rrrrr}
 & 1 & 4 & -6 & -6 & -12 \\
2 & & 2 & 12 & 12 & 12 \\
\hline
 & 1 & 6 & 6 & 6 & 0
\end{array}
$$

18. Prove Theorem 6–5 for the case where all terms with odd powers of x have negative coefficients and all terms with even powers of x have positive coefficients.

C 19. Prove: If f is a polynomial with integer coefficients, then $f(\sqrt{2})$ and $f(-\sqrt{2})$ differ only in the sign of the irrational part; that is, if $f(\sqrt{2}) = a + b\sqrt{2}$, with $a, b \in J$, then $f(-\sqrt{2}) = a - b\sqrt{2}$. Now use this result to prove the following: If f is a polynomial with integer coefficients and $\sqrt{2}$ is a zero of f, then $-\sqrt{2}$ is also a zero.

6–5 Approximating Irrational Zeros

You have learned how to find the rational zeros of polynomials with integer or rational coefficients. The question remains, "How can you find other real zeros?" In particular, how can you locate an irrational zero?

Of course graphs aid in the search for other zeros. Polynomials are all *continuous* functions, in the sense that there are no breaks in their graphs. Closeness of arguments guarantees closeness of values. Thus when graphing a polynomial, it is reasonable to plot several points and connect them with a smooth curve. This idea is formally expressed in Theorem 6–6, which is given here without proof.

Theorem 6–6 If f is a polynomial function with domain **R** and $f(x_1) = y_1$ and $f(x_2) = y_2$, then for every y_0 between y_1 and y_2 there is an x_0 between x_1 and x_2 such that $f(x_0) = y_0$.

Example 1 Sketch a graph of the function $f(x) = x^3 - 4x + 1$.

SOLUTION Compute several values of f, using successive integers as arguments. Here is a table of values found by using synthetic substitution.

x	−3	−2	−1	0	1	2	3
f(x)	−14	1	4	1	−2	1	16

Now you can plot these points and sketch the graph.

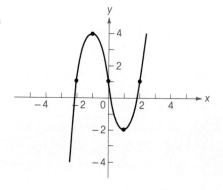

A sketch of the graph of a polynomial function yields an initial estimate of its zeros. In example 1, there seems to be a zero between -3 and -2, a second zero between 0 and 1, and a third one between 1 and 2. In fact, this much could be learned just by looking at the table of values. For example, $f(-3) = -14$ and $f(-2) = 1$. Since $f(x)$ is negative at the argument -3 and positive at -2, and since f is continuous, there must be a zero of f between -3 and -2. This important property is stated as a corollary to Theorem 6–6.

Corollary 6–6 **Intermediate-Value Property**

If f is a polynomial function and $f(a)$ and $f(b)$ have opposite signs, then there must be at least one zero of f between a and b.

Notice that there might be more than one zero in the interval $[a, b]$. For instance, in example 1, $f(-3) < 0$ and $f(2) > 0$, and there appear to be three zeros in $[-3, 2]$. (How do you know that this function has only three zeros?) Also, if $f(a)$ and $f(b)$ have the same sign, there may still be one or more zeros of f between a and b.

The intermediate-value property is the key to a method of approximating a zero to any desired degree of accuracy.

Example 2 For the polynomial function $f(x) = x^3 - 4x + 1$, which was graphed in example 1, estimate the smallest zero to the nearest tenth.

SOLUTION The graph on page 228 shows a zero closer to -2 than to -3. Compute values of f as follows, using synthetic substitution.

x	-3	$\cdots$	-2.2	-2.1	-2
$f(x)$	-14	$\cdots$	-0.848	0.139	1

Since $f(-2.2) < 0$ and $f(-2.1) > 0$, there is a zero of f between -2.2 and -2.1. By synthetic substitution, $f(-2.15) = -0.338$. Thus the zero is between -2.15 and -2.1. So, to the nearest tenth, the zero is -2.1.

If the instructions in example 2 required locating a zero correct to the nearest hundredth, you could subdivide the interval $[-2.15, -2.1]$ in steps of 0.01 and look for the change of sign in the values. Theoretically you can continue this process to attain any desired accuracy.

Exercises

A Use the intermediate-value property to show that each polynomial function has zeros as indicated.

1. $f(x) = x^3 - 6x^2 + 9x - 17$ has a zero between 4 and 5.
2. $f(x) = x^4 - 20x^2 + 48x + 3$ has a zero between -1 and 0.

3. $f(x) = -4x^3 + 8x^2 + 7x - 5$ has a zero between 0 and 1 and also a zero between 2 and 3.

4. $f(x) = x^3 - 5x + 1$ has a zero between -3 and -2, a second zero between 0 and 1, and a third between 2 and 3.

Sketch a graph of each polynomial function. From the graph determine the number of real zeros of the polynomial. Estimate the greatest zero to the nearest tenth.

5. $f(x) = x^3 - 4x - 1$
6. $f(x) = x^3 - 3x + 1$

B 7. $f(x) = x^3 + 3x^2 - 6x - 3$
8. $f(x) = -3x^3 + x^2 + 8x + 2$

9. $f(x) = x^4 - 2x^3 + x^2 - 1$
10. $f(x) = -x^4 + 4x^3 + 4x - 12$

11. Show that $f(x) = 12x^3 - 23x^2 - 10x + 25$ has two zeros between 1 and 2.
    [HINT: Compute $f(1.5)$.]

12. Show that $f(x) = x^4 - 4x^3 + x^2 + 6x + 2$ has two zeros between -1 and 0 and two zeros between 2 and 3.

C Use the intermediate-value property to approximate the given principal root to two decimal places.

13. $\sqrt[3]{7}$ [HINT: Consider $f(x) = x^3 - 7$.]
14. $\sqrt[4]{3}$

15. If $f(a) < 0$ and $f(b) > 0$, there could be exactly two distinct zeros of the polynomial function between a and b. Explain how this could happen.

Use the intermediate-value property to prove each of the following.

16. If f is a polynomial function and $f(0) > 0$ and $f(1) < 1$, then $f(c) = c$ for some c in $[0, 1]$. Interpret this statement graphically.

17. If f and g are polynomial functions, $f(a) > g(a)$, and $f(b) < g(b)$, then $f(c) = g(c)$ for some c in $[a, b]$.

6-6 Maximum and Minimum Values

The graph of $f(x) = x^3 - 4x + 1$ in section 6-5 suggests that on the interval $[-2, 0]$ the highest point on the graph is at $(-1, 4)$. Is this, in fact, the highest point?

Example 1 Find the argument in $[-2, 0]$ to the nearest hundredth at which the polynomial function $f(x) = x^3 - 4x + 1$ reaches a maximum value.

(cont. on p. 231)

SOLUTION Use a trial-and-error approach. Start somewhere to the left of the apparent high point, and proceed in steps of 0.1 to calculate function values until they begin to decrease.

x	-1.5	-1.4	-1.3	-1.2	-1.1	-1
$f(x)$	3.625	3.856	4.003	4.072	4.069	4

From this table it appears that the maximum point occurs for x between -1.3 and -1.1 and is probably close to -1.2.

Calculate some additional values near -1.2.

x	-1.17	-1.16	-1.15	-1.14
$f(x)$	4.078	4.079	4.079	4.078

Use a calculator to show that

$f(-1.16) = 4.079104$ and $f(-1.15) = 4.079125$.

To the nearest hundredth, the maximum occurs at $x = -1.15$.

Example 2 Four congruent square corners are cut from a sheet of aluminum that measures 6 inches by 5 inches. The aluminum sheet is then folded to make an open box. How large should each square be in order to obtain the box with the largest possible volume?

SOLUTION As with similar problems in Chapter 4, you need to create a function which expresses the quantity to be maximized in terms of one variable. Let x represent the side of each square to be removed. Then the dimensions of the open box are x, $6 - 2x$, and $5 - 2x$. The function that gives the volume of the box is

$f(x) = x(6 - 2x)(5 - 2x)$, or

$f(x) = 4x^3 - 22x^2 + 30x$.

The description of the box requires that x be between 0 and 2.5 (why?). Sketch a graph of f on the interval $[0, 2.5]$.

From the graph on the next page you can see that the function appears to be a maximum near $x = 1$.

(cont. on p. 232)

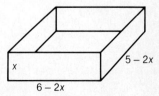

Calculate some additional values so as to approximate the maximum value of f accurate to two decimal places.

x	0.8	0.9	1
f(x)	11.968	12.096	12

These values indicate a maximum near $x = 0.9$.

x	0.89	0.9	0.91	0.92
f(x)	12.094	12.096	12.096	12.094

Calculator values are $f(0.9) = 12.09600$ and $f(0.91) = 12.096084$.

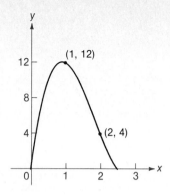

ANSWER The sides of the squares cut from each corner should measure 0.91 inches, accurate to two decimal places.

Exercises

A By trial and error, find to the nearest hundredth the argument at which each function has the indicated maximum or minimum.

1. $f(x) = x^3 - 3x$; maximum on $[-2, 0]$
2. $f(x) = x^3 - 3x - 4$; minimum on $[-1, 2]$
3. $f(x) = 9x + 2 - x^3$; minimum on $[-2, 0]$
4. $f(x) = x^4 - 2x^3 - 7x^2 + 10x + 10$; maximum on $[0, 1]$

In exercises 5–19, give answers correct to the nearest hundredth.

An open box is made, as in example 2, by cutting squares from each corner of a sheet of copper. What dimensions will the largest such box have if the original piece of copper measures

5. 9 cm by 7 cm?
6. 5 inches by 5 inches?

B 7. If p and q are two positive numbers such that $p + q = 5$, and if the sum of the cube of p and the square of q is as small as possible, what are the numbers p and q?

8. Write the number 10 as the sum of two positive numbers in such a way that the product of one by the square of the other is as large as possible.

9. The length, width, and height of a room with a square floor total 31 feet. What dimensions does the room have if it was built to have the largest possible volume?

10. A rectangular bin has a square base and no top. The combined area of the sides and the bottom is 50 square feet. Find the dimensions of the bin that has maximum volume.

11. The radius r and the height h of a silo in the shape of a right circular cylinder have a sum of 100 feet. What is the greatest possible volume of the silo?

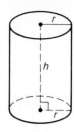

12. A factory that employs x-hundred workers can produce $x^3 - 11x^2 + 30x$ electric automobiles per day. If at most 500 workers are available, how many hundreds of them should be employed to make the factory's daily production a maximum?

13. A rectangle $PQRS$ is inscribed in the first quadrant beneath the line $4y = 12 - 3x$, as shown. Find the coordinates of point P for which the area of the rectangle is a maximum.

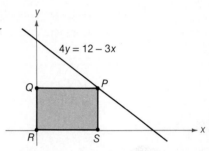

14. A rectangle $PQRS$ is inscribed under the parabola $y = 1 - x^2$, as shown. Find the coordinates of point P for which the area of the rectangle is a maximum.

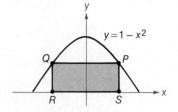

C 15. A point $P(c, c^2)$ on the graph of $y = x^2$ has abscissa c such that $0 \le c \le 2$. Point Q is the foot of the perpendicular from P to the x-axis and A is the point $(2, 0)$. What is the greatest possible area for $\triangle PAQ$?

16. A point $P(c, c^2 + 1)$ is on the graph of the parabola $y = x^2 + 1$. Let S denote the square of the distance from P to $A(4, 1)$.
   a. What is the least possible value of S?
   b. What point on the parabola is closest to A?

17. Find the minimum value of the square of the distance from $B(3, 0)$ to a point on the graph of $y = x^3$.

18. Find the dimensions of the right circular cylinder of greatest volume that can be inscribed in a right circular cone of height 12 cm and radius 5 cm. [HINT: Let r represent the radius and h the height of the cylinder. Use similar triangles to express h in terms of r. The volume of a cylinder is $V = \pi r^2 h$. Express V as a function of r alone.]

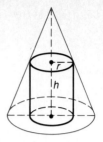

19. The strength of a wooden beam of rectangular cross section is proportional to the width of the beam and to the square of its depth (Strength $= kwd^2$). Find the dimensions of the strongest beam that can be sawed from a round log whose diameter is 2 feet.

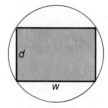

20. Write a program to have the computer determine, to the nearest thousandth, the arguments at which a given function achieves its maximum and minimum values on a given interval. Run your program using the function $f(x) = 0.25x^4 - 1.5x^2 + 2.01$ on the interval $[-1, 2]$.

6–7 Polynomial and Fractional Inequalities

In Chapter 4 you learned how to solve quadratic inequalities by completing the square or by graphing. In working with polynomial inequalities of higher degree such as $(x^2 - 4)(2x + 3) > 0$, completing the square is not possible and the graph may be difficult to draw. In fact, if you could solve the above inequality algebraically, the solution would help you to graph the function $f: x \rightarrow (x^2 - 4)(2x + 3)$.

We solve such inequalities by a number-line analysis of the signs of $f(x)$. Since polynomial functions are continuous, the intermediate-value property implies that the function values change signs in some interval only if that interval contains a zero of the function. Since $f(x) = (x^2 - 4)(2x + 3)$ has -2, $-\frac{3}{2}$, and 2 as its only zeros, mark these on a number line. (Arguments x appear beneath the line, and information about the values $f(x)$ above the line.) In the intervals $(-\infty, -2)$, $(-2, -\frac{3}{2})$, $(-\frac{3}{2}, 2)$, or $(2, \infty)$, $f(x)$ cannot change signs. The three zeros of f divide the number line into four intervals, or zones, and within any given zone, no sign change can occur. To discover the sign of $f(x)$ within each zone, all you need to do is test any argument x and see

whether $f(x)$ is positive or negative on that zone. The results are summarized in the table below.

x	-3	-2	$-\frac{5}{3}$	$-\frac{3}{2}$	0	2	3
$f(x)$	-15	0	0.4	0	-12	0	45
Sign of $f(x)$	$-$		$+$		$-$		$+$

Now you can complete the number-line analysis, as in the diagram below.

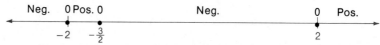

Finally, note from the analysis that the original inequality, $(x^2 - 4)(2x + 3) > 0$, has the following solution set:

$$\{x: -2 < x < -\tfrac{3}{2} \text{ or } x > 2\} = (-2, -\tfrac{3}{2}) \cup (2, \infty).$$

Example 1 Solve the inequality $3x^3 + 4x^2 - 5x - 2 \le 0$.

SOLUTION In example 2 of section 6–3, the polynomial $f(x) = 3x^3 + 4x^2 - 5x - 2$ was found to have -2, $-\tfrac{1}{3}$, and 1 as its zeros. Testing one argument x from each of the four resulting zones reveals the following sign pattern for $f(x)$.

Neg.   0   Pos.   0   Neg.   0   Pos.
       -2 $-\tfrac{1}{3}$ 1

ANSWER The solution set for $f(x) \le 0$ is $(-\infty, -2] \cup [-\tfrac{1}{3}, 1]$.

The number-line analysis depends upon expressing the given inequality with a single polynomial on one side and zero on the other.

Example 2 Solve the inequality $4x(x^4 + x^3 - x - 1) \ge 1 - x^3$.

SOLUTION Subtract $1 - x^3$ from both sides of the inequality and simplify to get

$$4x^5 + 4x^4 + x^3 - 4x^2 - 4x - 1 \ge 0.$$

In example 3 of section 6–4, you found that this polynomial can be factored as

$$f(x) = 4(x - 1)(x + \tfrac{1}{2})^2(x^2 + x + 1).$$

So the only real zeros of f are $-\tfrac{1}{2}$ and 1. Test the three resulting zones on a number line to get the following results.

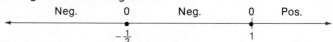

ANSWER The original inequality is true whenever $f(x) \ge 0$, so the solution set is $\{x: x = -\tfrac{1}{2} \text{ or } x \ge 1\}$.

Notice from the last example that the signs do not always alternate from one zone to the next adjacent one.

Example 3 Solve the inequality $\dfrac{x + 5}{x - 3} \leq 0$.

SOLUTION The function $f: x \to \dfrac{x + 5}{x - 3}$ is not continuous at $x = 3$, so you might expect a change in the sign of values of f near $x = 3$. The sign of $f(x)$ can also change as x passes through the zero $x = -5$. These are the only places where $f(x)$ can change sign. Place this information on a number line, using the symbol $*$ as a reminder that $f(3)$ is undefined.

There are 3 intervals in which $f(x)$ cannot change sign: $(-\infty, -5)$, $(-5, 3)$, and $(3, \infty)$. Test each zone with a representative argument and you will find that $f(x)$ has the signs indicated on the following diagram.

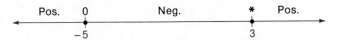

ANSWER The solution set for the given inequality is $\{x : -5 \leq x < 3\} = [-5, 3)$.

Example 4 Solve the inequality $\dfrac{2x + 5}{x - 3} < 1$.

SOLUTION First rewrite the inequality with all nonzero terms on the same side.

$$\dfrac{2x + 5}{x - 3} - 1 < 0$$

$$\dfrac{2x + 5 - x + 3}{x - 3} < 0$$

$$\dfrac{x + 8}{x - 3} < 0$$

```
        Pos.   0        Neg.       *    Pos.
     ←─────────●────────────────────●─────────→
              -8                    3
```

ANSWER The number-line analysis pictured above shows the solution set to be
$\{x : -8 < x < 3\}$.

Example 5 Solve the inequality $\dfrac{x}{x-3} \geq \dfrac{1}{x+5}$.

SOLUTION $\dfrac{x}{x-3} - \dfrac{1}{x+5} \geq 0$ *Rewrite with all nonzero terms on the left side.*

$\dfrac{x^2 + 5x - x + 3}{(x-3)(x+5)} \geq 0$

$\dfrac{(x+3)(x+1)}{(x-3)(x+5)} \geq 0$

The number-line zones are determined by the zeros, -3 and -1, and by the numbers where the fraction is undefined, -5 and 3. Test a sample argument in each zone to verify that the signs of the function are as shown on the number line below. Also check to see which endpoints of intervals are to be included.

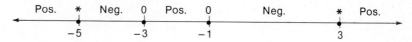

ANSWER The solution set is $(-\infty, -5) \cup [-3, -1] \cup (3, \infty)$.

Exercises

A Solve each inequality by using a number-line analysis.

1. $(x+1)(x-5) < 0$
2. $(3x-5)(x+2) \geq 0$
3. $(x^2 - 4)(x+1) \geq 0$
4. $(2-x)(4-9x^2) \leq 0$
5. $4x^2 - 10x > 0$
6. $2x^2 + 11x - 21 < 0$
7. $(2x^2 - 5x - 3)(x+2) \geq 0$
8. $(x^2 - x - 2)(9 - 2x) \leq 0$
9. $\dfrac{x+2}{x+4} < 0$
10. $\dfrac{x-7}{x+5} > 0$
11. $\dfrac{x-1}{x+1} \geq 0$
12. $\dfrac{x+3}{x-3} \leq 0$
13. $\dfrac{2x}{x-1} \leq 0$
14. $\dfrac{3x}{x+1} \geq 0$
15. $\dfrac{(x+1)(2x-5)}{(3x-6)} \geq 0$
16. $\dfrac{(x-2)(3+x)}{(5+3x)} \leq 0$

B 17. $(3x-1)(x^2 - 6x + 20) \leq 0$
18. $(5-2x)(x^2 + x + 1) > 0$
19. $2x^3 - 5x^2 - 12x > 0$
20. $5x^3 - 13x^2 - 6x < 0$

21. $2x^3 - 6x > 7x^2 - 21$
22. $x^3 - 20 \geq 4x^2 - 5x$
23. $x^3 - x^2 + 4x < 4$
24. $x^3 + 6x^2 \geq -5x$
25. $(x^2 - 16)(x^2 + 2x - 3) \leq 0$
26. $(x^2 - 25)(9 - 4x^2) > 0$
27. $\dfrac{5 - 2x}{x^2 - 4} > 0$
28. $\dfrac{x - 3}{25x^2 - 9} < 0$
29. $\dfrac{1}{x - 1} > 3$
30. $\dfrac{2}{2x + 7} > \dfrac{4}{3}$
31. $\dfrac{4x}{x - 2} \geq -1$
32. $\dfrac{x + 2}{3 - x} \leq -2$
33. $\dfrac{25}{9 - x^2} \leq 2$
34. $\dfrac{3x^2 - 5}{x} > 2x$
35. $\dfrac{5}{x + 1} \leq \dfrac{2}{3 - 2x}$
36. $\dfrac{2}{3x - 7} \geq \dfrac{3}{x}$

C 37. $\dfrac{x + 5}{x^2 - 4} < \dfrac{1}{x}$
38. $\dfrac{3x}{x^2 - 3x + 2} < \dfrac{4}{x - 2}$

39. For what values of x does the graph of $y = \dfrac{1}{x}$ lie below the graph of $y = 2x + 1$?

6–8 Graphing Rational Functions

Functions like $f(x) = \dfrac{3x + 1}{x - 2}$, $g(x) = \dfrac{x^2 + 3x - 4}{x^2 - x - 6}$, and $h(x) = \dfrac{1}{x^2 - x - 2}$, each of which is a ratio of two polynomial functions, are called **rational functions**. In a calculus course, you will learn techniques for graphing such functions, but you can sketch such graphs approximately now.

Consider $f(x) = \dfrac{3x + 1}{x - 2}$. The domain of f is $R - \{2\}$. When x is close to 2, $|f(x)|$ will be very large. For instance, you can verify that $(1.99, -697)$ and $(2.01, 703)$ are both points on the graph of f. The graph rises (or falls) steeply near the line $x = 2$; this line is a vertical asymptote of the graph of f.

You can find the horizontal asymptote of the graph of f by asking what happens to $f(x)$ when $|x|$ gets very large. If values of $f(x)$ come arbitrarily close to a real number k as x increases without bound, we say that "$f(x)$ has **limit** k as x approaches infinity," and write $\lim\limits_{x \to \infty} f(x) = k$. If you divide numerator and denominator of $\frac{3x+1}{x-2}$ by x to obtain $f(x) = \frac{3 + \frac{1}{x}}{1 - \frac{2}{x}}$, you can see that $f(x)$ comes very close to 3 as x increases without bound. Thus $\lim_{x \to \infty} \frac{3x+1}{x-2} = 3$.

Also, $\lim_{x \to -\infty} \frac{3x+1}{x-2} = 3$. So the left and right extremes of the graph of f will show points coming closer and closer to the line $y = 3$; this line is the horizontal asymptote of the graph of f.

Next you can use methods from the previous section to show that $f(x) > 0$ when $x < -\frac{1}{3}$ or $x > 2$. So, on these intervals, the graph of f will be above the x-axis. Putting all this information together and plotting a few specific points, including the x- and y-intercepts, you can sketch a graph of f. Note from the graph that Rng $f = \mathbf{R} - \{3\}$.

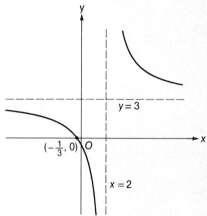

Here is a summary of aids to sketching the rational function $f(x) = \frac{p(x)}{q(x)}$.

① Determine the domain of f.

② Sketch any vertical asymptotes of f.

③ Find the zeros of f by solving $p(x) = 0$.

④ Evaluate $\lim_{x \to \infty} f(x)$ and $\lim_{x \to -\infty} f(x)$, and sketch the horizontal asymptotes.

⑤ Determine the intervals where $f(x) > 0$.

⑥ Plot some points, including the x- and y-intercepts and some points to test the behavior of f near the asymptotes.

Example 1 Sketch the graph of the rational function $f(x) = \frac{x^2 + 3x - 4}{x^2 - x - 6}$. Determine the range of f.

SOLUTION After factoring $f(x)$ as $f(x) = \frac{(x+4)(x-1)}{(x+2)(x-3)}$, you can see that

① Dom $f = \{x : x \neq -2 \text{ and } x \neq 3\}$;

② the vertical asymptotes of f are the lines $x = -2$ and $x = 3$; and

③ the zeros of f are -4 and 1.

(cont. on p. 240)

④ To find the horizontal asymptotes, rewrite $f(x)$ as

$$f(x) = \frac{(x^2 + 3x - 4)\left(\frac{1}{x^2}\right)}{(x^2 - x - 6)\left(\frac{1}{x^2}\right)} = \frac{1 + \frac{3}{x} - \frac{4}{x^2}}{1 - \frac{1}{x} - \frac{6}{x^2}}.$$

For $|x|$ large, whether x is positive or negative, $f(x)$ will be very close to 1; that is, $\lim_{x \to \infty} f(x) = 1$ and $\lim_{x \to -\infty} f(x) = 1$. Hence the horizontal asymptote is the line $y = 1$.

⑤ The signs of $f(x)$ can change only at the zeros (-4 and 1) or at the numbers where the function is not defined (-2 and 3). Check the signs in each of the resulting zones on a number line to find that $f(x)$ is positive on $(-\infty, -4) \cup (-2, 1) \cup (3, \infty)$.

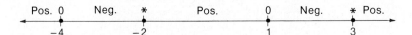

⑥ Plot a few points, including the x- and y-intercepts and some points to test the behavior near the asymptotes. The final graph is shown below.

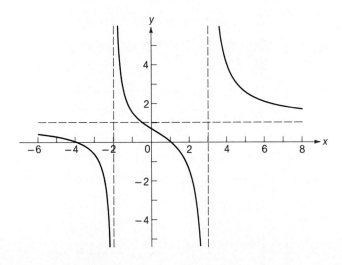

You can see from the graph that $\text{Rng } f = \mathbf{R}$. Note that $1 \in \text{Rng } f$ even though $y = 1$ is a horizontal asymptote.

Most of the aids to sketching rational functions can be used in graphing any function of the form $f(x) = \dfrac{1}{g(x)}$, even if g is not a polynomial function.

Example 2 Sketch a graph of $f(x) = \dfrac{1}{|x| - 1}$.

SOLUTION Start by sketching the function $g(x) = |x| - 1$. Since the zeros of g are 1 and -1, ① Dom $f = \mathbf{R} - \{-1, 1\}$, and ② f will have as vertical asymptotes the lines $x = 1$ and $x = -1$. ③ $f(x) = \dfrac{1}{g(x)}$ will have no zeros, regardless of the choice of g (why?). ④ When $|x|$ is large, $f(x)$ will be close to 0.

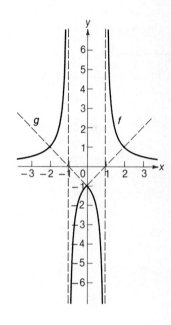

$$\lim_{x \to \infty} \frac{1}{|x| - 1} = 0 \text{ and } \lim_{x \to -\infty} \frac{1}{|x| - 1} = 0$$

So the x-axis will be a horizontal asymptote of the graph of f.

⑤ Where $g(x) > 0$, $f(x) = \dfrac{1}{g(x)}$ will be positive. So the graph of f will be above the x-axis wherever the graph of g is above the x-axis. ⑥ Mark the points where $g(x) = 1$ or $g(x) = -1$. These points will also be on the graph of f. (Why?)

Plotting a few points, and putting all this information together, you can sketch the graph.

Exercises

A Sketch a graph of each function. Specify the domain and the vertical and horizontal asymptotes. Also determine the range of each function.

1. $f(x) = \dfrac{x + 2}{x + 3}$
2. $g(x) = \dfrac{x + 3}{x - 1}$
3. $h(x) = \dfrac{x}{x - 1}$
4. $f(x) = \dfrac{4x + 1}{x}$
5. $g(x) = \dfrac{2x - 5}{x - 2}$
6. $h(x) = \dfrac{x + 1}{2x + 1}$

Determine the zeros and the domain of each function. Also find $\lim_{x \to \infty} f(x)$ and write the equations of any horizontal asymptotes.

7. $f(x) = \dfrac{(2x + 4)(x - 1)}{(x + 3)(x - 2)}$

8. $f(x) = \dfrac{(x + 5)(x - 5)}{(x + 4)(x - 3)}$

9. $f(x) = \dfrac{x^2 - 5x + 6}{x + 1}$

10. $f(x) = \dfrac{x^2 + 7x + 10}{x - 3}$

11. $f(x) = \dfrac{x - 4}{x^2 - 3x - 4}$

12. $f(x) = \dfrac{(x - 2)(x + 3)(x + 5)}{(x^2 - 4)(x - 1)}$

B Sketch a graph of each function and from the graph determine the range of the function.

13. $g: x \to \dfrac{1}{x^2 - 1}$

14. $h: x \to \dfrac{1}{x^2 - 4}$

15. $r(x) = \dfrac{1}{x^2 + 3x + 2}$

16. $h(x) = \dfrac{1}{x^2 - x - 2}$

17. $F(x) = \dfrac{x^2 - 4}{x^2 - 1}$

18. $G(x) = \dfrac{x^2 + 3x}{x^2 - 4}$

19. $f: x \to \dfrac{1}{|x| - 2}$

20. $g: x \to \dfrac{1}{|x - 2|}$

C Determine graphically the approximate solution of each inequality.

21. $\dfrac{2x - 5}{x - 2} > 4 - 3x$

22. $\dfrac{4x + 1}{x} < 4 - |x|$

23. $\dfrac{x^2 - 4}{x^2 - 1} \geq 4 - x^2$

24. $\dfrac{1}{x^2 - x - 2} \leq |x + 1|$

25. Determine graphically the number of points of intersection of the graph of
$$f(x) = \dfrac{2x^2 + 6x}{x^2 - 4}$$
with the ellipse $x^2 + 4y^2 = 16$.

26. Write a program to have a computer print the values $f(10), f(100), f(1000), \ldots, f(10^{10})$ for a given function. Run your program for the three functions in exercises 7, 9, and 11 and use the results to confirm your findings about the horizontal asymptotes of these functions.

Special   Interval Bisection

The **interval-bisection method** for approximating a zero of a function adapts the method you learned in section 6–5, using the intermediate-value property, to the computer. Instead of computing values of the given function f at each of many arguments throughout the interval, a repetitive procedure is followed in which only the midpoint of the interval is used. If $f(x)$ changes sign on interval $[a, b]$, the midpoint value $f\left(\dfrac{a+b}{2}\right)$ is computed. Depending on the sign of this value, you learn which half of $[a, b]$ contains the zero to be located. Now repeat the midpoint-value check, using the new interval $\left[a, \dfrac{a+b}{2}\right]$ or $\left[\dfrac{a+b}{2}, b\right]$. In this way you continue finding smaller and smaller intervals that contain the zero of f. Continue bisecting intervals until an interval of suitably short length is found. For instance, if $|b - a| \leq 1$, then after ten bisections the length of the new interval will be less than $\dfrac{1}{2^{10}} < 0.001$. This process quickly converges toward the zero that is being approximated.

Here is a computer program written in BASIC to have a computer carry out the interval-bisection method. In example 2 of section 6–5, the intermediate-value property was used to show that the polynomial function $f(x) = x^3 - 4x + 1$ has -2.1 (nearest tenth) as a zero. For comparison, the same function is used with the program written below.

```
10 DEF FNF (X) = X ↑ 3 − 4*X + 1
20 PRINT "WHAT INTERVAL";
30 INPUT A,B
40 LET C = (A + B)/2
50 IF ABS(A−B) < 0.0001 THEN 110
60 IF FNF(A) * FNF(C) < = 0 THEN 90
70 LET A = C
80 GOTO 40
90 LET B = C
100 GOTO 40
110 PRINT "THE ZERO IS APPROXIMATELY"; C
120 END
```

Run
WHAT INTERVAL? −3, −2
THE ZERO IS APPROXIMATELY −2.1149

In this program, the definition of the function $f(x)$ is typed in line 10. Lines 20 and 30 have the user input the endpoints a and b of some interval known to contain the zero. Lines 40 to 100 form a loop in which the computer repeatedly finds $f(c)$, where $c = \dfrac{a+b}{2}$ is the interval midpoint, and uses the intermediate-value property (line 60) to discover which half of the interval $[a, b]$ contains a zero of f. The endpoints a and b are then redefined to establish the new interval, and the search continues. The program instructs the computer to keep bisecting the intervals until (line 50) $|a - b| < 0.0001$. If greater accuracy is needed in the approximation, then line 50 can be changed. In line 110, the computer prints the final approximation, the midpoint of the last interval used. Notice that for this program to run, you must be sure when you input the initial endpoints a and b that the corresponding function values $f(a)$ and $f(b)$ have opposite signs.

Exercises

Use the interval-bisection method computer program to approximate the indicated zero of the polynomial f.

1. $f(x) = x^3 - 4x + 1$; the zero between 0 and 1.

2. $f(x) = x^3 - 6x^2 + 9x - 17$; the zero between 4 and 5.

3. Modify the given computer program so that if $f(x)$ does not change sign on $[a, b]$ the computer will ask the user to input a different interval.

4. Modify the given program so that the user will be asked to specify (as an input) the desired accuracy of the approximation to be found.

Chapter Review Exercises

6–1, page 209

Use synthetic substitution to find $f(-2)$ and $f(3)$ for each polynomial function.

1. $f(x) = x^3 - 2x^2 + 3x + 4$
2. $f(x) = 2x^4 - x^2 - 3x + 5$

6–2, page 215

3. Use the factor theorem to show that $x + 3$ is a factor of $x^5 + 243$.

4. Use synthetic division to find the quotient and the remainder when $2x^4 - 3x^2 + 7x + 1$ is divided by $x + 2$.

6–3, page 220

Make a list of the possible rational zeros of each polynomial.

5. $g(x) = 9x^3 - 6x^2 + 2x + 3$
6. $h(x) = \frac{1}{2}x^5 - \frac{1}{3}x^3 + 1$

7. Find all the rational zeros of the polynomial function $f(x) = 4x^3 + 4x^2 - 7x + 2$.

6–4, page 225

8. Which of the following polynomial functions can have no negative zeros?
   a. $p(x) = x^3 + 3x^2 + 1$
   b. $r(x) = x^4 - 3x^3 + 2x^2 - x + 1$
   c. $t(x) = 2x^3 + x - 7$

9. Find all the real zeros of $f(x) = 4x^4 - 4x^3 - 7x^2 + 8x - 2$.

6–5, page 228

10. Use the intermediate-value property to show that $f(x) = x^3 + 6x^2 - 5x - 3$ has a zero between 1 and 2.

11. Use a graph of $g(x) = 2x^3 - 9x + 1$ to determine the number of real zeros of g. Estimate the greatest zero (nearest tenth).

6–6, page 230

12. Find, to the nearest hundredth, the argument in $[-1, 0]$ at which the function $f(x) = 2x^3 - 3x + 1$ reaches a maximum value.

13. Rectangular mailing cartons are being manufactured so that each carton has a square base and the sum of the length, width, and height of each box is 20 in. What dimensions should each box have if it is to have the greatest possible volume?

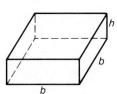

6–7, page 234

Solve each inequality by using a number-line analysis.

14. $(x^2 - 1)(2 - x) < 0$
15. $\dfrac{7 - 2x}{x^2 - 9} \geq 0$
16. $\dfrac{3}{4 + 2x} \geq \dfrac{1}{x}$

6–8, page 238

Graph each function.

17. $f(x) = \dfrac{2x - 1}{x + 5}$
18. $g: x \rightarrow \dfrac{1}{x^2 + x - 6}$
19. $h: x \rightarrow \dfrac{1}{|x| - 4}$

Chapter Test

1. Use synthetic substitution to find $g(-1)$ and $g(3)$ if $g(x) = x^4 - 11x^2 - 5x + 2$.

2. Use synthetic division to find the quotient and the remainder when $x^3 + \frac{1}{3}x^2 - 9x - 3$ is divided by $x + 2$.

3. Use the factor theorem to show that $x - 3$ is a factor of $x^5 - 243$.

4. Find the polynomial function P of lowest degree that has 0, -1, and 2 as zeros if $P(1) = 4$.

5. All of the zeros of $f(x) = 2x^4 + 3x^3 - 12x^2 - 7x + 6$ are rational numbers. Find them.

6. Use the intermediate-value property to show that the function $f(x) = 6x^3 - 4x^2 - 27x + 18$ has a negative zero. Find that zero to the nearest tenth.

7. If $g(x) = x^3 + 4x^2 + kx - 3$ and $g(-3) = 36$, find k.

8. From the graph of g as shown, sketch the graph of $f(x) = \dfrac{1}{g(x)}$.

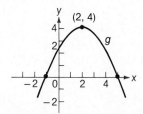

9. Two nonnegative numbers are to have a sum of 4. How should they be chosen if the cube of the first added to twice the square of the second is to yield the least sum possible? Use trial and error, giving your answer to the nearest tenth.

10. Locate the asymptotes and then graph $f(x) = \dfrac{2x + 3}{x + 1}$.

11. Use your solution of problem 10 to determine the number of solutions of the equation $\dfrac{2x + 3}{x + 1} = 2 - |x|$.

The remaining problems concern the function $f: x \to \dfrac{2x(x + 3)}{(x + 2)(x - 2)}$.

12. Determine the domain of f.

13. Find the zeros of f.

14. Find $\lim\limits_{x \to \infty} f(x)$.

15. Write equations for all the asymptotes of f.

16. Solve $f(x) \geq 0$.

17. Sketch a graph of f.

Inverse Functions

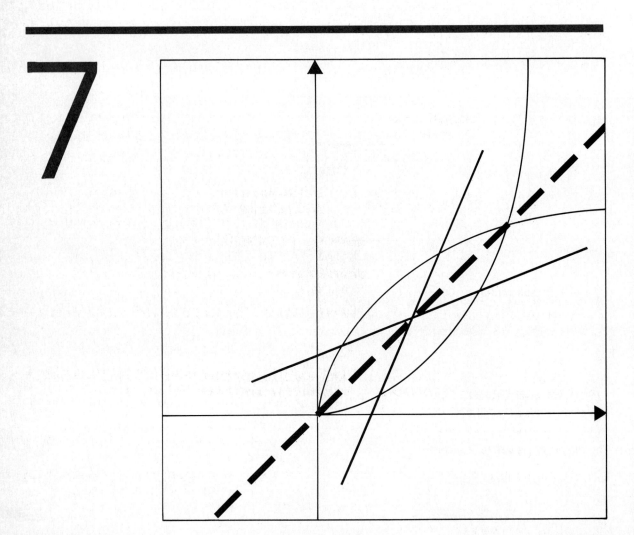

7–1 Finding Inverse Functions

Fahrenheit and Celsius temperatures are related by the formula
$$F = \tfrac{9}{5}C + 32.$$
That is, the function $T: C \to \tfrac{9}{5}C + 32$ converts Celsius degrees into Fahrenheit degrees. For example, if $C = 20°$, then $T(C) = 68°$, or $20°C = 68°F$.

We often want to convert from Fahrenheit to Celsius temperature. In the language of functions, this means finding the *inverse* function. For a linear function such as T, this is easy. First solve the formula for C,
$$C = \tfrac{5}{9}(F - 32),$$
and then create the function
$$T': F \to \tfrac{5}{9}(F - 32).$$
This function T' converts Fahrenheit degrees to Celsius degrees and so is the inverse function.

Notice that without regard to the temperature application,
$$T: x \to \tfrac{9}{5}x + 32 \text{ and } T': x \to \tfrac{5}{9}(x - 32)$$
are functions such that T' "undoes" T. That is, T takes a number, multiplies it by $\tfrac{9}{5}$, then adds 32. But T' subtracts 32 from the number and then multiplies by $\tfrac{5}{9}$.

For many other functions it is easy to identify an inverse function which takes a range element of the given function back to the original argument.

Example 1 Find an inverse for each function.

    a. $f: x \to \tfrac{x}{2}$ **b.** $F(x) = \tfrac{1}{x}$ **c.** $h: x \to \tfrac{1}{x} + 1$

SOLUTION   **a.** Since f divides every argument by 2, an inverse function must multiply each of its arguments by 2. Hence an inverse of f is
$$g: x \to 2x.$$

    b. Taking the reciprocal of $\tfrac{1}{x}$ gives you x. Therefore F is its own inverse.

    c. Function h first takes the reciprocal of each argument and then adds 1. By subtracting 1 and then taking the reciprocal, the original argument is obtained. Thus an inverse of h is
$$g: x \to \tfrac{1}{x - 1}.$$

In the real-number system the subject of inverses is tied directly to the existence of identity elements. Recall that 0 is the identity for addition, or the additive identity, because
$$a + 0 = 0 + a = a \text{ for every } a \in \mathbf{R}.$$

Also, for each $a \in \mathbf{R}$ there corresponds a number x such that
$$x + a = a + x = 0.$$
The number x, as you know, is the additive inverse of a, and is denoted by the special symbol $-a$.

Similarly for function composition, for any function f and any number x in the domain of f, the function $I: x \to x$ has the property
$$[f \circ I](x) = f[I(x)] = f(x) = I[f(x)] = [I \circ f](x), \text{ or}$$
$$f \circ I = f = I \circ f.$$

Thus I is the identity element for the operation of function composition, and for this reason $I: x \to x$ is called the identity function.

Following the pattern in the real-number system for defining additive inverses, we now require that if a function f is to have an inverse element g for the operation of function composition, then it should be true that $[g \circ f](x) = I(x)$ and $[f \circ g](x) = I(x)$.

Definition For a given function f, a function g is an **inverse** of f if and only if $[g \circ f](x) = x$ for every $x \in \text{Dom } f$ and $[f \circ g](x) = x$ for every $x \in \text{Dom } g$.

Example 2 Verify that the function $f: x \to 2x - 3$ has $g: x \to \dfrac{(x+3)}{2}$ as its inverse.

SOLUTION
$$[f \circ g](x) = f[g(x)] = f\left[\dfrac{x+3}{2}\right]$$
$$= 2\left(\dfrac{x+3}{2}\right) - 3$$
$$= x$$
$$[g \circ f](x) = g[f(x)] = g(2x - 3)$$
$$= \dfrac{(2x-3)+3}{2}$$
$$= x$$

The results in example 2 indicate that the definition is consistent with the idea that an inverse should be an "undoing" function.

As with real numbers, a given function f has at most one inverse. Consequently, that function can be assigned its own symbol, f^{-1}, which is called *the* inverse of f. With the new symbol, we can write the following restatement of the definition of an inverse function for composition:

If $x \in \text{Dom } f$, then $[f^{-1} \circ f](x) = x$, and if $x \in \text{Dom } f^{-1}$, then $[f \circ f^{-1}](x) = x$.

It is incorrect to write $f^{-1} \circ f = I$, because the domain of I is **R** whereas the domain of $f^{-1} \circ f$ is the domain of f, which might be a subset of **R**. Recall that equal functions must have the same domain.

By the definition, if g is the inverse of f, then f is the inverse of g. In other words,

$$(f^{-1})^{-1} = f.$$

Be careful not to confuse f^{-1} with the function $\frac{1}{f}$; they are completely different functions.

You have seen that inverses for some functions can be obtained almost by inspection. With more complicated functions it is desirable to have an algebraic method that is based on the definition of an inverse.

To illustrate this approach, consider again the function $f: x \rightarrow 2x - 3$, and let g denote its inverse.

Since $f(x) = 2x - 3$, you know that $[f \circ g](x) = f[g(x)] = 2g(x) - 3$.

But from the definition, you also know that

$$[f \circ g](x) = x.$$

$$2g(x) - 3 = x$$

$$g(x) = \frac{x + 3}{2} \qquad \text{(See example 2.)}$$

Example 3 Find $f^{-1}(x)$ if $f(x) = \frac{4x - 3}{x + 2}$.

SOLUTION Let g be the inverse of f. Then by the definition of an inverse,

$$[f \circ g](x) = x.$$

$$f(x) = \frac{4x - 3}{x + 2}$$

$$[f \circ g](x) = f[g(x)] = \frac{4g(x) - 3}{g(x) + 2}$$

Hence, $\dfrac{4g(x) - 3}{g(x) + 2} = x.$

Now solve for $g(x)$.

$$xg(x) + 2x = 4g(x) - 3$$

$$g(x) = \frac{-2x - 3}{x - 4}$$

$$f^{-1}(x) = \frac{2x + 3}{4 - x}$$

Notice that in example 3 Dom $f = \mathbf{R} - \{-2\}$ and Dom $f^{-1} = \mathbf{R} - \{4\}$. To find the ranges of f and f^{-1}, it is helpful to sketch their graphs.

Note that the graph of f has the line $x = -2$ as a vertical asymptote and the line $y = 4$ as a horizontal asymptote. Likewise, the vertical line $x = 4$ and the horizontal line $y = -2$ are asymptotes of f^{-1}.

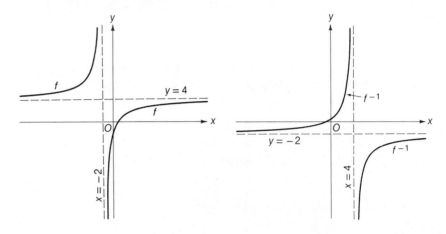

Observe from the graphs that Rng $f = \mathbf{R} - \{4\}$ and Rng $f^{-1} = \mathbf{R} - \{-2\}$, so that Dom $f = $ Rng f^{-1} and Rng $f = $ Dom f^{-1}.

Exercises

A Determine by the method of example 1 the inverse of each function.

1. $f(x) = x - 7$
2. $g(x) = 5x$
3. $h(x) = \frac{2}{3}x$
4. $k(x) = 2x + 3$
5. $f: x \to x$
6. $g(x) = \frac{2}{x}$
7. $h: x \to \frac{x}{3} - 4$
8. $F: x \to \frac{1}{x + 5}$

Use the definition to find the inverse of each function.

9. $f(x) = 2x + 7$
10. $g(x) = 4 - 3x$
11. $f: x \to \frac{1}{3x - 1}$
12. $g: x \to \frac{2}{x + 3}$

B 13. $h(x) = \frac{3x}{x + 2}$
14. $F(x) = \frac{2x}{2 - 5x}$
15. $f: x \to \frac{4x - 3}{5 - 3x}$
16. $g: x \to \frac{3x + 1}{2x + 7}$

Find the domain and range of both f and f^{-1}.

17. $f: x \to \frac{3}{x + 2}$
18. $f: x \to \frac{2x}{x - 1}$

C For the given function g, find all arguments t such that $g(t) = g^{-1}(t)$.

19. $g(x) = \dfrac{1}{2x - 3}$

20. $g(x) = \dfrac{2x}{x - 1}$

Find a function f such that the given equation is satisfied for every $x \in$ Dom f.
[HINT: Use a method similar to the algebraic method of finding an inverse.]

21. $[g \circ f](x) = \dfrac{1}{x - 2}$; $g(x) = \dfrac{2x}{x - 1}$

22. $[H \circ f](x) = x^2 - 3$; $H(x) = \dfrac{1 - 2x}{x + 2}$

23. Consider the set S of all real-valued functions with domain $\mathbf{R}$, together with the operations of function addition and function multiplication.

  a. Identify an identity element for function addition and determine whether each element of S has an inverse function for function addition.

  b. Identify an identity element for function multiplication and determine whether each element of S has an inverse for function multiplication.

7–2 One-to-One Functions

In the previous section you saw how the inverse of a function can be found algebraically from a formula for $f(x)$. The basic principle used was that for every $x \in$ Dom f^{-1}, $[f \circ f^{-1}](x) = x$. This procedure assumes that the given function has an inverse.

Not every function has an inverse. For example, consider h defined by

$$h(x) = |x|.$$

If g were an inverse of h, it would then follow that

$$3 = [g \circ h](3) = g[h(3)] = g(|3|) = g(3), \text{ and}$$
$$-3 = [g \circ h](-3) = g[h(-3)] = g(|-3|) = g(3).$$

But $g(3) = 3$ and $g(3) = -3$ is a contradiction, which implies that h cannot have an inverse.

Now suppose that f is a function which always assigns different values to different arguments. If each of the ordered pairs is reversed, the resulting set of ordered pairs will be a function, since each argument will have exactly one value. Call this function g. Thus if $(a, b) \in f$, then $(b, a) \in g$.

It follows that
$$[g \circ f](a) = g[f(a)] = g(b) = a.$$
Likewise, if $(c, d) \in g$, then $(d, c) \in f$, and
$$[f \circ g](c) = f[g(c)] = f(d) = c.$$
By the definition of an inverse, g is an inverse of f.

The demonstration above proves that a function has an inverse if it has the property that no two arguments have the same value. Such functions are called one-to-one functions. The definition is stated in an equivalent form that is particularly suitable for proofs.

Definition A function f is **one-to-one** if and only if for all arguments a and b $f(a) = f(b)$ implies $a = b$.

You have seen that if a function is one-to-one, then it has an inverse. Now consider the converse: if a function has an inverse, then it is one-to-one.

Assume that g is an inverse of f and that $f(a) = f(b)$. Since $f(a)$ and $f(b)$ are in the domain of g,
$$g[f(a)] = g[f(b)], \text{ or } [g \circ f](a) = [g \circ f](b).$$
Since $[g \circ f](a) = a$ and $[g \circ f](b) = b$, it follows that $a = b$. Thus $f(a) = f(b)$ implies $a = b$, which proves that f is one-to-one.

These results are summarized in the following theorem which provides the standard method of determining whether or not a function has an inverse.

Theorem 7–1 A function has an inverse if and only if it is **one-to-one**.

In view of the central importance of Theorem 7–1, you should be well acquainted with the ways of determining whether or not a function is one-to-one. The definition suggests that a good strategy for proving a function f is one-to-one will be to assume that $f(a) = f(b)$ and then prove that $a = b$.

Example 1 Prove that the function g defined by $g(x) = \dfrac{3x}{x + 2}$ is one-to-one.

SOLUTION Assume that $g(a) = g(b)$. Then
$$\frac{3a}{a + 2} = \frac{3b}{b + 2}.$$
$3ab + 6a = 3ab + 6b$, or $a = b$

Since $g(a) = g(b)$ implies $a = b$, the function is one-to-one.

To prove that a function is *not* one-to-one, you need only exhibit two distinct arguments that have the same value.

Example 2 Prove that $f: x \to \dfrac{1}{x^2 + 1}$ is not one-to-one.

SOLUTION $f(2) = \dfrac{1}{2^2 + 1} = \dfrac{1}{5}$ and $f(-2) = \dfrac{1}{(-2)^2 + 1} = \dfrac{1}{5}$.

Since $f(2) = f(-2)$, f is not one-to-one.

A graph can be helpful in determining whether or not a given function is one-to-one. If any horizontal line intersects the graph in at least two points, then the x-coordinates of those points are distinct arguments that yield the same function value. This implies that the function is *not* one-to-one. On the other hand, if no horizontal line intersects the graph in two or more points, then the function is one-to-one. This test for a one-to-one function is called the **horizontal-line test**.

Example 3 Use the horizontal-line test to determine whether the given function is one-to-one.

a. $f: x \to \dfrac{1}{(x - 2)^2}$ **b.** $h: x \to \dfrac{1}{2 - x}$

SOLUTION
a. You can see from the graph below that the horizontal line $y = 1$, and many others, intersect the graph in two points. So f is *not* one-to-one.

b. The graph below shows that no horizontal line intersects the graph more than once. So h is one-to-one.

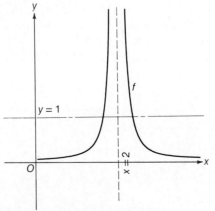
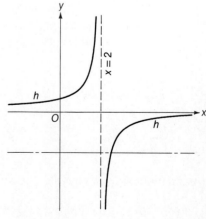

If f is a function of the form $f: x \rightarrow x^{2n-1}$, where $n \in \mathbb{N}$, that is, $f(x)$ is an odd power of x, then no horizontal line intersects the graph of the function in more than one point. Thus the horizontal-line test confirms that odd powers of x define one-to-one functions.

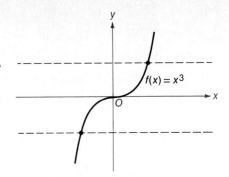

If it is difficult to graph a given function, then the horizontal-line test is not particularly helpful. If, however, the function can be recognized as a composite of one-to-one functions, the following theorem is useful.

Theorem 7–2 If f and g are one-to-one functions, then $f \circ g$ is one-to-one.

PROOF Use the definition, that is, assume that $[f \circ g](a) = [f \circ g](b)$ and show that $a = b$. The assumption is equivalent to

$$f[g(a)] = f[g(b)].$$

Since f is one-to-one, $g(a) = g(b)$, and since g is one-to-one, $a = b$.

Thus $f \circ g$ is also one-to-one.

Example 4 Prove that $H: x \rightarrow \left(\dfrac{3x}{x+2}\right)^3$ is one-to-one.

SOLUTION If $f: x \rightarrow x^3$ and $g: x \rightarrow \dfrac{3x}{x+2}$, then $H = f \circ g$.

As noted above, f is one-to-one, and in example 1 it was proved that g is one-to-one. Thus H is a composite of two one-to-one functions, and by Theorem 7–2, H is one-to-one.

Exercises

A 1. Specify three ways of showing that a function is one-to-one.

2. Without graphing, how can you show that a function is *not* one-to-one?

Use the definition as in example 1 to prove that each function is one-to-one.

3. $f(x) = \dfrac{1}{x}$
4. $g(x) = 3x - 4$
5. $h(x) = 5 - 2x$
6. $k(x) = \dfrac{1}{2x+1}$

7. $F: x \to \dfrac{3}{4 - 3x}$
8. $H: x \to \dfrac{x + 2}{3x}$
9. $g: x \to \dfrac{x}{4x - 1}$
10. $h: x \to \dfrac{2x - 3}{x + 4}$
11. $f(x) = \sqrt{x + 3}$
12. $g(x) = 1 + \sqrt{3 - 4x}$

Sketch a graph of each function and use the horizontal line test to determine whether or not the given function is one-to-one.

13. $f(x) = |x| + 2$
14. $h(x) = \dfrac{1}{x - 5}$
15. $g(x) = (x - 1)^3$
16. $k(x) = x^5 - 4$
17. $f(x) = \dfrac{1}{x + 3}$
18. $g(x) = x^2 + 4$

Determine whether each function has an inverse. Justify your conclusion.

19. $f(x) = x^4 - 3$
20. $g(x) = 2x + 7$
21. $h(x) = \dfrac{1}{4 - x}$
22. $f(x) = 3 - x^2$

B 23. $F(x) = \dfrac{1}{2x^2 + 1}$
24. $G(x) = 4 - |x - 1|$
25. $h(x) = x^3 - 3x$
26. $f(x) = \dfrac{2x + 1}{x}$
27. $f(x) = \dfrac{1}{x^3 + 2}$
28. $k(x) = \sqrt{x - 2}$

29. Prove that if f, g, and h are one-to-one functions, then $f \circ g \circ h$ is one-to-one.
30. If $f(x) = x^3 + k$ and $(-1, 3) \in f^{-1}$, evaluate k.
31. If the graph of g contains $(2, -3)$ and $g^{-1}(x) = \dfrac{kx}{x + 1}$, find k.

Use Theorem 7-2 to show that each function is one-to-one.

32. $f(x) = \dfrac{1}{x^3}$
33. $g: x \to (x - 3)^5$
34. $F(x) = \left(\dfrac{3x}{2x - 1}\right)^7$
35. $h(x) = \dfrac{1}{x^3 - 4}$

C 36. Specify whether each of the following is True or False.
   a. If f is an odd function, then f is one-to-one.
   b. If f and g are one-to-one, then $f + g$ is one-to-one.
   c. If $g = \{(1, -1), (2, -4), (-1, 3)\}$ and $f: x \to x^2$, then $f \circ g$ is one-to-one.
   d. If f has an inverse, then f^{-1} is one-to-one.

Determine whether each function has an inverse.

37. $g: x \to \dfrac{1}{(x + 4)(x - 2)}$ on the domain $(-\infty, -1]$.

38. $k(x) = \begin{cases} 2x - 3 & \text{if } x \neq 2 \\ 7 & \text{if } x = 2 \end{cases}$

7–3 Increasing and Decreasing Functions

The functions whose graphs are shown below are members of an important subclass of one-to-one functions called increasing functions.

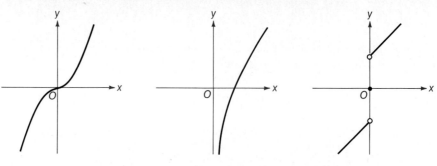

Definition A function f is **increasing** if and only if for all a and b in Dom f,
$$a < b \text{ implies } f(a) < f(b).$$

This definition spells out the property that each of the functions pictured above exhibits: as x increases, the function values become larger. That is, the graph goes upward to the right. What is easy to see graphically can also be determined algebraically.

Example 1 Prove that if $f(x) = 2x + 3$, then f is an increasing function.

SOLUTION Since f is a linear function with slope 2, it is clear from a graph that f is increasing. To prove this algebraically, use the definition. Suppose that a and b are any two numbers such that $a < b$. Then show that $f(a) < f(b)$.

$$a < b$$
$$2a < 2b$$
$$2a + 3 < 2b + 3$$

Since $f(a) = 2a + 3$ and $f(b) = 2b + 3$, the conclusion is that $f(a) < f(b)$ and f is an increasing function.

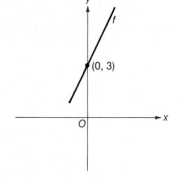

If the graph of a function goes downward to the right, then the function is called a decreasing function.

Definition A function f is **decreasing** if and only if for all a and b in Dom f,
$$a < b \text{ implies } f(a) > f(b).$$

A word of caution: Do not jump to the conclusion that if a function is not increasing then it must be a decreasing function.

Example 2 Prove that the function $g: x \to \dfrac{-1}{x}$ with domain $\mathbf{R} - \{0\}$ is neither increasing nor decreasing.

SOLUTION To prove that g is not increasing, find two numbers a and b such that $a < b$ but $g(a) \geq g(b)$. Let $a = -2$ and $b = 2$. Then $g(-2) = \tfrac{1}{2}$ and $g(2) = -\tfrac{1}{2}$, so that $g(-2) > g(2)$. Thus g is not an increasing function.

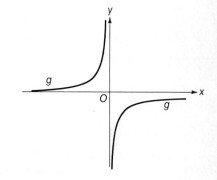

Note also that $g(2) < g(3)$, so g is not a decreasing function either.

Example 3 If $f(x) = \dfrac{1}{x^2 + 1}$ with domain $[0, \infty)$, prove that f is a decreasing function.

SOLUTION Assume that $0 \leq a < b$. Then since a and b are nonnegative,
$$a^2 < b^2.$$
$$a^2 + 1 < b^2 + 1$$
$$\frac{1}{a^2 + 1} > \frac{1}{b^2 + 1}$$
Hence $f(a) > f(b)$ and, by the definition, f is decreasing.

Example 3 shows that sometimes a function which is neither increasing nor decreasing on its natural domain can be made a decreasing function if its domain is appropriately restricted. This leads to the concept of a function that is increasing or decreasing *on an interval*.

Definition A function f is **increasing on an interval** if for all arguments a and b in that interval,
$$a < b \text{ implies } f(a) < f(b).$$

The interval in which a and b lie can be closed like $[c, d]$; open or half open like (c, d) or $(c, d]$; or unbounded like intervals of the form $(-\infty, d)$ or (c, ∞).

Example 4 For the function $f(x) = |x^2 - 4|$, sketch a graph and find the intervals on which f is increasing.

SOLUTION First sketch the graph.
It is evident from the graph that f is increasing on $[-2, 0]$ and on $[2, \infty)$.

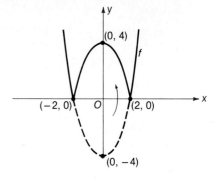

A glance at the graphs below will show that any function of the form $f: x \to x^n$, $n \in \mathbb{N}$, is increasing on $[0, \infty)$. Also, you can see from the graph at the right below that if n is an odd integer, then the function I^n is increasing on its entire domain, $(-\infty, \infty)$. These statements can be proved algebraically.

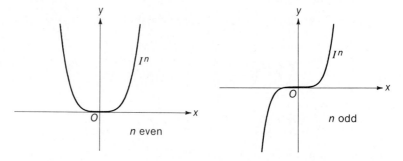

A useful sequence of theorems can be proved about compositions of increasing and decreasing functions. One of them is in the next example; the others are in the exercises.

Example 5 Prove that if f is an increasing function and g is a decreasing function, then $f \circ g$ is a decreasing function.

SOLUTION Assume that f is increasing and that g is decreasing. Assume also that a and b are any two numbers in the domain of $f \circ g$ such that $a < b$.

$a < b$

$g(a) > g(b)$ *g is a decreasing function.*

$f[g(a)] > f[g(b)]$ *f is an increasing function.*

Thus $a < b$ implies $[f \circ g](a) > [f \circ g](b)$, and $f \circ g$ is a decreasing function.

A fundamental property of increasing and decreasing functions is that they all have inverses.

Theorem 7–3 If a function is increasing or decreasing then the function has an inverse.

The proof is left to the exercises.

Exercises

A Classify each function as increasing, decreasing, or neither.

1. $f(x) = 2x + 1$
2. $g(x) = x^2 - 3$
3. $h(x) = |x| + 2$
4. $k(x) = 4 - 3x$
5. $f: x \rightarrow \dfrac{1}{x - 3}$
6. $g: x \rightarrow x^3 - 5$
7. $f(x) = x^2 + x$
8. $h(x) = \dfrac{1}{3x - 1}$

9. Define: A function f is decreasing on an interval.

From the graphs identify the intervals on which the function is decreasing.

10.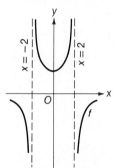
11.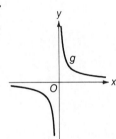
12.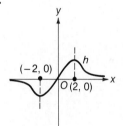
13.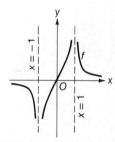

Sketch a graph of each function and find the intervals on which the function is increasing.

14. $g: x \rightarrow (x - 2)^2$
15. $k: x \rightarrow |x + 3|$
16. $f: x \rightarrow |5 - 2x|$
17. $h: x \rightarrow 4 - (x + 1)^2$

B 18. $G(x) = 1 - |x + 3|$
19. $F(x) = x^2 - 4x$
20. $h(x) = \dfrac{1}{x^2 - 9}$
21. $g(x) = x^2 - 4x + 5$

22. $k(x) = |x^3 + 1|$
23. $f(x) = \dfrac{1}{1 - |x|}$
24. $g(x) = \left|\dfrac{x + 1}{x}\right|$
25. $k(x) = \dfrac{1}{x^2 + 2x + 2}$

Prove that the given function is neither increasing nor decreasing.

26. $f: x \rightarrow x^2 + 4x + 3$
27. $h: x \rightarrow \dfrac{1}{|x + 3|}$

Use the definition to prove that the given function is either increasing or decreasing.

28. $f(x) = 2x + 1$

29. $g(x) = 4 - 3x$

30. $h(x) = mx + b; m < 0$

31. $k(x) = mx + b; m > 0$

32. $g(x) = \dfrac{1}{2x - 3}$; Dom $g = (\tfrac{3}{2}, \infty)$

33. $f(x) = \dfrac{2}{4 - x}$; Dom $f = (-\infty, 4)$

34. If f and g are increasing functions, prove that $f \circ g$ is also increasing.

35. If f and g are decreasing functions, prove that $f \circ g$ is increasing.

Rewrite the given function as a composite of two functions and use the results stated in example 5 together with exercises 34 and 35 to determine whether the given function is increasing or decreasing.

36. $f: x \to (3x - 1)^3$

37. $g: x \to (2 - 7x)^3$

38. $F: x \to (-x - x^3)^5$

39. $h: x \to (x^3 + 1)^7$

C 40. If $g: x \to 4 - 3x$, $h: x \to x^3$, and f is decreasing, what can be said of $h \circ f \circ g$? Explain your answer.

41. Prove the following part of Theorem 7–3: If f is increasing, then f has an inverse. [HINT: Assume that f is an increasing function and prove that f is one-to-one.]

42. For each of the following, specify True or False and justify your answer briefly.
  a. If f is one-to-one, then either f is increasing or it is decreasing.
  b. If f and g are increasing, then $f + g$ is increasing.
  c. If f and g are increasing, then $f \cdot g$ is increasing.
  d. If h is decreasing and $F: x \to [2h(x) + 3]^3$, then F is decreasing.

43. Given the function $f: x \to x + \dfrac{1}{x}$.
  a. Prove algebraically that f is decreasing on $(0, 1]$.
  b. Assuming that f is increasing on $[1, \infty)$, sketch a graph of f.

7–4 Graphing Inverses

The statement was made at the beginning of this chapter that a function can have at most one inverse, which is the reason f^{-1} is an unambiguous symbol for the inverse function. We now prove that the inverse is unique.

Theorem 7–4 A given function can have at most one inverse.

PLAN FOR PROOF Show that if g and h are inverses of f, then $g = h$.

PROOF Assume that g and h are inverses of f. Choose any $x \in$ Dom g and let $g(x) = y$. Then since g is an inverse of f, $f[g(x)] = x$. Also $f[g(x)] = f(y)$. So $f(y) = x$.

It follows that $h[f(y)] = h(x)$. And since h is an inverse of f, $h[f(y)] = y$. Hence $h(x) = y$. This proves that for every $x \in$ Dom g, $g(x) = h(x)$.

In the same way it is proved that for every $x \in$ Dom h, $h(x) = g(x)$. Therefore g and h are the same function.

Creating the inverse of a given one-to-one function can be done by simply reversing the elements of each ordered pair of the function. For instance, if

$$f = \{(1, 3), (2, -8), (4, 5)\}, \text{ then}$$

$$f^{-1} = \{(3, 1), (-8, 2), (5, 4)\}.$$

The fact that f^{-1} is obtained by reversing the ordered pairs of f has three direct consequences.

1. Given numbers a and b, $f(a) = b$ if and only if $f^{-1}(b) = a$.

2. Dom $f =$ Rng f^{-1}, and Rng $f =$ Dom f^{-1}.

3. The graph of f^{-1} is the reflection of the graph of f in the line $y = x$.

To justify the third of these observations, note that the line $y = x$ is the perpendicular bisector of the segment that joins the points (a, b) and (b, a). The proof of this fact is left to the exercises.

Since (b, a) lies on the graph of f^{-1} if and only if (a, b) lies on the graph of f, it must be the case that the graph of f^{-1} is the mirror image of the graph of f in the line $y = x$.

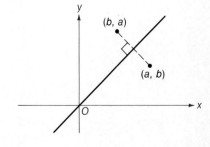

Example 1 For the function f whose graph is shown, sketch the graph of f^{-1} and find the domain and range of f^{-1}.

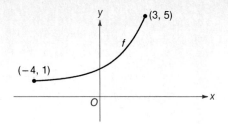

SOLUTION The graph of f^{-1} is obtained by reflecting the graph of f in the line $y = x$.

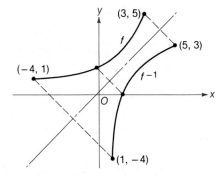

The graph indicates that Dom $f = [-4, 3]$ and Rng $f = [1, 5]$. Hence, Dom $f^{-1} = [1, 5]$ and Rng $f^{-1} = [-4, 3]$.

Reflecting the graph of f in the line $y = x$ to obtain a graph of f^{-1} is made easier if you first find the coordinates (a, b) of two or three points on the graph of f and then plot the corresponding points (b, a) on the graph of f^{-1}. Note that any points where the graph of f intersects the line $y = x$ will also be on the graph of f^{-1}. Finally, you should use the fact that if f has a horizontal asymptote $y = a$, then f^{-1} will have the line $x = a$ as a vertical asymptote. A corresponding observation applies to any vertical asymptote of f.

Example 2 If $f(x) = \dfrac{1}{x - 3}$, sketch a graph of f^{-1} and find its range and domain.

SOLUTION The graph of f is the graph of the reciprocal function $g: x \to \dfrac{1}{x}$ shifted 3 units to the right.

Clearly, f is one-to-one and therefore has an inverse. Since

Dom $f = R - \{3\}$ and Rng $f = R - \{0\}$,

it follows that

Dom $f^{-1} = R - \{0\}$ and
Rng $f^{-1} = R - \{3\}$.

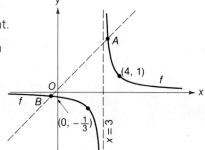

262 Chapter 7: Inverse Functions

To sketch the graph of f use the fact that the x-axis and the line $x = 3$ are asymptotes. Thus the y-axis and the line $y = 3$ will be asymptotes for f^{-1}. Since the graph of f intersects the line $y = x$ at points A and B, these points are also on the graph of f^{-1}.

With fixed points A and B identified and the asymptotes drawn, reflect the graph of f (dashed line) in the identity line to obtain the graph of f^{-1} (solid line).

Inspection of the graph of f^{-1} will confirm earlier conclusions about the domain and range of the inverse function.

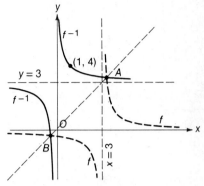

The following theorem can be helpful in sketching the graph of f^{-1} from the graph of f.

Theorem 7-5
a. If f is increasing, then f^{-1} is increasing.
b. If f is decreasing, then f^{-1} is decreasing.

PROOF a. By hypothesis, f is increasing. Thus f has an inverse. Assume that a and b are any arguments in the domain of f^{-1} such that $a < b$. Now show that $f^{-1}(a) < f^{-1}(b)$.

Use an indirect approach. Suppose that
$$f^{-1}(a) \geq f^{-1}(b).$$
Then since f is increasing,
$$f[f^{-1}(a)] \geq f[f^{-1}(b)].$$
But since $f[f^{-1}(x)] = x$ for every $x \in \text{Dom } f^{-1}$, it follows that
$$a \geq b.$$
This is a contradiction of the assumption that $a < b$, and the contradiction implies
$$f^{-1}(a) \not\geq f^{-1}(b),$$
so that $f^{-1}(a) < f^{-1}(b)$
and f^{-1} is an increasing function.

The proof of b is left to the exercises.

Exercises

A
1. If $f = \{(2, -1), (3, 7), (5, -8)\}$, determine f^{-1}.
2. If $g = \{(-4, 7), (\pi, 1), (2, -1)\}$, determine g^{-1}.
3. If the lines $x = -2$ and $y = 5$ are asymptotes of a one-to-one function g, what are the asymptotes of g^{-1}?
4. If the x-axis and the line $x = 4$ are asymptotes of a one-to-one function g, what are the asymptotes of g^{-1}?

For the given one-to-one function f, find the range of f^{-1}.

5. $f: x \to \sqrt{x + 2}$
6. $f: x \to \dfrac{1}{x + 7}$
7. $f: x \to \dfrac{1}{3 - x}$
8. $f: x \to \sqrt{x - 4}$

Sketch a graph of the given function. Use that graph to graph the inverse. Specify the domain and range of the inverse.

9. $f(x) = 2x + 5$
10. $g(x) = 4 - 2x$
11. $h(x) = 6 - x$
12. $f(x) = x + 2$
13. $k(x) = x^3$
14. $F(x) = 1 - x^3$
15. $H(x) = (x + 2)^3$
16. $G(x) = (3 - x)^3$

Sketch a graph of f^{-1} from the given graph of f.

17.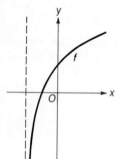
18.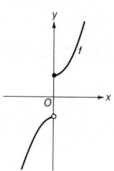
19.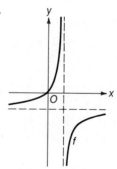
20.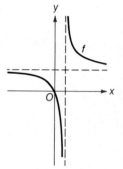

B Find the range of g by solving algebraically for $g^{-1}(x)$ and using Rng g = Dom g^{-1}.

21. $g(x) = \dfrac{x}{3x - 1}$
22. $g(x) = \dfrac{4x + 1}{x - 3}$

Find all points which are common to the graphs of f and f^{-1}.

23. $f(x) = 3x - 1$
24. $f(x) = 5 - 2x$
25. $f(x) = \dfrac{3x}{2x - 1}$
26. $f(x) = \dfrac{x + 9}{3x + 1}$

Sketch a graph of the given function f. Use that graph to sketch a graph of f^{-1}. Specify the domain and range of f and f^{-1}.

27. $f(x) = \dfrac{1}{x + 2}$
28. $f(x) = \dfrac{1}{5 - x}$
29. $f(x) = \dfrac{x + 2}{x}$
30. $f(x) = \dfrac{3 - x}{x}$

31. $f(x) = (x + 2)^2$; Dom $f = [-2, \infty)$

32. $f(x) = \dfrac{1}{x^2 + 1}$; Dom $f = (-\infty, 0]$

33. Given points $P(a, b)$ and $Q(b, a)$, prove that the line $y = x$ is the perpendicular bisector of $\overline{PQ}$.

34. Prove Theorem 7–5b: If a function f is decreasing, then f^{-1} is decreasing.

C 35. Evaluate m in $g(x) = mx + 7$ if the graph of g is also the graph of g^{-1}.

36. Specify True or False for each statement, and justify your conclusion briefly.

   a. If f is decreasing and $a < b$, then $f^{-1}(a) > f^{-1}(b)$.
   b. If $f = f^{-1}$ and Dom $f = \mathbf{R}$, then $f = I$.
   c. If f is increasing and g is increasing, then $f + g$ has an inverse.
   d. If f is decreasing and $f(a) < f(b)$, then $a > b$.

37. Prove that if f is an odd one-to-one function, then f^{-1} is odd.

7–5 Inverses of Power Functions

Functions of the form I^n, $n \in \mathbf{N}$, are commonly called power functions. In this section, we consider inverses of the power functions.

You have already seen that the function I^3 is an increasing function and therefore has an inverse. Suppose that g is the function defined by $g(x) = \sqrt[3]{x}$; then for all $x \in \mathbf{R}$,

$$[I^3 \circ g](x) = (\sqrt[3]{x})^3 = x = \sqrt[3]{x^3} = [g \circ I^3](x).$$

Therefore, $g: x \to \sqrt[3]{x}$ is the inverse of I^3.

The function I^2 presents a problem; it does not have an inverse since it is not a one-to-one function. However, on the interval $[0, \infty)$ it *is* increasing. So, restricting the domain yields the function

$$f = \{(x, y) : y = x^2, x \in [0, \infty)\},$$

which is increasing and therefore has an inverse.

That inverse is

$g : x \to \sqrt{x}$, since for every $x \geq 0$

$$[g \circ f](x) = \sqrt{x^2} = |x| = x \text{ and}$$
$$[f \circ g](x) = (\sqrt{x})^2 = x.$$

Thus $f^{-1}(x) = \sqrt{x}$ and Dom $f^{-1} =$ Rng $f = [0, \infty)$.

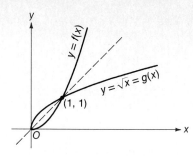

This function g is called the square-root function, and is frequently built into computer languages under the name SQR.

What you have observed about the functions I^2 and I^3 can be extended to all the power functions. If n is odd, then I^n is increasing on $(-\infty, \infty)$ and therefore has an inverse.

If n is even, then I^n is increasing on $[0, \infty)$ and consequently the function

$$f = \{(x, y) : y = x^n, x \in [0, \infty)\}$$

has an inverse.

These results are summarized in the following theorem.

Theorem 7–6 Let $n \in \mathbf{N}$, let $D = \begin{cases} \mathbf{R} & \text{if } n \text{ is odd} \\ [0, \infty) & \text{if } n \text{ is even} \end{cases}$, and let f be the function with domain D defined by $f(x) = x^n$. Then f has an inverse with domain D defined by
$$f^{-1}(x) = \sqrt[n]{x}.$$

The graphs of $\sqrt[n]{x}$, or $x^{1/n}$, can be sketched in the usual manner by reflecting the graphs of the corresponding power functions in the line $y = x$.

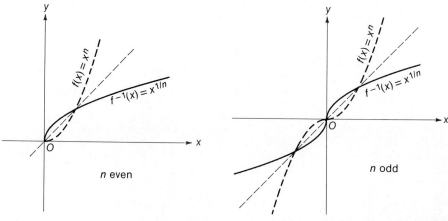

Example 1 Sketch a graph of $f(x) = 1 + \sqrt[3]{x-2}$.

SOLUTION If $g(x) = x + 1$, $h(x) = \sqrt[3]{x}$, and $k(x) = x - 2$, then f is the composite $g \circ h \circ k$. The graph of f is obtained by first sketching the graph of $h(x) = \sqrt[3]{x}$ and then shifting this graph 2 units to the right and 1 unit upward.

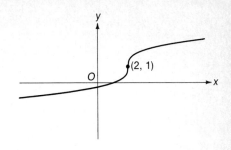

Example 2 If $f: x \to \sqrt{x+4}$, find the domain and range of f^{-1}, determine $f^{-1}(x)$; and sketch a graph of f^{-1}.

SOLUTION Since Dom $f = [-4, \infty)$ and Rng $f = [0, \infty)$, it follows that

Dom $f^{-1} = [0, \infty)$ and Rng $f^{-1} = [-4, \infty)$.

To find $f^{-1}(x)$, let $g = f^{-1}$ and note that

$[f \circ g](x) = x$ and $[f \circ g](x) = \sqrt{g(x) + 4}$.

$$\sqrt{g(x) + 4} = x$$
$$g(x) + 4 = x^2$$
$$f^{-1}(x) = x^2 - 4$$

Notice that since Dom $f^{-1} = [0, \infty)$, the graph of f^{-1} includes only one side of the parabola $y = x^2 - 4$.

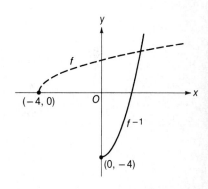

Example 3 Locate between consecutive integers all the solutions of the equation $(x-1)^3 = \sqrt{x+4}$.

SOLUTION From the graphs of

$f: x \to (x-1)^3$ and

$g: x \to \sqrt{x+4}$,

you can see that there is only one such solution and it is between $x = 2$ and $x = 3$.

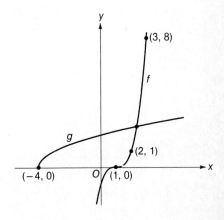

Example 4 Solve for x: $\sqrt{x + 5} \leq \sqrt{3 - 2x}$.

SOLUTION For this inequality the domain is
$$[-5, \infty) \cap (-\infty, \tfrac{3}{2}] = [-5, \tfrac{3}{2}].$$
On this domain, both $\sqrt{x + 5}$ and $\sqrt{3 - 2x}$ are nonnegative. For nonnegative a and b, $a < b$ if and only if $a^2 < b^2$. So squaring both sides of the given inequality yields an equivalent inequality on the domain $[-5, \tfrac{3}{2}]$.

$$\sqrt{x + 5} \leq \sqrt{3 - 2x}$$
$$x + 5 \leq 3 - 2x$$
$$3x \leq -2$$
$$x \leq -\tfrac{2}{3}$$

ANSWER The solution is the set $\left(-\infty, -\tfrac{2}{3}\right] \cap \left[-5, \tfrac{3}{2}\right]$, or $\left[-5, -\tfrac{2}{3}\right]$.

Exercises

A Specify the domain and range and sketch a graph of each function.

1. $f: x \to x^{1/5}$
2. $g: x \to \sqrt[4]{x}$
3. $h(x) = \sqrt{-x}$
4. $k: x \to 2 + \sqrt[3]{x}$
5. $f: x \to 1 - \sqrt{x}$
6. $g: x \to \sqrt{x + 2}$
7. $h(x) = \sqrt[3]{x - 1}$
8. $k(x) = \sqrt[3]{8 - x}$
9. $f(x) = 1 + \sqrt{x - 3}$
10. $g(x) = 2 - \sqrt{x + 4}$
11. $h(x) = 3 + \sqrt[3]{x + 8}$
12. $k(x) = |\sqrt[3]{x + 1}|$

For each graph, specify a function that has the properties pictured.

13.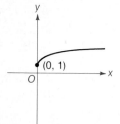
14.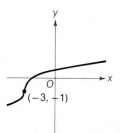
15.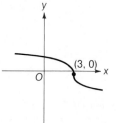
16.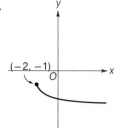

Specify the domain and range and sketch a graph of each function.

17. $f(x) = \dfrac{1}{\sqrt[3]{x}}$
18. $g(x) = \dfrac{1}{\sqrt{x + 4}}$
19. $h(x) = \dfrac{1}{\sqrt{x - 9}}$
20. $k(x) = \dfrac{1}{\sqrt{x + 2}}$
21. $f(x) = \dfrac{-1}{\sqrt{x + 3}}$
22. $g(x) = \dfrac{1}{\sqrt{4 - x}}$

B For each function f, find the domain and range of f^{-1}, determine $f^{-1}(x)$, and sketch a graph of f^{-1}.

23. $f: x \to \sqrt{x+3}$

24. $f: x \to \sqrt[3]{x+1}$

25. $f: x \to x^3 + 1$

26. $f: x \to 2 - \sqrt{x}$

27. $f: x \to \dfrac{1}{\sqrt{x-4}}$

28. $f: x \to \dfrac{1}{\sqrt{x+2}}$

By considering graphs of appropriate functions, solve each inequality.

29. $x^{1/3} < x$

30. $x^{1/3} > x^2$

31. $x^{1/2} < x^2$

32. $x^{1/4} < x$

By considering graphs of appropriate functions, locate between consecutive integers all solutions of each equation.

33. $\sqrt{x+4} = x^2$

34. $\sqrt[3]{x} = 1 - x^2$

C 35. $x^2 = 1 - \sqrt[3]{x-4}$

36. $\dfrac{1}{\sqrt{2-x}} = -(x-3)^3$

37. $(x-3)^4 + \sqrt{x+2} = 4$

38. $\dfrac{1}{x^2 - 1} + 1 - \sqrt[3]{x} = 0$

Solve the following inequalities.

39. $\sqrt{x+5} \le \sqrt{3-2x}$

40. $\sqrt{4x+1} < \sqrt{x+3}$

Use appropriate graphs as an aid in solving the following inequalities.

41. $\sqrt{x+8} \ge |x+2|$

42. $\sqrt{x-5} + 7 \le x$

43. Solve to the nearest tenth: $x^3 + \sqrt{x+4} = 0$

7–6 Inverses of Composite Functions

Suppose you know the inverses of functions f and g. Now consider how to find the inverse of the composite $f \circ g$. The idea is quite simple. Suppose that F denotes the process "put on shoes" and G denotes "put on socks." Each morning in the course of getting dressed you perform a ritual that can be expressed concisely as $F \circ G$, taking care to make G the inner function. At night you carry out the inverse process $(F \circ G)^{-1}$. Clearly $G^{-1} \circ F^{-1}$ will produce the desired result, which suggests that

$$(F \circ G)^{-1} = G^{-1} \circ F^{-1}.$$

Notice the crucial reversal in the order of operations.

Consider now an algebraic example. Let $h: x \to x^3 + 1$. Using the definition of the inverse and the method shown in section 7–1, you can confirm that $h^{-1}(x) = \sqrt[3]{x-1}$.

Now observe that if $f: x \to x + 1$ and $g: x \to x^3$, then $h = f \circ g$. Since

$f^{-1}(x) = x - 1$ and $g^{-1}(x) = \sqrt[3]{x}$,

$$[g^{-1} \circ f^{-1}](x) = g^{-1}(x - 1)$$
$$= \sqrt[3]{x - 1}.$$

Thus $h^{-1}(x) = [f \circ g]^{-1}(x) = [g^{-1} \circ f^{-1}](x)$. In other words,

$$(f \circ g)^{-1} = g^{-1} \circ f^{-1},$$

a conclusion that corresponds to the "shoes-and-socks" analogy.

Theorem 7–7 If f and g have inverses, then $f \circ g$ has an inverse and $(f \circ g)^{-1} = g^{-1} \circ f^{-1}$.

PROOF
$$[(f \circ g) \circ (g^{-1} \circ f^{-1})](x) = f(g \circ g^{-1}[f^{-1}(x)])$$
$$= f[f^{-1}(x)]$$
$$= x$$

Similarly,

$$[(g^{-1} \circ f^{-1}) \circ (f \circ g)](x) = x.$$

Therefore, $(f \circ g)^{-1} = g^{-1} \circ f^{-1}$.

Example 1 If $f: x \to \dfrac{1}{(x - 2)^3}$, determine $f^{-1}(x)$.

SOLUTION Let $g: x \to \dfrac{1}{x}$ and $h: x \to x - 2$. Then

$$f = g \circ I^3 \circ h \text{ and}$$
$$f^{-1} = h^{-1} \circ (I^3)^{-1} \circ g^{-1}.$$

But $h^{-1}: x \to x + 2$, $(I^3)^{-1}: x \to \sqrt[3]{x}$, and $g^{-1}: x \to \dfrac{1}{x}$.

Hence $f^{-1}(x) = \sqrt[3]{\dfrac{1}{x}} + 2$.

Like the even power functions, there are many other functions which are not one-to-one on their domains and therefore do not have inverses. At times, however, it is important to be able to achieve an effect much like taking an inverse of such a function. The key is to restrict the domain so that the function is one-to-one on the restricted domain. The function thus modified will have an inverse.

Example 2 Limit the domain of the function $G: x \to x^2 - 4$ to obtain a function g which has an inverse. Then specify the domain and range of g^{-1}, sketch its graph, and determine a formula for $g^{-1}(x)$.

(cont. on p. 271)

SOLUTION The graph of G is a parabola with vertex $(0, -4)$. Clearly G is decreasing on $(-\infty, 0]$ and increasing on $[0, \infty)$. Define g by restricting the domain to $(-\infty, 0]$:

$$g = \{(x, y) : y = x^2 - 4, x \le 0\}.$$

Thus, Dom $g = (-\infty, 0]$ and Rng $g = [-4, \infty)$. Since g is decreasing, g has an inverse with Dom $g^{-1} = [-4, \infty)$ and Rng $g^{-1} = (-\infty, 0]$. Let $h: x \to x - 4$ and $f: x \to x^2$, $x \le 0$. Then

$$g = h \circ f \text{ and } g^{-1} = f^{-1} \circ h^{-1} \quad \text{(Theorem 7-7).}$$

But $h^{-1}: x \to x + 4$ and $f^{-1}: x \to -\sqrt{x}$ (why not $\sqrt{x}$?). So

$$g^{-1}(x) = -\sqrt{x + 4} \text{ with Dom } g^{-1} = [-4, \infty).$$

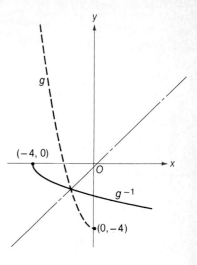

In example 2, by defining g with domain $(-\infty, 0]$, we chose only one of many possible ways of restricting the domain of the original function G in order to obtain a one-to-one function. Other possible restricted domains are $[0, \infty)$, $[10, \infty)$, and $[-100, -10]$.

Exercises

A For each function f express f as a composite of two or more functions. Then apply Theorem 7-7 to obtain a formula for $f^{-1}(x)$.

1. $f(x) = x^3 - 1$
2. $f(x) = 2x + 7$
3. $f(x) = \dfrac{1}{x + 4}$
4. $f(x) = \sqrt[3]{x - 5}$
5. $f(x) = \sqrt[3]{x + 3}$
6. $f(x) = \dfrac{1}{3 - x}$
7. $f(x) = (x - 2)^5$
8. $f(x) = \sqrt[3]{x} - 7$

B 9. $f(x) = (x - 2)^3 - 3$
10. $f(x) = \dfrac{1}{2x + 9}$
11. $f(x) = \dfrac{1}{(x + 2)^3} + 1$
12. $f(x) = \dfrac{1}{\sqrt[3]{x - 4}} - 3$
13. $f(x) = \dfrac{1}{2 + \sqrt{x - 5}}$
14. $f(x) = \dfrac{2}{1 + \sqrt{4 - x}}$

For each function F, restrict the domain to obtain a function f that has an inverse. Specify the domain and range of f^{-1}, sketch its graph, and determine a formula for $f^{-1}(x)$.

15. $F(x) = x^2 + 1$
16. $F(x) = x^4 - 3$
17. $F(x) = \dfrac{1}{9 - x^2}$

C 18. $F(x) = \dfrac{1}{x^2 - 4}$
19. $F(x) = (x - 2)^2$
20. $F(x) = \dfrac{1}{x^2 + 1}$

Special Groups

You know the following facts about integers: the integers are closed under addition; addition is an associative operation; 0 is an integer, and every integer has an additive inverse which is itself an integer. These 4 facts imply that the set of integers together with the operation of addition constitute a fundamental algebraic structure known as a group.

Definition A set S together with an operation # is a **group** if and only if
a. S is closed under #;
b. # is an associative operation on S;
c. S contains an identity for #; and
d. every member of S has a # inverse.

The symbol $\langle S, \# \rangle$ can be used to denote such a group.

By checking the 4 requirements in the definition, you can confirm that the set of positive numbers **P** under the operation of multiplication is also an example of a group.

In the examples above, $\langle \mathbf{J}, + \rangle$ and $\langle \mathbf{P}, \cdot \rangle$, the underlying sets were sets of numbers and the operations were arithmetic operations. Such narrow restrictions are not necessary. To show this, consider the set $\mathscr{L}$ of all linear functions and the operation of function composition.

Recall that a linear function is defined as a function f such that, for some pair of constants m and b, $m \neq 0$, $f(x) = mx + b$ for all $x \in \mathbf{R}$. Now let g and h be any two linear functions such that $g(x) = nx + c$ and $h(x) = px + d$, with $n \neq 0$ and $p \neq 0$. Then
$$[g \circ h](x) = g[h(x)] = g[px + d]$$
$$= n(px + d) + c$$
$$= (np)x + (nd + c).$$

Since $n \neq 0$ and $p \neq 0$, $np \neq 0$. Thus $g \circ h$ is a linear function with $m = np$ and $b = nd + c$. This proves that the set of all linear functions is closed under function composition.

Next note that the identity $I: x \to x$ is a linear function with $m = 1$ and $b = 0$.

Now suppose that $g: x \to nx + c$, with $n \neq 0$, is any linear function. Then $h: x \to \dfrac{x}{n} - \dfrac{c}{n}$ is a linear function with $m = \dfrac{1}{n}$ and $b = -\dfrac{c}{n}$. You can confirm that for every $x \in \mathbf{R}$

$$[g \circ h](x) = [h \circ g](x) = x,$$

and thus show that $g^{-1} = h$. This will prove that every linear function has an inverse under function composition and that this inverse is itself a linear function.

Finally, recall that function composition is associative on the set of all functions so that this operation is associative on $\mathscr{L}$. It follows that $\langle \mathscr{L}, \circ \rangle$ consisting of the linear functions under function composition is a group.

The study of groups has its origins in the work of Evariste Galois, a brilliant young French mathematician who died in 1832. Death did not come unexpectedly to Galois. Facing a duel, prompted by his political feelings, he spent the night before hastily outlining the results of his mathematical research (a margin note said, "I have no time"). The next day he was shot and killed in the duel.

Attached to his final writings was a letter which read in part:

"Ask Jacobi or Gauss publicly to give their opinion, not as to the truth, but as to the importance of the theorems. Subsequently there will be, I hope, some people who will find it to their profit to decipher all this mess."

Over the past one hundred fifty years, Galois's last hope has been more than fulfilled as mathematicians have studied his ideas with intense interest and found them enormously productive. The theorems he referred to were directed to the theory of equations but in fact contained the central concepts of what has come to be known as group theory. Though Galois received disgracefully little attention during his lifetime, he is now recognized as the father of group theory. Applications of this vital branch of modern mathematics are of particular importance in nuclear physics.

Perhaps the most remarkable aspect of Galois's achievement was that he was killed five months before his twenty-first birthday.

Exercises

Determine whether each structure is a group.

1. $\langle J, \cdot \rangle$
2. $\langle W, + \rangle$
3. $\langle S, \cdot \rangle$, where $S = \{x \in \mathbf{R} : 0 < x \leq 1\}$
4. $\langle T, \odot \rangle$, where $T = \{1, 2, 3, 4\}$ and $\odot$ is defined for all $a, b \in T$ by $a \odot b$ is the remainder when ab is divided by 5.
5. Let $S = \{1, 2, 3, \cdots, 10, 11, 12\}$ and define clock addition, $\oplus$, for all $c, d \in S$ by $c \oplus d$ is the time of day or night d hours after c o'clock.

    a. Complete the table for $\oplus$.
    b. Illustrate that $\oplus$ is an associative operation on S by considering $7 \oplus (3 \oplus 10)$.
    c. What numbers are their own inverses under $\oplus$?
    d. Is $\langle S, \oplus \rangle$ a group? Justify your answer.

$\oplus$	1	2	3	$\cdots$	12
1					
2					
3					
$\vdots$					
12					

6. Consider the following set of motions carried out on an equilateral triangle ABC.

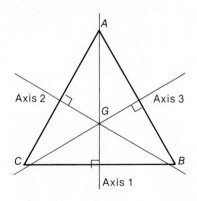

R_1 Rotate about G 120° clockwise
R_2 Rotate about G 240° clockwise
R_3 Rotate about G 360° clockwise
S_1 Spin about axis 1 one-half turn
S_2 Spin about axis 2 one-half turn
S_3 Spin about axis 3 one-half turn

Let $T = \{R_1, R_2, R_3, S_1, S_2, S_3\}$ and let $*$ be the operation defined for all motions X and Y in T by $X * Y$ is the motion Y followed by the motion X. Notice that $R_1 * R_2$ has the same effect as R_3, and that $R_2 * S_1$ has the same effect as S_2 alone.

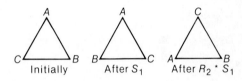

Initially   After S_1 After $R_2 * S_1$

These results are indicated by writing

$R_1 * R_2 = R_3$ and $R_2 * S_1 = S_2$.

a. Determine $R_1 * S_2$.
b. Complete the table below for $*$.

$*$	R_1	R_2	R_3	S_1	S_2	S_3
R_1			R_3			
R_2					S_2	
R_3						
S_1						
S_2						
S_3						

c. Is there a $*$ identity?
d. What is the $*$ inverse for R_2?
e. Illustrate that $*$ is associative on T by considering $R_2 * (S_2 * R_1)$.
f. Is $\langle T, * \rangle$ a group? Justify your answer.

Chapter Review Exercises

7–1, page 247

1. Use the definition of an inverse to find $g^{-1}(x)$ if $g(x) = \dfrac{2x}{x+1}$.

7–2, page 251

2. Which of the following functions are one-to-one?

    a. $f(x) = (x-3)^2$ **b.** $g(x) = |x+1|$ **c.** $h(x) = \dfrac{1}{x-5}$

3. Prove that the function $f: x \to \dfrac{1}{2x-7}$ is one-to-one.

7–3, page 256

4. Classify each function as increasing, decreasing, or neither.

    a. $f(x) = 3 - 2x$ **b.** $h(x) = x^3 + 4$ **c.** $g(x) = x^2 - 4$

5. Use the definition to prove that $g: x \to 5x - 9$ is increasing.

7–4, page 261

6. Graph $f: x \to \dfrac{1}{x^2 - 4}$. Identify intervals where f is increasing.

7. If f and g are decreasing functions, prove that $f + g$ is decreasing.

8. Find the domain and range of f^{-1} if $f(x) = \dfrac{1}{1-2x}$.

9. The graph of g^{-1} contains $(-1, 3)$ and $g(x) = \dfrac{x}{2x+k}$. Find $g(2)$.

7–5, page 265

10. Sketch a graph of $f(x) = \sqrt{x-4}$. Use this graph to graph f^{-1} and specify the domain and range of f^{-1}.

11. Specify the domain of the inequality $\sqrt{x+1} < \sqrt{1-2x}$.

12. If $h(x) = |1 + \sqrt[3]{x}|$, specify the domain and range of h and sketch its graph.

7–6, page 269

13. If $g(x) = \dfrac{1}{\sqrt[3]{x+8}}$, express g as a composite and use Theorem 7–7 to obtain a formula for $g^{-1}(x)$.

14. If $F(x) = x^2 - 4x$, restrict the domain so as to obtain a function f that has an inverse. Specify the domain and range of f^{-1}, sketch its graph, and find a formula for f^{-1}.

Chapter Test

1. If $f = \{(1, 3), (-2, \sqrt{3}), (7, 12)\}$, what is f^{-1}?

2. Given that h is a one-to-one function, solve for x: $h(2x - 1) = h(3)$.

3. If $f(x) = \dfrac{1}{x - 2}$, what lines will be asymptotes of f^{-1}?

4. Use the definition of an inverse to find $f^{-1}(x)$ if $f(x) = \dfrac{3x}{x - 4}$.

5. Which of the following functions are one-to-one?

   a. $f(x) = \dfrac{1}{x - 3}$ b. $g(x) = \sqrt{x + 1}$ c. $h(x) = 2 + |x|$

6. Sketch a graph of the function $f: x \to \sqrt[3]{x - 2} + 1$. Label coordinates of two points on the graph.

7. If $h(x) = (x + 4)^3$, express h as a composite and then obtain a formula for $h^{-1}(x)$.

8. If g is a decreasing function and $h: x \to [g(x)]^5$, is h increasing, decreasing, or neither? Justify your answer briefly.

9. Sketch a graph of $f(x) = 3 + \sqrt{x}$, use this graph to graph f^{-1}, and specify the domain and range of f^{-1}.

10. Sketch a graph of the function $g: x \to \dfrac{1}{9 - x^2}$. Identify any intervals on which g is decreasing.

11. Consider the graph of $f: x \to \sqrt{x - 4}$ and solve the inequality $\sqrt{x - 4} < 3$.

12. For the function $F(x) = (x + 2)^2$, restrict the domain so as to obtain a function f that has an inverse. Specify the domain and range of f^{-1}, sketch its graph, and find $f^{-1}(x)$.

Exponential and Logarithmic Functions

8

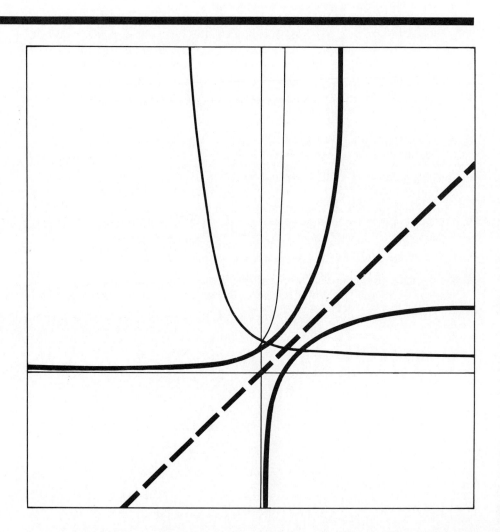

8-1 Exponential Functions

In August 1980, newspapers reported an analysis of India's food production. These reports highlighted the fact that India was growing enough food to feed its population of 630 million people.

While food production was rising, population growth was slowing down. The population growth rate had declined from 2.2% in the late 1960s to just under 2%. It was expected to fall to about 1.6% after 1985.

The difference between the growth rates of 2.2% and 1.6% is of greater consequence than you might think. For example, with a 2.2% growth rate over the next 20 years, India's 1980 population of 630 million would grow to over 973 million in the year 2000. With a growth rate of 1.6%, the population would be about 865 million at the end of the century. A decrease of only 0.6% in the growth rate means 108 million fewer people to feed in 2000.

Mathematically, constant growth rates can be described by exponential functions. These functions are useful in describing a host of other phenomena, such as radioactive decay, chemical reactions, heating phenomena, and transient electrical states.

To begin a study of exponential functions, consider the function $f: x \rightarrow 2^x$. Plot a few points of the graph that correspond to integer arguments. Since $2^{p/q} = (2^{1/q})^p$ is defined for all integral values of p and q except $q = 0$, the domain of f includes all rational numbers.

In fact, by using only the square root key on a calculator, you can compute decimal approximations for values generated from arguments such as $\frac{1}{2}, \frac{5}{2}, -\frac{3}{2}$, and even values like $f(\frac{3}{4})$ and $f(-\frac{1}{8})$. [How would you obtain $2^{1/8}$ using the square root function?] By also plotting points such as $(\frac{3}{2}, 2\sqrt{2})$, $(\frac{1}{4}, \sqrt[4]{2})$, and $\left(-\frac{1}{2}, \frac{\sqrt{2}}{2}\right)$, you can see that they fall nicely on a smooth curve through the previously plotted points.

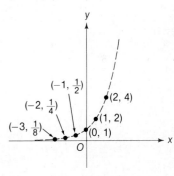

x	2^x
-3	$\frac{1}{8}$
-2	$\frac{1}{4}$
-1	$\frac{1}{2}$
0	1
1	2
2	4

But there are infinitely many points on this curve for which you do not have a method of calculating values, namely, the points that correspond to irrational values of x such as π and $\sqrt{3}$.

Even though $f: x \rightarrow 2^x$ is, at this time, defined only for rational arguments, the function is interesting. The graph shows that it is an increasing function and has positive values. Recall that

$$2^{a+b} = 2^a \cdot 2^b, \quad 2^{a-b} = 2^a \div 2^b, \quad \text{and} \quad 2^{ab} = (2^a)^b.$$

It follows that for this exponential function f, and for all $a, b \in \mathbf{Q}$,

1. $f(a + b) = f(a) \cdot f(b)$,
2. $f(a - b) = f(a) \div f(b)$, and
3. $f(ab) = [f(a)]^b$.

There is nothing unique about 2 as the number selected as the base in the function f; you can create similar functions using any number b as long as $b > 0$ but $b \neq 1$. (If $b = 1$, all values of the function are 1.) So rather than dealing only with the function $f: x \rightarrow 2^x$, it will save time to consider the whole class of exponential functions of the form

$$g: x \rightarrow b^x,\ b > 0 \text{ and } b \neq 1.$$

The domain of such a function is the set of rational numbers, since b^x has not yet been defined if x is an irrational number.

A calculator with a $\boxed{Y^x}$ key makes the computation of exponential function values very easy.

Example 1 Evaluate 13^4 by using a calculator.

SOLUTION Use the key sequence $\boxed{1}\boxed{3}\ \boxed{Y^x}\ \boxed{4}\ \boxed{=}$ to find that $13^4 = 28{,}561$.

Example 2 Use a calculator to evaluate $0.435^{13/29}$ to 3 decimal places.

SOLUTION Note the use of parentheses in the key sequence

$\boxed{0}\boxed{.}\boxed{4}\boxed{3}\boxed{5}\ \boxed{Y^x}\ \boxed{(}\ \boxed{1}\boxed{3}\ \boxed{\div}\ \boxed{2}\boxed{9}\ \boxed{)}\ \boxed{=}$,

which yields 0.689, accurate to 3 decimal places.

A fundamental property of exponential functions is that

if $b > 1$, then $g: x \rightarrow b^x$ is increasing, and

if $0 < b < 1$, then $g: x \rightarrow b^x$ is decreasing.

You can convince yourself of the validity of this statement by considering graphs on the domain $\mathbf{Q}$ of

$f: x \rightarrow 2^x$ and $g: x \rightarrow \left(\frac{1}{2}\right)^x$.

This means that to solve an equation like $5^x = 5^3$, you simply observe that the function $f(x) = 5^x$ is increasing and therefore one-to-one, and conclude that 3 is the only solution.

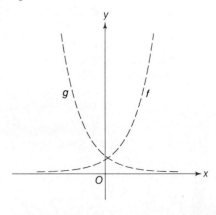

Example 3 Solve the equation $2^x = 4\sqrt{2}$ on the domain **Q**.

SOLUTION First, express $4\sqrt{2}$ as a power of 2.
$$2^x = 2^2 \cdot 2^{1/2} = 2^{5/2}$$
Let $g: x \to 2^x$. Then $g(x) = g(\frac{5}{2})$. Since g is increasing, g is one-to-one. Therefore $x = \frac{5}{2}$.

To solve exponential inequalities, use the fact that the exponential functions are either increasing or decreasing.

Example 4 Solve on the domain **Q**: $\left(\frac{1}{3}\right)^x < 81$.

SOLUTION Both sides of the inequality are powers of 3, so the inequality can be rewritten as
$$(3^{-1})^x < 3^4, \text{ or } 3^{-x} < 3^4.$$
Let $h(x) = 3^x$ with domain **Q**; then $3^{-x} < 3^4$ is equivalent to $h(-x) < h(4)$. Since h is increasing, it follows that $-x < 4$. The solution set is
$$\{x \in \mathbf{Q}: x > -4\}.$$

Example 5 The population of India was 616.4 million in January, 1979, and was 630 million a year later. Assume that the annual rate of population increase remains constant for the remainder of the century. Estimate India's population in January 2000.

SOLUTION Let x be the number of years elapsed since January, 1979. Let P be the function with domain $\{x \in \mathbf{N}: 0 \leq x \leq 21\}$ that assigns to an argument x a value which is the population of India in the year $1979 + x$.
$$P(0) = 616.4 \times 10^6$$
$$P(1) = 630 \times 10^6$$
What is $P(21)$? To say that the annual rate of population growth remains constant through 2000 A.D. means that for some constant k,
$$\frac{P(1)}{P(0)} = \frac{P(2)}{P(1)} = \cdots = \frac{P(x+1)}{P(x)} = \cdots = \frac{P(21)}{P(20)} = k.$$
$$P(1) = kP(0)$$
$$P(2) = kP(1) = k[kP(0)] = k^2 P(0)$$
$$P(3) = kP(2) = k[k^2 P(0)] = k^3 P(0)$$
$\cdots$

(cont. on p. 280)

$$P(x) = k^x P(0)$$

Now use known values for $P(0)$ and $P(1)$ to determine k.

$$k = \frac{P(1)}{P(0)} = \frac{630 \times 10^6}{616.4 \times 10^6} \doteq 1.022$$

$$P(x) \doteq (1.022)^x (616.4 \times 10^6)$$

In the year 2000, the population will be

$$P(21) \doteq (1.022)^{21}(616.4 \times 10^6)$$
$$\doteq 973.5 \times 10^6.$$

Example 6 Iromie deposits $200 in a bank which pays interest annually at a rate of 6%. How much money will be in her account at the end of 8 years?

SOLUTION At the end of one year Iromie's account will contain $200(1.06)$ dollars; at the end of two years it will contain $[200(1.06)][1.06]$, or $200(1.06)^2$ dollars. At the end of t years the amount $A(t)$ in the account will be

$$A(t) = 200(1.06)^t.$$

After 8 years Iromie will have in her account

$$A(8) = 200(1.06)^8,$$

or approximately $318.77.

Exercises

A Evaluate each expression to 3 decimal places by using a calculator.

1. a. 2.13^4 b. 3.23^{-3} c. $\sqrt[3]{2}$ d. $3.18^{2/3}$
2. a. 1.73^5 b. $\sqrt[7]{32.6}$ c. 41^{-6} d. $1.026^{4.3}$

Solve each equation on the domain Q.

3. $2^x = 16$
4. $3^x = 27$
5. $3^x = 3\sqrt{3}$
6. $2^x = 8\sqrt{2}$
7. $5^y = 1$
8. $1 - 13^y = 0$
9. $4^y = 2$
10. $9^y = \sqrt{3}$
11. $\left(\frac{1}{3}\right)^x = 9$
12. $\left(\frac{1}{4}\right)^x - 64 = 0$
13. $2^{-x} = 32$
14. $3^{-x} = \frac{1}{81}$
15. $3^{t-2} = 243$
16. $5^{2t-1} = 125$

B 17. $\left(\frac{1}{2}\right)^{1-2t} = 16$
18. $\left(\frac{3}{4}\right)^{3-t} - \frac{9}{16} = 0$
19. $\frac{4^x}{3^{x+1}} = \frac{9}{64}$
20. $\left(\frac{2}{3}\right)^x = \sqrt{\frac{3}{2}}$

Solve each inequality on the domain Q.

21. $3^x < 27$ **22.** $2^x \geq 64$ **23.** $\left(\frac{1}{2}\right)^u \geq 8\sqrt{2}$

24. $3^{-u} < 9\sqrt{3}$ **25.** $4^{2v-1} < 64$ **26.** $\left(\frac{1}{5}\right)^v > 125$

27. If the population of Mexico increases at a rate of 3% each year, and if the population in 1975 was 28 million, what will be the population in the year 2000?

28. If the population of India was 630 million in 1980, and if the annual growth rate was 2.3% between 1950 and 1980, what was the population in 1950?

29. If $100 is invested in an account which pays interest annually, how much money will an investor have in the account at the end of 10 years if the interest rate is 6%? 9%? 12%?

30. In 1980, a small pizza with salami, peppers, mushrooms, tomatoes, sprinkled with walnuts and peppermint candy cost $2.65. If the annual rate of inflation remains at 10%, what would be the price of the same pizza in 1995?

31. Suppose that a ham sandwich costs $1.20 on your eighteenth birthday. If the annual rate of inflation remains constant at 9%, what would you expect to pay for the same sandwich when you retire at age 65?

32. An executive buys an automobile for $23,500. During its long life, the car is sold repeatedly, each time bringing the seller 72% of the price paid for it. How much did the 14th owner pay for the car?

C 33. If a basketball player ordinarily sinks $\frac{4}{5}$ of his foul shots, then the probability that in n free throws he will make all of his shots is given by the function $f(n) = \left(\frac{4}{5}\right)^n$.

What is the greatest number of foul shots he can take in a game and have at least a 25% chance of sinking all of his free throws? [HINT: Use a trial-and-error solution.]

34. A rubber ball has the property that when it is dropped it rebounds to 0.7 the height from which it was dropped. If a ball is dropped from a height of 8 feet,
  a. To what height does it rebound on the 15th bounce?
  b. After how many bounces will the rebound be less than 1 foot?

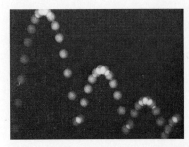

Use a number-line analysis to solve for x on the domain Q.

35. $(x + 1)(2^x - 1) < 0$ **36.** $\dfrac{9 - 3^x}{x^2 - 16} \leq 0$

37. Assume that $700 is invested in a bank that pays 6% interest annually. To determine by trial and error the number of years that must pass until the value of the account exceeds $8500, write a computer program that will print out the value of the investment at the end of each year from 5 to 15 years.

8–2 Exponential Functions on *R*

In section 8–1, exponential functions were introduced but the domain was limited to **Q**. This was necessary because previously no meaning had been attached to irrational exponents. Thus it was not possible to find function values for arguments like π and $\sqrt{2}$. Now we shall define irrational exponents and thereby be able to extend the domain of the exponential functions to include all real numbers. This will be done in such a way that

a. all of the exponential properties apply on the domain **R**, and

b. the graph of any exponential function will be a smooth continuous curve.

Before giving the general definition for irrational exponents, we will show how 2^π is defined. The completeness axiom, which states that every nonempty set of real numbers which has an upper bound has a *least* upper bound, provides the basis for the definition.

Consider the set $S = \{2^x : x \in \mathbf{Q} \text{ and } x < \pi\}$. Notice that $2^{3.1} \in S$ because $3.1 \in \mathbf{Q}$ and $3.1 < \pi$; on the other hand $2^4 \notin S$, since $4 \not< \pi$. But $2^4 = 16$ is an upper bound of *S* as the following argument shows.

Let $2^x \in S$. Then $x \in \mathbf{Q}$ and $x < \pi < 4$. But since x and $4 \in \mathbf{Q}$, you know that $2^x < 2^4$. Thus every member of S is less than 16, and 16 is therefore an upper bound of S.

Since S is nonempty and has an upper bound, the completeness axiom implies that S has a least upper bound, and this is the number which will be designated 2^π.

Definition If k is an irrational number, then b^k is the least upper bound of

a. $\{b^x : x \in \mathbf{Q} \text{ and } x < k\}$ if $b > 1$;

b. $\{b^x : x \in \mathbf{Q} \text{ and } x > k\}$ if $0 < b < 1$.

Two parts of the definition are needed because for $0 < b < 1$, b^x is decreasing on domain **Q** in such a way that $\{b^x : x \in \mathbf{Q} \text{ and } x < k\}$ does not have an upper bound. Notice that for $b < 0$, irrational powers are not defined.

Even though 2^π is irrational, the fact that it is the least upper bound of $\{2^x : x \in \mathbf{Q} \text{ and } x < \pi\}$ implies that 2^π can be placed in as small an interval as you wish between rational powers of 2, as the figure suggests.

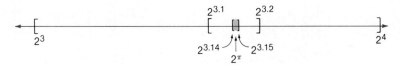

With a calculator, 2^π can be approximated very easily; simply use the $\boxed{Y^x}$ key to verify that $2^\pi \doteq 8.8249778$.

Example 1 Use a calculator to evaluate $6.2^{\sqrt[3]{17}}$ accurate to 3 decimal places.

SOLUTION Use the $\boxed{Y^x}$ key twice as follows:

$$6.2\ \boxed{Y^x}\ \boxed{(}\ 17\ \boxed{Y^x}\ 3\ \boxed{1/X}\ \boxed{)}\ \boxed{=}\ .$$

To 3 decimal places, the result is 109.009.

From the definition of b^k for irrational exponents it can be proved that if $b > 0$ and x and y are any real numbers, then

$$b^x \cdot b^y = b^{x+y}; \qquad b^x \div b^y = b^{x-y}; \qquad \text{and } (b^x)^y = b^{xy}.$$

This means that the irrational exponents have been defined in such a way that all of the exponent laws applied to positive bases also hold for all real number exponents.

Now we are ready to extend the exponential functions so that the domain of each is the set of all real numbers **R**, not just the set of rational-numbers.

Definition If $b > 0$ and $b \neq 1$, the **exponential function with base b**, denoted $\exp_b$, is defined for every real number x by

$$\exp_b(x) = b^x,$$

Thus, $\exp_2(3) = 2^3 = 8$, and, again using the $\boxed{Y^x}$ key of a calculator,

$$\exp_{1.7}(-\sqrt{2}) = 1.7^{-\sqrt{2}} \doteq 0.4721674.$$

The definition of irrational exponents makes it possible to prove that if x and y are any real numbers with $x < y$, then

$$b^x < b^y \text{ if } b > 1 \quad \text{and} \quad b^x > b^y \text{ if } 0 < b < 1.$$

This result directly implies the following theorem.

Theorem 8–1 For every exponential function $\exp_b$,

    a. If $b > 1$, then $\exp_b$ is increasing.

    b. If $0 < b < 1$, then $\exp_b$ is decreasing.

Example 2 Solve for x on **R**: $\exp_{1/2}(x - 1) \leq 2^\pi$.

SOLUTION Write the inequality as

$$\left(\tfrac{1}{2}\right)^{x-1} \leq \left(\tfrac{1}{2}\right)^{-\pi}.$$

Since $\tfrac{1}{2} < 1$, $\exp_{1/2}$ is decreasing and $x - 1 \geq -\pi$, or $x \geq 1 - \pi$. The solution set is $[1 - \pi, \infty)$.

In section 8–1 the graph of $y = 2^x$ was shown, using a dotted line to suggest gaps at irrational values of x. Now that $\exp_2$ has been defined on **R**, these gaps can be filled in so that the graph of $y = \exp_2(x)$ is a smooth continuous curve.

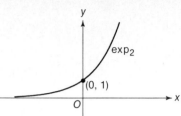

It should be evident that the range of $\exp_2$ is the set of positive real numbers, and that the graph approaches the x-axis as an asymptote. This is true for all exponential functions.

The graph of $\exp_{1/2}$ is easily obtained from the graph of $\exp_2$ by noting that
$$\left(\tfrac{1}{2}\right)^x = (2^{-1})^x = 2^{-x}.$$

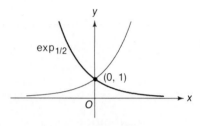

Thus, $\exp_{1/2}(x) = f(-x)$, where $f\colon x \to \exp_2(x)$. This means that the graph of $\exp_{1/2}$ is the reflection in the y-axis of the graph of $\exp_2$.

The graphs of all exponential functions have the x-axis as an asymptote. Note also that all such graphs contain the point $(0, 1)$, since $\exp_b(0) = 1$ for every base b.

Example 3 Sketch the graph of $g\colon x \to \left(\tfrac{2}{3}\right)^{x-2}$.

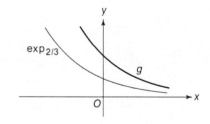

SOLUTION If $h\colon x \to x - 2$, then $g = \exp_{2/3} \circ h$. The graph of g is obtained by shifting the graph of $\exp_{2/3}$ to the right 2 units.

Example 4 The graph of an exponential function contains the point $(4, 19)$. Evaluate $f(-3)$ to four decimal places.

SOLUTION $f(x) = \exp_b(x) = b^x$

$f(4) = b^4 = 19$

$(b^4)^{1/4} = 19^{1/4}$

Thus $b = 19^{1/4}$ and $f(x) = (19^{1/4})^x$.

$f(-3) = (19^{1/4})^{-3} = 19^{-3/4} \doteq 0.1099$

The properties of the exponential functions can be summarized as follows:

1. Dom $\exp_b = $ **R** and Rng $\exp_b = (0, \infty)$.

2. If $b > 1$, $\exp_b$ is an increasing function.
   If $0 < b < 1$, $\exp_b$ is a decreasing function.

3. For all $x, y \in R$, $\exp_b(x + y) = \exp_b(x) \cdot \exp_b(y)$;

   $$\exp_b(x - y) = \frac{\exp_b(x)}{\exp_b(y)}; \text{ and } \exp_b(xy) = [\exp_b(x)]^y.$$

4. The characteristic shapes of the graphs are as shown below.

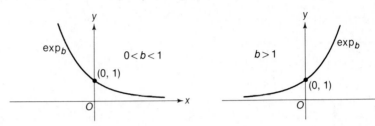

Exercises

A Sketch graphs of the given functions on the same axes.

1. $\exp_3$ and $\exp_{1/3}$
2. $\exp_{2/3}$ and $\exp_{3/2}$
3. $f(x) = 2^x$ and $g(x) = \left(\frac{5}{2}\right)^x$
4. $h(x) = 3^x$ and $k(x) = 4^x$

Solve for x.

5. $\exp_3(x) = 81$
6. $\exp_2(x) = 2\sqrt{2}$
7. $\exp_9(x) = 3^\pi$
8. $\exp_4(x) = 2^{\sqrt{3}}$

B 9. $\exp_3(x) \le \sqrt{3}$
10. $\exp_4(x) > 2^\pi$
11. $\left(\frac{1}{3}\right)^x > 3^{\sqrt{2}}$
12. $\left(\frac{2}{3}\right)^x \le \frac{4}{9}$

Sketch a graph of the given function. Label the coordinates of two points on the graph.

13. $f(x) = 3^{-x}$
14. $g(x) = 2^{x+1}$
15. $h: x \to \left(\frac{2}{3}\right)^{x-1}$
16. $k(x) = \left(\frac{3}{2}\right)^{x-3}$
17. $f(x) = 2 + \exp_3(x + 2)$
18. $k(x) = 1 - \exp_{1/2}(x)$

The graph of an exponential function f contains point A.

19. Evaluate $f(-3)$ if A is $(2, 8)$.
20. Evaluate $f(2)$ if A is $\left(3, \frac{8}{27}\right)$.

For the given function f, sketch the graph of f^{-1}. Specify the domain and range of f^{-1}.

21. $f(x) = 2^x$
22. $f(x) = \left(\frac{3}{4}\right)^x$

Find by trial and error an approximate solution of each equation correct to one decimal place.

23. $\exp_3(x) = 87$
24. $\left(\frac{1}{2}\right)^x = 5$

25. The price of a telescope lens in dollars is expressed by the function $300 \exp_{1.1} \circ f$, where $f(x)$ is the circumference of the lens in inches. What will be the price of a lens with a radius of 3 inches?

C 26. For each statement specify True or False and justify your answer.
   a. If f is increasing, then $\exp_2 \circ f$ is increasing.
   b. If f is an odd function, then $\exp_2 \circ f$ is an odd function.
   c. There are no solution of the equation $\left(\frac{1}{2}\right)^x = \text{int}(x)$.

Sketch a graph of each function. Label coordinates of two points on the graph.

27. $h: x \to \dfrac{1}{3^x + 1}$

28. $k: x \to |\exp_{1/2}(x) - 3|$

By considering the graphs of appropriate functions determine the number of solutions of the given equation and locate the solutions between consecutive integers.

29. $x^3 + 2 - \left(\frac{1}{3}\right)^x = 0$

30. $2^x + 1 + \dfrac{1}{x^2 - 9} = 0$

8–3 Logarithmic Functions

Exponential functions play a prominent role in describing the way quantities like temperature, radioactivity, population, and investments change with time. Thus there is a need to solve equations that involve expressions such as 2^x or $(1.04)^t$. The preceding exercise set included problems asking for trial-and-error solutions of equations like $3^x = 87$. A trial-and-error solution is perfectly valid but it is quite laborious if accuracy to several decimal places is required. The class of functions called logarithmic functions provides a method for obtaining a decimal approximation of the solution of $3^x = 87$ with no more difficulty than solving $x^3 = 87$.

Recall that all exponential functions are either increasing (if $b > 1$) or decreasing (if $0 < b < 1$). Consequently every exponential function is one-to-one and has an inverse. These inverses are called logarithmic functions.

Definition Let $b \neq 1$ be any positive real number. The **logarithm function to the base b**, denoted $\log_b$, is the inverse of the exponential function with base b, that is,

$$\log_b = (\exp_b)^{-1}.$$

If $\log_b y = x$, then x is called the logarithm of y to the base b. From your knowledge of inverse functions you can see that

$$\log_b(y) = x \text{ if and only if } \exp_b(x) = y.$$

This means that $\log_b(y) = x$ is equivalent to $b^x = y$.

To evaluate $\log_2(32)$, observe that $2^5 = 32$, and it follows that $\log_2(32) = 5$. Similarly, $\log_{10}(0.01) = -2$ because $10^{-2} = 0.01$.

Example 1 Evaluate $\log_3(27\sqrt{3})$.

SOLUTION Let $k = \log_3(27\sqrt{3})$.
$3^k = 27\sqrt{3}$
$3^k = 3^{7/2}$
$k = \frac{7}{2}$

Example 2 Solve for x: $\log_2(x) = -3$.

SOLUTION $\log_2(x) = -3$
$x = \exp_2(-3)$
$x = 2^{-3} = \frac{1}{8}$

Calculators with a $\boxed{Y^x}$ key will immediately display exponential function values for any base. However, function values for only two of the logarithmic functions can be obtained directly from most calculators. One of these functions is $\log_{10}$; its function values are called **common logarithms**. When referring to common logarithms it is customary to omit the base and write $\log(x)$. To evaluate $\log(2)$ on a calculator, press 2 $\boxed{\text{LOG}}$ to get the display 0.30103.

Example 3 Solve for x: $10^x = 7$.

SOLUTION The equation $10^x = 7$ is equivalent to $x = \log(7)$. Use a calculator to find $x \doteq 0.845$, accurate to 3 decimal places.

Three basic properties of logarithmic functions are obtained from corresponding properties of exponential functions.

1. Since Dom $\exp_b = \mathbf{R}$, Rng $\exp_b = (0, \infty)$, and $\log_b = (\exp_b)^{-1}$, it follows that Dom $\log_b = (0, \infty)$ and Rng $\log_b = \mathbf{R}$.

2. Since $\exp_b$ is increasing for $b > 1$, Theorem 7–5 implies that $\log_b$ is increasing if $b > 1$. Similarly, since $\exp_b$ is decreasing if $0 < b < 1$, $\log_b$ is decreasing if $0 < b < 1$.

3. From the graphs of the exponential functions, you can see that every logarithmic function is asymptotic to the y-axis and contains the point $(1, 0)$.

Thus the graph of the logarithmic function for $b > 1$ is as shown at the left below. For $0 < b < 1$, the graph is as shown at the right.

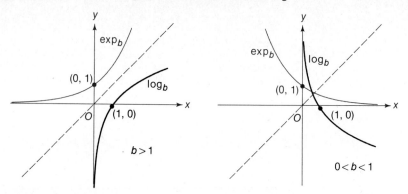

Most values of a logarithmic function are irrational numbers even when the bases and arguments are natural numbers. For example, $\log_3 (2) \doteq 0.6309298$. However, with rational bases it is easy to generate a collection of rational arguments with rational values which can be useful in curve sketching. Since (r, b^r) is an element of $\exp_b$, you know that (b^r, r) is an element of $\log_b$. So to find points with rational coordinates on the graph of $\log_b$, select integer values of r and calculate b^r.

Example 4 Find four points with rational coordinates on the graph of $f(x) = \log_3 (x)$ and sketch the graph.

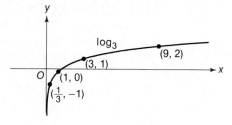

SOLUTION A typical point is $(3^x, x)$. So for $x = -1, 0, 1,$ and 2, the corresponding points on the graph are $\left(\frac{1}{3}, -1\right)$, $(1, 0)$, $(3, 1)$, and $(9, 2)$.

Up to now, we have always used parentheses when writing function values such as $\exp_2 (5)$, $\log_7 (x)$, or $\log (3)$. This was to emphasize that $\exp_b$ and $\log_b$ are functions. Customarily, when there is no chance of confusion, the parentheses are omitted. We write $\exp_2 5$, $\log_7 x$, and $\log 3$.

Exercises

A Evaluate each expression.

1. $\log_7 7$
2. $\log_3 1$
3. $\log_3 27$
4. $\log_2 16$
5. $\log_\pi 1$
6. $\log 10$
7. $\log_5 \frac{1}{125}$
8. $\log_3 \frac{1}{81}$

9. $\log_{1/2} 4$
10. $\log_{1/5} 125$
11. $2^{\log_2 5}$
12. $10^{\log 3}$

Sketch graphs of the given functions on the same axes.

13. $\log_3$ and $\log_{1/3}$
14. $\log_{3/4}$ and $\log_{4/3}$
15. $f: x \to \log_2(-x)$ and $g: x \to -\log_2 x$
16. $f: x \to \log_2 x$ and $g: x \to \log_3 x$

Sketch a graph of the given function. Label coordinates of two points on each graph.

17. $f(x) = \log_2(x - 3)$
18. $g(x) = \log_3(x + 2)$
19. $h(x) = 3 - \log_3 x$
20. $k(x) = 2 + \log_2(x - 1)$

Solve each equation.

21. $\log_4 x = 3$
22. $\log_3 x = -2$
23. $\log_2(x + 1) = 2$
24. $\log_7 x = 0$
25. $\log_3(3^x) = -2$
26. $\log_4(4^x) = 3$

Use a calculator to solve for x. Express answers accurate to 3 decimal places.

27. $7x = \log 3$
28. $(3x)(\log 5) = \log 21$

B 29. $10^x = 21.6$
30. $10^x = 0.283$
31. $3x^2 - x(\log 2) = 0$
32. $2x^2 - \log 7 = 3x$

Assume g is a logarithmic function whose graph contains point A.

33. If A is $(32, 5)$, find $g(\sqrt{8})$.
34. If A is $(9, -2)$, find $g\left(\frac{1}{27}\right)$.

Sketch a graph of each function. Label the coordinates of two points on each graph.

35. $f(x) = 2 + \log_2(-x)$
36. $h(x) = |\log_3(x + 1)|$

37. The monkey population of a zoo t years after the zoo opens is given by $P(t) = P_0[1 + \log_3(t + 1)]$. If there are 12 monkeys when the zoo opens, what will be the monkey population after 8 years?

38. In a certain country, the cost-of-living index increased logarithmically between 1967 and 1977 so that t years after 1967 the index was given by $C(t) = 1 + \log(2t + 1)$. If a ticket to a movie cost $5 in 1967, what should a movie ticket have cost in 1974?

Specify the domain and range of f^{-1}, and find a formula for $f^{-1}(x)$.

C 39. $f(x) = \log_3(x + 2)$
40. $f(x) = \left(\frac{1}{3}\right)^x + 2$
41. $f(x) = 2^{x+1} - 3$
42. $f(x) = \dfrac{1}{3^x + 1}$

43. Solve graphically the inequality $2\log_3(x) + 1 - x \le 0$.

Sketch a graph of the given function. Label coordinates of two points.

44. $f(x) = 1 + \log_2|x|$
45. $g(x) = \log_2(x^2 - 4)$

8–4 Properties of Logarithms

If $\exp_b$ is any exponential function and x and y are any real numbers, then you know that

$$b^{x+y} = b^x b^y, \quad b^{x-y} = b^x \div b^y, \quad \text{and} \quad b^{xy} = (b^x)^y.$$

Thus $2^{\pi+3} = 2^{\pi} 2^3$, $2^{\pi-3} = 2^{\pi} \div 2^3$, and $2^{3\pi} = (2^3)^{\pi} = 8^{\pi}$.

These properties of exponentials lead to a corresponding set of properties for logarithms.

Theorem 8–2 For all $c, d \in (0, \infty)$ and $r \in \mathbf{R}$, and for every logarithmic function $\log_b$,

a. $\log_b cd = \log_b c + \log_b d$,

b. $\log_b \left(\dfrac{c}{d}\right) = \log_b c - \log_b d$, and

c. $\log_b c^r = r \log_b c$.

PROOF Let c and d be positive numbers such that

$$\log_b c = p \quad \text{and} \quad \log_b d = q.$$

Converting to equivalent exponential equations yields

$$c = b^p \quad \text{and} \quad d = b^q.$$

Now apply the properties of exponents.

a. $cd = b^{p+q}$

b. $\dfrac{c}{d} = b^{p-q}$

c. $c^r = b^{rp}$

Converting back to equivalent logarithmic equations gives

a. $\log_b cd = p + q = \log_b c + \log_b d$

b. $\log_b \left(\dfrac{c}{d}\right) = p - q = \log_b c - \log_b d$

c. $\log_b c^r = rp = r \log_b c$

Corollary 8–2 follows immediately from Theorem 8–2c by letting $r = -1$.

Corollary 8–2 $\log_b \dfrac{1}{c} = -\log_b c$

Example 1 Simplify $\log_6 9 + \log_6 4$.

SOLUTION $\log_6 9 + \log_6 4 = \log_6 (9 \cdot 4)$ Use Theorem 8–2a.
$$= \log_6 36$$
$$= 2$$

Example 2 Simplify $\frac{2}{3} \log_7 125 - 2 \log_7 35$.

SOLUTION $\frac{2}{3} \log_7 125 - 2 \log_7 35 = \log_7 125^{2/3} - \log_7 35^2$
$$= \log_7 \left(\frac{25}{35^2}\right) = \log_7 \left(\frac{1}{49}\right)$$
$$= \log_7 (7^{-2})$$
$$= -2$$

The properties of logarithms can often be applied to advantage in graphing.

Example 3 Sketch a graph of the function $f: x \rightarrow \log \frac{1}{x + 1}$.

SOLUTION By Corollary 8–2, $f(x) = -\log (x + 1)$, so first sketch a graph of $h(x) = \log (x + 1)$, which is the graph of $\log x$ translated 1 unit to the left. Then to get a graph of f, reflect the graph of h in the x-axis.

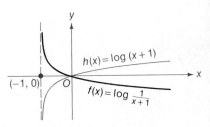

The properties of logarithms can also be helpful in solving equations and inequalities. The following examples show the importance of identifying the domain in logarithmic equations.

Example 4 Solve for x: $\log x + \log (x - 1) = \log (2x + 10)$.

SOLUTION Domain $= (0, \infty) \cap (1, \infty) \cap (-5, \infty) = (1, \infty)$.
$$\log x + \log (x - 1) = \log (2x + 10)$$
$$\log x(x - 1) = \log (2x + 10)$$

Since log is a one-to-one function,
$$x^2 - x = 2x + 10.$$
$$x^2 - 3x - 10 = 0$$
$$(x - 5)(x + 2) = 0$$
$$x = 5 \text{ or } x = -2$$

But -2 is not in the domain, so $x = 5$ is the only solution.

Example 5 Solve for x: $2 \log_2 x - \log_2 (x - 1) = 2$.

SOLUTION Domain $= (0, \infty) \cap (1, \infty) = (1, \infty)$

$$2 \log_2 x - \log_2 (x - 1) = 2$$
$$\log_2 x^2 - \log_2 (x - 1) = 2$$
$$\log_2 \left(\frac{x^2}{x - 1}\right) = 2$$
$$\frac{x^2}{x - 1} = 2^2$$
$$x^2 - 4x + 4 = 0$$
$$(x - 2)^2 = 0$$

Since 2 is in the domain, the only solution is $x = 2$.

Example 6 Solve for x: $\log_2 x^3 \leq 9$.

SOLUTION The domain is $(0, \infty)$.

$$\log_2 x^3 \leq 9$$
$$3 \log_2 x \leq 9$$
$$\log_2 x \leq 3$$

Since $\exp_2$ is an increasing function,

$$\exp_2 (\log_2 x) \leq \exp_2 (3)$$
$$x \leq 2^3.$$

The solution set is $(-\infty, 8] \cap (0, \infty) = (0, 8]$.

Exercises

A Simplify each expression.

1. $\log 20 + \log 5$
2. $\log_4 32 + \log_4 2$
3. $\log_5 75 - \log_5 3$
4. $\log_3 54 - \log_3 6$
5. $\log_7 4 + \log_7 \left(\frac{1}{4}\right)$
6. $\log 50 - \log \left(\frac{1}{2}\right)$
7. $2 \log_6 \sqrt{54} - \log_6 9$
8. $\log_9 \sqrt[3]{36} - \frac{1}{3} \log_9 4$
9. $\log \sqrt{1000} \div \log \sqrt[3]{100}$
10. $\log_3 \sqrt{27} \div \log_3 \sqrt[3]{81}$

State the domain and solve for x.

11. $\log_5 (x + 2) = \log_5 (2x + 1)$
12. $\log (x + 3) = \log (4x - 6)$

13. $\log_2 (x^2 - 4) = \log_2 3x$
14. $2 \log_3 x = \log_3 16$
15. $\log_3 (x - 2) = -1$
16. $\log_4 (3 - 2x) = 2$
17. $\log_2 (x - 2) + \log_2 x = \log_2 8$
18. $\log (x + 1) + \log (2x + 1) = \log 6$
19. $\log (x + 1) + \log (x - 2) = 1$
20. $\log_2 (x - 2) + \log_2 x = 3$

B Sketch a graph of each function. Label coordinates of two points on the graph. [HINT: Use the properties of logarithmic functions as in example 3.]

21. $f: x \rightarrow \log_2 \sqrt{x - 3}$
22. $f: x \rightarrow \log_2 \dfrac{1}{x + 3}$
23. $g(x) = \log_3 \dfrac{1}{x - 2}$
24. $g(x) = \log_3 x^3$

State the domain and solve for x.

25. $\log_2 (3x - 2) - \log_2 (x + 1) = 3$
26. $\log_3 (2x + 1) - 2 \log_3 x = 1$
27. $\log_3 (x - 4) < 4$
28. $\log_2 (x + 5) > 1$
29. $\log_{1/2} (3 - x) < -1$
30. $\log_{1/3} (2x + 1) > -2$
31. $\log (x + 6) < \log (3 - 2x)$
32. $\log (x^2 - 2) \leq \log (3x + 2)$

If $\log_3 2 = a$ and $\log_3 10 = b$, evaluate each of the following expressions in terms of a and b.

33. **a.** $\log_3 6$ **b.** $\log_3 200$ **c.** $\log_3 5$ **d.** $\log_3 \sqrt[3]{108}$
34. **a.** $\log_3 20$ **b.** $\log_3 24$ **c.** $\log_3 15$ **d.** $\log_3 \sqrt{48}$

C Sketch a graph of the given function. Label coordinates of two points on the graph.

35. $h(x) = 2 - \log_3 \left(\dfrac{x}{3}\right)$
36. $k(x) = \log_2 (x - 1)^3$

Solve for x.

37. $(\log_2 x)^2 - \log_2 x = 0$
38. $(\log x)^2 - 3 \log x - 4 = 0$
39. $2 \log_2 x - 7 = \dfrac{4}{\log_2 x}$
40. $\dfrac{\log_2 x - 3}{3 - x} < 0$

41. Prove that $\log_3 5$ is an irrational number. [HINT: Use an indirect proof, assuming that $\log_3 5 = \dfrac{p}{q}$, where p and q are integers with $q \neq 0$.]

8–5 Exponential Equations

Logarithmic functions were introduced by suggesting that they would be valuable in solving exponential equations. In this section we show just how this is done.

The equation
$$3^x = 87$$
is equivalent to
$$x = \log_3 87.$$

Although $\log_3 87$ is a compact way of expressing the solution, a decimal approximation to $\log_3 87$ cannot be found directly either by using tables or a calculator. However, by using the next theorem it is possible to express $\log_3 87$ in terms of easily approximated logarithms.

Theorem 8–3 **Change of Base**

If a, b, and c are positive numbers with $b \neq 1$ and $c \neq 1$, then
$$\log_b a = \frac{\log_c a}{\log_c b}.$$

PROOF Suppose $b^x = a$.

Take the logarithm to the base c of both sides.

$$\log_c b^x = \log_c a$$
$$x \log_c b = \log_c a$$
$$x = \frac{\log_c a}{\log_c b}$$

But the given equation $b^x = a$ is equivalent to $x = \log_b a$. So by substitution,

$$\log_b a = \frac{\log_c a}{\log_c b}.$$

Evaluating $\log_3 87$ can now be done by letting $a = 87$, $b = 3$, and $c = 10$ in Theorem 8–3 and writing

$$\log_3 87 = \frac{\log 87}{\log 3},$$

which can easily be approximated by calculator as 4.065.

The change-of-base theorem implies that logarithmic function values for just one base are needed in order to compute logarithms to any base. Calculators usually incorporate two logarithmic functions. In addition to common logarithms, calculators provide values of the **natural logarithm function**. The base of this function is an irrational number which is approximately 2.71828 and is designated by the letter e. The reason why e is singled out as the base for natural logarithms requires a knowledge of calculus. For now, you need only know that e is a constant that appears in scientific writing with a frequency comparable to that of π. Ordinarily the notation ln is used to denote $\log_e$.

Example 1 Solve for x: $2^{x-3} = 5$.

SOLUTION

$\ln(2^{x-3}) = \ln 5$ *Take the natural logarithm of both sides.*

$(x - 3) \ln 2 = \ln 5$

$x - 3 = \dfrac{\ln 5}{\ln 2}$

$x = \dfrac{\ln 5}{\ln 2} + 3$

By using a calculator with an $\boxed{\text{LN X}}$ key you find that

$x \doteq 5.322$.

You could have solved the equation in example 1 in exactly the same way using common logarithms rather than natural logarithms.

Exponential inequalities are solved by using the same method used for exponential equations.

Example 2 Solve for x: $\left(\dfrac{1}{7}\right)^x \leq 38$.

SOLUTION Since ln is an increasing function,

$\ln\left(\dfrac{1}{7}\right)^x \leq \ln 38$.

$x \ln\left(\dfrac{1}{7}\right) \leq \ln 38$

$x \geq \dfrac{\ln 38}{\ln\left(\dfrac{1}{7}\right)}$ *Reverse the inequality since $\ln\left(\dfrac{1}{7}\right) < 0$.*

$x \geq -\dfrac{\ln 38}{\ln 7} \doteq -1.869$

The solution set is approximately $[-1.869, \infty)$.

The paths of music and mathematics have been intertwined at least as far back as the days of Pythagoras. Exponential functions are related to the tuning of musical instruments.

If two musical notes are an octave apart, the higher note has a frequency twice that of the lower note. Between notes on a piano that are an octave apart there are 11 keys, which separate the octave into 12 semitone intervals. A piano is tuned so that the ratio of frequencies for adjacent notes is a constant.

Example 3 If middle C on a piano is tuned to a frequency of 260 hertz (cycles per second), find the frequency of the F note that is five notes above middle C and locate the key that is tuned to a frequency of 655 hertz.

SOLUTION Let k be the ratio of the frequencies of adjacent notes, and let $f(n)$ be the frequency of a note n semitones above middle C. Then

$$f(n) = 260k^n.$$

To evaluate k use the fact that the next C above middle C (one octave) has a frequency of $(2)(260)$ hertz, or 520 hertz.

$$f(12) = 520 = 260k^{12}$$
$$k^{12} = 2$$
$$k = 2^{1/12}$$
$$f(n) = 260(2^{1/12})^n = 260(2^{n/12})$$

To find the frequency of F, evaluate $f(5)$.

$$f(5) = 260(2^{5/12}) \doteq 347 \text{ hertz}$$

To locate the note tuned to 655 hertz, solve the exponential equation

$$260(2^{n/12}) = 655.$$
$$2^{n/12} = \frac{655}{260}$$
$$\log 2^{n/12} = \log\left(\frac{655}{260}\right)$$
$$\frac{n}{12} \log 2 = \log 655 - \log 260$$
$$n = 12\left(\frac{\log 655 - \log 260}{\log 2}\right)$$
$$n \doteq 15.996 \doteq 16$$

The 16th key above middle C is called E.

Exercises

A Use a calculator to evaluate each expression to 3 decimal places.

1. $\ln 5$
2. $\ln 2 \cdot \ln 3$
3. $\log_3 2$
4. $\log_2 7$

5. $\log_{2/3} 5$ **6.** $\log_{\pi} 15$ **7.** $\log_7 \frac{5}{3}$ **8.** $\log(\log_3 2)$

Simplify each expression without using a calculator.

9. $e^{\ln 3}$ **10.** $10^{\log 14}$ **11.** $10^{3 \log 2}$

12. $e^{1/2 \ln 8}$ **13.** $\log_3 5 \div \log_3 10$ **14.** $\dfrac{\log_2 3}{\log 3}$

Solve each equation. Express answers accurate to 3 decimal places.

15. $7^x = 31$ **16.** $2^x = 9$ **17.** $\left(\frac{2}{3}\right)^x = 10$ **18.** $(0.72)^x = 9$

19. $e^t = 13$ **20.** $e^{t+2} = 7$ **21.** $3^{t-1} = 0.82$ **22.** $5^{2t-3} = 8$

B 23. $7^{u^2-1} = 5$ **24.** $\left(\frac{1}{2}\right)^{3u^2} = 0.75$

25. If the graph of an exponential function contains (2, 7), where does the graph intersect the line $y = 6$?

26. The graph of an exponential function contains $(-3, 5)$. Evaluate $f(6)$.

Solve each inequality.

27. $3^x \leq 5$ **28.** $2^x > 10$ **29.** $\left(\frac{1}{3}\right)^x > 24$

30. $\left(\frac{3}{4}\right)^x \leq 7$ **31.** $2^{1-x} \leq 75$ **32.** $3^{x^2-2} \geq 8$

33. If middle A, 9 semitones above middle C, is tuned to 430 hertz, what should be the frequency of middle C?

34. If middle F, 5 notes above middle C, is tuned to 350 hertz, how many semitones above middle C is a note tuned to 588 hertz?

35. Prove the following corollary to Theorem 8–3: If a and b are positive with $a \neq 1$ and $b \neq 1$, then $\log_b a = \dfrac{1}{\log_a b}$.

36. If $f(x) = 3^x$ and $g(x) = 3 \cdot 2^x$, find the intersection point of the graphs of f and g.

C 37. Prove that if f and g are logarithmic functions, then there exists a constant k such that $f(x) = k \cdot g(x)$.

38. Two notes constitute a musical fourth if the ratio of their frequencies is 4 : 3. How many semitone intervals separate the notes of a musical fourth?

39. A professor of music theory wishes to experiment with a new scale that will have 18 intervals per octave instead of 12. She wishes to have a piano-like instrument constructed that will span 3 octaves. If the lowest note on the keyboard is to be tuned to 120 hertz and the highest to 960 hertz, how many of the keys will have frequencies lower than 600 hertz?

40. Suppose that for tax purposes the value of a car is considered to decrease 25% each year. If a new Zephyr Ostentation is valued at $19,000, how many years must elapse before the car is valued at less than $500?

41. A child requires 40 milligrams of a certain anesthetic in order to maintain an acceptable level of anesthesia during surgery. The amount of anesthetic in the child's system is given by $A = M(0.791)^t$, where M is the weight in milligrams of the anesthetic administered and t is the time in hours after its administration. If 75 milligrams of the anesthetic is administered at 9:00 A.M., by what time should the operation be completed if the child is not to experience unacceptable discomfort?

42. Solve for x to one decimal place: $x^3 = \log_3 (x + 1)$.

Solve for x.

43. $\dfrac{3^x}{3^x - 10} < 1$

44. $\dfrac{3}{4 - 2^x} = 2^x$

45. Prove that if $a > b > 1$, then $\log_b a + \log_a b > 2$.

8–6 Applications of Exponential Functions

A quantity A *grows exponentially* if its growth is described by a function of the form

$$A(t) = Cb^t,$$

where C and b are constants with $b > 0$ and $b \neq 1$. One common example of a quantity that grows exponentially is the value of a savings account in which money is invested at a constant interest rate that is compounded over some regular time interval.

To illustrate, suppose that on January 1, 1985, a person has $800 in a savings account in a bank which pays 6 percent annual interest, compounded quarterly. If A represents the amount of money in the account at a given time, then three months later (one interest period), the amount in the account is

$$A = 800\left(1 + \frac{0.06}{4}\right) = 800(1.015).$$

At the end of 2 interest periods,
$$A = [800(1.015)][1.015] = 800(1.015)^2,$$
and after n interest periods,
$$A = 800(1.015)^n.$$
Since there are 4 interest payments a year, $n = 4t$ where t is in years. Thus A can be expressed as a function of t, the time in years since January 1, 1985.
$$A(t) = 800(1.015)^{4t} = 800(1.015^4)^t$$
$$\doteq 800(1.06136)^t$$
Thus A is a function of the form $A(t) = Cb^t$, and therefore the person's investment grows exponentially.

In the general case, the formula for the amount in the account is
$$A(t) = P\left(1 + \frac{1}{n} \cdot \frac{r}{100}\right)^{nt},$$
where P is the amount originally deposited, the principal;
r is the annual rate of interest in percent;
n is the number of times per year that interest is paid; and
t is the time in years since the account was started.

Example 1 If $200 is deposited in an account that pays $7\frac{1}{4}\%$ compounded monthly, what is the account balance after 40 months?

SOLUTION $A(t) = P\left(1 + \frac{1}{n} \cdot \frac{r}{100}\right)^{nt}$

$$A\left(\frac{40}{12}\right) = 200\left(1 + \frac{7.25}{1200}\right)^{12(40/12)} = 200\left(\frac{1207.25}{1200}\right)^{40}$$

Use a calculator to find that, to the nearest penny, $A = \$254.49$.

Return now to the $800 investment considered earlier. A very different approach to deriving a formula for $A(t)$ is possible by simply assuming that the money invested grows exponentially, that is, that
$$A(t) = Cb^t.$$
To evaluate C, use the fact that when $t = 0$, $A = 800$.
$$A(0) = 800 = Cb^0$$

Thus C = 800 and
$$A(t) = 800b^t.$$

The interest is compounded quarterly, so during the first quarter year the value of the investment increases by $\left(\frac{0.06}{4}\right)(800)$, or \$12. Hence at the end of 3 months, the account value is \$812. Use this fact to determine b.

$$A\left(\tfrac{1}{4}\right) = 812 = 800b^{1/4}$$
$$b = \left(\frac{812}{800}\right)^4 \doteq 1.06136$$

So the amount $A(t)$ after t years is
$$A(t) \doteq 800(1.06136)^t,$$

which is the same result obtained earlier by a different method. Notice that 6% interest compounded quarterly corresponds to an effective annual interest rate of approximately 6.136%.

The rate of change in the value of the investment is directly proportional to the value of the investment. In this case, on January 1, 1985, the investment was increasing (0.06136)(800) or \$49.09 per year. After t years the investment is increasing in value $(0.06136)A(t)$ dollars per year. That is, the rate of change in the value of the investment is a constant times the amount $A(t)$ of the investment. This example can be generalized in the form of an important principle which we shall not prove but which should be noted carefully.

The Exponential-Growth Principle

Exponential growth occurs whenever the rate of change of a given quantity is directly proportional to the existing amount of that quantity.

Radioactive elements like radium and uranium disintegrate at a rate which is directly proportional to the quantity of radioactive material present. Consequently a given mass of radioactive material disintegrates exponentially and the amount present at any time t is given by

$$M(t) = M_0 b^t,$$

where M_0 is the amount present at $t = 0$. Thus, levels of radioactivity in the waste products from nuclear power reactors are described in terms of exponential functions.

A vivid way of describing the rate at which a radioactive isotope disintegrates is to state its half-life. The **half-life** of a particular isotope is the time required for a given quantity of the isotope to decompose to one-half the original amount.

Thus to say that the half-life of carbon 14 is 5600 years means that if there are 8 grams of the isotope present today; then after 5600 years, 4 grams of carbon 14 will remain and after 11,200 years, 2 grams will remain.

Example 2 Assume that the half-life of a radioactive isotope is 16 years. How much of the isotope will remain after 11 years if the initial amount is 50 grams?

SOLUTION Let $M(t)$ be the mass present after t years, so that

$$M(t) = M_0 b^t.$$

Since $M(0) = 50$, you know that $50 = M_0 b^0$, or $M_0 = 50$, and

$$M(t) = 50b^t.$$

The half-life is 16 years so

$$M(16) = \frac{50}{2} = 50b^{16}.$$
$$b^{16} = \left(\frac{1}{2}\right), \text{ or } b = \left(\frac{1}{2}\right)^{1/16}$$
$$M(t) = 50\left[\left(\frac{1}{2}\right)^{1/16}\right]^t = 50\left(\frac{1}{2}\right)^{t/16}$$

After 11 years the amount remaining is

$$M(11) = 50\left(\frac{1}{2}\right)^{11/16} \doteq 31.046 \text{ grams.}$$

The term half-life is not confined to describing radioactive decay. For instance, to say that a substance ingested into a person's body has a half-life of 4 days means that after 4 days the amount of the substance remaining has been reduced by one half. If M_0 is the quantity of a material initially present that decreases exponentially, and if the half-life is k units, then after t time units the quantity of the material remaining, $M(t)$, is given by

$$M(t) = M_0\left(\frac{1}{2}\right)^{t/k}.$$

Example 3 When an overheated car is brought into a garage, the temperature of the engine block is 350°F. Ten minutes later the temperature has dropped to 280°F. If the mechanic will not work on the engine until its temperature has dropped to 150°F, how much time must pass before work on the engine can proceed? Consider the air temperature in the garage to be 80°F.

SOLUTION It is reasonable to assume that if D is the difference between the engine-block temperature and the garage temperature, then the rate at which D is changing (the rate of cooling) at any given moment is directly proportional to D itself. Thus the exponential-growth principle is applicable, and it can be assumed that the engine block cools exponentially.

(cont. on p. 302)

The cooling can therefore be described by a function D such that the temperature difference t minutes after the 350°F reading was noted is

$$D(t) = Cb^t.$$

It is known that $D(0) = 350 - 80 = 270$.

Thus $D(t) = 270b^t$.

It is also known that $D(10) = 280 - 80 = 200$.

Hence $D(10) = 200 = 270b^{10}$

$$b = \left(\frac{200}{270}\right)^{1/10} \doteq 0.970435$$

So $D(t) = 270(0.970435)^t$.

Now solve the equation $D(t) = 150 - 80 = 70$.

$$270(0.970435)^t = 70$$
$$0.970435^t = \frac{70}{270}$$
$$\log (0.970435)^t = \log \frac{70}{270}$$
$$t = \frac{\log 70 - \log 270}{\log 0.970435}$$
$$t \doteq 44.98$$

About 45 minutes after the car was brought into the garage, the engine should have cooled to about 150°F.

Exercises

A   **1.** An investor places $1000 in a savings account which pays 6% interest. What is the value of the account at the end of 10 years if the interest is compounded semiannually? Monthly? Daily?

2. An investor places $1000 in a money-market fund which pays interest compounded quarterly. What is the value of the account after 20 years if the interest rate is 7%? 10%? 13%?

3. A student earns $750 during the summer and deposits the money in a savings account that pays $5\frac{1}{4}\%$ interest compounded quarterly. After how long will the account balance be $1000?

4. Find the time required for an investment to double in value if the interest is compounded monthly and the interest rate is
   a. 5%.   **b.** 10%.   **c.** 15%.

5. The half-life of a certain radioactive isotope is 24 years. What mass of the isotope will remain from an initial mass of 60 grams after
   a. 96 years?   **b.** 12 years?

6. The half-life of a radioactive material is 28 seconds. Find the percent of radioactive material remaining after 150 seconds.

7. The half-life of an isotope is 27.3 hours. How much time is required for 5.3 grams to decay so that only 1.9 grams remain?

8. After 100 days, 30 grams of radioactive polonium remain from an initial mass of 50 grams. Determine the half-life of polonium.

B 9. After 250 minutes, 180 grams of radioactive material is found to have disintegrated to 15 grams. How many grams of the material were there after 60 minutes?

10. A new-born baby can be expected to double its weight in 5 months. Assuming exponential growth in the early months of life, what would you expect a baby weighing 7 pounds at birth to weigh after 2 months?

11. When she was born, Ramona's grandparents started a savings account for her so that when she became 18 she would receive $5000 for college expenses. If the savings bank paid 6% interest compounded monthly, how much did the grandparents deposit initially?

12. Suppose you are forced by misfortune to borrow $500 from an unscrupulous money lender who charges 100% interest compounded daily. If you wait a year to repay the loan, how much does the money lender expect to receive for each dollar loaned? What would he expect if the interest were compounded each hour? Without computing, estimate the required payment if the interest were compounded each second.

C 13. To start breeding rats for sale to research biologists, a scientific supply company buys a number of genetically selected rats. After 6 months there are 786 rats, and at the end of 9 months there are 1720 rats. Assuming that the population has grown exponentially, estimate the number of rats originally procured.

14. At 7:00 A.M. the temperature of a bottle of milk removed from a refrigerator is 36°F and at 8:30 A.M. its temperature has risen to 43°F. What would you predict its temperature to be at 10:45 A.M. if the room temperature stays constant at 72°F?

15. In 1970 the population of Matanuska, Alaska, was 6509 and by 1980 it had grown to 17,766. Assuming exponential growth, when would you expect Matanuska's population to reach 30,000?

16. The Fibonacci family wishes to conserve energy by shutting off the heat in the house at 9:00 P.M. Assume that on a particular night the outdoor temperature remains constant at 10°F and that exponential cooling takes place in the house after the heat is turned off. The indoor temperature at 9:00 P.M., when the heat is shut off, is 68°F. At 10:00 P.M. it is 61°F.
   a. What indoor temperature would you predict at 10:45 P.M.?
   b. It is decided that the temperature should remain above 55°F until 11:00 P.M. when the family's favorite TV program ends. When should the heat be shut off?

17. After exercise, the difference between a person's pulse rate and the normal pulse rate decreases exponentially. Immediately after a road race, Betsy's pulse rate is 140 beats per minute. Five minutes later it has dropped to 116. If Betsy's normal pulse rate is 68, how long should it take her pulse to reach 90?

Chapter Review Exercises

8–1, page 277

Evaluate each expression, accurate to 3 decimal places.

1. 3.17^5
2. $\sqrt[3]{5.1}$
3. $21^{4/7}$
4. $263^{-2/3}$

Solve for x on the domain Q.

5. $3^{x+1} = 9\sqrt{3}$
6. $\left(\frac{1}{2}\right)^x \leq 64$

7. The school-age population of a certain city is now approximately 63,000 and is decreasing at a rate of 3% per year. If this rate of decrease continues, what will be the school-age population 8 years from now?

8–2, page 282

Sketch a graph of each function and label the coordinates of two points on each graph.

8. $f: x \to 1 + \left(\frac{1}{3}\right)^x$
9. $g: x \to \exp_2(x-1) - 3$

Solve for x on the domain R.

10. $\exp_8 x = 2^\pi$
11. $\left(\frac{1}{5}\right)^x \leq 125^{\sqrt{2}}$

12. The graph of an exponential function contains the point $\left(-3, \frac{1}{64}\right)$. Where does the graph intersect the line $x = 2$?

13. If $f = \exp_2 \circ g$ and g is decreasing, prove or disprove that f is decreasing.

8–3, page 286

Evaluate each expression.

14. $\log_2 32$
15. $5 \log_7 1$
16. $\log(\exp_{10} 5)$
17. $2^{\log_2 7}$

Sketch a graph of each function. Label coordinates of two points on each graph.

18. $f(x) = \log_2 (x - 3)$

19. $g(x) = 1 + \log_3 (-x)$

Solve each equation or inequality for x.

20. $\log_2 (x - 3) = 4$

21. $\log_3 x \leq 5$

22. If $f: x \to \log_3 (x - 2)$, sketch a graph of f^{-1}, specify the domain and range of f^{-1}, and find a formula for $f^{-1}(x)$.

8–4, page 290

Use the properties of logarithmic functions as an aid in sketching a graph of each function.

23. $f(x) = \log_3 \sqrt{x + 1}$

24. $g(x) = \log_2 \left(\dfrac{4}{x - 3}\right)$

State the domain and solve for x.

25. $\log_2 (x + 1) + \log_2 (x + 3) = 3$

26. $\log_3 (2x - 1) \leq \log_3 (x + 4)$

8–5, page 294

Solve for x correct to 3 decimal places.

27. $10^x = 463$
28. $e^x = 33.4$
29. $3^x = 5$
30. $\left(\dfrac{1}{2}\right)^x = 3.2$

31. If the graph of an exponential function contains the point $(-2, 9)$, where does the graph intersect the line $y = 5$?

32. Determine the number of solutions of the following equation and locate the smallest solution between consecutive integers.

$$2^x - 1 = \log_3 (x + 4)$$

8–6, page 298

33. When a valve at the bottom of a tank that contains 120 liters of water is opened, the amount of water remaining in the tank t minutes later is given by

$$Q(t) = 120(0.79)^t.$$

How long after opening the valve will the tank contain 15 liters of water?

34. At 2:00 P.M. there are 680 bacteria in a culture and 3 hours later there are 3420. Assuming exponential population growth, determine the time at which there should be 10,000 bacteria present in the culture.

Special Carbon Dating

Living organisms contain approximately the same proportion of radioactive carbon 14 as does the earth's atmosphere. However, when an organism dies it ceases to take in carbon 14, so that its level of radioactivity drops as the carbon 14 present at death decays to nitrogen 14. By comparing the level of radioactivity of a dead specimen, such as a skeleton, with the level of radioactivity in a living organism, scientists can measure the ratio of the amount of carbon 14 in the specimen to the amount present at the time of the organism's death. Since it is known that carbon 14 decays exponentially, and that the half-life of carbon 14 is 5568 years, it is possible to estimate the time elapsed since the organism died.

This technique of measuring time is called **carbon dating** and is effective when applied to specimens that have died anywhere from 400 to 40,000 years ago. Carbon dating is of particular value to archaeologists in dating ancient artifacts and to paleontologists in finding the ages of remains of prehistoric animals.

Example When the skeleton of a woolly mammoth is discovered in the Siberian tundra, it is found that the level of radioactivity in the skeleton is 14% of that of a living animal. Estimate how long ago the mammoth lived.

SOLUTION Let $M(t)$ be the amount of carbon 14 present in the skeleton t years after the death of the mammoth. Then

$$M(t) = M_0 \left(\frac{1}{2}\right)^{t/5568},$$

where M_0 was the amount of carbon 14 present at the time of death. Assuming that $M(t)$ is proportional to the level of radioactivity, the problem is to solve the equation

$$\frac{M(t)}{M_0} = \left(\frac{1}{2}\right)^{t/5568},$$

given that

$$\frac{M(t)}{M_0} = 0.14.$$

$$0.14 = \left(\frac{1}{2}\right)^{t/5568}$$

$$\ln 0.14 = \left(\frac{t}{5568}\right) \ln \frac{1}{2}$$

$$t = 5568 \left(\frac{\ln 0.14}{\ln 0.5}\right)$$

$$\doteq 15{,}793$$

The mammoth lived about 16,000 years ago.

Exercises

1. Excavating the ruins of an ancient temple, archaeologists find that the level of radioactivity of one of the beams is 0.77 times that of a living tree. Estimate how long ago the temple was built.

2. The city of Pompeii was destroyed by the volcanic eruption of Mt. Vesuvius in 79 A.D. If archaeologists this year were to find the skeleton of a child who was apparently killed during the eruption, how should the level of radioactivity of the skeleton compare with that of a living person?

Chapter Test

Evaluate each expression.

1. $\log_2 \sqrt[3]{4}$
2. $10^{\log 13}$

Solve for x.

3. $\log_3 (x + 2) = 4$
4. $2^x + 1 \leq 33$

Sketch a graph of each function. Label the coordinates of two points on each graph.

5. $f: x \to \left(\frac{1}{3}\right)^x + 2$
6. $g: x \to \log_2 (x + 5)$

7. State the domain and solve for x: $\log (x + 3) + \log x = 1$.

8. If you pay $12 for a pair of jeans today, what should you expect to pay for the same jeans 7 years from now, assuming a constant inflation rate of 11% per year?

For each graph shown below, specify a function that has the properties indicated.

9.
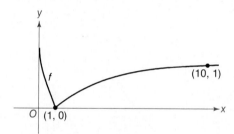

10.
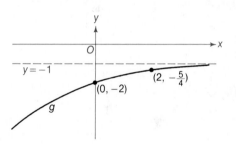

11. Solve for x accurate to 3 decimal places: $\dfrac{8}{3^x + 2} = 3$.

12. If $f(x) = 4 + \exp_3 x$, find $f^{-1}(x)$ and determine the domain and range of f^{-1}.

13. Use graphs to solve for x: $\log x \leq 1 - x^2$.

14. An electrical condenser discharges in such a way that the voltage across its terminals is expressed by the function

$$V(t) = V_0 (3)^{-t/40}$$

where t is the time in seconds. How many seconds are required for the voltage to drop from 120 to 10 volts?

Circular Functions

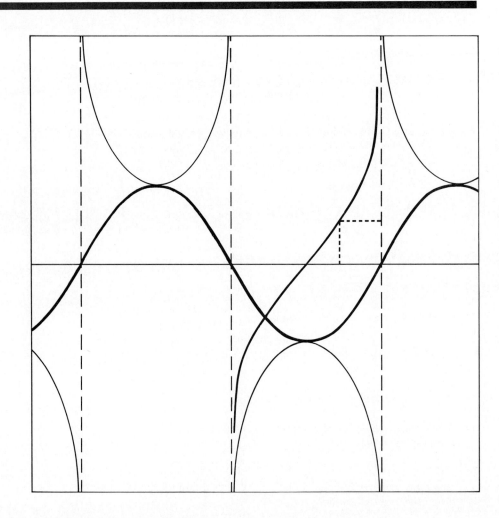

9

9-1 The Wrapping Function

Imagine that you have a piece of string 2 units long. Attach one end to the point (1, 0) on the unit circle, $x^2 + y^2 = 1$. Keeping the string taut, wrap it around the circle in the counter-clockwise direction. Call the other end of the string the terminal point and label it $W(2)$, for "wrapping 2 units." How can you find the coordinates of the point $W(2)$? From the figure the point appears to be about $(-0.4, 0.9)$.

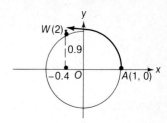

If the string is t units long, then the terminal point is called $W(t)$. Since the circumference of the unit circle is 2π, $W(\pi) = (-1, 0)$. If a number t is negative, cut the string to a length $|t|$ and wrap it clockwise around the circle. For example, $W\left(-\frac{\pi}{2}\right) = (0, -1)$. Also note that $W(0) = (1, 0)$. This process relates a unique point $W(t)$ to each real number t and thus defines a function that is called the **wrapping function W**.

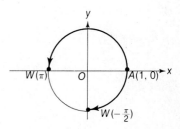

The domain of W is the set of real numbers; the range of W is the set of points on the unit circle. If t is a multiple of $\frac{\pi}{2}$, then $W(t)$ is easy to locate on the unit circle.

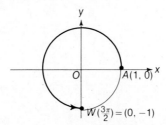
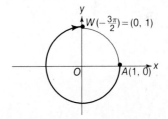

There are some other special arguments t for which you can determine exact values of W. $W\left(\frac{\pi}{6}\right)$ is the endpoint of an arc with length $\frac{\pi}{6}$ measured from (1, 0). Label this point P. The length of $\overset{\frown}{AP}$ is $\frac{1}{12}$ the circumference of the circle, so $\overset{\frown}{AP}$ has degree measure 30° and $m\angle AOP$ is 30°. So $\triangle POB$ is a 30°-60°-90° triangle with $OP = 1$, $BP = \frac{1}{2}$, and $OB = \frac{\sqrt{3}}{2}$.

$$W\left(\frac{\pi}{6}\right) = \left(\frac{\sqrt{3}}{2}, \frac{1}{2}\right)$$

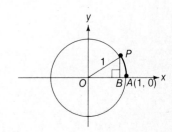

9-1: The Wrapping Function 309

To find $W(\frac{\pi}{4})$, note from the diagram that $\widehat{AP}$ is one fourth the length π of the arc of the semicircle.

$$m\angle AOP = \tfrac{1}{4}(180°) = 45°$$

So $\triangle POB$ is a 45°-45°-90° triangle with $OP = 1$, $BP = \frac{\sqrt{2}}{2}$, and $OB = \frac{\sqrt{2}}{2}$.

$$W\left(\frac{\pi}{4}\right) = \left(\frac{\sqrt{2}}{2}, \frac{\sqrt{2}}{2}\right)$$

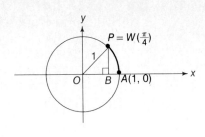

For $W(\frac{\pi}{3})$, $\widehat{AP}$ is one third the length π of the arc of the semicircle.

$$m\angle AOP = \tfrac{1}{3}(180°) = 60°$$

So $\triangle POB$ is a 30°-60°-90° triangle with $OP = 1$, $BP = \frac{\sqrt{3}}{2}$, and $OB = \frac{1}{2}$.

$$W\left(\frac{\pi}{3}\right) = \left(\frac{1}{2}, \frac{\sqrt{3}}{2}\right)$$

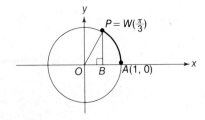

If $t = \frac{5\pi}{6}$, the two arcs shown in black are congruent; they both have length $\frac{\pi}{6}$. By symmetry,

$$W\left(\frac{5\pi}{6}\right) = \left(-\frac{\sqrt{3}}{2}, \frac{1}{2}\right),$$

$$W\left(\frac{7\pi}{6}\right) = \left(-\frac{\sqrt{3}}{2}, -\frac{1}{2}\right), \text{ and}$$

$$W\left(\frac{11\pi}{6}\right) = \left(\frac{\sqrt{3}}{2}, -\frac{1}{2}\right).$$

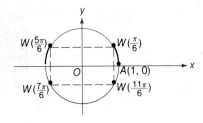

It will be useful to call the arguments $\frac{\pi}{6}$, $\frac{5\pi}{6}$, $\frac{7\pi}{6}$, and $\frac{11\pi}{6}$ the $\frac{\pi}{6}$ family of arguments.

In similar fashion you can locate the terminal points for the $\frac{\pi}{4}$ family of arguments $\frac{\pi}{4}$, $\frac{3\pi}{4}$, $\frac{5\pi}{4}$, and $\frac{7\pi}{4}$, as well as the $\frac{\pi}{3}$ family of arguments $\frac{\pi}{3}$, $\frac{2\pi}{3}$, $\frac{4\pi}{3}$, and $\frac{5\pi}{3}$. In every case you need only remember the coordinates of $W(t)$ for $t = \frac{\pi}{6}$, $\frac{\pi}{4}$, or $\frac{\pi}{3}$ in order to identify $W(t)$ for t in any of the families.

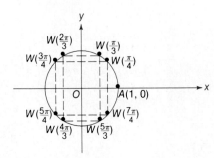

The quadrant in which $W(t)$ is located determines the signs of the coordinates.

If t is a negative multiple of one of the arguments $\frac{\pi}{6}$, $\frac{\pi}{4}$, or $\frac{\pi}{3}$, then finding $W(t)$ still depends upon finding the correct quadrant and the correct family of arguments for $W(t)$.

Example 1 Find the coordinates of $W\left(-\frac{2\pi}{3}\right)$.

SOLUTION Since $-\frac{2\pi}{3}$ is negative, wrap clockwise from $(1, 0)$ and count off 2 units of $\frac{\pi}{3}$ each. The terminal point $W\left(-\frac{2\pi}{3}\right)$ is the same point as $W\left(\frac{4\pi}{3}\right)$.

$$W\left(-\frac{2\pi}{3}\right) = W\left(\frac{4\pi}{3}\right) = \left(-\frac{1}{2}, -\frac{\sqrt{3}}{2}\right)$$

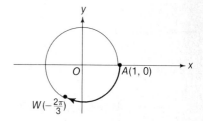

Sometimes t is a multiple of one of the standard arguments but also greater than 2π or less than -2π. Then you must wrap more than once around the unit circle to find the correct quadrant.

Example 2 Find the coordinates of $W\left(\frac{35\pi}{6}\right)$.

SOLUTION Rewrite $\frac{35\pi}{6}$ as $4\pi + \frac{11\pi}{6}$. Locate $W\left(\frac{35\pi}{6}\right)$ by wrapping around the circle counterclockwise 2 complete revolutions, or 4π units, and an additional $\frac{11\pi}{6}$ units. Hence

$$W\left(\frac{35\pi}{6}\right) = W\left(\frac{11\pi}{6}\right) = \left(\frac{\sqrt{3}}{2}, -\frac{1}{2}\right).$$

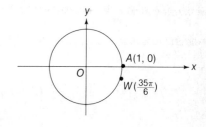

Exercises

A Find the coordinates of the terminal point.

1. $W\left(\frac{\pi}{2}\right)$
2. $W(-\pi)$
3. $W\left(\frac{\pi}{6}\right)$
4. $W(5\pi)$
5. $W\left(\frac{4\pi}{3}\right)$
6. $W\left(-\frac{3\pi}{2}\right)$
7. $W\left(\frac{\pi}{4}\right)$
8. $W(-6\pi)$
9. $W(8\pi)$
10. $W\left(-\frac{\pi}{2}\right)$
11. $W\left(\frac{7\pi}{6}\right)$
12. $W\left(-\frac{7\pi}{6}\right)$
13. $W\left(-\frac{\pi}{6}\right)$
14. $W\left(\frac{7\pi}{3}\right)$
15. $W\left(\frac{5\pi}{4}\right)$
16. $W\left(-\frac{\pi}{4}\right)$

17. $W\left(-\frac{11\pi}{6}\right)$
18. $W\left(\frac{11\pi}{3}\right)$
19. $W\left(-\frac{5\pi}{4}\right)$
20. $W\left(-\frac{17\pi}{3}\right)$

B In what quadrant is each point located?

21. $W(3)$
22. $W(-3)$
23. $W(-7)$
24. $W(6)$

Estimate the coordinates of the terminal point.

25. $W(1.4)$
26. $W(3)$
27. $W(-1)$
28. $W(-1.4)$

Determine five different arguments t so that $W(t)$ has the given coordinates.

29. $(-1, 0)$
30. $\left(\frac{\sqrt{2}}{2}, \frac{\sqrt{2}}{2}\right)$

31. The up-and-down motion of an industrial piston drives one end E of a connecting rod in a circle of radius 0.6 meter. Introduce a coordinate system with origin at the center of the circle so that E has coordinates (x, y) and find the maximum and minimum values of both x and y.

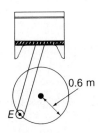

32. For t_0 as indicated in the figure, sketch the positions of $W(t_0 + \pi)$, $W(-t_0)$, $W(t_0 + 2\pi)$, and $W(t_0 - 2\pi)$.

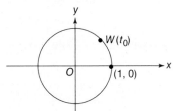

33. If the coordinates of $W(t_0)$ are (a, b), what are the coordinates of points P_1, P_2, and P_3?

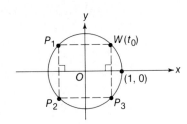

34. If $W(t_0) = \left(\frac{3}{5}, \frac{4}{5}\right)$, what are the coordinates of $W(t_0 + \pi)$, $W(t_0 + 2\pi)$, and $W(-t_0)$?

C For what integers n is the given statement true?

35. $W\left(\frac{8\pi}{3}\right) = W\left(\frac{2\pi}{3} + 2\pi n\right)$
36. $W\left(\frac{27\pi}{4}\right) = W\left(\frac{3\pi}{4} + 2\pi n\right)$

37. Explain why $W(x + 2\pi n) = W(x)$ for all real numbers x and all integers n.

9–2 Sine and Cosine Functions

The trigonometric ratios sine and cosine were defined in antiquity to help traders and explorers navigate by the stars in their travels. In modern times, scientists and engineers have been assisted in their exploration of technical fields by the so-called circular functions, which go by the old trigonometric names sine and cosine.

In section 9–1, whenever you found the coordinates of a particular terminal point $W(t)$, you also found cos (t) and sin (t). They are simply the coordinates of the point $W(t)$.

Definition Let t be any real number and $W(t) = (x, y)$ be the point that the wrapping function assigns to t. Then x is the **cosine of t** and y is the **sine of t**.

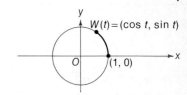

It is customary to abbreviate the function cosine as *cos* and the function sine as *sin*. Since t can be any real number, the domain of cosine and sine is **R**. By referring to the diagrams on page 310, you can verify that the sine and cosine values for arguments t in the $\frac{\pi}{3}$, $\frac{\pi}{4}$, and $\frac{\pi}{6}$ families are as given in the table below. Notice that we customarily write sin (t) as sin t and cos (t) as cos t. If $t \in [0, 2\pi]$ is a number other than one of the special arguments, you can estimate the coordinates (cos t, sin t) for $W(t)$. In the previous section you saw that cos $2 \doteq -0.4$ and sin $2 \doteq 0.9$. It used to be common practice to find values of sin t and cos t from tables, and such a table for decimal arguments is given in the Appendix. Nowadays, most people rely on a calculator, which provides more accurate approximations very quickly.

For any value of t, the point (cos t, sin t) lies on the unit circle, $x^2 + y^2 = 1$. Thus, by substitution, we obtain the following very useful theorem.

t	cos t	sin t	t	cos t	sin t
0	1	0			
$\frac{\pi}{6}$	$\frac{\sqrt{3}}{2}$	$\frac{1}{2}$	$\frac{7\pi}{6}$	$-\frac{\sqrt{3}}{2}$	$-\frac{1}{2}$
$\frac{\pi}{4}$	$\frac{\sqrt{2}}{2}$	$\frac{\sqrt{2}}{2}$	$\frac{5\pi}{4}$	$-\frac{\sqrt{2}}{2}$	$-\frac{\sqrt{2}}{2}$
$\frac{\pi}{3}$	$\frac{1}{2}$	$\frac{\sqrt{3}}{2}$	$\frac{4\pi}{3}$	$-\frac{1}{2}$	$-\frac{\sqrt{3}}{2}$
$\frac{\pi}{2}$	0	1	$\frac{3\pi}{2}$	0	-1
$\frac{2\pi}{3}$	$-\frac{1}{2}$	$\frac{\sqrt{3}}{2}$	$\frac{5\pi}{3}$	$\frac{1}{2}$	$-\frac{\sqrt{3}}{2}$
$\frac{3\pi}{4}$	$-\frac{\sqrt{2}}{2}$	$\frac{\sqrt{2}}{2}$	$\frac{7\pi}{4}$	$\frac{\sqrt{2}}{2}$	$-\frac{\sqrt{2}}{2}$
$\frac{5\pi}{6}$	$-\frac{\sqrt{3}}{2}$	$\frac{1}{2}$	$\frac{11\pi}{6}$	$\frac{\sqrt{3}}{2}$	$-\frac{1}{2}$
π	-1	0	2π	1	0

Theorem 9–1 **Pythagorean Identity**

For all real numbers t, $\sin^2 t + \cos^2 t = 1$.

We usually write $(\sin t)^2$ as $\sin^2 t$ and $(\cos t)^2$ as $\cos^2 t$.

Example 1 If $\sin t = \frac{1}{3}$ and $W(t)$ is in quadrant II, find $\cos t$ exactly.

SOLUTION By substitution in the Pythagorean identity and subtraction,

$$\cos^2 t = 1 - \frac{1}{9} = \frac{8}{9}.$$

$$\cos t = \pm\sqrt{\frac{8}{9}} = \pm\frac{2\sqrt{2}}{3}$$

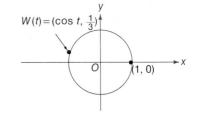

Since $W(t)$ is in quadrant II, $\cos t$ must be negative.

ANSWER $\cos t = -\dfrac{2\sqrt{2}}{3}$

Most calculators accept arguments in radians, degrees, or grads. By tradition, the real-number argument t is referred to as t radians. Thus for a calculator in the radian mode,

$\sin\left(\frac{\pi}{3}\right) = \sin\left(\frac{\pi}{3} \text{ radians}\right) \doteq 0.8660$, and

$\sin(2) = \sin(2 \text{ radians}) \doteq 0.9093$.

But if the calculator is in degree mode, you will find $\sin(2°) \doteq 0.0349$ and if the calculator is in grad mode, you will find $\sin(2 \text{ grads}) \doteq 0.0314$.

In this chapter, we work exclusively in radians. In Chapter 10, the work is in degrees. The unit grad is used principally by civil engineers and is not discussed here.

Example 2 Use a calculator to find each of the following accurate to 4 decimal places:
$\cos 2$, $\sin(0.2)$, $\sin(1.0)$, $\sin\left(\frac{\pi}{6}\right)$, $\cos(15.2)$, $\sin(-2.4)$.

SOLUTION
$\cos 2 \doteq -0.4161$ $\sin(0.2) \doteq 0.1987$
$\sin(1.0) \doteq 0.8415$ $\sin\left(\frac{\pi}{6}\right) \doteq 0.5000$
$\cos(15.2) \doteq -0.8737$ $\sin(-2.4) \doteq -0.6755$

Exercises

A Evaluate exactly without using a calculator or a table.

1. $\sin\left(\dfrac{2\pi}{3}\right)$
2. $\cos\left(\dfrac{5\pi}{4}\right)$
3. $\cos\left(\dfrac{5\pi}{6}\right)$
4. $\sin\left(\dfrac{11\pi}{6}\right)$
5. $\cos\left(\dfrac{5\pi}{3}\right)$
6. $\sin\left(\dfrac{7\pi}{4}\right)$
7. $\cos\left(\dfrac{11\pi}{6}\right)$
8. $\sin\left(\dfrac{5\pi}{3}\right)$
9. $\sin\left(\dfrac{3\pi}{4}\right)$
10. $\cos\left(\dfrac{2\pi}{3}\right)$

Approximate by using a calculator. Leave answers accurate to the nearest hundredth.

11. $\sin(3.1)$
12. $\cos(1.4)$
13. $\cos\left(\dfrac{\pi}{3}\right)$
14. $\sin\left(\dfrac{2\pi}{5}\right)$
15. $\cos(10)$
16. $[\sin(1)]^2 + [\cos(1)]^2$

B Evaluate exactly. Do not use decimal approximations.

17. $\sin\left(-\dfrac{\pi}{4}\right)$
18. $\cos\left(-\dfrac{\pi}{6}\right)$

19. $\dfrac{\sin\left(\dfrac{11\pi}{6}\right) - \cos\left(\dfrac{\pi}{6}\right)}{\sin\left(-\dfrac{\pi}{6}\right) + \cos\left(\dfrac{5\pi}{6}\right)}$

20. $\dfrac{1 - \sin\left(\dfrac{7\pi}{4}\right)}{1 + \cos\left(\dfrac{3\pi}{4}\right)}$

21. $\left(\sin\dfrac{11\pi}{6}\right)^2 + \left(\cos\dfrac{11\pi}{6}\right)^2$
22. $\sin\left(\dfrac{\pi}{2} + \dfrac{\pi}{3}\right)$

23. If $\sin t = \dfrac{3}{5}$ and $W(t)$ is in quadrant II, find $\cos t$ exactly.
24. If $\cos t = -\dfrac{1}{3}$ and $W(t)$ is in quadrant II, find $\sin t$ exactly.
25. If $\sin t = -\dfrac{2}{3}$ and $W(t)$ is in quadrant III, find $\cos t$ exactly.
26. If $\cos t = \dfrac{4}{5}$ and $W(t)$ is in quadrant IV, find $\sin t$ exactly.

C 27. For a simple alternating current generator the electric current in amperes is given by $I = \sin(\pi t)$, where t is in seconds.
Find the current I when $t = 0, \dfrac{1}{4}, \dfrac{1}{2}, \dfrac{3}{4},$ and 1.

28. **a.** Write a program to have a computer print the value of $\dfrac{\sin x}{x}$ for $x = 1, \dfrac{1}{5},$ $\dfrac{1}{25}, \dfrac{1}{125}, \dfrac{1}{625},$ and $\dfrac{1}{3125}$.

  b. Run the program.

  c. What do your results from **b** appear to say about the limit of $\dfrac{\sin x}{x}$ as x approaches zero?

29. A mass moving up and down at the end of a spring has height in meters given by

$$h(t) = 5 \cos\left(\frac{\pi t}{6}\right),$$

where t is in seconds. What is the smallest positive value of t for which $h(t) = 5$?

9–3 Graphs of Sine and Cosine

You now have enough values for the sine and cosine functions to sketch their graphs on the interval $[0, 2\pi]$. From the table on page 313, plot the values of sin t. You could plot more points by using a calculator for approximations. But instead, picture a particle moving around the unit circle, and keep track of its height, that is, the y-coordinate of the particle. If $t > 0$, the particle starts at height 0 and as t increases, $y = \sin t$ increases gradually to a maximum of 1 and then falls back to 0 at $t = \pi$. If $t > \pi$, the y-coordinate becomes negative, reaches a minimum of -1, and then returns to 0 when the particle reaches $A(1, 0)$. The particle continues to travel around the circle and its y-coordinate, sin t, goes up to 1, back to 0, down to -1, and back to 0. If $t < 0$, the particle goes the other direction around the circle and gives the same up-and-down pattern.

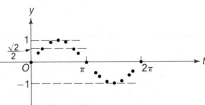

Thus, a graph of the sine function looks like the following.

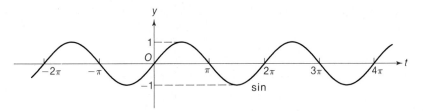

You can graph the cosine function in the same manner.

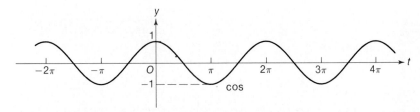

You have probably already noticed that the range of both these circular functions, sine and cosine, is the set of real numbers in the closed interval from -1 to 1, that is,

$$\text{Rng sin} = \text{Rng cos} = [-1, 1].$$

Both of these circular functions repeat themselves regularly in a manner unlike any other previous elementary functions you have studied. Such functions are called periodic functions.

Definition A nonconstant function f is said to be **periodic** if there is a nonzero number p such that $f(x) = f(x + p)$ for all x in Dom f.

Any such p is called *a period* of f. The smallest positive p is called *the period* of f.

It should be clear from the diagrams on the previous page that the periods of the cosine and sine functions are 2π.

Knowing that the period of sine is 2π allows you to write

$$\sin(x + 4\pi) = \sin[(x + 2\pi) + 2\pi],$$
$$= \sin(x + 2\pi),$$
$$= \sin x.$$

Continuing this process again and again for sine and cosine establishes that

$$\sin(x + 2\pi n) = \sin x \quad \text{and} \quad \cos(x + 2\pi n) = \cos x, \text{ for any integer } n.$$

Some periodic functions are not circular functions.

Example 1 Graph the function $f(x) = x - [x]$ and determine its period.

SOLUTION It is easy to verify that the graph is as shown at the right, and that the period is 1. Thus

$f(x + 1) = f(x)$ for all $x \in \mathbb{R}$.

Example 2 Evaluate $\sin\left(\dfrac{23\pi}{6}\right)$ exactly.

SOLUTION $W\left(\dfrac{23\pi}{6}\right) = W\left(\dfrac{11\pi}{6} + 2\pi\right) = W\left(\dfrac{11\pi}{6}\right)$

Thus $\sin\left(\dfrac{23\pi}{6}\right) = \sin\left(\dfrac{11\pi}{6}\right)$,

which you already know equals $-\dfrac{1}{2}$.

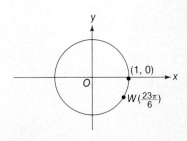

Example 3 Evaluate $\cos\left(-\dfrac{21\pi}{4}\right)$ exactly.

SOLUTION Write $-\dfrac{21\pi}{4}$ as $\dfrac{3\pi}{4} + (-6\pi)$. Since -6π is -3 times the period 2π, it follows that

$$\cos\left(\dfrac{3\pi}{4} + [-6\pi]\right) = \cos\left(\dfrac{3\pi}{4}\right), \text{ which equals } -\dfrac{\sqrt{2}}{2}.$$

ANSWER $\cos\left(-\dfrac{21\pi}{4}\right) = -\dfrac{\sqrt{2}}{2}$

Exercises

A 1. What is the domain of the sine function?
2. What is the domain of the cosine function?

Evaluate exactly without using a calculator or a table.

3. $\cos\left(2\pi + \dfrac{3\pi}{4}\right)$
4. $\sin\left(2\pi + \dfrac{2\pi}{3}\right)$
5. $\sin\left(\dfrac{5\pi}{6} - 2\pi\right)$
6. $\cos\left(\dfrac{3\pi}{2} + 4\pi\right)$
7. $\sin\left(\dfrac{7\pi}{6} + 2\pi\right)$
8. $\cos\left(\dfrac{4\pi}{3} + 2\pi\right)$
9. $\cos\left(-\dfrac{\pi}{6} + 6\pi\right)$
10. $\sin\left(\dfrac{5\pi}{3} + 6\pi\right)$
11. $\sin\left(\dfrac{4\pi}{3} + 4\pi\right)$
12. $\cos\left(-\dfrac{2\pi}{3} + 4\pi\right)$
13. $\cos\left(-\dfrac{11\pi}{2}\right)$
14. $\sin\left(\dfrac{37\pi}{3}\right)$
15. $\sin\left(-\dfrac{19\pi}{3}\right)$
16. $\cos\left(\dfrac{17\pi}{4}\right)$
17. $\cos\left(\dfrac{25\pi}{3}\right)$
18. $\sin\left(\dfrac{29\pi}{3}\right)$
19. $\sin\left(\dfrac{101\pi}{2}\right)$
20. $\cos\left(-\dfrac{103\pi}{2}\right)$

B 21. For the periodic function f whose graph is shown:
   a. What is the period of f?
   b. Evaluate $f(10.1)$.
   c. Evaluate $f(9.8)$.

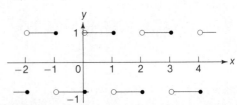

318 Chapter 9: Circular Functions

22. For the periodic function g whose graph is shown:
   a. What is the period of g?
   b. Is 6 a period of g?
   c. Is g(100) = g(1)?

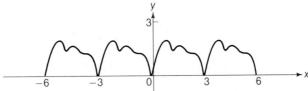

23. a. Let $0 < t < \frac{\pi}{2}$. Show that the right triangles in the figures are congruent.
   b. Why does this congruence show that cos (t) = cos (−t)?
   c. Why does this congruence show that sin (t) = −sin (−t)?

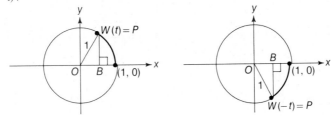

24. Repeat exercise 23 for $\frac{\pi}{2} < t < \pi$ with W(t) in quadrant II.

25. Exercises 23 and 24 establish that cosine is an *even* function and that sine is an *odd* function. How are these facts reflected in the graphs of sine and cosine?

26. If the period of the function f is p, prove that f(x − p) = f(x) for all x in the domain of f.

C 27. a. Write a computer program to have a computer print the value of $\frac{\cos x - 1}{x}$ for x = 1, 0.1, 0.01, 0.001, 0.0001.
   b. Run the program.
   c. What do your results from b appear to say about the limit of $\frac{\cos x - 1}{x}$ as x approaches zero?

28. Find all $x \in [0, 2\pi]$ where f(x) = sin x is an increasing function.

29. Find all $x \in [0, 2\pi]$ where f(x) = cos x is a decreasing function.

Solve each equation on the interval $[0, 2\pi]$.

30. sin x = 1

31. cos x = 0

32. cos x = 0.5

33. sin x = $\frac{\sqrt{3}}{2}$

9–4 Equations and Inequalities

Imagine a perfect pendulum, one with no resistance, swinging back and forth over a coordinate line. The end E of the pendulum is above a point with coordinate P on the line. Let t be the time elapsed since the pendulum started swinging. For a particular pendulum the coordinate P may be given as a function of t by

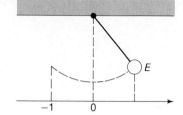

$$P(t) = \cos t.$$

Surprisingly, the amount of mass at E does not affect the coordinate P.

When the pendulum starts swinging, $t = 0$ and $P(0) = \cos(0) = 1$. So the pendulum starts with $P = 1$, as in the figure, and starts swinging to the left. How long does it take for the pendulum to return to its starting position where $P = 1$? Inspection of the graph of $P(t) = \cos t$ shows that $P(t) = 1$ for $t = 2\pi$.

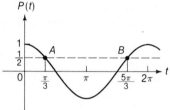

What are the first two arguments t for which $P(t) = 0.5$? You already know that $\cos \frac{\pi}{3} = 0.5$ (point A on the graph) and $\cos \frac{5\pi}{3} = 0.5$ (point B on the graph). Because of the periodicity of cosine, these values are repeated every 2π units.

Suppose the motion of a pendulum is described by the function

$$P(t) = 2 \cos t.$$

For what $t \in [0, 2\pi]$ is $P(t) = 1$?

$$2 \cos t = 1$$
$$\cos t = \frac{1}{2}$$
$$t = \frac{\pi}{3}, \frac{5\pi}{3}$$

These two examples illustrate the importance of being able to solve equations of the form $P = a \cos bt$.

Example 1 Solve for $x \in [0, 2\pi]$: $\sin x = 1$.

SOLUTION A graph of $y = \sin x$ shows that there is only one solution,

$$x = \frac{\pi}{2}.$$

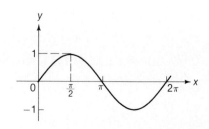

Example 2 Solve for $x \in [0, 2\pi]$: $2 \cos x + 1 = 0$.

SOLUTION First, solve the given equation for $\cos x$.

$$\cos x = -\frac{1}{2}$$

From a graph of cos, observe that there are two solutions and they are between $\frac{\pi}{2}$ and $\frac{3\pi}{2}$. Recall the special arguments in the $\frac{\pi}{3}$ family to identify the solutions as $x = \frac{2\pi}{3}$ and $x = \frac{4\pi}{3}$.

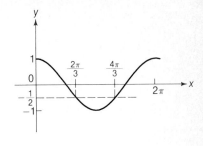

If $\sin x$ or $\cos x$ is not equal to one of the special numbers ± 1, $\pm \frac{1}{2}$, $\pm \frac{\sqrt{2}}{2}$, $\pm \frac{\sqrt{3}}{2}$, or 0, then a table or calculator must be used. But a sketch is always useful in approximating the solutions.

Example 3 Approximate the smallest positive x for which $\cos x = 0.4$.

SOLUTION From a graph of cosine you can see that the solution x_0 is a little less than $\frac{\pi}{2}$, which is approximately 1.57. By calculator, $\cos 1.1 \doteq 0.45$ and $\cos 1.2 \doteq 0.36$. Therefore x_0 is about halfway between 1.1 and 1.2.

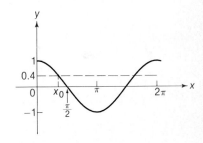

ANSWER $x_0 \doteq 1.15$

On a calculator set in the radian mode, the sequence 0.4 [INV] [COS] displays 1.1592795, which is a better approximation of x_0 in example 3. Since neither the sine nor the cosine function is one-to-one on **R**, they do not have inverses. You must proceed with caution when using the key [INV] with sin or cos. This matter will be discussed in section 9–10.

Some inequalities involving circular functions are particularly simple. The solution of $\cos x < 2$ is the set of real numbers, since the cosine of any real number is in the interval $[-1, 1]$. Likewise, the solution of $\sin x < -1$ is the empty set, since the sine of any real number is always greater than or equal to -1. For more difficult inequalities, graphing usually helps.

Example 4 Solve for $x \in [0, 2\pi]$: $\sin x \geq \dfrac{\sqrt{3}}{2}$.

SOLUTION On the graph of sin shown, the answer is the interval $[r, s]$, since for all $x \in [r, s]$ the graph of sin is at or above $\dfrac{\sqrt{3}}{2}$. For the arguments $\dfrac{\pi}{3}$ and $\dfrac{2\pi}{3}$, you know that $\sin \dfrac{\pi}{3} = \sin \dfrac{2\pi}{3} = \dfrac{\sqrt{3}}{2}$.

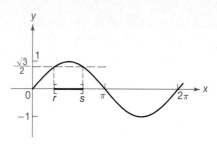

The solution is the set of all x such that $\dfrac{\pi}{3} \leq x \leq \dfrac{2\pi}{3}$.

Example 5 One end of a beam is built into a rigid wall. The other end can vibrate, and its vertical displacement at time t is given by $Y = 2 \cos t$. For what t in $[0, 2\pi]$ is $Y \geq 1$?

SOLUTION Solve the inequality $2 \cos t \geq 1$ for $\cos t$:

$$\cos t \geq \tfrac{1}{2}.$$

The graph of cosine shows that there are two intervals on which the graph of cos is at or above $\tfrac{1}{2}$. The solutions of $\cos t = \tfrac{1}{2}$ are the special arguments $\dfrac{\pi}{3}$ and $\dfrac{5\pi}{3}$. The solution set for the inequality is

$$\left\{t: 0 \leq t \leq \tfrac{\pi}{3} \text{ or } \tfrac{5\pi}{3} \leq t \leq 2\pi\right\}.$$

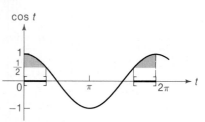

Exercises

A Solve for all $x \in [0, 2\pi]$.

1. $\cos x = 1$
2. $\sin x = 0$
3. $\sin x = \tfrac{1}{2}$
4. $\cos x = -\tfrac{1}{2}$
5. $\cos x = -\dfrac{\sqrt{3}}{2}$
6. $\sin x = \dfrac{\sqrt{3}}{2}$
7. $\sin x = -1$
8. $\cos x = \dfrac{\sqrt{2}}{2}$
9. $\sin x = -\dfrac{\sqrt{2}}{2}$
10. $\sin x = 1$
11. $\sin x < 2$
12. $\cos x > -2$

B Solve for all $x \in [0, 2\pi]$.

13. $3 \sin x = \sin x + 1$
14. $2 \cos x + \sqrt{2} = 0$
15. $\cos x > 0.5$
16. $\sin x \leq -1$
17. $2 \sin x - \sqrt{2} < 0$
18. $\cos x < \dfrac{\sqrt{3}}{2}$

19. $\sin x \geq \dfrac{\sqrt{3}}{2}$ **20.** $\cos^2 x > 0$

Use a graph to determine the number of solutions in $[0, \pi]$. Use a calculator to approximate the smallest positive solution to the nearest hundredth.

21. $\cos x = 0.9$ **22.** $\sin x = 0.1$ **23.** $\sin x = -0.9$

24. $\cos x = -0.1$ **25.** $\sin x = 0.6$ **26.** $\sin x = -0.4$

27. In the first 6 seconds that a generator is operating, the current in amperes is given in terms of time t by $I(t) = \cos t$. During what time intervals is the current I greater than 0.5 ampere?

28. A mass is moving up and down at the end of a spring and its height in meters at time t seconds is given by $H(t) = 2 \sin t + 1$.
   a. What is the height H when $t = 0$?
   b. How long does it take before H is 0 for the second time?
   c. After approximately how long is $H = 0.4$ meter for the first time?

29. The number r is a solution of the equation $\sin x = k$, $k > 0$. By referring to the figure, specify in terms of r the other three solutions in the interval $[0, 3\pi]$.

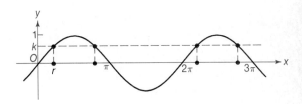

30. Use the illustration in exercise 29 to specify the first two intervals in which $\sin x > k$.

31. The motion of a pendulum is described by $P(t) = a \cos t$.
   a. If $a = 2$, find all $t \in [0, 2\pi]$ for which $P = 1$.
   b. If $a = 4$, find all $t \in [0, 2\pi]$ for which $P = 2\sqrt{2}$.

C Find approximations to the nearest hundredth for all solutions in $[0, 2\pi]$.

32. $\sin x < 0.7$ **33.** $\cos x > -0.3$ **34.** $2 \sin^2 x + \sin x = 1$

35. The motion of a pendulum is described by $P(t) = a \cos bt$.
   a. If $a = 2$ and $b = \pi$, find all $t \in [0, 2]$ for which $P = 0$.
   b. If $a = 2$ and $b = \pi$, find the time required for the pendulum to make one complete swing back and forth (the *period* of the pendulum).

36. a. Write a computer program to print a table of values of the function $f(x) = 2 \sin x$ for $x = 0, 0.4, 0.8, 1.2, \ldots, 6.4$.
   b. Run the program and use the printout to sketch the function $f(x)$ on the interval $[0, 2\pi]$.
   c. Use your sketch to estimate the solution of $f(x) > 0.5$ on the interval $[0, 2\pi]$.

9–5 Graphs Derived from Sine and Cosine

In example 5 of section 9–4 the vibrating end of a beam had a vertical displacement Y at time t given by

$$Y = 2 \cos t.$$

The graph of $Y(t)$ is found by stretching vertically the graph of $Y = \cos t$ by a factor of 2.

$$Y(0) = 2 \cos 0 = 2$$
$$Y\left(\tfrac{\pi}{3}\right) = 2 \cos \tfrac{\pi}{3} = 1$$
$$Y(\pi) = 2 \cos \pi = -2.$$

The graphs of $Y = \cos t$ and of $Y = 2 \cos t$ are shown on the same axes.

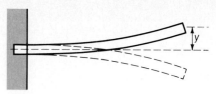

You already know how to shift and stretch the graph of a function f to get the graph of $g(x) = af(x + h) + k$.

Example 1 Graph $g(x) = \sin x + 1$ on the interval $[0, 2\pi]$.

SOLUTION Shift the graph of sine up 1 unit. Note that Rng g is $[0, 2]$ and that $\tfrac{3\pi}{2}$ is the only zero of g.

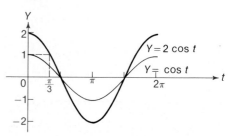

Example 2 Graph $f(x) = -2 \cos x$ on the interval $[0, 2\pi]$.

SOLUTION Multiply each value of $\cos x$ by 2. Then reflect the resulting graph in the x-axis.

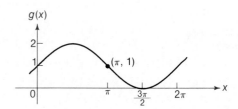

Example 3: Sketch a graph of the function $f(x) = 2 \cos\left(x - \tfrac{\pi}{4}\right) - 1$.

SOLUTION First sketch a graph of $g(x) = 2 \cos x$.

Then shift this graph to the right $\tfrac{\pi}{4}$ units.

(cont. on p. 325)

324 Chapter 9: Circular Functions

Finally shift this graph down 1 unit to obtain the graph of f. It is useful to label maximum and minimum points as shown in the final graph below.

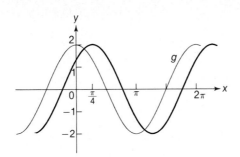
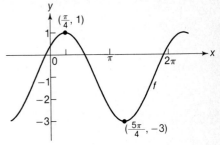

Example 4 The path of a bouncing ball is approximated by the graph of the function $g(x) = |\sin(x + \frac{\pi}{2})|$. Sketch the graph and determine the period of g.

SOLUTION Write the function as the composite

$$g = \text{abs} \circ \sin \circ (I + \tfrac{\pi}{2}).$$

First sketch the graph of sin; then shift this graph to the left $\frac{\pi}{2}$ units, as shown.

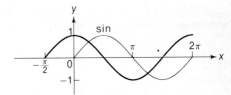

Finally, take the absolute value of the function graphed to obtain the graph of g. The period of g is π.

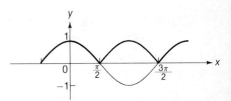

Exercises

A Graph two periods of the given function and label the coordinates of one point not on an axis.

1. $f(x) = \sin x - 1$
2. $f(x) = \cos x + 2$
3. $f(x) = 2 \cos x + 1$
4. $f(x) = -2 \sin x$
5. $g(x) = -\sin x + 2$
6. $f(x) = 2 \cos x - 1$
7. $f(x) = \cos\left(x - \frac{\pi}{2}\right)$
8. $f(x) = \sin(x - \pi)$

B 9. $g(x) = 2 \sin\left(x + \frac{\pi}{4}\right)$
10. $g(x) = -\cos\left(x + \frac{\pi}{2}\right)$

11. $f(x) = 2 \sin(x - \pi)$
12. $g(x) = |2 \sin x|$
13. $f(x) = 1 - \cos x$
14. $g(x) = 1 + \sin x$
15. $f(x) = 3 - \sin x$
16. $g(x) = |\cos x - 1|$
17. $f(x) = \sin\left(x - \dfrac{\pi}{2}\right)$
18. $g(x) = \cos(x - \pi)$
19. $h(x) = 2 \cos\left(x - \dfrac{3\pi}{2}\right) + 1$
20. $h(x) = 1 - 2 \sin\left(x + \dfrac{\pi}{2}\right)$

21. The Celsius temperature C at a point on a metal plate varies with time t and is given by $C(t) = 10 \cos t - 5$. Graph $C(t)$ for $0 \le t \le 9$ and find those values of t for which the temperature is positive.

22. The voltage across the terminals of a circuit at time t is given by $V(t) = 1 - 2 \sin t$. Graph V for $0 \le t \le 4\pi$, and determine all t for which $V(t) \ge 0$.

Graph each function, label the coordinates of a point not on an axis, and specify the range of the function.

23. $f(t) = \sin(t - 1)$
24. $g(t) = 2 \cos(t + 1)$
25. $g(x) = \tfrac{1}{2} \sin(-x)$
26. $f(x) = \tfrac{1}{2} \cos(-x)$
27. $f(x) = |1 - 2 \sin(x - \pi)|$
28. $g(x) = |2 \cos(x + \pi)|$
29. $g(t) = \left|1 + \dfrac{\cos t}{2}\right|$
30. $f(t) = \left|1 - \dfrac{\sin t}{2}\right|$

31. The magnitude of a shock wave in the earth's crust is given as a function of time t by $M(t) = |2 \cos t - 1|$. For what t in the interval $[0, 2\pi]$ is M greater than 1?

32. A mass at the end of a spring has a height at time t given by $h(t) = 10 - 2 \cos t$. Graph the function h for $0 \le t \le 12$ and determine the times when h is a maximum.

33. What is the period of the function $g(t) = 1 + |2 \cos t|$?

C Graph each function.

34. $f(x) = 1 - |2 \cos x - 1|$
35. $f(x) = 1 + 2\left|\sin\left(x - \dfrac{\pi}{2}\right)\right|$
36. $g(x) = \dfrac{1}{\sin x}$
37. $g(x) = \dfrac{1}{\cos x}$
38. $H(t) = \sin\left(\dfrac{t}{2}\right)$
39. $h(t) = \sin(2t)$

9–6 Secant and Cosecant

You have studied the reciprocals of some functions in Chapter 6. To find values of the function $f(x) = \frac{1}{\sin x}$ and sketch its graph, you could use a calculator to compute and graph many points, and then connect these points with a smooth curve to obtain a graph of f. However, you might miss some important issues like periodicity, zeros, asymptotes, and range.

The method outlined in Chapter 6 depends upon graphing $g(x) = \sin x$ and then asking some questions about this graph.

1. What are the zeros of $g(x) = \sin x$? They are 0, π, 2π, etc. On a graph of g, draw vertical dashed lines at these arguments.

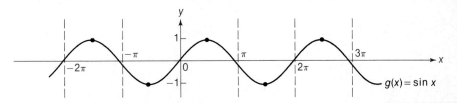

2. Where is $\sin x = \pm 1$? At these points, the graphs of the functions $\sin$ and $\frac{1}{\sin}$ are the same. Mark these points, as shown above.

3. Where is $\sin x$ positive? Where is it negative? The function $\frac{1}{\sin}$ has the same sign as the function $\sin$.

4. Where is $|\sin x|$ small? There, the reciprocal will be large. Near the zeros, $\sin x$ is small. So, near the zeros $\frac{1}{\sin x}$ is either large positive or large negative. Therefore the vertical dashed lines are vertical asymptotes.

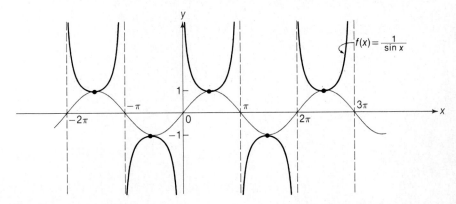

The reciprocals of the sine and cosine functions are used so often that they have their own special names.

9–6: Secant and Cosecant   327

Definition The **secant function**, written sec, is defined for all x, $\cos x \neq 0$, by

$$\sec x = \frac{1}{\cos x}.$$

The **cosecant function**, written csc, is defined for all x, $\sin x \neq 0$, by

$$\csc x = \frac{1}{\sin x}.$$

From the graph of $\csc x = \frac{1}{\sin x}$ on the previous page, you can see that the csc function has no zeros, has period 2π, and its range is $(-\infty, -1] \cup [1, \infty)$.

Example 1 Find $\sec \frac{\pi}{6}$ and $\csc \frac{5\pi}{4}$.

SOLUTION $\sec \frac{\pi}{6} = \dfrac{1}{\cos \frac{\pi}{6}} = \dfrac{1}{\frac{\sqrt{3}}{2}} = \dfrac{2\sqrt{3}}{3}$

$\csc \frac{5\pi}{4} = \dfrac{1}{\sin \frac{5\pi}{4}} = \dfrac{1}{-\frac{\sqrt{2}}{2}} = -\sqrt{2}$

Example 2 Sketch the secant function; find its domain, range, zeros, and asymptotes.

SOLUTION $f(x) = \sec x = \dfrac{1}{\cos x}$

Use a sketch of cos and repeat steps 1–4 from page 327.

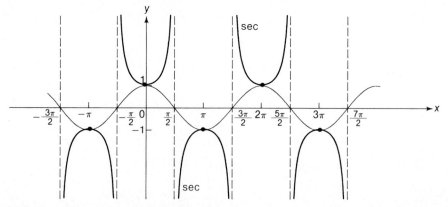

From the graph you can see that the period of sec is 2π; the domain is $\mathbf{R} - \left\{\frac{\pi}{2} + \pi k, k \in \mathbf{J}\right\}$, and there are no zeros. There are vertical asymptotes at $x = \frac{\pi}{2} + \pi k$, $k \in \mathbf{J}$.

Example 3 Use graphs to estimate the smallest positive solution of csc $x = x + 1$.

SOLUTION Sketch $f(x) = $ csc x and
$g(x) = x + 1$.

If you do not use the same scales on the two axes, be careful to graph the line with the correct slope.

The graphs first intersect at $x = r$, which you can estimate visually to be about 0.7.

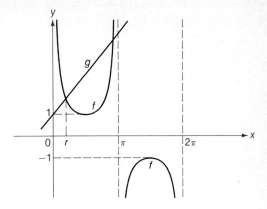

Exercises

A Evaluate exactly.

1. sec $\frac{\pi}{4}$
2. csc $\left(-\frac{\pi}{4}\right)$
3. sec 0
4. csc $\frac{\pi}{2}$
5. csc $\frac{5\pi}{6}$
6. sec $\frac{4\pi}{3}$
7. sec π
8. csc $\left(-\frac{\pi}{2}\right)$
9. csc $\left(-\frac{2\pi}{3}\right)$
10. sec $\left(-\frac{11\pi}{6}\right)$
11. sec $\frac{41\pi}{4}$
12. csc $\frac{23\pi}{6}$

Evaluate to the nearest hundredth.

13. sec (-3.1)
14. csc (6.2)
15. csc $(\pi + 1)$
16. sec $(1 - \pi)$

B Graph each function.

17. $f(x) = 2$ sec x
18. $f(x) = -$csc x
19. $g(t) = 1 - $ sec t
20. $g(t) = $ csc $(t - \pi)$
21. $h(x) = |1 - $ csc $x|$
22. $h(x) = |2$ sec $x - 1|$

If cos $x = 0.6$ and $0 < x < \frac{\pi}{2}$, evaluate exactly:

23. sin x
24. sec x
25. csc x

Use graphs to estimate the smallest positive solution.

26. csc $x = x$
27. sec $x = -\frac{1}{2}x$

For what $x \in [0, 2\pi]$ is each of the following true?

28. sec $x > 0$
29. csc $x > 0$
30. $1 + $ sec $x > 0$
31. $2 + $ csc $x \geq 0$

C Assume (correctly) that cosine is an even function and that sine is an odd function.

32. Prove that secant is an even function.
33. Prove that cosecant is an odd function.
34. Prove or disprove that cos ∘ sin is an even function.

9-7 Tangent and Cotangent

Recall that any point P on the unit circle has coordinates $(\cos t, \sin t)$. Suppose P is in the first quadrant. Let point R be the intersection of $\overrightarrow{OP}$ and the tangent line through $A(1, 0)$. By similar triangles,

$$\frac{AR}{OA} = \frac{PB}{OB} = \frac{\sin t}{\cos t}.$$

Since $OA = 1$, the tangent segment AR has length $\frac{\sin t}{\cos t}$. This is the origin of the name for the tangent function. Its reciprocal, the cotangent function, can be obtained in a similar way (see exercise 30).

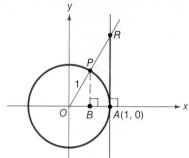

Definition The **tangent function**, written tan, is defined for all x, $\cos x \neq 0$, as

$$\tan x = \frac{\sin x}{\cos x}.$$

The **cotangent function**, written cot, is defined for all x, $\sin x \neq 0$, as

$$\cot x = \frac{\cos x}{\sin x}.$$

Example 1 Evaluate exactly: $\tan \frac{\pi}{6}$, $\cot \frac{2\pi}{3}$, and $\tan \frac{3\pi}{4}$.

SOLUTION

$$\tan \frac{\pi}{6} = \frac{\sin \frac{\pi}{6}}{\cos \frac{\pi}{6}} = \frac{\frac{1}{2}}{\frac{\sqrt{3}}{2}} = \frac{\sqrt{3}}{3}$$

$$\cot \frac{2\pi}{3} = \frac{\cos \frac{2\pi}{3}}{\sin \frac{2\pi}{3}} = \frac{-\frac{1}{2}}{\frac{\sqrt{3}}{2}} = -\frac{\sqrt{3}}{3}$$

$$\tan \frac{3\pi}{4} = \frac{\sin \frac{3\pi}{4}}{\cos \frac{3\pi}{4}} = \frac{\frac{\sqrt{2}}{2}}{-\frac{\sqrt{2}}{2}} = -1$$

The tangent function is almost as commonly used as the sine and cosine functions. Any calculator that has sin and cos keys will also have a tan key that can be used to find approximations. Be sure the calculator is in radian mode.

Example 2 Find approximations to the nearest thousandth of tan 1, tan(−1.6), and cot 6.1.

SOLUTION By calculator:

$$\tan 1 \doteq 1.557,$$

$$\tan(-1.6) \doteq 34.233, \text{ and}$$

since 6.1 is in the domain of both tan and cot,

$$\cot 6.1 = \frac{1}{\tan 6.1} \doteq -5.398.$$

To graph the tangent function, begin by plotting some points that you can calculate easily by using $\tan x = \frac{\sin x}{\cos x}$: $\tan 0 = 0$, $\tan \frac{\pi}{4} = 1$, $\tan\left(-\frac{\pi}{4}\right) = -1$, $\tan \frac{\pi}{3} = \sqrt{3} \doteq 1.7$, and $\tan\left(-\frac{\pi}{3}\right) = -\sqrt{3} \doteq -1.7$.

As x increases and approaches $\frac{\pi}{2}$, $\sin x$ approaches 1 and $\cos x$ approaches 0, so that their quotient, $\tan x$, becomes infinitely large. At $x = \frac{\pi}{2}$, tan is not defined. As x decreases and approaches $-\frac{\pi}{2}$, $\tan x$ is negative and $|\tan x|$ becomes infinitely large.

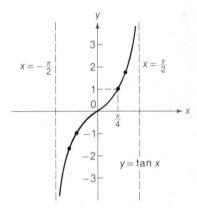

The graph of tan on the interval $\left(-\frac{\pi}{2}, \frac{\pi}{2}\right)$ is as shown at the right.

What is the graph of tan on its entire domain? Is tan periodic? To answer these questions, look again at the unit circle, and observe that the slope of $\overrightarrow{OP}$ is tan t. So as t increases from 0 toward $\frac{\pi}{2}$, the slope of $\overrightarrow{OP}$ also increases. Likewise, as t decreases from 0 to $-\frac{\pi}{2}$, that slope is negative and decreasing. If $W(t)$ is in quadrants II or III, by similar arguments you can conclude that the slope of $\overrightarrow{OP}$ follows the same pattern. A more complete sketch of the graph of tan is as shown on the next page.

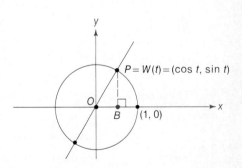

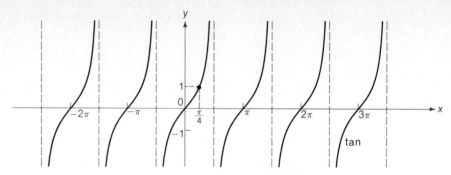

From the graph you can see that the range of tan is **R**. The period of tan is π. The set of zeros is $\{k\pi, k \in J\}$. Vertical asymptotes occur at $x = \frac{\pi}{2} + k\pi$, $k \in J$, and the domain of tan is $\mathbf{R} - \left\{\frac{\pi}{2} + k\pi, k \in J\right\}$.

To sketch graphs involving the tangent function, it is usually convenient to plot points corresponding to arguments $\frac{k\pi}{4}$, $k \in J$.

To sketch a graph of $f(x) = \cot x = \frac{\cos x}{\sin x}$, note that the zeros of f occur at arguments x for which $\cos x = 0$ and asymptotes occur for arguments x for which $\sin x = 0$. Also, $\cot \frac{\pi}{4} = 1$. The graph is shown below.

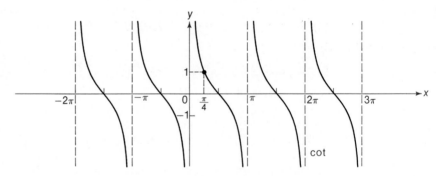

The tangent and cotangent functions are related to the secant and cosecant functions by Pythagorean-type expressions.

Theorem 9–2 **a.** For all x such that $\cos x \neq 0$, $\tan^2 x + 1 = \sec^2 x$.
b. For all x such that $\sin x \neq 0$, $\cot^2 x + 1 = \csc^2 x$.

The proofs follow immediately from the definitions and Theorem 9–1, which asserts that $\sin^2 x + \cos^2 x = 1$.

Example 3 If $\tan x = 2$ and $\pi < x < \frac{3\pi}{2}$, find $\sec x$ and $\csc x$ exactly.

SOLUTION Use $\tan^2 x + 1 = \sec^2 x$ and $\tan x = 2$ to find $\sec x = \pm\sqrt{5}$. Since $\pi < x < \frac{3\pi}{2}$, $\cos x$ and $\sec x$ must be negative, so that $\sec x = -\sqrt{5}$.

From the definition, $\cot x = \frac{1}{\tan x} = \frac{1}{2}$. Substitute in $\cot^2 x + 1 = \csc^2 x$ to find $\csc^2 x = \frac{5}{4}$ and $\csc x = \pm\frac{\sqrt{5}}{2}$. Since $\pi < x < \frac{3\pi}{2}$, $\sin x$ and $\csc x$ must be negative, so that $\csc x = -\frac{\sqrt{5}}{2}$.

Exercises

A Evaluate each expression exactly, if possible.

1. $\tan \frac{\pi}{2}$
2. $\cot \frac{\pi}{2}$
3. $\tan \frac{2\pi}{3}$
4. $\cot \frac{3\pi}{4}$
5. $\cot \frac{\pi}{6}$
6. $\cot\left(-\frac{\pi}{6}\right)$
7. $\cot \frac{11\pi}{3}$
8. $\tan\left(-\frac{\pi}{3}\right)$
9. $\tan\left(-\frac{5\pi}{6}\right)$
10. $\cot \frac{11\pi}{6}$
11. $\tan \frac{9\pi}{4}$
12. $\cot \frac{4\pi}{3}$

Evaluate to the nearest thousandth.

13. $\tan \frac{\pi}{5}$
14. $\cot(\pi + 1)$
15. $\cot(-1.5)$
16. $\tan 3$

B Evaluate exactly.

17. $\csc \frac{59\pi}{6} + \sec\left(-\frac{41\pi}{4}\right) - \cot \frac{13\pi}{4}$
18. $\cot\left(-\frac{3\pi}{4}\right) - \tan \frac{11\pi}{6} - \csc \frac{7\pi}{2}$

Graph the function on the interval $[0, \pi]$ and label the coordinates of one point not on an axis.

19. $f(x) = \tan x - 1$
20. $f(x) = |2 \tan x|$
21. $g(x) = \tan\left(x + \frac{\pi}{2}\right)$
22. $g(x) = \tan\left(x - \frac{3\pi}{2}\right)$
23. $f(t) = -\tan t$
24. $f(t) = 1 + \tan t$
25. $v(x) = |1 - \tan x|$
26. $h(x) = 1 - \tan\left(x + \frac{\pi}{2}\right)$

27. Prove Theorem 9–2(a): $\tan^2 x + 1 = \sec^2 x$.

28. Prove Theorem 9–2(b): $\cot^2 x + 1 = \csc^2 x$.

29. Give the domain, range, period, and zeros of the cotangent function.

30. In the figure, t is a number between 0 and π, $W(t)$ has coordinates $(\cos t, \sin t)$, and point S is the intersection of $\overrightarrow{OP}$ with the line $y = 1$. Find the coordinates of S in terms of t.

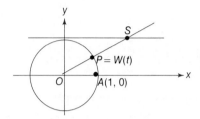

31. Find $\sec x$ exactly if $\tan x = 2$ and $\sin x > 0$.

32. Find $\tan x$ exactly if $\sec x = 1.4$ and $\sin x < 0$.

33. If $\sin x = \frac{7}{25}$ and $0 < x < \frac{\pi}{2}$, find exact values of $\cos x$, $\tan x$, $\sec x$, $\csc x$, and $\cot x$.

34. If $\tan x = \frac{1}{2}$ and $0 < x < \frac{\pi}{2}$, find exact values of $\cos x$, $\sin x$, $\sec x$, $\csc x$, and $\cot x$.

35. The height of a rocket is given in kilometers by $h(t) = \tan t$, where t is in minutes. Use a graph to estimate the time t that it takes for $h(t)$ to be at least 2 kilometers.

C 36. Graph $y = \tan x$ and $y = \frac{1}{2}x + 1$ on the same axes and thereby estimate the solutions of $2 \tan x = x + 2$ on the interval $\left[-\frac{\pi}{2}, \frac{3\pi}{2}\right]$.

37. Use a calculator to find the smallest positive solution of the equation in exercise 36 to the nearest hundredth.

Estimate graphically the smallest positive solution of each equation.

38. $\tan x = \frac{1}{x}$ 39. $x + \sec x = 0$ 40. $\cot x = \log x$ 41. $2^x = \csc x - 1$

42. What is the domain of $f = \log \circ \tan$? Graph the function f.

43. The equation $\tan \alpha = \alpha$ occurs in the study of the diffraction of light. Find a solution to the nearest hundredth of this equation, given that $\alpha \in \left(\frac{\pi}{2}, 2\pi\right)$.

44. Prove that $f: x \to 2 \sin x + \tan x$ is periodic and determine the period of f.

9–8 The Graph of g(x) = f(ax)

The voltage a particular generator produces is given as a function of time t in seconds by $g(t) = \sin 2t$. The range of g is the same as for sine, namely $[-1, 1]$, and the shape of the graph must resemble a sine curve. You know that one complete period of sin occurs when its argument, here $2t$, covers the interval $[0, 2\pi]$. Therefore g will complete one period when $0 \leq 2t \leq 2\pi$, or $0 \leq t \leq \pi$. This suggests that the graph of g repeats itself every π units. Verify that the sketch at the right is correct by evaluating $g(t)$ at some special arguments: $g\left(\dfrac{\pi}{4}\right) = \sin \dfrac{\pi}{2} = 1$, $g\left(\dfrac{\pi}{2}\right) = \sin \pi = 0$. It is easy to show algebraically that π is a period of g:

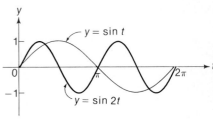

$$g(t + \pi) = \sin 2(t + \pi) = \sin 2t = g(t).$$

The graph of $g(t) = f(2t)$ is a horizontal compression of the graph of $f = \sin$.

Theorem 9–3 If f is a periodic function with the period p, then $g(x) = f(ax)$ is also periodic and the period of g is $\dfrac{p}{|a|}$, $a \neq 0$.

The proof of this theorem is left to the exercises. It has two parts:

1. Prove that $\dfrac{p}{|a|}$ is a period of g.

2. Prove that $\dfrac{p}{|a|}$ is the smallest positive period of g.

Example 1 Find the period for each of the functions $f(x) = \cos 3x$, $g(x) = \sin \dfrac{x}{2}$, and $h(x) = \tan(-\pi x)$.

SOLUTION For $f(x) = \cos 3x$, the period of cos is 2π and $|a| = 3$. Thus the period of f is $\dfrac{2\pi}{3}$.

For $g(x) = \sin \dfrac{x}{2}$, the period of sin is 2π and $|a| = \dfrac{1}{2}$.

Thus the period of g is $\dfrac{2\pi}{\frac{1}{2}}$, or 4π.

Since the period of tan is π, the period of h is $\dfrac{\pi}{|-\pi|} = 1$.

Example 2 Given the graph of f as shown at the right, sketch a graph of $g(x) = f(3x)$.

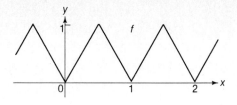

SOLUTION By inspection, the period p of f is 1. Since $a = 3$, the period of $g(x) = f(3x)$ is $\frac{p}{|a|}$, or $\frac{1}{3}$. So the graph of g is found by compressing horizontally the graph of f.

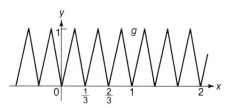

Example 3 Graph one period of the function $g: x \to 1 + \cos 3x$.

SOLUTION The period of the function $y = \cos 3x$ is $\frac{2\pi}{3}$. First, sketch a graph of this function. Now shift this graph upward 1 unit. Check the result by graphing a convenient point such as $\left(\frac{\pi}{6}, 1\right)$.

$g\left(\frac{\pi}{6}\right) = 1 + \cos\left(3 \cdot \frac{\pi}{6}\right) = 1$

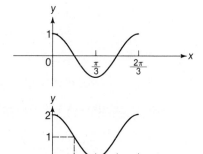

Example 4 Graph one period of the function $g(x) = 2 \tan\left(-\frac{x}{2}\right)$.

SOLUTION Since $\tan\left(-\frac{x}{2}\right) = \tan\left[\left(-\frac{1}{2}\right)x\right]$, you know that $a = -\frac{1}{2}$ and the period of g is that of tan, π, divided by $|a|$, or $\frac{1}{2}$. The period of g is 2π. Sketch a typical period of $f(x) = \tan\left(\frac{1}{2}x\right)$ between the asymptotes $x = -\pi$ and $x = \pi$.

Now stretch this graph vertically by a factor of 2 and reflect it in the y-axis.

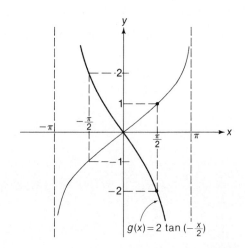

336 Chapter 9: Circular Functions

If $g(x) = f(ax)$, then the graph of g is obtained from the graph of f by horizontal stretching if $|a| < 1$ and by horizontal compression if $|a| > 1$.

Example 5 An economist finds from company records that sales, in thousands of dollars per month, fluctuate periodically and are described by

$$S(t) = 10 + \sin pt.$$

January sales are $S(1)$, February $S(2)$, March $S(3)$, and so on. If the sales cycle repeats itself each year, what is the value of the positive constant p and for what month are sales the greatest?

SOLUTION The period of $S(t)$ is $\dfrac{2\pi}{|p|}$. Since p is positive and the cycle repeats itself every 12 months, set $\dfrac{2\pi}{p} = 12$, so that $p = \dfrac{\pi}{6}$.

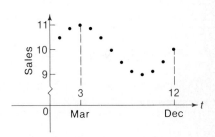

Now you can write $S(t) = 10 + \sin \dfrac{\pi t}{6}$.

The maximum value of $S(t)$ occurs when $\dfrac{\pi t}{6}$ equals $\dfrac{\pi}{2}$, since $\sin \dfrac{\pi t}{6}$ then equals 1. Hence, $t = 3$ and the best sales are in March. One period of the graph of sales by month is a set of 12 dots, not the entire sine curve.

Exercises

A Find the period of each function.

1. $g(x) = \sin 5x$
2. $g(x) = 3 \cos 2x$
3. $g(x) = \sin(-3x)$
4. $g(x) = -\tan(4\pi x)$
5. $f(x) = 2 \cos\left(\dfrac{\pi x}{2}\right)$
6. $f(x) = 3 + \sin \dfrac{x}{4}$
7. $h(x) = \sec \dfrac{x}{2} + 1$
8. $h(x) = 1 - 3 \csc(-2x)$

Graph one period of each function and label the coordinates of one point not on either axis.

9. $g(x) = 2 \sin 3x$
10. $g(x) = 3 \cos 2x$
11. $g(x) = 1 - \sin 3x$
12. $g(x) = \tan \dfrac{\pi x}{2}$
13. $f(x) = -\tan \dfrac{x}{3}$
14. $f(x) = 1 + \cos 3x$
15. $f(x) = \sin \pi x$
16. $f(x) = \tan 2\pi x$

B **17.** $f(x) = \sin(-2x)$ **18.** $f(x) = \cos\left(-\dfrac{x}{2}\right)$ **19.** $g(x) = 1 + \cos(-\pi x)$

20. $g(x) = 1 - \tan\left(-\dfrac{x}{3}\right)$ **21.** $h(x) = 1 - \sec\dfrac{x}{2}$ **22.** $h(x) = |\csc \pi x|$

Determine graphically the number of solutions of each equation on the given interval.

23. $\sin 3x = 0.4$; $[0, 2\pi]$ **24.** $\tan \dfrac{\pi x}{2} = -2$; $[0, 6]$

25. The vertical displacement V in feet of a mass at the end of a spring is given as a function of time t in seconds by $V(t) = A \cos kt$. The maximum displacement is 3 feet and the mass completes one oscillation in two seconds. Find the positive constants A and k.

26. The charge q in coulombs on a certain condenser is given by $q = 0.01(1 - \cos 10t)$, where t is the time elapsed in seconds. What is the maximum charge on the condenser and when does it first occur?

27. The current in amperes in an alternating-current circuit is given by $I = 10 \sin 120\pi t$, where time t is in seconds. How many complete sine-wave cycles are there in one second?

28. The horizontal displacement of the end of a swinging pendulum is given by $D(t) = -3 \cos \dfrac{\pi t}{2}$, where t is the time in seconds that the pendulum has been swinging. How long from $t = 0$ does it take for this pendulum to pass the vertical position the first time?

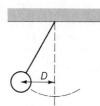

29. If $f(x) = \sin \pi x$, prove that $f(x - 2) = f(x)$ for all $x \in \mathbf{R}$.

30. An economist plots points as shown in the figure. He assumes that the points lie on a sine curve with equation $y = A + B \sin Ct$. Find approximate values of the constants A, B, and C.

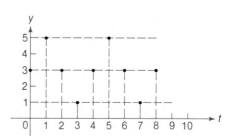

C **31. a.** Find a cosine function f whose graph passes through the points A, B, and C in the figure.
   b. Find a quadratic function g whose graph passes through points A, B, and C.
   c. Write a computer program to print out the values of f and g for $x = -1$ to $x = 1$ in steps of 0.1. Run the program.
   d. For what $x \in [-1, 1]$ is $f(x) - g(x)$ greatest?

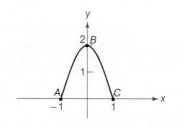

32. The first part of the proof of Theorem 9–3 is as follows. Give reasons for each step.

1. $g\left(x + \dfrac{p}{|a|}\right) = f\left[a\left(x + \dfrac{p}{|a|}\right)\right]$

2. $\phantom{g\left(x + \dfrac{p}{|a|}\right)} = f\left[ax + p\dfrac{a}{|a|}\right]$

3. $g\left(x + \dfrac{p}{|a|}\right) = \begin{cases} f(ax + p) \text{ if } a > 0 \\ f(ax - p) \text{ if } a < 0 \end{cases}$

4. $f(ax + p) = f(ax - p) = f(ax)$

5. Therefore $g\left(x + \dfrac{p}{|a|}\right) = f(ax) = g(x)$.

6. Therefore $\dfrac{p}{|a|}$ is a period of g.

33. The second part of the proof of Theorem 9–3 is as follows. Give a reason for each step.

Let k be any positive period of g.

1. $g\left(\dfrac{x}{a} + k\right) = g\left(\dfrac{x}{a}\right)$

2. $f\left[a\left(\dfrac{x}{a} + k\right)\right] = f\left[a\left(\dfrac{x}{a}\right)\right]$

3. $f(x + ak) = f(x)$

4. ak is a period of f.

5. $|a|k$ is a period of f.

6. $p \leq |a|k$

7. $\dfrac{p}{|a|}$ is the smallest positive period of g.

8. $\dfrac{p}{|a|}$ is *the* period of g.

34. a. Write a computer program for arguments x between 0 and 6.3 in steps of 0.2 to print 4 columns of numbers headed *sin 2x, sin x, cos x,* and *2 sin x cos x*.
b. Run the program.
c. Use the results of the run to help you graph $g(x) = 2 \sin x \cos x$ on $[0, 2\pi]$.

9–9 The Graph of g(x) = f(ax + b)

You know that the graph of $f(x) = \sin 2x$ is as shown at the right. If you shift this graph to the left $\frac{\pi}{2}$ units, the result is the graph of $g(x) = f\left(x + \frac{\pi}{2}\right) = \sin 2\left(x + \frac{\pi}{2}\right)$. Both f and g have the same period, namely π. Horizontal shifting of the graph does not change the period.

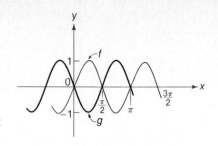

Theorem 9–4 If the function f is periodic with period p and $g(x) = f[a(x - h)]$, then g is periodic with period $\dfrac{p}{|a|}$, $a \neq 0$.

Example 1 Graph one period of the function $g(x) = \cos\left[\frac{1}{2}(x + \pi)\right]$.

SOLUTION Sketch the graph of $f(x) = \cos \frac{1}{2}x$. The period is 2π divided by $\frac{1}{2}$, or 4π.

Now shift this graph to the left π units, since $g(x) = f(x + \pi)$.

Check the coordinates of a convenient point directly from the given function, say at $x = \pi$.

$$g(\pi) = \cos\left[\tfrac{1}{2}(\pi + \pi)\right]$$
$$= \cos \pi = -1$$

So g contains the point $(\pi, -1)$.

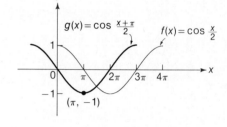

Example 2 Graph two periods of the function $f(x) = 3 \sin[-2(x - 1)]$.

SOLUTION The basic shape of the graph of f is that of sin stretched vertically by a factor of 3.

Since $a = -2$, the period of f is $\dfrac{2\pi}{|-2|} = \pi$. Since $a < 0$, reflect the graph of $y = 3 \sin 2x$ in the vertical axis and label appropriate points. Now shift this graph to the right 1 unit. Note that the first peak, M, is just to the right of the y-axis at $\left(-\frac{\pi}{4} + 1, 3\right)$.

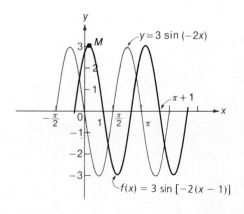

If f is a periodic function and completes a period of its graph on an interval, say $[r, s]$, then instead of factoring the argument $ax + b$, it is easier and more instructive to write $r \le ax + b \le s$, solve for x, and then proceed.

Example 3 Graph one period of the function $g(x) = \tan\left(2x - \frac{\pi}{2}\right)$.

SOLUTION Since the tangent function completes one period between $-\frac{\pi}{2}$ and $\frac{\pi}{2}$, g completes one period when $-\frac{\pi}{2} < 2x - \frac{\pi}{2} < \frac{\pi}{2}$, which is equivalent to $0 < x < \frac{\pi}{2}$. So sketch a tan-shaped curve on $\left(0, \frac{\pi}{2}\right)$, as shown. It is easy to verify that $g\left(\frac{3\pi}{8}\right) = 1$.

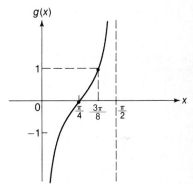

Exercises

A Graph one period of each function and label the coordinates of one point not on an axis.

1. $f(x) = \cos[2(x - \pi)]$
2. $f(x) = \sin[\pi(x + 1)]$
3. $f(x) = \tan(2x + \pi)$
4. $f(x) = \cos\left[-\frac{1}{2}(x - \pi)\right]$
5. $f(x) = \sin\dfrac{x - \pi}{2}$
6. $f(x) = 2\cos(\pi x + \pi)$
7. $f(x) = \cos\dfrac{x + \pi}{2}$
8. $f(x) = 6\cos(3x - \pi)$
9. $f(x) = -5\sin\left(3x - \dfrac{\pi}{2}\right)$
10. $f(x) = 2\tan\left(\dfrac{\pi - \pi x}{2}\right)$

B Graph each function and give its period, y-intercept, and range.

11. $f(x) = 1 - \tan\left(2\pi x + \dfrac{\pi}{2}\right)$
12. $f(x) = 2 - \sin(2x - \pi)$
13. $f(x) = 1 + 2\cos(2x - 3)$
14. $f(x) = 1 - 2\sin(2\pi - \pi x)$
15. $f(x) = |\sin(1 - x) - 1|$
16. $f(x) = |1 - 3\sin(\pi x - 1)|$

17. For what $x \in [0, 2\pi]$ is the function $f(x) = \sin\frac{1}{2}(x + 1)$ increasing?
18. For what $t \in [0, \pi]$ is the function $g(t) = \cos(2t + 1)$ decreasing?

The position of a particle moving along the x-axis is given as a function of time t by $P(t) = \tan\left(\pi t - \dfrac{\pi}{4}\right)$.

19. Where is the particle when $t = 0$?

20. When does the particle reach the origin?

21. As t approaches $\dfrac{3}{4}$ what happens to the particle?

The position of a particle moving back and forth on the x-axis has its x-coordinate given as a function of time t elapsed in seconds by $x(t) = 1 - \cos\left(\pi t - \dfrac{\pi}{2}\right)$.

22. Where is the particle when $t = 0$ and how long does it take for the particle to return to this position?

23. When the particle starts moving, at $t = 0$, does it move to the right or to the left on the x-axis?

In an alternating-current circuit, the current I in amperes may be expressed as a function of time t in seconds by $I = I_m \sin(120\pi t + \alpha)$, where I_m is the maximum current available and α is a constant which depends upon the circuit elements.

24. If $I_m = 10$ amperes and $\alpha = \dfrac{\pi}{6}$, find the smallest positive t for which $I = I_m$.

25. If $I_m = 6$ amperes and $\alpha = -\dfrac{\pi}{4}$, find the smallest positive t for which $I = I_m$.

26. If the function I is as sketched, determine I_m and α.

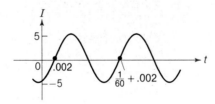

C **27.** Use graphs to help approximate the least positive solution of the equation $x + \tan(2x - 2) = 0$.

28. Use a calculator to help find the solution of the equation in exercise 27 to the nearest hundredth.

29. Sketch on the same axes the functions $f(x) = \sin x$ and $g(x) \doteq x - \dfrac{1}{6}x^3$. For x near zero is it reasonable to assume that $f(x) \doteq g(x)$?

30. Write a computer program to evaluate the expression $\sin x - x + \dfrac{1}{6}x^3$ for 50 different numbers $x \in [-1, 1]$.

Special Hyperbolic Functions

The circular functions sine and cosine were defined in terms of a wrapping function W and the unit circle, $x^2 + y^2 = 1$.

Suppose that the "unit hyperbola," $x^2 - y^2 = 1$, is used instead of the unit circle. For any number t you can locate a point $P(t)$ on the hyperbola by a wrapping function based on the starting point $A(1, 0)$.

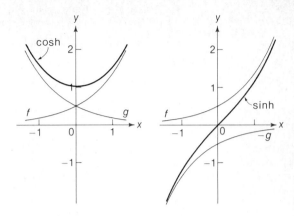

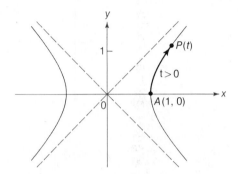

The function cosh describes the shape of a flexible cable supported at the same height at each end. This curve is called a **catenary**. It looks like, but is not, a parabola. In a suspension bridge, the uniform heavy load of the roadway does stretch the cable into the shape of a parabola.

The coordinates of $P(t)$ are called (cosh t, sinh t) where the symbols **cosh** and **sinh** represent the **hyperbolic cosine** and the **hyperbolic sine** functions.

Instead of the Pythagorean relationship $\sin^2 t + \cos^2 t = 1$, these hyperbolic functions are related by

$$\cosh^2 t - \sinh^2 t = 1.$$

These two hyperbolic functions can also be defined in terms of exponential functions with base e, the irrational number which is approximately 2.71828.

$$\sinh x = \frac{e^x - e^{-x}}{2} \qquad \cosh x = \frac{e^x + e^{-x}}{2}$$

Just as for the circular functions, the **hyperbolic tangent**, **tanh**, is defined by

$$\tanh x = \frac{\sinh x}{\cosh x}$$

By graphing $f(x) = \frac{e^x}{2}$ and $g(x) = \frac{e^{-x}}{2}$, you can verify that the graphs of cosh and sinh are as follows.

Exercises

1. Prove that $\cosh^2 t - \sinh^2 t = 1$.
2. Use the definition and the graphs of cosh and sinh to sketch a graph of tanh.
3. Write a definition of tanh in terms of exponential functions with the base e.
4. Prove that $e^x = \cosh x + \sinh x$, and that $e^{-x} = \cosh x - \sinh x$.

9-10 Inverse Circular Functions

To solve the equation log $x = 0.7$, you must isolate x by "undoing" the effect of the logarithm function on x. Since log is a one-to-one function and therefore has an inverse, this can be done by writing $x =$ exp 0.7.

What do you do if the equation to be solved is sin $x = 0.7$? Since sine is not a one-to-one function on its domain **R**, it has no inverse there. You have probably already discovered that on a calculator, you can use the (INV) and (SIN) keys to solve the equation. By calculator, `0.7` (INV) (SIN) (=) yields the result `0.7753975`.

A graph of sin shows that sin $x = 0.7$ has an infinite number of solutions. The calculator found only one approximation.

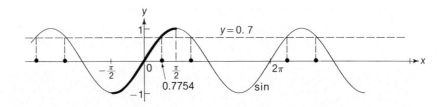

Notice from the graph that sine is an increasing function on the interval $\left[-\frac{\pi}{2}, \frac{\pi}{2}\right]$. The sine function with the restricted domain $\left[-\frac{\pi}{2}, \frac{\pi}{2}\right]$ is called Sin and since it is an increasing function, it has an inverse.

Definition The **inverse sine function**, written arcsin, or Sin^{-1}, is defined by

$$\arcsin x = y \text{ if and only if } \sin y = x,$$

where $-1 \leq x \leq 1$ and $-\frac{\pi}{2} \leq y \leq \frac{\pi}{2}$.

The domain of arcsin, namely, the interval $[-1, 1]$, is the range of Sin; the range of arcsin, namely $\left[-\frac{\pi}{2}, \frac{\pi}{2}\right]$, is the domain of the restricted sine function, Sin. Reflecting the graph of the Sin function in the line $y = x$ gives the graph of $y = f(x) = \arcsin x$. The arcsin of $\frac{1}{2}$ is that number in the interval $\left[-\frac{\pi}{2}, \frac{\pi}{2}\right]$ such that its sine is $\frac{1}{2}$; that is, arcsin $\frac{1}{2} = \frac{\pi}{6}$.

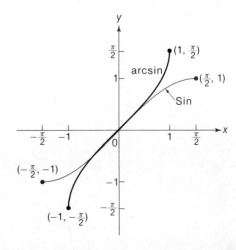

Example 1 Evaluate exactly $\arcsin \frac{\sqrt{3}}{2}$ and $\arcsin\left(-\frac{\sqrt{2}}{2}\right)$.

SOLUTION Since $\sin \frac{\pi}{3} = \frac{\sqrt{3}}{2}$ and $\frac{\pi}{3} \in \left[-\frac{\pi}{2}, \frac{\pi}{2}\right]$,

$$\arcsin \frac{\sqrt{3}}{2} = \frac{\pi}{3}.$$

Since $\sin\left(-\frac{\pi}{4}\right) = -\frac{\sqrt{2}}{2}$ and $-\frac{\pi}{4} \in \left[-\frac{\pi}{2}, \frac{\pi}{2}\right]$,

$$\arcsin\left(-\frac{\sqrt{2}}{2}\right) = -\frac{\pi}{4}.$$

Example 2 The coordinate of a particle moving back and forth along the *x*-axis is given by $x(t) = 3 \sin 2t - 1$ at time *t* seconds. Find how long it takes the particle to move from its initial position $x(0) = -1$ to the origin, $x = 0$.

SOLUTION Set $3 \sin 2t - 1 = 0$ and solve for *t*:

$$\sin 2t = \tfrac{1}{3}.$$
$$2t = \arcsin \tfrac{1}{3}$$

So $t = \tfrac{1}{2} \arcsin \tfrac{1}{3}$ seconds is the answer.

By calculator, $t \doteq 0.17$ seconds.

A calculator will find at most one solution of an equation such as $\sin x = -0.7$ and sometimes will not find any of the solutions you want.

Example 3 Solve $\sin x = -0.7$ for all $x \in [0, 2\pi]$ accurate to the nearest thousandth.

SOLUTION From a graph of $f(x) = \sin x$ on $[0, 2\pi]$, it is clear that there are two solutions, *r* and *s*, in the interval.

Solving $\sin x = -0.7$ for *x* gives

$$x = \arcsin(-0.7),$$

and a calculator finds the number -0.775 as an approximation of its only solution.

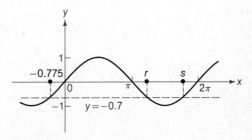

You must now use the symmetry of the sine function to determine

$r \doteq \pi + 0.775 \doteq 3.917$ and $s \doteq 2\pi - 0.775 \doteq 5.508$

as the answers to the problem.

On the restricted domain $[0, \pi]$, cosine is decreasing and is therefore one-to-one. On this domain, cosine is called Cos and has an inverse.

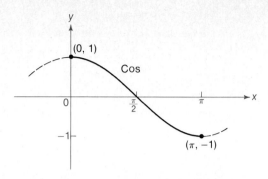

Definition The **inverse cosine function**, written arccos, or Cos^{-1}, is defined by

$$\arccos x = y \text{ if and only if } \cos y = x,$$

where $-1 \leq x \leq 1$ and $0 \leq y \leq \pi$.

The domain of arccos is $[-1, 1]$ and the range of arccos is $[0, \pi]$. The graph of arccos is shown below. You should verify the following values of arccos from the definition.

$$\arccos 1 = 0$$
$$\arccos \frac{\sqrt{2}}{2} = \frac{\pi}{4}$$
$$\arccos \left(-\frac{1}{2}\right) = \frac{2\pi}{3}$$
$$\arccos \frac{\sqrt{3}}{2} = \frac{\pi}{6}$$
$$\arccos \frac{1}{2} = \frac{\pi}{3}$$
$$\arccos \left(-\frac{\sqrt{3}}{2}\right) = \frac{5\pi}{6}$$

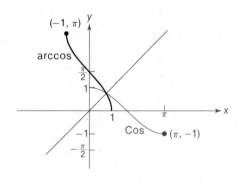

Example 4 Solve for x: $\arccos x > \frac{3}{4}$.

SOLUTION First solve the equation

$$\arccos x = 0.75.$$

Use a calculator to estimate x.

$$x = \cos 0.75 \doteq 0.7317$$

From a graph of arccos you can now see that the solution of the inequality is the interval $[-1, 0.7317)$.

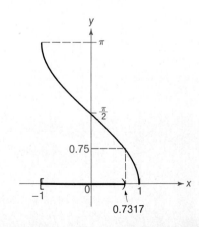

To define the arctan function, first restrict the domain of the tangent function to the interval $\left(-\frac{\pi}{2}, \frac{\pi}{2}\right)$ and call this restricted function Tan. Since Tan is one-to-one, it has an inverse.

Definition The **inverse tangent function**, written arctan, or Tan^{-1}, is defined by

$$\arctan x = y \text{ if and only if } \tan y = x,$$

where x is any real number and $-\frac{\pi}{2} < y < \frac{\pi}{2}$.

The domain of arctan is **R** and the range is $\left(-\frac{\pi}{2}, \frac{\pi}{2}\right)$. Graphs of the Tangent function and the arctangent function are shown below. Notice from the graph that $\lim_{x \to \infty} \arctan x = \frac{\pi}{2}$ and $\lim_{x \to -\infty} \arctan x = -\frac{\pi}{2}$.

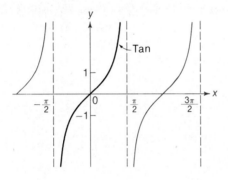
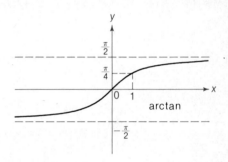

Example 5 Graph the function $f(x) = 1 + 2 \arctan x$ and estimate its zero to three decimal places. Find $\lim_{x \to \infty} f(x)$.

SOLUTION Stretch the graph of arctan vertically by a factor of 2 and then shift the graph up 1 unit.

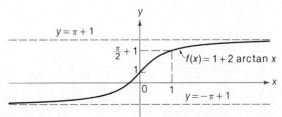

To find the zero of f, solve the equation $1 + 2 \arctan x = 0$ to get $x = \text{Tan}\left(-\frac{1}{2}\right)$. By calculator, $x \doteq -0.546$, accurate to 3 decimal places. From the graph, $\lim_{x \to \infty} f(x) = \pi + 1$.

Exercises

A Evaluate exactly or state that the expression is not defined.

1. $\arcsin\left(-\frac{1}{2}\right)$
2. $\arccos\left(-\frac{\sqrt{3}}{2}\right)$
3. $\arcsin 1$
4. $\cos^{-1}\frac{\pi}{2}$
5. $\arccos 0$
6. $\arcsin\frac{\pi}{3}$
7. $\arctan(-\sqrt{3})$
8. $1 - 2\arctan 1$
9. $\arctan(-1)$
10. $\arctan\sqrt{3}$
11. $\arcsin\frac{\sqrt{2}}{2}$
12. $\sin^{-1}\left(-\frac{\sqrt{3}}{2}\right)$

Use a calculator to approximate each expression.

13. $1 - \arctan 2.943$
14. $10^{\arctan \pi}$

Solve for a value of x to the nearest thousandth.

15. $\sin x = 0.4$
16. $4\cos x = 3$
17. $\arctan x = 1.5$
18. $\tan x = 1.5$
19. $3\cos x + 1 = 0$
20. $3\sin 2x - 1 = 0$

B Solve for x to the nearest thousandth on the interval $[0, 2\pi]$.

21. $\sin x = -0.4$
22. $\cos x = 0.7$
23. $\tan x = 2$
24. $\sin x = -0.8$
25. $\tan x = -2$
26. $\cos x = -0.3$

27. The charge q on a certain condenser is given in coulombs by $q = 0.01(1 - \cos 10t)$, where t is the time elapsed in seconds. Find the time required for q to equal 0.015 coulomb.

28. The current in amperes in an alternating-current circuit is given by $I = 10\sin(\pi t - 1)$, where t is the time elapsed in seconds. Find a value of $t \geq 0$, accurate to the nearest thousandth, when $I = 8$ amperes.

29. The vertical displacement of a mass oscillating at the end of a spring is given as a function of time t in seconds by $V(t) = 3\cos \pi t$ centimeters. If the mass starts oscillating when $t = 0$, how long does it take until $V(t)$ first equals 2 cm?

30. The vertical displacement of the end of a vibrating rod is given in millimeters by $Y(t) = 10\sin 50t$, where t is in seconds. How long does it take the end to move vertically 1 mm?

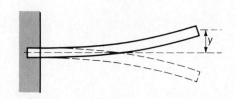

348 Chapter 9: Circular Functions

31. Use graphs of arcsin and arccos on the same axes to solve the inequality arcsin $x <$ arccos x.

For what real numbers x is each statement true?

32. $\sin(\arcsin x) = x$

33. $\arcsin(\sin x) = x$

Sketch each function f and label the coordinates of one point not on an axis.

34. $f(x) = 2 \arcsin x$

35. $f(x) = -\arctan x$

36. $f(x) = \arctan(x - 1)$

37. $f(x) = \arccos(x + 1)$

38. $f(x) = \dfrac{\pi}{2} + \arccos x$

39. $f(x) = 2 \arctan x$

40. $f(x) = |\arctan x|$

41. $f(x) = \dfrac{\pi}{2} + \arcsin x$

C Find $g^{-1}(t)$.

42. $g(t) = 2 \text{ Tan } t - 1$

43. $g(t) = 1 - \text{Sin } t$

44. $g(t) = \text{Cos } \pi t$

Solve for x.

45. $\arccos x > 2$

46. $10^{\tan x} = 3$

Chapter Review Exercises

9–1, page 310

Find the coordinates of the terminal point.

1. $W\left(-\dfrac{7\pi}{6}\right)$

2. $W\left(\dfrac{17\pi}{4}\right)$

3. Determine three different arguments t such that $W(t)$ has the coordinates $\left(-\dfrac{1}{2}, \dfrac{\sqrt{3}}{2}\right)$.

9–2, page 313

Evaluate exactly, without using a calculator or table.

4. $\sin \dfrac{7\pi}{6}$

5. $\cos \dfrac{7\pi}{4}$

6. If $\cos t = \dfrac{12}{13}$ and $W(t)$ is in quadrant IV, find $\sin t$ exactly.

9–3, page 316

Evaluate exactly, without using a calculator or table.

7. $\cos \dfrac{11\pi}{4}$

8. $\sin \dfrac{25\pi}{3}$

9–4, page 320

Solve for all $x \in [0, 2\pi]$.

9. $2 \cos x = -1$ **10.** $\sin x > 0.5$

11. The motion of a pendulum is described by $P(t) = a \cos t$. If $a = 6$, find all $t \in [0, 2\pi]$ for which $P = 3\sqrt{3}$.

9–5, page 324

Sketch a graph of the given function.

12. $f(x) = -3 \sin x$ **13.** $g(x) = 1 + 2 \sin\left(x + \dfrac{\pi}{4}\right)$

9–6, page 327

Evaluate exactly.

14. $\sec \dfrac{11\pi}{6}$ **15.** $\csc \dfrac{-4\pi}{3}$

16. If $\cos x = \dfrac{3}{5}$ and $0 < x < \dfrac{\pi}{2}$, find $\sec x$, $\sin x$, and $\csc x$ exactly.

9–7, page 330

Evaluate exactly.

17. $\tan \dfrac{-\pi}{4}$ **18.** $\cot \dfrac{\pi}{3}$

19. If $\tan x = 10$ and $\cos x < 0$, find $\sec x$ exactly.

9–8, page 335

20. Find the period of the function $f(x) = 2 + 3 \cos 4x$.

21. Graph one period of the function $g: x \to 3 \sin 2x$.

9–9, page 340

Graph one period of each function.

22. $h(x) = \sin(2x + \pi)$ **23.** $f(x) = 2 \tan\left(3x + \dfrac{\pi}{2}\right)$

9–10, page 344

Evaluate exactly, without using a calculator or table.

24. $\arcsin(-1)$ **25.** $\arctan(-\sqrt{3})$ **26.** $\arccos\left(-\dfrac{1}{2}\right)$

Chapter Test

1. Find the coordinates of the terminal point $W\left(\dfrac{25\pi}{6}\right)$.

Evaluate exactly, without using a calculator or table.

2. $\sin \dfrac{23\pi}{6}$

3. $\cos \dfrac{\pi}{6} - \sin \dfrac{5\pi}{3}$

4. $\sec \dfrac{11\pi}{3} + \tan \dfrac{-5\pi}{4}$

5. $\arctan(1) + \arcsin\left(-\dfrac{1}{2}\right)$

Solve for $x \in [0, 2\pi]$.

6. $1 + 2\sin x = 0$

7. $\sin x \geq 3$

8. $2\cos x > \sqrt{2}$

9. $\tan^2 x = 3$

10. Find the period of the function $F: x \to 2\sin(3\pi x - 1)$.

11. The graph of a periodic function is shown in the figure.
    a. What is the period of g?
    b. Is 9 a period of g?
    c. Evaluate $g(62)$.

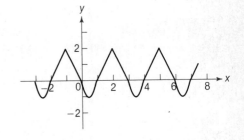

12. If $\sin x = \dfrac{1}{4}$ and $\dfrac{\pi}{2} < x < \pi$, find the exact value of each of the following:

    a. $\cos x$
    b. $\tan x$
    c. $\sec x$
    d. $\csc x$

Graph one period of each function and label the coordinates of one point not on either axis.

13. $f(x) = 3\cos x$

14. $g(x) = 2\cos 3x$

15. $h(x) = -2\cos\left(x + \dfrac{\pi}{2}\right) + 1$

16. A mass at the end of a spring has a height at time t given by $h(t) = 8 - 2\sin\dfrac{\pi t}{2}$.
    Graph the function h for t in the interval $0 \leq t \leq 10$ and determine the times when h is a minimum.

Trigonometric Functions

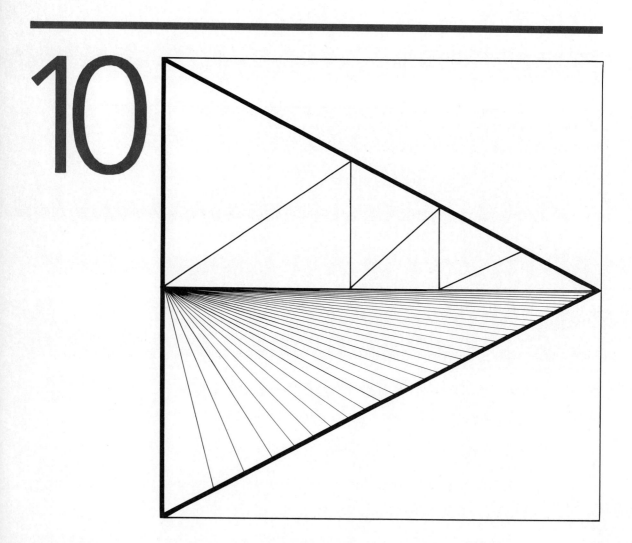

10–1 Angles and Angular Velocity

You know that $\sin \frac{\pi}{6} = \sin 30° = \frac{1}{2}$, and that on calculators it is important to distinguish between radian and degree measure. What is the connection between radians and degrees?

For any angle AOB with degree measure θ, a coordinate system can be established with vertex O, and the positive x-axis as ray $\overrightarrow{OA}$. This is called the **standard position** of an angle AOB; $\overrightarrow{OA}$ is called the **initial side** and $\overrightarrow{OB}$ the **terminal side**.

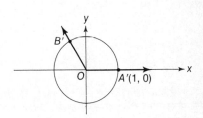

An angle in standard position intersects the unit circle at two points, A'(1, 0) and B' on $\overrightarrow{OB}$. The length of the arc A'B' is called the **radian measure** of ∠AOB. Thus, for example, a 90° angle has a radian measure of $\frac{\pi}{2}$.

In general, an angle of θ° has a measure of $\frac{\theta\pi}{180}$ radians.

Example 1 What is the radian measure of an angle with measure 100°?

SOLUTION Since $\theta° = \frac{\theta\pi}{180}$ radians,

$$100° = 100 \cdot \frac{\pi}{180} = \frac{5\pi}{9} \text{ radians.}$$

With the advent of calculators it has become standard practice to leave angle measures in degrees in decimal form, like 57.3°. It is still important, however, to be able to change degrees, minutes, and seconds to decimal form or to radians.

Example 2 What is the radian measure of an angle with measure 37° 20′ 42″?

SOLUTION $37°20′42″ = 37°\left(20 + \frac{42}{60}\right)' = 37°(20.7)'$

$= \left(37 + \frac{20.7}{60}\right)° = 37.345°$

$= 37.345 \cdot \frac{\pi}{180}$ radians

$\doteq 0.6518$ radian

10–1: Angles and Angular Velocity 353

By reversing the process, you find that 1 radian is $\frac{180}{\pi}$ degrees, or about 57.3°.

It is particularly useful to know radian and degree equivalents for the special arguments of Chapter 9, as listed below.

$\frac{\pi}{6}$	$\frac{\pi}{4}$	$\frac{\pi}{3}$	$\frac{\pi}{2}$	$\frac{2\pi}{3}$	$\frac{3\pi}{4}$	$\frac{5\pi}{6}$	π
30°	45°	60°	90°	120°	135°	150°	180°

In geometry, the measures of angles are all *between* 0° and 180°. (Some books define 0° and 180° angles as well.) Placing an angle in standard position on a coordinate system permits an extension of the definition so that the measure of an angle may be any number of degrees or radians. The idea is to think of rotating the initial side about the origin until it coincides with the terminal side.

The diagrams below show angles of 210°, or $\frac{7\pi}{6}$ radians (counterclockwise rotation), and −45°, or $-\frac{\pi}{4}$ radians (clockwise rotation).

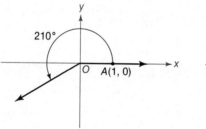
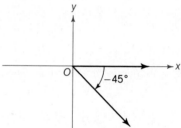

Example 3 Draw an angle of −270° in standard position.

SOLUTION Rotate a ray clockwise from the positive *x*-axis through −270° or its equivalent, $-\frac{3\pi}{2}$ radians.

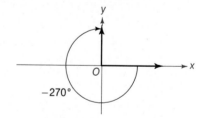

It is conventional and efficient not always to distinguish between geometric angles, measures of rotations, and the measure of an angle. Usually the context makes clear which of the three is meant.

If an object moves in a straight line at constant linear speed v, then the distance s it travels in time t is related to v by $v = \frac{s}{t}$, provided v and t are expressed in the same time units. Suppose that the object is on the rim of a uniformly rotating wheel of radius r. Then the distance s that it travels along an arc with central angle θ is θr. Its linear velocity is then $v = \frac{s}{t} = \frac{\theta r}{t}$. The quantity $\frac{\theta}{t}$ is called the **angular velocity** of the point and is often represented by the Greek letter ω (omega). Thus linear speed and angular velocity are related by the equation $v = \omega r$. Angular velocity is measured in radians per unit time.

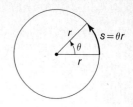

If the rate of rotation is given as W revolutions per unit time, then the angular velocity ω is $2\pi W$ radians per unit time.

Example 4 An LP record is turning at the standard $33\frac{1}{3}$ revolutions per minute (rpm). What is the linear speed of a point 6 inches from the center of the record?

SOLUTION The angular velocity ω of the point is $2\pi\left(33\frac{1}{3}\right)$ radians per minute and r is 6 inches. Therefore, from $v = \omega r$, the linear speed v is $2\pi\left(33\frac{1}{3}\right)(6)$, or about 1257 inches per minute.

Example 5 A car traveling 40 mph pulls a trailer that has wheels with radius 10 in. What is the angular velocity in radians per minute of a point on the wheel? How many revolutions per minute does the wheel make?

SOLUTION The linear speed v of the point is 40 mph and r is 10 inches. Before substituting in the equation $v = \omega r$, change the units of linear speed to inches per minute.

$$v = \left(40 \frac{\text{miles}}{\text{hour}}\right)\left(\frac{12 \cdot 5280 \text{ inches}}{1 \text{ mile}}\right)\left(\frac{1 \text{ hours}}{60 \text{ minute}}\right)$$

$$= 42{,}240 \frac{\text{inches}}{\text{minute}}$$

$$\omega = \frac{v}{r} = \frac{42{,}240}{10} = 4224 \frac{\text{radians}}{\text{minute}}$$

The number of revolutions per minute of the wheel is $\frac{4224}{2\pi}$, or about 672 rpm.

Exercises

A Find the radian measure for each degree measure.

1. 60°
2. 45°
3. −135°
4. −120°
5. 49° 30′
6. 146.3°
7. 21° 50′ 11″
8. 75° 4′ 40″
9. 35° 20′
10. 68° 50″
11. 600°
12. −225°

10–1: Angles and Angular Velocity   355

Sketch the angle with given measure in standard position.

13. 2 radians **14.** $\frac{1}{2}$ radian **15.** 260° **16.** −90°
17. −135° **18.** 60° **19.** 405° **20.** 3 radians

Express each angle measure as a decimal number of degrees, to the nearest hundredth.

21. 1.7 radians **22.** 0.2 radian **23.** 47°11′ **24.** 21°20′19″

B 25. An automobile is traveling at 30 mph and its wheels are 27 in. in diameter. What is the angular velocity of a point on the wheel?

26. A wheel is rotating at 2π radians per second and its radius is 3 inches. What is the linear velocity of a point on the rim of the wheel?

27. A bicycle rider goes 15 mph and the bicycle wheels are 29 in. in diameter. What is the angular velocity of a point on a wheel?

28. A large industrial flywheel with radius 40 feet turns at 20 rpm. What is the linear velocity of a point
  a. on the rim of the wheel?
  b. half way out one spoke of the wheel?

29. Suppose a discus thrower turns at the rate of 1 revolution per second, whirling the discus in a circle with radius 3 feet. What is the linear velocity of the discus when it leaves the athlete's hand?

C 30. A belt connects two pulleys as shown. If the radii of the pulleys are 20 cm and 30 cm, respectively, and the angular velocity of the smaller pulley is 5 radians per second, what is the angular velocity of the larger pulley?

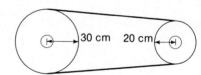

10–2 Trigonometric Functions

Light traveling from point A in one medium to point C in a different medium always takes the path that requires the least time. If its velocity in the first medium is v and in the second medium is u, then the light is bent at point B on the boundary of the two media so as to satisfy Snell's Law:

$$\frac{\sin \theta}{v} = \frac{\sin \phi}{u}.$$

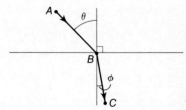

If $u = 0.95v$, then $\sin \phi = 0.95 \sin \theta$ and $\phi = \arcsin[0.95 \sin \theta]$. If a light ray reaches point B so that $\theta = 30°$, or $\frac{\pi}{6}$ radians, it leaves point B at an angle ϕ such that

$$\phi = \arcsin (0.95 \sin 30°)$$
$$= \arcsin (0.475)$$
$$\doteq 28.4°.$$

When the domains of the circular functions are degree measures, as in the previous example, the functions are called **trigonometric functions**. In order to evaluate a trigonometric function at one of the familiar arguments such as 30°, 45°, 60°, and so on, you may nevertheless want to change from degree to radian measure.

Example 1 Evaluate $\cos 120° + \tan 45°$.

SOLUTION
$$\cos 120° = \cos \frac{2\pi}{3} = -\frac{1}{2}$$
$$\tan 45° = \tan \frac{\pi}{4} = 1$$
$$\cos 120° + \tan 45° = \frac{1}{2}$$

For angles whose degree measure θ is greater than 360 or negative, use the periodicity of the trigonometric functions, which follows from the periodicity of the circular functions.

$$\left. \begin{array}{l} \sin(\theta + 360k) = \sin \theta \\ \cos(\theta + 360k) = \cos \theta \\ \tan(\theta + 180k) = \tan \theta \end{array} \right\} \text{ where } k \in J.$$

As above, we do not always include the degree symbol after an angle measure if the meaning is clear from the context.

Example 2 Evaluate $\csc 570°$ exactly.

SOLUTION
$$\csc 570° = \csc(360° + 210°) = \csc 210°$$
$$= \frac{1}{\sin 210°} = \frac{1}{\sin \frac{7\pi}{6}}$$
$$= \frac{1}{-\frac{1}{2}} = -2$$

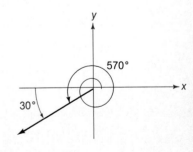

With a calculator you can enter the argument for the trigonometric function directly without relying upon periodicity. Be careful, however, to have the calculator in degree mode. For those without calculators, there is a table of values for trigonometric functions in the Appendix. Any time directions call for calculator approximations, you may use the table instead.

Example 3 Use a calculator to evaluate approximately $\cos 59°$, $\tan(-20°)$, and $\csc 568°$.

SOLUTION
$$\cos 59° \doteq 0.515$$
$$\tan(-20°) \doteq -0.364$$
$$\csc 568° = \frac{1}{\sin 568°} \doteq -2.130$$

Equations and inequalities that involve trigonometric functions may be solved by using the same methods as in Chapter 9.

Example 4 Find all θ in the interval $[0°, 360°]$ such that $3\cos^2\theta + \sin\theta = 3$.

SOLUTION Since the Pythagorean relationship $\sin^2\theta + \cos^2\theta = 1$ is true for degree measure as well as for radians, substituting in the equation for $\cos^2\theta$ gives

$$3(1 - \sin^2\theta) + \sin\theta = 3.$$
$$-3\sin^2\theta + \sin\theta = 0$$
$$\sin\theta\,(1 - 3\sin\theta) = 0$$
$$\sin\theta = 0 \quad or \quad \sin\theta = \tfrac{1}{3}$$

From $\sin\theta = 0$, $\theta = 0°$, $180°$, or $360°$.
From $\sin\theta = \tfrac{1}{3}$, $\theta = \arcsin\tfrac{1}{3} \doteq 19.5°$.
By symmetry, θ is also $180° - 19.5° \doteq 160.5°$.

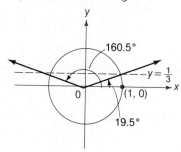

ANSWER $\theta = 0°, 180°, 360°$ $\;or\;$ $\theta \doteq 19.5°, 160.5°$.

Exercises

A Evaluate exactly.

1. $\sin 45°$
2. $\cot 30°$
3. $\sin 60°$
4. $\sec 120°$
5. $\sin 180°$
6. $\sin 30°$
7. $\tan 45°$
8. $\csc 135°$
9. $\sec 150°$
10. $\cos 315°$
11. $\cos(-210°)$
12. $\sin(-120°)$

Approximate by using a calculator.

13. $\sin 17°$
14. $\tan 200°$
15. $\sec 100°$
16. $\cot 500°$
17. $\cos(-37°)$
18. $\sin(-149°)$

Solve each equation for all θ in the interval [0°, 360°].

19. $2\cos\theta = 1$ **20.** $2\sin\theta = \sqrt{3}$ **21.** $\cot\theta = 1$

22. $\sec\theta = -1$ **23.** $\sin^2\theta = \sin\theta$ **24.** $\cos^2\theta = \frac{3}{4}$

B 25. In order to stay on a circular track with radius r, a motorcycle racer traveling with speed v must bank his cycle at an angle θ. These three variables are related by the formula

$$\tan\theta = \frac{v^2}{rg},$$

where g is the acceleration of gravity, 9.8 m/sec². If the radius of the track is 80 meters and $v = 20$ m/sec, find the banking angle θ.

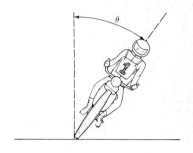

26. Use Snell's Law,

$$\frac{\sin\theta}{v} = \frac{\sin\phi}{u},$$

with $u = 0.95\,v$ and θ = 20°, to estimate φ in degrees.

27. A formula used in determining Doppler effect (the change in pitch of sound due to movement of the source of the sound) for a sound wave traveling in a vacuum is

$$\sin\theta = \frac{\sqrt{1-\beta^2}\sin\alpha}{1+\beta\cos\alpha}.$$

Find θ if β = 0.4 and α = 12°.

28. When light passes through a transparent solid prism with apex angle A, the light rays are bent and leave the prism at an angle δ with the original direction, as shown in the diagram. The index of refraction of the prism is given by

$$n = \frac{\sin\left(\frac{A+\delta}{2}\right)}{\sin\frac{A}{2}}.$$

Find δ if $n = 1.6$ and $A = 60°$.

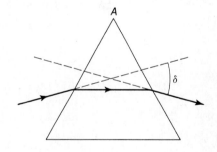

Solve for all θ in the interval [0°, 360°].

29. $\cos\theta = 0.2$ **30.** $\sin\theta = -0.91$

C 31. $12\sin^2\theta + \sin\theta = 6$ **32.** $2\sin\theta(3\sin\theta + 1) = \sin^2\theta$

33. $4\sin^3\theta = 3\sin\theta$ **34.** $2\cos^2\theta = \cos\theta + 1$

35. $\sin^2\theta + 3\sin\theta + 1 = 0$ **36.** $\tan^2\theta = 11 + \sec\theta$

10–3 Parametric Equations

An airplane on a bearing of 30° is 5 miles away from an airport tower. (*Bearing* specifies the direction of travel as an angle measured clockwise from north.) What are its coordinates on the axis system shown? The tower is at the origin. Since $\sin \angle AOR = \sin 60° = \frac{AR}{5}$, $AR = 5 \sin 60° = \frac{5\sqrt{3}}{2}$.

Similarly, $\cos 60° = \frac{OR}{5}$, and $OR = 5 \cos 60° = \frac{5}{2}$. Therefore, the plane's coordinates are $\left(\frac{5}{2}, \frac{5\sqrt{3}}{2}\right)$.

Describing the position of a point by determining its coordinates from other information is an important process. Suppose Q is the point r units from the origin on the terminal side of angle θ. What are the coordinates (x, y) of Q in terms of r and θ? Segment OQ intersects the unit circle at P, whose coordinates are $(\cos \theta, \sin \theta)$. In the figure, $\triangle OBP$ and $\triangle OTQ$ are similar right triangles. Therefore,

$$\frac{y}{\sin \theta} = \frac{r}{1} \text{ and } \frac{x}{\cos \theta} = \frac{r}{1}.$$

This proves the following useful theorem when Q is in the first quadrant.

Theorem 10–1 If point Q is r units from the origin on the terminal side of an angle θ in standard position, then its coordinates are

$$x = r \cos \theta,$$
$$y = r \sin \theta.$$

The proofs for Q in other quadrants are left to the exercises.

Example 1 Find the coordinates of the point Q in the figure if $OQ = 5$.

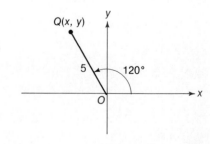

SOLUTION By Theorem 10–1,

$$x = 5 \cos 120° = 5\left(-\frac{1}{2}\right) = -\frac{5}{2}.$$

$$y = 5 \sin 120° = 5\left(\frac{\sqrt{3}}{2}\right) = \frac{5\sqrt{3}}{2}$$

The coordinates of Q are $\left(-\frac{5}{2}, \frac{5\sqrt{3}}{2}\right)$.

By solving the equations in Theorem 10–1 for $\sin \theta$ and $\cos \theta$, and then using the definitions of the other four functions, it is easy to verify the following corollary.

Corollary 10–1 If an angle θ is in standard position with $Q(x, y)$ on its terminal side r units from the origin, $r \neq 0$, then

$$\sin \theta = \frac{y}{r} \qquad \cos \theta = \frac{x}{r}$$

$$\left.\begin{array}{l}\tan \theta = \dfrac{y}{x} \\ \\ \sec \theta = \dfrac{r}{x}\end{array}\right\} x \neq 0 \qquad \left.\begin{array}{l}\cot \theta = \dfrac{x}{y} \\ \\ \csc \theta = \dfrac{r}{y}\end{array}\right\} y \neq 0.$$

Example 2 If $Q(-3, 5)$ lies on the terminal side of an angle θ, find $\theta \in [0°, 360°]$.

SOLUTION By Corollary 10–1,

$$\tan \theta = -\frac{5}{3}.$$

$$\theta = \arctan\left(-\frac{5}{3}\right)$$

A calculator gives the solution $\theta \doteq -59°$, but this angle is not in the interval $[0°, 360°]$. From the diagram it is clear that the desired angle is $\theta \doteq 180° - 59° = 121°$.

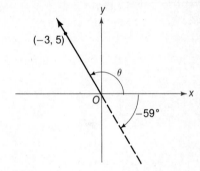

For any point $Q(x, y)$ on the terminal side of an angle θ, the slope of the line $\overrightarrow{OQ}$ is $\dfrac{y}{x}$, $x \neq 0$, and $\dfrac{y}{x} = \tan \theta$. Therefore, if Q is not on the y-axis, the slope of $\overrightarrow{OQ}$ is $\tan \theta$ and the slope of any line parallel to $\overrightarrow{OQ}$ is also $\tan \theta$.

Example 3 Find an equation of the line ℓ that contains $(3, 2)$ and intersects the x-axis at an angle of $60°$.

SOLUTION The slope of the line is $\tan 60°$, or $\sqrt{3}$. Therefore an equation of the line is

$$y - 2 = \tan 60° \, (x - 3), \text{ or}$$

$$y = \sqrt{3}x - 3\sqrt{3} + 2.$$

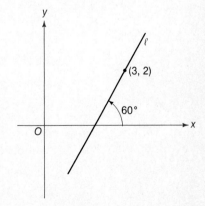

If both the x- and y-coordinates of a point Q are given as functions of another variable θ, say $x = f(\theta)$ and $y = g(\theta)$, then the locus of points Q is said to be described by the **parametric equations** $x = f(\theta)$ and $y = g(\theta)$. The argument θ is called the **parameter**.

Example 4 What is the locus of points $Q(x, y)$ described by the parametric equations

$x = 3 \cos \theta$ and
$y = 3 \sin \theta$?

SOLUTION Eliminate the parameter θ by observing that

$x^2 + y^2 = 9 \cos^2 \theta + 9 \sin^2 \theta$, or

$x^2 + y^2 = 9$ for all points Q.

Thus the locus is the circle with center (0, 0) and radius 3.

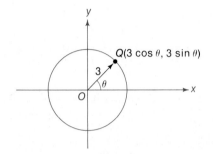

Example 5 Sketch the locus of points $P(x, y)$ described by the parametric equations

$x = \sin \theta$,
$y = 2 \sin^2 \theta$.

SOLUTION By substituting x for sin θ in the second equation, $y = 2 \sin^2 \theta$ may be written as $y = 2x^2$. The graph of this equation is a parabola, as shown in the figure. But since $x = \sin \theta$, you know from the range of the sine function that x is restricted to the interval $[-1, 1]$. Therefore the sketch is only that portion of the parabola with $-1 \le x \le 1$ and $0 \le y \le 2$.

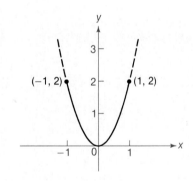

Example 6 Show that the parametric equations $x = 5 \cos \theta$ and $y = 3 \sin \theta$ describe an ellipse.

SOLUTION Solve the parametric equations for cos θ and sin θ: $\cos \theta = \dfrac{x}{5}$, $\sin \theta = \dfrac{y}{3}$.

Substitution in the Pythagorean relationship $\sin^2 \theta + \cos^2 \theta = 1$ yields

$$\frac{y^2}{9} + \frac{x^2}{25} = 1.$$

You should recognize this as the standard equation of an ellipse centered at the origin with semimajor axis 5 and semiminor axis 3.

Exercises

A For each figure, find the coordinates of Q if OQ = 5.

1.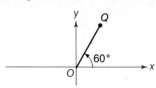
2.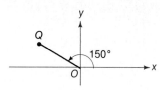

If Q lies on the terminal side of θ in standard position, find θ ∈ [0°, 360°].

3. $Q(-4, 4)$
4. $Q(2\sqrt{3}, 2)$
5. $Q(2, 3)$
6. $Q(-1, 3)$

Find an equation of the line ℓ containing P that makes an angle θ with the x-axis.

7. $P(2, -1)$, θ = 45°
8. $P(3, 2)$, θ = 30°
9. $P(-1, 2)$, θ = 100°
10. $P(0, 1)$, θ = 70°

B 11. Prove Theorem 10–1 if point Q is in

    a. Quadrant II.   **b.** Quadrant III.   **c.** Quadrant IV.

Describe the locus of points P(x, y) and express its equation in terms of x and y only.

12. $x = 5 \cos θ$
    $y = 5 \sin θ$
13. $x = 7 \cos θ$
    $y = 7 \sin θ$
14. $x = 5 \cos θ$
    $y = 2 \sin θ$
15. $x = 3 \cos θ$
    $y = 7 \sin θ$

16. Find parametric equations of the ellipse with equation
$$\frac{x^2}{a^2} + \frac{y^2}{b^2} = 1.$$

17. By eliminating the parameter show that the parametric equations $x = -1 + t$ and $y = 2 + 3t$ describe a line. Write the equation of the line in slope-intercept form.

Describe geometrically the locus of points P(x, y).

18. $x = \sin θ$
    $y = 2 \sin^2 θ - 1$
19. $x = \cos θ$
    $y = 3 \cos^2 θ$
20. $x = \cos^2 θ$
    $y = \sin^2 θ$
21. $x = 2$
    $y = 3 \sin θ$

C 22. If $x = a \cos θ + h$ and $y = a \sin θ + k$, where a, h, and k are constants, a > 0, show that the locus of points P(x, y) is a circle.

23. Show that the parametric equations $x = a \cos θ + h$ and $y = b \sin θ + k$, where a and b are positive constants and h and k are any constants, describe an ellipse or a circle.

24. Show that the parametric equations $x = a \sec \theta$ and $y = b \tan \theta$, where a and b are positive constants, describe a hyperbola.

In exercises 25 and 26, Q lies on the upper half of the circle $x^2 + y^2 = 16$. Find the coordinates of $P(x, y)$ in terms of θ and sketch the locus of points P.

25. ℓ is the line $y = 4$.
$\overline{QP} \| \ell$
$\overline{PR} \perp \ell$
$0° < \theta < 180°$

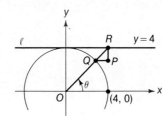

26. $QP = PR$
$0° < \theta < 180°$

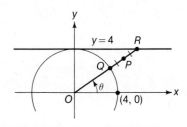

10–4 Right-Triangle Trigonometry

A tree casts a shadow 100 ft. long, and the angle of elevation of the top of the tree from the end of the shadow is θ. In previous mathematics courses you learned to express the height H of the tree as a function of the angle θ by writing $\tan \theta$ as the ratio $\dfrac{H}{100}$ of the side opposite θ to the side adjacent to θ. Thus the height H is given as a function of θ by
$H(\theta) = 100 \tan \theta$.

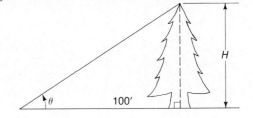

All of the familiar expressions relating lengths of the sides of a right triangle follow from Corollary 10–1, with Q in Quadrant I, $x \neq 0$, $y \neq 0$, $r \neq 0$.

$$\sin \theta = \frac{y}{r}$$
$$\cos \theta = \frac{x}{r}$$
$$\tan \theta = \frac{y}{x}$$
$$\cot \theta = \frac{x}{y}$$
$$\sec \theta = \frac{r}{x}$$
$$\csc \theta = \frac{r}{y}$$

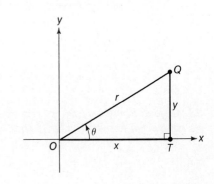

These ratios of lengths of the sides of a right triangle give values of the trigonometric functions of an acute angle even if the angle is not in standard position on a coordinate system.

Notice that we usually say "find $\angle A$," or "find A," rather than the more cumbersome "find the measure of $\angle A$." Similarly, sometimes we say "find the side . . ." as shorthand for "find the length of the side."

Example 1   Find the acute angles of the right triangle with sides whose lengths are 3, 4, and 5 units.

SOLUTION   From the definition and the figure,

$$\sin A = \frac{BC}{AB} = \frac{3}{5} = 0.6.$$

$$A = \arcsin 0.6 \doteq 36.87°$$

Since $\angle B$ is the complement of $\angle A$,

$$B = 90° - A \doteq 90° - 36.87°$$
$$\doteq 53.13°.$$

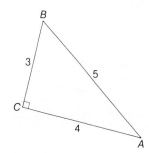

For any right triangle ABC where C is the right angle,

$$\sin A = \frac{BC}{AB} = \cos B.$$

Angles A and B are complements of each other. The prefix *co* of cosine is derived from *complement* since cosine of θ equals the sine of the complement of θ. Similarly,

$$\tan A = \frac{BC}{AC} = \cot B, \text{ and}$$

$$\sec A = \frac{AB}{AC} = \csc B.$$

Example 2   Rowland's kite is exactly 200 ft. above the ground. Express the number of feet of string L in terms of the angle of inclination θ from Rowland.

SOLUTION   From the figure,

$$\sin \theta = \frac{200}{L}.$$

Hence, $L = \dfrac{200}{\sin \theta} = 200 \csc \theta$, where $0° < \theta < 90°$.

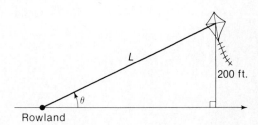

10–4: Right-Triangle Trigonometry   **365**

Example 3 One hundred meters from the base of a building the angle of elevation of the top of the building is 30°. From the same spot the angle of elevation of the top of a flagpole on the building is 45°. What is the height of the flagpole?

SOLUTION Let H be the height of the flagpole and B the height of the building. Then
$$\tan 30° = \frac{B}{100} \quad \text{and} \quad \tan 45° = \frac{H + B}{100}.$$

So $B = 100 \tan 30° = \dfrac{100}{\sqrt{3}}$, and

$H + B = 100 \tan 45° = (100)(1) = 100.$

Thus, $H = 100 - \dfrac{100}{\sqrt{3}} \doteq 42.3$ meters.

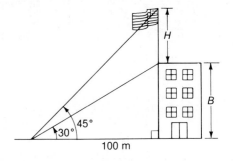

Example 4 A surveyor needs to measure the distance from a point P across a stream to a point Q on the abutment of a bridge. He makes angle measurements at P and R, as shown in the figure; these two points are 38 ft. apart. What is the distance PQ?

SOLUTION On a drawing which shows the data, extend $\overline{PR}$ and construct a perpendicular $\overline{QT}$ from Q to line PR.

$\angle QRT = 63°$

From $\triangle QRT$, $\tan 63° = \dfrac{QT}{RT}$.

From $\triangle QPT$, $\tan 49° = \dfrac{QT}{38 + RT}$.

From the first equation,
$RT \doteq (0.5095) QT.$

Substitute in the second equation:

$1.150 \doteq \dfrac{QT}{38 + (0.5095) QT}.$

$QT \doteq 105.5$ feet

Since $\sin 49° = \dfrac{QT}{PQ} \doteq \dfrac{105.5}{PQ},$

$PQ \doteq \dfrac{105.5}{\sin 49°} \doteq \dfrac{105.5}{0.7547}.$

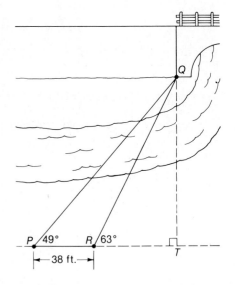

ANSWER The distance PQ is approximately 140 feet.

Exercises

A In exercises 1–20, △ABC is a right triangle with right angle at C.

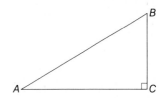

1. If $AC = 4$ and $AB = 7$, find sin A.
2. If $AC = 5$ and $BC = 4$, find sin A.
3. If $AC = 4$ and $BC = 4$, find cos B.
4. If $AB = 13$ and $BC = 5$, find cos B.
5. If $AC = BC$, find tan B.
6. If csc $B = 3$, find tan A.
7. If tan $A = \frac{3}{4}$ and $BC = 7$, find AC.
8. If cot $B = 2$ and $BC = 1$, find AB.
9. If $AC = 2$ and sec $B = 2$, find AB.
10. If $AB = 5$ and tan $A = 3$, find AC.
11. If $AC = 5$ and $BC = 7$, find ∠B.
12. If $AB = 9$ and $AC = 7$, find ∠A.
13. If ∠$B = 30°$ and $BC = 3$, find AB.
14. If ∠$A = 30°$ and $BC = 3$, find AB.
15. If ∠$B = 45°$ and $AB = 2$, find AC.
16. If ∠$A = 60°$ and $BC = 3$, find AB.
17. If $AC = 2\sqrt{3}$ and $AB = 4$, find ∠B.
18. If $BC = 4$ and $AC = 5$, find ∠A.
19. If ∠$B = 62°30'$ and $AB = 7$, find AC.
20. If ∠$A = 41°35'$ and $AC = 4.5$, find BC.

B 21. Find the perimeter of a regular pentagon inscribed in a circle with radius 10 in.

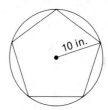

22. A rectangular field is 102 m wide and 234 m long. Find the angle between the longer side and a diagonal.

23. The equal sides of an isosceles trapezoid make angles of 70° with the longer base. If the lengths of the bases are 10 m and 14 m, what is the perimeter of the trapezoid?

24. The angle of elevation of a flagpole from a point level with its base is 45°. From a point 20 ft. further away, the angle of elevation is 30°. What is the height of the flagpole to the nearest 0.1 ft.?

25. The angle of elevation of a TV tower is 65°. From a point 42 ft. further away, the angle of elevation is 52°. How high is the tower?

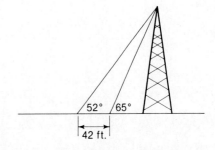

26. A ladder L meters long leans against a wall. The foot of the ladder is Q meters from the wall and at that point the ladder makes an angle θ with level ground.
   a. Express Q as a function of θ.
   b. Express θ as a function of Q.

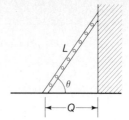

27. A right circular cone has a base with radius 7 ft., and the angle between the vertical height and the slant height is θ. Express the volume of the cone as a function of θ.

28. A radar tracking station picks up an airplane flying at an elevation of 2000 ft. In terms of the ground-level distance from the station to the plane, express the angle of elevation of the radar beam to the plane.

C 29. A billboard 11 ft. high on top of a base 20 ft. high is viewed from the level road, as pictured. Express θ, the angle subtended by the billboard, in terms of x, the distance of the viewer from the base.

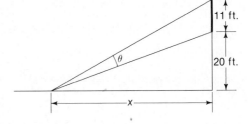

30. A billboard rests on top of a base. From a point level with the bottom of the base, the angle of elevation of the bottom of the billboard is 26° and of the top is 31°. From a point 21 meters further away, the angle of elevation of the top of the billboard is only 17°. How tall is the billboard?

10–5 The Law of Sines

Trigonometry can be used to determine the unknown sides and angles of triangles that are not right triangles. This is often done by using the Law of Sines. Notice that in naming the parts of a triangle, it is customary to label the side opposite an angle with the corresponding lower case letter.

Theorem 10–2 **The Law of Sines**

In any triangle ABC, $\dfrac{a}{\sin A} = \dfrac{b}{\sin B} = \dfrac{c}{\sin C}$.

(cont. on p. 369)

PROOF Place △ABC on a coordinate system in two different ways as shown.

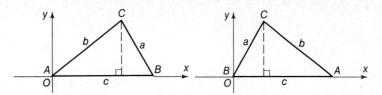

In each figure, determine the y-coordinate of vertex C by using Theorem 10–1.

In the triangle at the left, the y-coordinate of C is $b \sin A$.
In the triangle at the right, the y-coordinate of C is $a \sin B$.

Equate these y-coordinates:

$$b \sin A = a \sin B.$$

Divide both sides by $\sin A \sin B$:

$$\frac{b}{\sin B} = \frac{a}{\sin A}.$$

To complete the proof, place side BC of the triangle on the x-axis to determine that

$$\frac{b}{\sin B} = \frac{c}{\sin C}.$$

Example 1 In △ABC, ∠A = 30°, ∠B = 45°, and b = 2. Find a.

SOLUTION Use the Law of Sines to get

$$\frac{a}{\sin 30°} = \frac{2}{\sin 45°}.$$

Therefore $\frac{a}{\frac{1}{2}} = \frac{2}{\frac{\sqrt{2}}{2}}$, and $a = \sqrt{2}$.

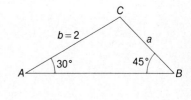

Example 2 In △ABC, ∠A = 45°, a = 10, and c = 3. Find ∠B.

SOLUTION Use the Law of Sines to get

$$\frac{10}{\sin 45°} = \frac{3}{\sin C}.$$

Then $\sin C = \frac{3\sqrt{2}}{20} \doteq 0.212$, and by calculator
$C \doteq 12.25°$.

$\angle B \doteq 180° - (12.25° + 45°) = 122.75°$

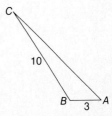

10–5: The Law of Sines   369

In order to use the Law of Sines in solving triangles, you must know three parts of the triangle, including one angle and the side opposite that angle. If you know two sides and the angle opposite one of them, there may be two noncongruent triangles that satisfy the given conditions.

Example 3 In △ABC, ∠A = 30°, b = 10, and a = 6. Find ∠B.

SOLUTION Substitution in the Law of Sines yields

$$\frac{6}{\sin 30°} = \frac{10}{\sin B}, \text{ or } \sin B = \frac{5}{6} \doteq 0.8333.$$

Then there are two angles, $B_1 \doteq 56.4°$ and $B_2 \doteq 180° - 56.4° = 123.6°$, that satisfy the given condition.

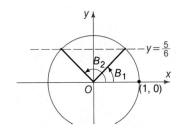

You can see why there are two solutions by trying to draw △ABC. Construct ∠A with measure 30°. On one of its sides, mark C such that AC = b = 10. Since a = 6, use a compass to draw an arc with radius 6 about C, intersecting the other side of ∠A at B_1 and at B_2.

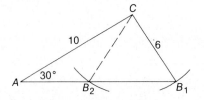 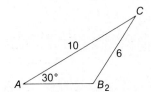

Both △ACB_1 and △ACB_2 represent the given information.

ANSWER There are two solutions, approximately 56.4° and 123.6°.

The situation in example 3 is customarily called the **ambiguous case** for the Law of Sines. The name is misleading, however, for the situation is no more ambiguous than the case of a quadratic equation with two solutions.

In the last example, had the problem read a = 2 instead of a = 6, an attempt to sketch the triangle would have yielded the figure at the right. Substitution in the Law of Sines gives

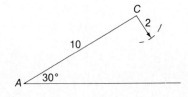

$$\frac{2}{\sin 30°} = \frac{10}{\sin B}, \text{ from which } \sin B = 2.5.$$

Clearly there is no △ABC and there is no solution to sin B = 2.5.

It is not always obvious from given information whether there will be zero, one, or two solutions. Fortunately, careful algebra will always determine all the answers even without a diagram.

Example 4 In $\triangle ABC$, $\angle A = 30°$, $b = 10$, and $a = 12$. Find $\angle B$.

SOLUTION From the Law of Sines,

$$\frac{12}{\sin 30°} = \frac{10}{\sin B}.$$

Therefore, $\sin B = \frac{5}{12} \doteq 0.417$, and $\angle B \doteq 24.6°$ or $155.4°$.

But the latter value is impossible, since the sum of two angles of the triangle would then be greater than $180°$ ($\angle A + \angle B \doteq 30° + 155.4°$).

Thus there is only one solution, namely, $\angle B \doteq 24.6°$.

This last example makes clear that you must check the measures of the angles you get from using the Law of Sines.

Exercises

A Use the Law of Sines to determine exactly, if possible, the requested side or angle of $\triangle ABC$.

1. $\angle A = 30°$, $\angle C = 45°$, $a = 3$. Find c.
2. $\angle A = 60°$, $\angle B = 45°$, $a = 3$. Find b.
3. $\angle A = 30°$, $c = 9$, $a = \sqrt{3}$. Find $\angle C$.
4. $\angle B = 30°$, $b = 4$, $a = 9$. Find $\angle A$.
5. $\angle A = 45°$, $a = 4$, $c = 2\sqrt{6}$. Find $\angle B$.
6. $\angle C = 45°$, $c = 4$, $a = 2\sqrt{2}$. Find $\angle A$.

Use the Law of Sines to determine approximately, if possible, the remaining sides and angles of $\triangle ABC$.

7. $\angle A = 40°$, $\angle C = 28°$, $a = 12.4$
8. $\angle B = 24°$, $a = 10$, $b = 5$
9. $\angle A = 66°$, $b = 12$, $a = 15$
10. $b = 98$, $a = 143$, $\angle B = 47°$
11. $\angle C = 34°$, $b = 6.9$, $c = 4.2$
12. $\angle A = 30°$, $\angle B = 40°$, $c = 10$

B 13. A ship's navigator sights a lighthouse on top of a cliff at an angle of elevation of $36°48'$. After sailing 2000 feet further away from the cliff, the navigator reads the angle of elevation as $29°6'$. How many feet above the sea is the lighthouse?

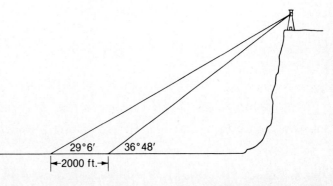

14. Barry and Costa are amateur radio operators, and Barry lives 102 km due north of Costa. If Barry picks up an S O S signal on a bearing of 130° and Costa receives the same signal on a bearing of 70°, how far from Costa is the sender of the S O S ?

15. In $\triangle ABC$, $\angle B = 45°$ and $b = 20$ ft. Express the length BC as a function of $\angle A$.

16. In $\triangle ABC$, $\angle B = 30°$ and $b = 20$ ft. Express the measure of $\angle A$ in terms of the length of side $\overline{AB}$.

17. Judy and her neighbor are arguing about how much frontage on the lake Judy owns. She owns a triangular plot, one side of which is lakefront. The other two sides of the triangle are 1000 ft. and 700 ft. long. The angle between the lakeshore and the 1000-foot side is 29°. How many feet of lake frontage does Judy own? Why might there be a dispute?

18. A buoy is sighted from a sailboat 30° to the port (left) side of the direction of travel. One mile later the buoy is 50° to port. How far is the buoy from the sailboat at the time of the second sighting?

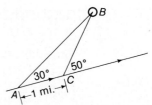

19. In order to measure the distance between two points P and Q on opposite sides of a gorge, a third point S is chosen. If $PS = 160$ ft., $QS = 210$ ft., and $\angle PQS = 43°$, find PQ.

20. George lives directly south of factory R. Factory S is on a line directly northwest of factory R. George hears the noon whistle from factory R four seconds past noon and the one from factory S six seconds past noon. Assume that the velocity of sound is 980 feet per second and find the distance between the two factories.

C 21. A straight road slopes upward 14° from the horizontal and a vertical telephone pole beside the road casts a shadow 60 feet down the road. If the angle of elevation of the sun is 55°, what is the height of the telephone pole?

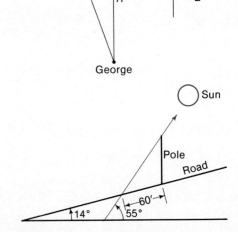

22. Write a computer program that will print out the length of side $\overline{BC}$ of $\triangle ABC$ if the input consists of $\angle A$, $\angle B$, and the length of side $\overline{AB}$.

23. Write a computer program with inputs $\angle A$, side $\overline{AB}$, and side $\overline{BC}$ that will print out all solutions for $\angle C$, or print that no solution for $\angle C$ exists.

10-6 The Law of Cosines

Two horizontal mine shafts of lengths 80 m and 100 m are dug into the side of a mountain; both strike a vein of ore. If the angle between the shafts is 60°, what is the length L of the vein of ore between them?

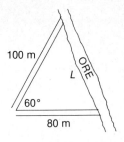

The question can be answered by using the Law of Cosines.

Theorem 10-3 **The Law of Cosines**

In any triangle ABC, $a^2 = b^2 + c^2 - 2bc \cos A$.

PROOF Place $\triangle ABC$ on coordinate axes as shown. By Theorem 10-1, the coordinates of C are $(b \cos A, b \sin A)$. Now use the distance formula to obtain an expression for BC, or a.

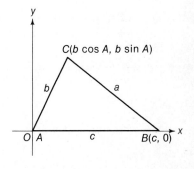

$a = \sqrt{(b \cos A - c)^2 + (b \sin A - 0)^2}$

$a^2 = (b \cos A - c)^2 + (b \sin A - 0)^2$

$\quad = b^2 \cos^2 A - 2bc \cos A + c^2 + b^2 \sin^2 A$

Since $b^2 \cos^2 A + b^2 \sin^2 A = b^2$,

$a^2 = b^2 + c^2 - 2bc \cos A$.

By renaming the sides and angles of $\triangle ABC$, two equivalent forms of the theorem can be obtained:

$$b^2 = a^2 + c^2 - 2ac \cos B, \text{ and}$$

$$c^2 = a^2 + b^2 - 2ab \cos C.$$

Return now to the problem at the beginning of this section. Use the Law of Cosines to write an expression for L^2.

$$L^2 = (100)^2 + (80)^2 - 2(100)(80) \cos 60°$$

$$= 10{,}000 + 6400 - 8000$$

$$= 8400$$

$$L \doteq 91.7 \text{ meters}$$

It is useful to remember the Law of Cosines in words as well as in symbols:

> The square of the length of one side of a triangle equals the sum of the squares of the lengths of the other two sides minus twice the product of the lengths of the other two sides and the cosine of the included angle.

Notice that if one of the angles of a triangle is a right angle, then the Pythagorean theorem may be considered a special case of the Law of Cosines.

Example 1 Use the Law of Cosines to find the largest angle of $\triangle ABC$ if $a = 3$, $b = 4$, and $c = 6$.

SOLUTION Since the largest angle of a triangle is opposite the longest side, you are to find $\angle C$. Substitute in the Law of Cosines and solve for $\cos C$:

$$c^2 = a^2 + b^2 - 2ab \cos C.$$
$$36 = 9 + 16 - 2(3)(4) \cos C$$
$$11 = -24 \cos C$$
$$\cos C = -\frac{11}{24} \doteq -0.4583$$
$$C \doteq 117.28°$$

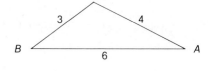

Recall from geometry that there are three basic ways of establishing the congruence of triangles. The abbreviations for those three ways—SSS, SAS, and ASA (or AAS)—can help you remember when to use the Law of Cosines or the Law of Sines and when to watch for the ambiguous case.

Use the Law of Cosines for
1. SSS, that is, when you know all three sides.
2. SAS, that is, when you know two sides and an included angle.

Use the Law of Sines for
1. ASA, that is, when you know two angles and an included side.
2. SSA. This is the case when there may be two solutions. (Remember, SSA does not provide triangle congruence as do the other three cases.)

Sometimes both laws are useful in one problem.

Example 2 In $\triangle ABC$, $a = 7$, $c = 10$, and $\angle B = 25°$. Find $\angle A$.

SOLUTION The given information fits the SAS form; use the Law of Cosines to find b.

$$b^2 = a^2 + c^2 - 2ac \cos B$$
$$b^2 = (7)^2 + (10)^2 - 2(7)(10) \cos 25°$$
$$\doteq 22.117$$
$$b \doteq 4.703$$

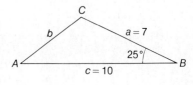

(cont. on p. 375)

Now use the Law of Sines to find $\angle A$.

$$\frac{a}{\sin A} = \frac{b}{\sin B}$$

$$\frac{7}{\sin A} \doteq \frac{4.703}{\sin 25°}$$

$$\sin A \doteq 0.6290$$

$$A \doteq 38.98°$$

(Why is $\angle A \doteq 38.98°$ and not $180° - 38.98°$?)

Exercises

A Find the specified part of $\triangle ABC$ exactly.

1. $a = 2$, $b = 3$, $c = 4$; find $\cos B$.
2. $a = 3$, $b = 4$, $\angle C = 30°$; find c.
3. $b = 3$, $c = 5$, $\angle A = 45°$; find a.
4. $a = 2$, $b = 1 + \sqrt{3}$, $c = \sqrt{6}$; find $\angle C$.

Find an approximate value of the specified part of $\triangle ABC$.

5. $a = 3$, $b = 4$, $\angle C = 40°$; find c.
6. $a = 3$, $b = 4$, $c = 6$; find $\angle B$.
7. $b = 4$, $a = 5$, $c = 7$; find $\angle C$.
8. $b = 4$, $c = 7$, $\angle A = 20°$; find $\angle B$.
9. $a = 5$, $b = 11$, $\angle C = 71°$; find c and $\angle A$.
10. $b = 9$, $c = 13$, $\angle A = 48°$; find $\angle B$.

B 11. Find the perimeter of a regular octagon inscribed in a circle with radius 6 inches.

12. A pilot sights two small towns, Amity and Bend, and determines that the angle between their lines of sight is 60°. The distance to Amity is 6 km and to Bend is 9 km. How far apart are Amity and Bend?

13. The sides of a triangular piece of land measure 110 ft., 187 ft., and 224 ft. Find the angles between the sides.

14. A plane is to fly from Baker to Como, a distance of 200 km. The plane flies 20 km on a course that is 11° off its correct heading before the pilot discovers the error. By what angle θ must he now change course in order to head for Como?

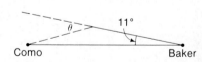

15. In $\triangle ABC$, $AB = 12$, and $BC = 20$. Express the angle θ between these two sides as a function of the length AC.

16. A hot-air balloon is sighted simultaneously from two towns that are 5 miles apart. The angle of elevation from one of the towns is 40° 27′ and from the other town is 58° 41′. Assume that the balloon is directly over a line between the two towns. Determine its air distance to the closer town and its altitude.

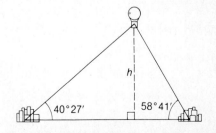

17. In $\triangle ABC$, $AB = 12$, $BC = 18$, and $\angle B = 40°$. Find the length of the median through vertex B.

C 18. A controller in an airport tower observes on her radar screen that two planes are flying toward one another on a straight line. One plane, flying at 500 mph, has a bearing of 41° from the tower and is 18 miles from the tower. The other plane, flying at 410 mph, has a bearing of 52° and is 24 miles from the tower. How long do the pilots now have to change courses and avoid collision? Give your answer to the nearest 0.1 sec.

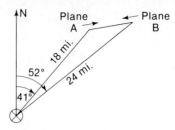

19. In $\triangle ABC$, $a = \sqrt{3}$, $b = \sqrt{3}$, and $c \geq 3$. What is the minimum possible measure for $\angle C$?

20. Derive the following formula for the perimeter P of a regular n-sided polygon inscribed in a circle with radius r:

$$P = nr\sqrt{2\left(1 - \cos\left[\frac{360°}{n}\right]\right)}.$$

21. From the contour map of Mt. Phillips and the data given on the diagram, determine the straight-line distance from A to D.

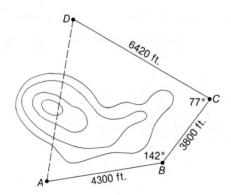

22. A radio tower is supported by three guy wires anchored on a circle with radius 100 ft. and centered at the base of the tower. The three anchor points, A, B, and C, are equally spaced around the circle. If the connecting point T is 200 ft. above the ground, what is the angle between any two of the wires, that is, $\angle ATC$?

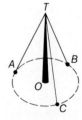

23. In example 2 of this section, the end of the solution reads

$\sin A \doteq 0.6290,$

$A \doteq 38.98°.$

Since $A \doteq 141.02°$ is also a solution of $\sin A \doteq 0.6290$, why must $\angle A$ be acute?

10–7 Area

The area of any triangle is given by $K = \frac{1}{2}(\text{base})(\text{height}) = \frac{1}{2}bh$.

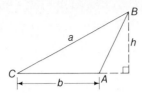 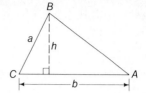

By Theorem 10–1 or by right-triangle trigonometry, the altitude of each triangle shown above is $h = a \sin C$. Therefore the area of $\triangle ABC$ is

$$K = \tfrac{1}{2}b(a \sin C)$$
$$= \tfrac{1}{2}ab \sin C.$$

This proves the following theorem.

Theorem 10–4 The area of a triangle is equal to one-half the product of the lengths of two sides and the sine of the included angle.

Example 1 Find the area of a triangular plot of land if two of its sides measure 110 ft. and 273 ft. and the included angle measures 40° 15′.

SOLUTION $K = \tfrac{1}{2}(110)(273) \sin 40° 15′$

$\doteq 9701.6$ square feet

Example 2 Express as a function of n the area K of an n-sided regular polygon inscribed in a circle with radius r.

SOLUTION Let the vertices of the polygon be $P_1, P_2, P_3, \cdots, P_n$. The area $K(n)$ of the polygon then equals n times the area of $\triangle OP_1P_2$.

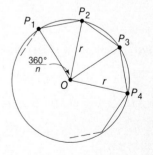

Since $\angle P_1OP_2 = \dfrac{360°}{n}$, by Theorem 10–4

area $\triangle OP_1P_2 = \tfrac{1}{2}(r)(r) \sin \dfrac{360°}{n}$.

$$K(n) = n\left(\frac{r^2}{2}\right) \sin \frac{360°}{n}$$

As the number of sides n of the regular polygon in example 2 increases, the area $K(n)$ should approach the area of a circle. The following results were obtained by calculator. They tend to confirm the expected result.

$$K(10) \doteq (2.938926)r^2$$
$$K(100) \doteq (3.139526)r^2$$
$$K(1000) \doteq (3.141573)r^2$$
$$\lim_{n \to \infty} K(n) = \pi r^2$$

Example 3 Find the area of a segment of a circle subtended by an angle of 120° in a circle with radius 3 cm.

SOLUTION The segment is the shaded region in the figure. Its area K is the difference between the area of sector $ABCD$ and the area of $\triangle ABC$.

$$\text{Area of sector} = \frac{120°}{360°} \pi (3)^2$$
$$= 3\pi$$
$$\text{Area of } \triangle ABC = \tfrac{1}{2}(3)(3) \sin 120°$$
$$= \frac{9\sqrt{3}}{4}$$

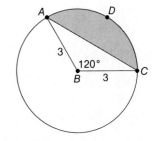

Therefore the area of the segment is

$$K = 3\pi - \frac{9\sqrt{3}}{4}.$$
$$K \doteq 5.53 \text{ cm}^2$$

Exercises

A Find the area of $\triangle ABC$.

1. $a = 7, b = 6, \angle C = 30°$
2. $a = 4, c = 9, \angle B = 45°$
3. $a = 4, c = 7, \angle B = 40°$
4. $c = 3, b = 4, \angle A = 63°$

B Heron's formula for the area K of $\triangle ABC$ is

$$K = \sqrt{s(s-a)(s-b)(s-c)}, \text{ where } s = \frac{a+b+c}{2}.$$

Use Heron's formula to find the area of $\triangle ABC$ in exercises 5–8.

5. $a = 4, b = 7, c = 5$
6. $a = 8, b = 6, c = 4$
7. $a = 4.2, b = 3.7, c = 4.9$
8. $a = 3.1, b = 2.9, c = 3.4$

9. Find the area of a segment of a circle subtended by a central angle of 30° in a circle with radius 5 cm.

10. In a circle with radius 2, a central angle of 45° subtends a segment. Find the area of that segment.

11. Give a formula for the area of parallelogram PQRS in terms of x, y, and θ. Use the figure at the right.

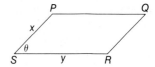

12. Show that the area of ▱PQRS is greatest when θ = 90°.

Calculate the area of each parallelogram. Refer to the figure for exercises 11 and 12.

13. x = 8.9, y = 14.3, θ = 47°   14. x = 21, y = 16, θ = 64°

15. The sides of a rhombus are each L inches long. If the angle between two sides is θ, express the area of the rhombus as a function of θ.

16. In exercise 15, what value of θ will give the maximum area?

17. The three sides of an isosceles triangle measure x, 2x, and 2x units. If the area of the triangle is px^2 square units, what is the value of the constant p?

18. Find the area of a regular 20-sided polygon inscribed in a circle with radius 5.

19. Find the area of the four-sided figure shown if the length of the indicated diagonal is 198 ft.

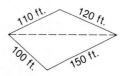

20. In a circle with radius 6, a chord of length 8 is drawn. Find the ratio of the areas of the two resulting segments.

21. For a circle with radius 5, express in terms of x the area of the smaller segment created by a chord which is x units from the center of the circle.

C 22. Find the area between two parallel chords one inch apart if the longer chord is 2 inches from the center of a circle with radius 5 inches.

23. Mr. Goodwin begins walking in a direction 60° east of north and goes 100 ft., where he turns 70° to the right and goes 120 ft., where he turns 100° to the right and goes 60 ft. How far is he from the starting point and what is the area of the land enclosed if he returns to the starting point by a straight-line path?

24. Use Heron's formula and the areas of $\triangle OAC$, $\triangle OAB$, and $\triangle OBC$ to find the radius of the circle inscribed in $\triangle ABC$ in terms of the sides of the triangle.

25. The roof of a house is in the shape of a pyramid. The dimensions of its base are 40 ft. and 26 ft., and the peak is known to be 10 ft. above the center of the base of the roof. To shingle the roof of the house, how many square feet must be covered?

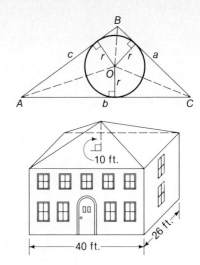

10–8 Polar Coordinates

The purpose of any coordinate system is to provide a frame of reference in order to locate points in a plane. You have been using exclusively the Cartesian coordinate system based on two perpendicular number lines. We now introduce another system known as the **polar coordinate system**.

Begin with a fixed point O, the **pole**, and a ray $\overrightarrow{OA}$, the **polar axis**. Establish a scale on the polar axis by marking off one unit.

To each point P in the plane, assign a pair of numbers (r, θ), where r is the distance OP and θ is the measure of $\angle AOP$. The numbers r and θ are called the **polar coordinates** of point P.

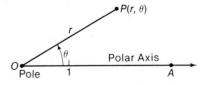

Because the rotations with magnitudes θ and $\theta + 360°$ (or $\theta + 2\pi$) are the same, the polar coordinates of a point are not unique. For example, $(2, 45°)$ and $(2, 405°)$ represent the same point. Furthermore r can be negative. When $r < 0$, measure $|r|$ units along the ray opposite the terminal ray. Thus $(-1, 120°) = (1, -60°)$. The pole O has coordinates $(0, \theta)$ for any θ.

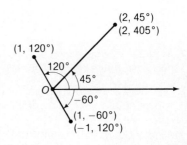

380 Chapter 10: Trigonometric Functions

The graph of an equation in polar coordinates r and θ is the set of points whose polar coordinates satisfy the equation.

Example 1 Sketch the graph of the polar equation $r = 2(1 + \cos \theta)$.

SOLUTION For several convenient numbers θ, find r and collect the results in a table as shown. Since cosine is an even function, $\cos(-\theta) = \cos \theta$, and the graph is symmetric about the line that contains the polar axis. The curve which results from joining the points (r, θ) is called a cardioid.

r	θ
4	0°
$2 + \sqrt{3}$	30°
$2 + \sqrt{2}$	45°
3	60°
2	90°
1	120°
$2 - \sqrt{2}$	135°
$2 - \sqrt{3}$	150°
0	180°

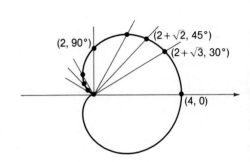

Polar coordinate graph paper helps in the sketching of polar equations, as shown in the next example.

Example 2 Sketch the graph of the polar equation $r = 2 \sin 3\theta$.

SOLUTION

θ	0°	15°	30°	45°	60°	75°	90°	105°	120°	135°	150°	165°	180°
r	0	$\sqrt{2}$	2	$\sqrt{2}$	0	$-\sqrt{2}$	-2	$-\sqrt{2}$	0	$\sqrt{2}$	2	$\sqrt{2}$	0

Sketching points (r, θ) for $0 \leq \theta \leq 60°$ gives the loop in the first quadrant.

For $60° < \theta < 120°$, $r < 0$, and you get the bottom loop.

For $120° < \theta < 180°$, you get the third loop.

For $\theta > 180°$ or $\theta < 0°$, the points coincide with points already graphed. So the whole graph is obtained from (r, θ) for which $\theta \in [0°, 180°)$. The graph is called a three-leaf rose.

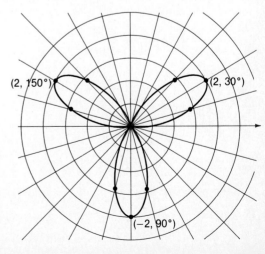

Example 3 Sketch a graph of the polar equation $r = \theta$ where θ is nonnegative and measured in radians.

SOLUTION Approximate r for some convenient values of θ with a calculator. As θ increases, $r = \theta$ increases, and the graph is a spiral, as shown.

θ	0	$\frac{\pi}{6}$	$\frac{\pi}{4}$	$\frac{\pi}{3}$	$\frac{\pi}{2}$	$\frac{3\pi}{4}$	π	$\frac{3\pi}{2}$	2π
r	0	0.52	0.79	1.05	1.57	2.36	3.14	4.71	6.28

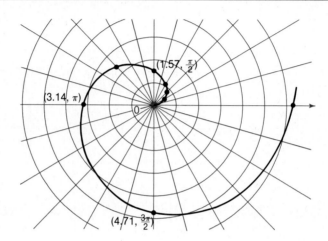

Exercises

A Sketch a graph of each polar equation.

1. $r = 3$
2. $\theta = \frac{\pi}{2}$
3. $r = \cos \theta$
4. $r = 3 \sin \theta$
5. $\theta = -\frac{\pi}{3}$
6. $r = 2$
7. $r = 1 - \cos \theta$ (cardioid)
8. $r = 1 - \sin \theta$ (cardioid)
9. $r = 2 \sin 2\theta$ (four-leaf rose)
10. $r = 2 \cos 2\theta$ (four-leaf rose)
11. $r = 1 + 2 \cos \theta$ (limaçon)
12. $r = 2 - \sin \theta$ (limaçon)

B 13. $r = \theta$ (spiral of Archimedes)
14. $r = 3^\theta$ (logarithmic spiral)
15. $r = \frac{2}{\theta}$ (hyperbolic spiral)
16. $r = 2 \cos 3\theta$ (three-leaf rose)
17. $r = \sin 5\theta$ (five-leaf rose)
18. $r^2 = 4 \sin 2\theta$ (lemniscate)

Find the points of intersection of the polar graphs.

19. $r = 1$
    $r = 2 \sin \theta$

20. $r = \sin \theta$
    $r = \cos \theta$

21. Find a formula for the distance between two points specified in polar coordinates as $P_1(r_1, \theta_1)$ and $P_2(r_2, \theta_2)$.

C Graph each polar equation.

22. $r = 2 \csc \theta + 3$ (conchoid)

23. $r^2 = \dfrac{2}{\theta}$ (lituus)

24. $r = 2(\sec \theta - \cos \theta)$ (cissoid)

10–9 Changing Coordinate Systems

The great advantage of the polar coordinate system is that some curves, like spirals, cardioids, and roses, have much simpler equations in polar form than in the more familiar Cartesian form. The circle $x^2 + y^2 = 4$ can, for example, be expressed very simply in polar form as $r = 2$.

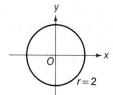

By Theorem 10–1, point $P(r, \theta)$ in the polar coordinate system has Cartesian coordinates

$$x = r \cos \theta \text{ and } y = r \sin \theta.$$

So the polar point $\left(\sqrt{2}, \dfrac{3\pi}{4}\right)$ has Cartesian coordinates

$$x = \sqrt{2} \cos \dfrac{3\pi}{4} = -1, \text{ and}$$

$$y = \sqrt{2} \sin \dfrac{3\pi}{4} = 1.$$

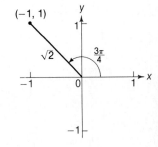

Since a point has many polar coordinates, one must be careful in changing form the Cartesian system to the polar system. From $x = r \cos \theta$ and $y = r \sin \theta$, it is clear that

$$x^2 + y^2 = r^2 \text{ and } \tan \theta = \dfrac{y}{x}.$$

Solving these equations for r and θ yields all possible polar coordinates. We usually select the pair with $r > 0$ and $|\theta|$ as small as possible.

Example 1 Find polar coordinates for the point P with Cartesian coordinates $(-\sqrt{3}, 1)$.

SOLUTION Substitute $x = -\sqrt{3}$ and $y = 1$ in
$r^2 = x^2 + y^2$ and $\tan \theta = \dfrac{y}{x}$:
$r^2 = (-\sqrt{3})^2 + 1 = 4$, and $\tan \theta = -\dfrac{1}{\sqrt{3}}$. So $r = 2$ and $\theta = \dfrac{5\pi}{6}$ are polar coordinates of the point P. But note that $\arctan\left(-\dfrac{1}{\sqrt{3}}\right)$ equals $-\dfrac{\pi}{6}$. So $r = -2$ and $\theta = -\dfrac{\pi}{6}$ are also polar coordinates of P.

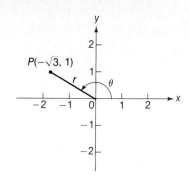

The equations $\begin{cases} x = r \cos \theta \\ y = r \sin \theta \end{cases}$ and $\begin{cases} r^2 = x^2 + y^2 \\ \tan \theta = \dfrac{y}{x} \end{cases}$ can also be used to change an equation of a curve from one coordinate system to the other.

Example 2 Show that the polar equation $r = 4 \sin \theta$ describes a circle.

SOLUTION Since $\sin \theta = \dfrac{y}{r} = \dfrac{y}{\sqrt{x^2 + y^2}}$, $r = 4 \sin \theta$ becomes

$$\sqrt{x^2 + y^2} = \dfrac{4y}{\sqrt{x^2 + y^2}}.$$

$$x^2 + y^2 = 4y$$

$$x^2 + y^2 - 4y = 0$$

$$x^2 + y^2 - 4y + 4 = 4$$

$$x^2 + (y - 2)^2 = 4$$

This is an equation of the curve in Cartesian coordinates and is easily recognized as an equation of a circle.

Example 3 Express $(x^2 + y^2)^{3/2} = y$ in polar form and sketch its graph.

SOLUTION In $(x^2 + y^2)^{3/2} = y$, replace $x^2 + y^2$ with r^2 and y with $r \sin \theta$ to yield
$(r^2)^{3/2} = r \sin \theta$.
Therefore, $r^3 = r \sin \theta$, and
$r^2 = \sin \theta$ is a polar equation of the curve.

(cont. on p. 385)

Now find sin θ for various values of θ from 0 to π. If θ is in the interval (π, 2π), sin θ is negative and there is no corresponding value of r. The curve is sketched in the figure at the right.

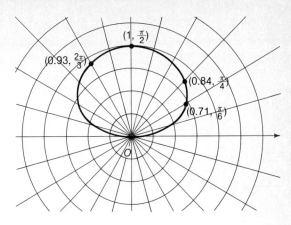

Exercises

A Find the Cartesian coordinates of each point given in polar form.

1. $\left(2, \dfrac{\pi}{4}\right)$
2. $\left(4, -\dfrac{\pi}{6}\right)$
3. $\left(1, -\dfrac{2\pi}{3}\right)$
4. $(2, \pi)$

Find polar coordinates of each point given in Cartesian form.

5. $(4, 4)$
6. $(\sqrt{3}, -1)$
7. $(-\sqrt{3}, -1)$
8. $(-2, 2\sqrt{3})$

Write in polar form the equation given in Cartesian form.

9. $x^2 + y^2 = 4x$
10. $x^2 + y^2 = 4y$
11. $2x + y = 6$
12. $x = 2$
13. $x^2 + y^2 = 9$
14. $x^2 = 16 - y^2$

B Transform the given polar equation into Cartesian form.

15. $r = 2 \sin \theta$
16. $r = 1 + \cos \theta$
17. $r = 1 - \sin \theta$
18. $r = \cos \theta$
19. $r = 2 \tan \theta$
20. $r = \dfrac{2}{1 - \cos \theta}$

Transform the given polar equation into Cartesian form and sketch the curve.

21. $r = 2 \sec \theta$
22. $r = \dfrac{2}{\cos \theta + \sin \theta}$
23. $r = \dfrac{4}{1 + \sin \theta}$
24. $r = \dfrac{2}{\cos \theta + 1}$

C Write the given equation in polar form and sketch the curve.

25. $(x^2 + y^2)^{3/2} = 4y$
26. $(x^2 + y^2 + y)^2 = x^2 + y^2$

10–9: Changing Coordinate Systems 385

Special Conic Sections in Polar Coordinates

It is rather easy to find a polar equation of a conic section with focus at the origin, eccentricity e, and directrix $\ell: x = k, k > 0$.

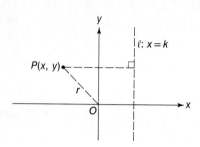

By the eccentricity definition of a conic section (see section 5–8),

$PO = e \cdot P\ell$.

Since

$PO = r$ and $P\ell = |k - x| = |k - r \cos \theta|$,

substitution yields

$r = e|k - r \cos \theta|$.

Solving for r in terms of θ gives

$$r = \frac{ek}{e \cos \theta + 1} \text{ or } r = \frac{ek}{e \cos \theta - 1}$$

as the polar form of the equation of the conic section. You should verify the following information.

For an *ellipse* with $e = \frac{1}{2}$ and $k = 2$:

$$r = \frac{1}{\frac{1}{2} \cos \theta + 1}.$$

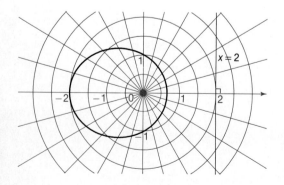

For a *parabola* with $e = 1$ and $k = 2$:

$$r = \frac{2}{\cos \theta + 1}.$$

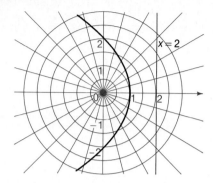

For a *hyperbola* with $e = \frac{3}{2}$ and $k = 2$:

$$r = \frac{3}{\frac{3}{2} \cos \theta + 1}.$$

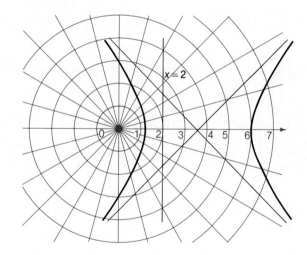

Exercises

Sketch each polar equation.

1. $r = \dfrac{4}{\cos \theta + 1}$

2. $r = \dfrac{2}{\frac{1}{2} \cos \theta - 1}$

3. $r = \dfrac{1}{2 \cos \theta + 1}$

Chapter Review Exercises

10–1, page 353

Find the radian measure for each degree measure.

1. 400°

2. −22°6′45″

Sketch the angle with given measure in standard position.

3. 4 radians

4. −280°

5. A record turns at 45 rpm. What is the linear velocity of a point 2 inches from the center of the record?

10–2, page 356

Evaluate exactly.

6. sin 210°

7. sec 120°

Solve each equation for all $\theta \in [0°, 360°]$.

8. $\tan^2 \theta = 1$

9. $2 = \sin \theta + 2 \cos^2 \theta$

10–3, page 360

10. Find the coordinates of the point Q in the figure if OQ = 8.

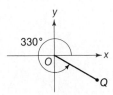

11. If $R(-3\sqrt{3}, -3)$ lies on the terminal side of an angle θ in standard position and $0° < \theta < 360°$, what is the measure of θ?

12. Sketch the locus of points $P(x, y)$ described by the parametric equations $x = \cos \theta$, $y = 4 \cos^2 \theta$.

10–4, page 364

In exercises 13–15, △RST is a right triangle with right angle T.

13. If RT = 6 and RS = 10, find sin R.

14. If $\tan S = \frac{3}{5}$ and RT = 15, find ST.

15. If sec R = 2 and RS = 8, find RT.

16. The angle of elevation of a water tower from a point level with its base is 60°. From a point 15 ft. further away, the angle of elevation is 30°. What is the height of the water tower to the nearest 0.1 ft.?

10–5, page 368

Use the Law of Sines to determine, if possible, the requested side or angle of △ABC.

17. ∠A = 60°, ∠B = 45°, b = 5. Find a.

18. ∠B = 25°, a = 12, b = 6. Find ∠A.

19. ∠C = 34°, a = 20, c = 11. Find ∠A.

20. Ship A is 80 miles due east of ship B when their captains pick up distress signals from ship C and decide that one of the ships must go to the rescue. Ship A picks up the signal coming from 80° west of north, and can travel at 21 mph. Ship B picks up the signal coming from 70° east of north, and can travel 14 mph. Which ship, A or B, can reach ship C first?

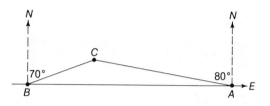

10–6, page 373

Use the Law of Cosines to find the specified part of △ABC.

21. a = 5, b = 6, c = 10. Find cos ∠C.

22. a = 7, c = 8, cos ∠B = 0.9. Find b.

23. Find the perimeter of a regular pentagon inscribed in a circle of radius 10 cm.

10–7, page 377

24. Find the area of a triangular plot of land if two of its sides measure 405 ft. and 382 ft. and the included angle measures 50°30′.

25. Find the area of the pentagon described in exercise 23.

10–8, page 380

Sketch a graph of the given polar equation.

26. r = 3   **27.** r = 3 cos θ   **28.** r = cos 3θ

10–9, page 383

Find the Cartesian coordinates of the point whose polar coordinates are given.

29. $\left(4, \frac{\pi}{4}\right)$ **30.** $\left(6, -\frac{\pi}{3}\right)$

Find polar coordinates of the point whose Cartesian coordinates are given.

31. (−5, 5)   **32.** (−4√3, −4)

33. Express the Cartesian equation $x^2 + y^2 = 6y$ in polar form and sketch its graph.

Chapter Test

1. Express $47°30'$ in radian measure.

2. Evaluate exactly: $\sin 150° + \sec(-420°)$.

Solve for all θ in the interval $[0°, 360°]$.

3. $2\cos\theta = 1$

4. $10\sin^2\theta = 7 + \cos\theta$

5. The linear velocity of a point on a wheel of diameter 26 in. is 88 ft./sec. How many revolutions per minute is the wheel turning?

6. The point $Q(-5, 5\sqrt{3})$ lies on the terminal side of an angle θ in standard position and $0° < \theta < 360°$. What is the measure of θ?

7. Sketch the locus of points $P(x, y)$ described by the parametric equations $x = \sin\theta$, $y = 4 - \sin^2\theta$.

8. Find the sine of the smallest angle of right triangle ABC if the sides have measures 7, 24, and 25.

9. A workman accidentally drops a wrench from the top of a tower 300 ft. high. From a point 1000 ft. away from the base of the tower, the angle of elevation of the wrench is θ. What is the height h of the wrench when $\theta = 15°$?

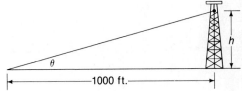

10. Sketch a graph of the polar equation $r = 2 + \cos\theta$.

11. The diagonals of a parallelogram measure 30 m and 50 m and form an angle whose measure is $60°$. Find the area of the parallelogram.

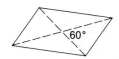

12. In $\triangle ABC$, $\angle A = 20°$, $a = 4$, and $c = 3$. Use the Law of Sines to determine $\angle C$.

13. In $\triangle ABC$, $\angle B = 14°$, $a = 9$, and $c = 7$. Use the Law of Cosines to determine b.

The pilot of an airplane decides to avoid a thunderstorm enroute from P to R, a distance of 120 miles. He changes course by $19°$ and then flies from P to Q to R. How many extra miles does he travel if

14. $PQ = 90$ miles?

15. $\angle QRP = 34°$?

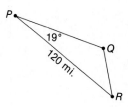

16. A forester is walking up a hill and observes that a Douglas fir tree near the top subtends an angle of $15°$. He walks 100 ft. further up the hill and then measures the subtended angle as $20°$. He then measures the distance from this second point to the base of the tree and finds that it is 185 ft. To the nearest foot, how tall is the tree?

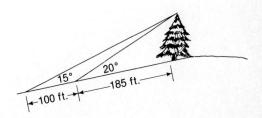

Circular Functions
Identities

11

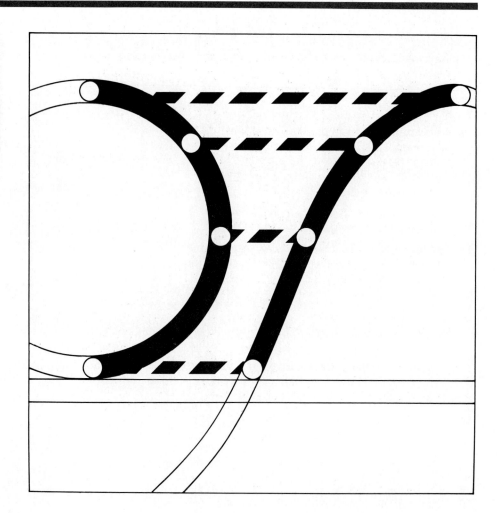

11–1 Identities

Any equation that is true for all numbers in its domain is called an **identity**. In the domain **R**, $a + b = b + a$ and $x(x + 5) = 5x + x^2$ are identities. You have already established two useful circular-function identities, namely, $\sin^2 x + \cos^2 x = 1$ and $\tan^2 x + 1 = \sec^2 x$.

Identities may be proved by using definitions, algebra, and previously proved identities.

Example 1 Prove: For all x such that $\sin x \neq 0$ and $\cos x \neq 0$, $\dfrac{\sec x}{\tan x} = \csc x$.

SOLUTION Use the definitions of the functions and algebra.

$$\frac{\sec x}{\tan x} = \frac{\dfrac{1}{\cos x}}{\dfrac{\sin x}{\cos x}} \qquad \text{Definitions of sec } x \text{ and tan } x$$

$$= \frac{1}{\sin x} \qquad \text{Multiply numerator and denominator by cos } x.$$

$$= \csc x \qquad \text{Definition of csc } x$$

Example 2 Prove that the following equation is an identity and give its domain:

$$\frac{1}{\sin x} - \sin x = \cot x \cos x.$$

SOLUTION A useful strategy for proving an identity is to start with the more complicated side and try to rewrite it in the form of the other side by establishing a sequence of equivalent expressions. In this example, start with the left side.

$$\frac{1}{\sin x} - \sin x = \frac{1 - \sin^2 x}{\sin x} = \frac{\cos^2 x}{\sin x}$$

$$= \frac{\cos x}{\sin x} \cos x$$

$$= \cot x \cos x$$

By the transitive property,

$$\frac{1}{\sin x} - \sin x = \cot x \cos x.$$

Since we want $\sin x \neq 0$, the domain is the set of all real numbers except zeros of the sine function, that is $\mathbf{R} - \{\pi k : k \in \mathbf{J}\}$.

11–1: Identities   391

Unless there is some special reason, we will usually not bother to state the domain of an identity because of the difficulty of determining it.

Sometimes the best strategy in proving an identity is to work with both sides *independently,* hoping to transform each into a common third expression.

Example 3 Prove the identity: $\tan x \sin x = \sec x - \cos x$.

SOLUTION Work with both sides independently, beginning with the left side.

$$\tan x \sin x = \frac{\sin x}{\cos x} \sin x$$

$$= \frac{\sin^2 x}{\cos x}$$

Now work with the right side of the identity.

$$\sec x - \cos x = \frac{1}{\cos x} - \cos x$$

$$= \frac{1 - \cos^2 x}{\cos x}$$

$$= \frac{\sin^2 x}{\cos x}$$

By the transitive property,

$$\tan x \sin x = \sec x - \cos x.$$

Identities are particularly useful for simplifying complicated expressions.

Example 4 Simplify the expression $\dfrac{1}{1 + \sin x} + \dfrac{1}{1 - \sin x}$.

SOLUTION
$$\frac{1}{1 + \sin x} + \frac{1}{1 - \sin x} = \frac{1 - \sin x + 1 + \sin x}{(1 + \sin x)(1 - \sin x)}$$

$$= \frac{2}{1 - \sin^2 x}$$

$$= \frac{2}{\cos^2 x}$$

$$\frac{1}{1 + \sin x} + \frac{1}{1 - \sin x} = 2 \sec^2 x$$

Identities are also useful in solving equations.

Example 5 Solve the equation $\sin x + \cos^2 x = 1$ on the interval $[0, 2\pi]$.

SOLUTION Use the Pythagorean identity $\sin^2 x + \cos^2 x = 1$ to replace $\cos^2 x$ by the equivalent expression $1 - \sin^2 x$.

$$\sin x + (1 - \sin^2 x) = 1$$
$$\sin x - \sin^2 x = 0$$
$$\sin x (1 - \sin x) = 0$$
$$\sin x = 0 \quad \text{or} \quad \sin x = 1$$

Therefore the solutions are

$$x = 0, \frac{\pi}{2}, \pi, \text{ and } 2\pi.$$

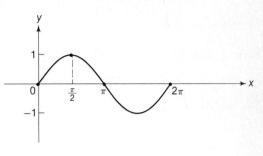

Exercises

A Prove each identity.

1. $\sin x \sec x = \tan x$
2. $\cos^2 x - \sin^2 x = 1 - 2 \sin^2 x$
3. $\sec x - \cos x = \sin x \tan x$
4. $1 = \csc x \cos x \tan x$
5. $\sec x + \tan x = \dfrac{\cos x}{1 - \sin x}$
6. $\cos^4 x - \sin^4 x = \cos^2 x - \sin^2 x$
7. $\dfrac{1 - \sin t}{\sin t \cot t} = \dfrac{\cos t}{1 + \sin t}$
8. $\sec \theta \tan \theta \csc \theta = \tan^2 \theta + 1$
9. $\sin \theta + \sin \theta \cot^2 \theta = \csc \theta$
10. $\tan t + \cot t = \tan t \csc^2 t$

B 11. $\sec x + \tan x + \cot x = \dfrac{1 + \sin x}{\cos x \sin x}$
12. $\dfrac{1 + \tan x}{\sin x} - \sec x = \csc x$
13. $\dfrac{1 - \tan^2 x}{1 - \cot^2 x} = \dfrac{\cos^2 x - 1}{\cos^2 x}$
14. $\dfrac{\cos A}{\cos A - \sin A} = \dfrac{1}{1 - \tan A}$
15. $\dfrac{1 + \cos x}{1 - \cos x} = \dfrac{1 + \sec x}{\sec x - 1}$
16. $\tan t + \cot t = \dfrac{1}{\cos t \sin t}$
17. $(\cos x + \sin x)(\cos x - \sin x) = 1 - 2 \sin^2 x$
18. $\dfrac{\sin x - \cos x}{\cos x} + \dfrac{\sin x + \cos x}{\sin x} = \sec x \csc x$

Simplify each expression.

19. $\sec^2 A - \tan^2 A$
20. $\sin^2 B + \cos^2 B$
21. $\tan r \cot r$
22. $\dfrac{\tan^2 A - \sin^2 A}{\tan^2 A \sin^2 A}$
23. $2(\csc^2 m - \cot^2 m)$
24. $\tan x \sin x + \cos x$

25. $\dfrac{\cos \theta \csc \theta}{\cot^2 \theta}$

26. $\dfrac{\csc \theta}{\tan \theta + \cot \theta}$

C 27. $\dfrac{\cot 2x}{\sec 2x - \tan 2x} - \dfrac{\cos 2x}{\sec 2x + \tan 2x} - \csc 2x$

Solve each equation on $[0, 2\pi]$.

28. $1 - \cos x = \sin^2 x$

29. $2\cos^2 x + \sin x = 1$

30. $\tan^2 \theta = \sec^2 \theta$

31. $\dfrac{\sin \theta}{\cos \theta} = \cot \theta$

11–2 Equations on R

A weight oscillating at the end of a spring has height $H(t)$ in meters above the ground at t seconds, where

$$H(t) = 2 + \sin t$$

and $t \geq 0$. Since the function H is periodic, the times at which the height is, say, 3 meters are infinite in number. They are the solutions of the equation

$$3 = 2 + \sin t, \text{ or } \sin t = 1.$$

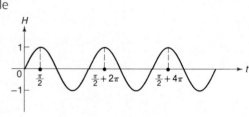

On the period $[0, 2\pi]$, there is a single solution $t = \dfrac{\pi}{2}$ sec., or $t \doteq 1.57$ sec. On the domain $[0, \infty)$, the solutions are $t = \dfrac{\pi}{2} + 2\pi k$, where k is a nonnegative integer. So $H = 3$ when $t \doteq 1.57$ sec., 7.85 sec., 14.14 sec., and so on.

The general strategy for solving equations that involve periodic functions is to start by finding all solutions in a single period and then add to each such solution an integer multiple of the period.

Example 1 Approximate all solutions of $5 \tan x = 12$.

SOLUTION
$5 \tan x = 12$
$\tan x = 2.40$
$x = \arctan 2.40$
$x \doteq 1.18$

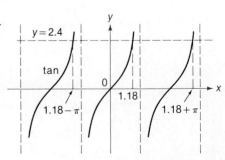

The solution set is approximately

$\{1.18 + \pi k : k \in J\}$.

394 Chapter 11: Circular Function Identities

If an equation has more than one solution in a period, then each such solution will generate an infinite number of additional solutions.

Example 2 Solve for x: $5 \cos x - 3 = 0$.

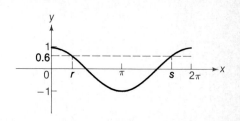

SOLUTION
$$5 \cos x - 3 = 0$$
$$\cos x = 0.60$$
$$x = \arccos 0.60$$
$$x \doteq 0.93$$

But the graph shows that there are two solutions, r and s, in the period $[0, 2\pi]$; the calculator found only one, $r \doteq 0.93$. From the symmetry of the graph of cosine it is clear that $s = 2\pi - r$, so that $s \doteq 5.35$. The solutions of the original equation are therefore

$$x \doteq 0.93 + 2\pi k \quad \text{and} \quad x \doteq 5.35 + 2\pi k, \quad k \in J.$$

Sometimes a solution of an equation found with a calculator is not in the standard interval.

Example 3 A traveling wave has height H meters at time $t \geq 0$ seconds that is given by $H(t) = 1 + \sin t$. When is the first time that $H = 0.28$ m?

SOLUTION The problem is to find the smallest positive t such that

$$1 + \sin t = 0.28.$$
$$\sin t = -0.72$$

By calculator,

$$t = \arcsin(-0.72) \doteq -0.80.$$

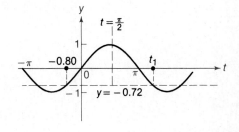

But a positive solution is required. From the graph you can see the symmetry of sin about the line $t = \frac{\pi}{2}$. Therefore,

$$t = t_1 \doteq \pi + 0.80 \doteq 3.94 \text{ sec.}$$

Example 4 Solve for x: $2 \sin^2 x + \cos x = 1$.

SOLUTION Replace $\sin^2 x$ in the given equation by $1 - \cos^2 x$:
$$2(1 - \cos^2 x) + \cos x = 1.$$
$$2 \cos^2 x - \cos x - 1 = 0$$

(cont. on p. 396)

$$(2\cos x + 1)(\cos x - 1) = 0$$

$$\cos x = -\tfrac{1}{2} \text{ or } \cos x = 1$$

So on the interval $[0, 2\pi]$, the solutions are

$$x = \frac{2\pi}{3}, \frac{4\pi}{3} \text{ and } x = 0, 2\pi.$$

These solutions in $[0, 2\pi]$ lead to the complete solution, which is

$$\left.\begin{array}{l} x = 2\pi k \\ x = \dfrac{2\pi}{3} + 2\pi k \\ x = \dfrac{4\pi}{3} + 2\pi k \end{array}\right\} \text{ for all } k \in J.$$

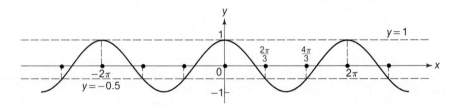

Exercises

A Solve for all solutions in **R**.

1. $\sin x = 0$
2. $\sin x = \dfrac{\sqrt{2}}{2}$
3. $\cos x = -\dfrac{\sqrt{2}}{2}$
4. $\cos x = -1$
5. $\sec x = -1$
6. $\sqrt{3} \tan x = 1$
7. $\csc x = -2$
8. $3\cos x + 1 = \cos x$
9. $\tan x + 1 = 0$
10. $\sin x = -\dfrac{\sqrt{3}}{2}$

11. If r and s are the solutions in $[0, 2\pi]$ of $\cos x = k$, where k is a constant, and $r = 1.4$, find s.

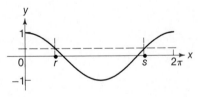

12. If r and s are the solutions in $[0, 2\pi]$ of $\sin x = k$, where k is a constant, and $r = 3.3$, find s.

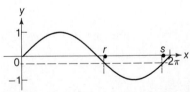

B The number x_0 is a solution of $f(x) = k$, where k is a constant. Find all the other solutions in **R**.

13. $f(x) = \sin x$ and $x_0 = -0.6$
14. $f(x) = \cos x$ and $x_0 = 0.2$

Find approximations to the nearest hundredth of all solutions in **R**.

15. $\cos x = 0.91$ **16.** $\sin x = 0.32$ **17.** $\tan x = 3.0$ **18.** $\cos x = -0.41$

19. $\sin x = -0.78$ **20.** $3 \csc x = 2$ **21.** $\sec x + 5 = 0$ **22.** $7 \cos x - 4 = 0$

23. $7 \sin x + 2 = 4 \sin x$ **24.** $5 \sin^2 x = 3$

25. $3 \sin^2 x = 5 \sin x + 2$ **26.** $5 \cos x + 6 \sin^2 x = 0$

Find exact values of all solutions in **R**.

27. $\sin^2 x = 1$ **28.** $2 \cos^2 x = \cos x$

29. $\sec^2 x = 4$ **30.** $3 \tan^2 x = 1$

31. $\sec^2 x - \sec x = 2$ **32.** $2 \sin x - 1 = 0$

33. $2 \cos^2 x + 5 \sin x + 1 = 0$ **34.** $2 \sin^2 x + \cos x = -1$

35. $\sin^2 x = \dfrac{1 - \cos x}{2}$ **36.** $2 \cos^2 x = 3(1 - \sin x)$

37. One end of a beam is built into a rigid wall. The other end can vibrate, and its vertical displacement at time t seconds is given by $y = 3 \cos t$ inches. Find the first 3 positive times t when the displacement is 1 inch. Express answers to the nearest hundredth of a second.

38. The voltage across the terminals of a circuit at time t seconds is given by $V(t) = 5 + 2 \sin t$ volts. Find, to the nearest hundredth, the 3 smallest positive values of t such that $V(t) = 4$ volts.

C Use the quadratic formula to find $\sin x$ or $\cos x$ and then solve for all solutions in **R**.

39. $2 \sin^2 x - 2 \sin x - 1 = 0$ **40.** $4 \cos^2 x - 4 \cos x - 1 = 0$

41. $3 \cos x (3 \cos x - 2) = 1$ **42.** $5 \sin x (5 \sin x - 4) = 1$

Solve for all x in **R**.

43. $2 \cos x \leq 1$ **44.** $2 \sin x + 1 \geq 0$ **45.** $2 \cos^2 x - \sqrt{3} \sin x < 1$

Determine whether the given statement is true or false.

46. For all $u, v \in \mathbf{R}$, $\cos (u + v) = \cos u + \cos v$.

47. For all $u, v \in \mathbf{R}$, $\sin (uv) = \sin u \sin v$.

48. A mass M is moving up and down at the end of a spring, and its height in inches above the ground at time t seconds is given by $H(t) = 3 - 2^{-t} \cos t$ inches.

  a. Estimate to the nearest hundredth of a second the smallest positive t for which $H(t) = 3.1$ inches.

  b. Sketch the function H for $t \geq 0$ and discuss what happens to the mass M as t becomes large.

11-3 Addition and Subtraction Identities

There is no obvious relationship between cos x, cos y, and cos (x + y) or cos (x − y). For instance, cos 60° + cos 45° ≠ cos 105° and cos 60° − cos 45° ≠ cos 15°. However, it is possible to calculate cos 15° if cos 60°, cos 45°, sin 60°, and sin 45° are known, as the next theorem shows.

Theorem 11–1   For all $x, y \in \mathbb{R}$, cos (x − y) = cos x cos y + sin x sin y.

PROOF   Let x and y be two real numbers and let $P = W(x)$ and $Q = W(y)$, where W is the wrapping function. This proof assumes that points O, P, and Q form a triangle, $\triangle OPQ$, in which $\angle POQ$ has measure $\alpha = x − y$.

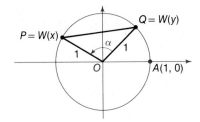

By the Law of Cosines,

$$(PQ)^2 = 1 + 1 - 2 \cos(x - y)$$
$$= 2 - 2 \cos(x - y).$$

Now use the coordinates of P(cos x, sin x) and Q(cos y, sin y) in the distance formula:

$$(PQ)^2 = (\cos x - \cos y)^2 + (\sin x - \sin y)^2$$
$$= \cos^2 x - 2 \cos x \cos y + \cos^2 y + \sin^2 x - 2 \sin x \sin y + \sin^2 y$$
$$= 2 - 2 \cos x \cos y - 2 \sin x \sin y.$$

Equate the two expressions for $(PQ)^2$.

$$2 - 2 \cos(x - y) = 2 - 2 \cos x \cos y - 2 \sin x \sin y.$$
$$\cos(x - y) = \cos x \cos y + \sin x \sin y$$

The six circular functions form three pairs of **cofunctions**: sine and cosine, secant and cosecant, tangent and cotangent. Each pair of cofunctions is related by complementary angles, x and $\frac{\pi}{2} - x$ in radian measure, or θ and $90° - \theta$ in degree measure.

Corollary 11–1a   **a.** $\cos\left(\frac{\pi}{2} - t\right) = \sin t$ **b.** $\sin\left(\frac{\pi}{2} - t\right) = \cos t$

               c. $\sec\left(\frac{\pi}{2} - t\right) = \csc t$ **d.** $\tan\left(\frac{\pi}{2} - t\right) = \cot t$

(cont. on p. 399)

PROOF **a.** In Theorem 11–1, substitute $x = \frac{\pi}{2}$ and $y = t$.

$$\cos\left(\frac{\pi}{2} - t\right) = \cos\frac{\pi}{2}\cos t + \sin\frac{\pi}{2}\sin t$$
$$= \sin t$$

b. In the identity in part **a**, replace t by $\left(\frac{\pi}{2} - t\right)$.

$$\cos\left[\frac{\pi}{2} - \left(\frac{\pi}{2} - t\right)\right] = \sin\left(\frac{\pi}{2} - t\right)$$
$$\cos t = \sin\left(\frac{\pi}{2} - t\right)$$

The proofs of **c** and **d** are left to the exercises.

From their graphs it is evident that cosine is an even function and sine is an odd function. These statements can now be proven algebraically.

In the identity for $\cos(x - y)$, let $x = 0$. Then
$$\cos(-y) = \cos y, \text{ for all } y \in \mathbf{R}.$$

In $\sin t = \cos\left(\frac{\pi}{2} - t\right)$, replace t with $-x$.

$$\sin(-x) = \cos\left(\frac{\pi}{2} + x\right) = \cos\left[x - \left(-\frac{\pi}{2}\right)\right]$$
$$= \cos x \cos\left(-\frac{\pi}{2}\right) + \sin x \sin\left(-\frac{\pi}{2}\right)$$
$$\sin(-x) = -\sin x, \text{ for all } x \in \mathbf{R}.$$

Corollary 11–1b **a.** Cosine is an even function; that is, for all $x \in \mathbf{R}$, $\cos(-x) = \cos x$.
b. Sine is an odd function; that is, for all $x \in \mathbf{R}$, $\sin(-x) = -\sin x$.

Theorem 11–2 For all $x, y \in \mathbf{R}$, $\cos(x + y) = \cos x \cos y - \sin x \sin y$.

PLAN FOR PROOF Observe that $\cos(x + y) = \cos[x - (-y)]$. Use Theorem 11–1 to write
$$\cos[x - (-y)] = \cos x \cos(-y) + \sin x \sin(-y)$$
and simplify by using Corollary 11–1b.

Theorem 11–3 For all $x, y \in \mathbb{R}$:

a. $\sin(x - y) = \sin x \cos y - \cos x \sin y$.

b. $\sin(x + y) = \sin x \cos y + \cos x \sin y$.

PROOF **a.** In $\sin t = \cos\left[\dfrac{\pi}{2} - t\right]$, let $t = x - y$.

$$\sin(x - y) = \cos\left[\dfrac{\pi}{2} - (x - y)\right]$$

$$= \cos\left[\left(\dfrac{\pi}{2} - x\right) + y\right]$$

$$= \cos\left(\dfrac{\pi}{2} - x\right)\cos y - \sin\left(\dfrac{\pi}{2} - x\right)\sin y$$

$$\sin(x - y) = \sin x \cos y - \cos x \sin y$$

The proof of **b** is left to the exercises and uses Corollary 11–1b.

The addition and subtraction identities in Theorems 11–1, 11–2, and 11–3 are valid for x and y measured in degrees as well as in radians. These identities are useful in simplifying expressions.

Example 1 Simplify: $\sin(x + 270°)$.

SOLUTION $\sin(x + 270°) = \sin x \cos 270° + \cos x \sin 270°$

$$= (\sin x)(0) + (\cos x)(-1) = -\cos x$$

The addition and subtraction identities can be used to evaluate exactly the circular functions at arguments which are the sums and differences of the special arguments.

Example 2 Evaluate $\cos \dfrac{\pi}{12}$ exactly.

SOLUTION Since $\dfrac{\pi}{12} = \dfrac{\pi}{3} - \dfrac{\pi}{4}$,

$$\cos \dfrac{\pi}{12} = \cos\left(\dfrac{\pi}{3} - \dfrac{\pi}{4}\right) = \cos \dfrac{\pi}{3} \cos \dfrac{\pi}{4} + \sin \dfrac{\pi}{3} \sin \dfrac{\pi}{4}$$

$$= \left(\dfrac{1}{2}\right)\left(\dfrac{\sqrt{2}}{2}\right) + \left(\dfrac{\sqrt{3}}{2}\right)\left(\dfrac{\sqrt{2}}{2}\right)$$

$$= \dfrac{\sqrt{2} + \sqrt{6}}{4}.$$

Example 3 Evaluate csc 105° exactly.

SOLUTION $\csc 105° = \csc(60° + 45°)$

$$= \frac{1}{\sin(60° + 45°)}$$

$$= \frac{1}{\left(\frac{\sqrt{3}}{2}\right)\left(\frac{\sqrt{2}}{2}\right) + \left(\frac{1}{2}\right)\left(\frac{\sqrt{2}}{2}\right)}$$

$$= \frac{4}{\sqrt{6} + \sqrt{2}}$$

By dividing the corresponding identities for sine and cosine you can prove the addition and subtraction identities for tangent. The proof is left to the exercises.

Theorem 11–4 For all real numbers x and y,

a. $\tan(x + y) = \dfrac{\tan x + \tan y}{1 - \tan x \tan y}$ and **b.** $\tan(x - y) = \dfrac{\tan x - \tan y}{1 + \tan x \tan y}$

wherever each expression is defined.

The addition and subtraction identities may also be useful in solving equations.

Example 4 Solve for x on $[0, \pi]$: $1 + \tan x = \tan\left(x + \dfrac{\pi}{4}\right)$.

SOLUTION Apply Theorem 11–4 to the right side:

$$1 + \tan x = \frac{\tan x + \tan \frac{\pi}{4}}{1 - \tan x \tan \frac{\pi}{4}} = \frac{\tan x + 1}{1 - \tan x}.$$

$1 - \tan^2 x = \tan x + 1$ *Multiply both sides by $1 - \tan x$.*

$0 = \tan^2 x + \tan x$

$0 = \tan x (\tan x + 1)$

So $\tan x = 0$ or $\tan x = -1$, and

$x = 0, \pi$ or $x = \dfrac{3\pi}{4}$.

Exercises

A Use the addition and subtraction identities to simplify each expression.

1. $\sin\left(x - \frac{\pi}{2}\right)$
2. $\cos(x - \pi)$
3. $\tan\left(x + \frac{\pi}{4}\right)$
4. $\tan(x - \pi)$
5. $\cos\left(\frac{\pi}{2} + x\right)$
6. $\csc\left(\frac{5\pi}{3} - x\right)$
7. $\sin(\pi - x)$
8. $\sec\left(\frac{\pi}{2} - x\right)$
9. $\cos 2x \cos x + \sin 2x \sin x$
10. $\sin 4 \cos 3 + \cos 4 \sin 3$
11. $\sin(60° + x) - \cos(30° + x)$
12. $(\sin x \cos y + \cos x \sin y)^2 + (\cos x \cos y - \sin x \sin y)^2$

Use the addition identities and $\frac{7\pi}{12} = \frac{\pi}{3} + \frac{\pi}{4}$ to find exact values.

13. $\sin \frac{7\pi}{12}$
14. $\tan \frac{7\pi}{12}$

Use the subtraction identities and $60° - 45° = 15°$ to find exact values.

15. $\sin 15°$
16. $\tan 15°$

B Find exact values.

17. $\sin 165°$
18. $\cos 75°$
19. $\cos \frac{5\pi}{12}$
20. $\sin \frac{\pi}{12}$
21. $\cos 255°$
22. $\sin 195°$

23. Find $\cos(x - y)$ if $\cos x = -\frac{12}{13}$, $\sin y = \frac{8}{17}$, x is in quadrant III and y is in quadrant II.

24. Find $\sin(x + y)$ if $\sin x = -\frac{4}{5}$, $\sec y = \frac{5}{3}$, $\cos x < 0$, and $\tan y > 0$.

For the given data find the exact value of $\cos(\alpha + \beta)$ if α and β are in the first quadrant.

25. $\sin \alpha = \frac{8}{17}$ and $\tan \beta = \frac{7}{24}$
26. $\tan \alpha = \frac{4}{3}$ and $\cot \beta = \frac{5}{12}$

Solve for all $x \in [0, 2\pi]$.

27. $\sin(x + \pi) = 1 + \cos\left(\frac{3\pi}{2} + x\right)$
28. $\sin\left(x + \frac{\pi}{6}\right) \sin\left(x - \frac{\pi}{6}\right) = \frac{3}{4}$

29. Simplify: $\dfrac{\cos(A + B) + \cos(A - B)}{\sin(A + B) + \sin(A - B)}$.

30. Prove: For all $t \in$ Dom csc, $\sec\left(\frac{\pi}{2} - t\right) = \csc t$.

31. Prove: For all $t \in$ Dom cot, $\tan\left(\dfrac{\pi}{2} - t\right) = \cot t$.

32. Prove Theorem 11–2: For all $x, y \in \mathbb{R}$,
$$\cos(x + y) = \cos x \cos y - \sin x \sin y.$$

33. Prove Theorem 11–3b: For all $x, y \in \mathbb{R}$,
$$\sin(x + y) = \sin x \cos y + \cos x \sin y.$$

34. Prove Theorem 11–4, the addition and subtraction identities for the tangent function.

35. Use the addition formula to prove that $\sin(x + 2\pi) = \sin x$ and thus verify that a period of sin is 2π.

Use addition and subtraction formulas to verify each identity.

36. $\tan(-x) = -\tan x$

37. $\cot(-x) = -\cot x$

38. $\tan(\pi - x) = -\tan x$

39. $\tan(\pi + x) = \tan x$

C 40. Prove: For all $x, y \in \mathbb{R}$, $\cos(x - y) - \cos(x + y) = 2\sin x \sin y$.

41. Prove: For all $x, y \in \mathbb{R}$, $\cos(x + y) + \cos(x - y) = 2\cos x \cos y$.

42. Express $\cos 3x \cos 5x - \sin 3x \sin 5x$ as the cosine of some expression.

Evaluate exactly.

43. $\tan\left[\arctan\dfrac{3}{4} + \arccos\left(-\dfrac{3}{5}\right)\right]$

44. $\cos\left[\arccos\dfrac{\sqrt{2}}{2} - \arccos\left(-\dfrac{3}{5}\right)\right]$

45. Prove that $\arctan\dfrac{2}{3} + \arctan\dfrac{1}{5} = \dfrac{\pi}{4}$.

46. Prove: If $x + y = \dfrac{\pi}{2}$, then $\sin^2 x + \sin^2 y = 1$.

47. Use Theorem 11–4 to help derive an expression for θ in terms of the slopes m_1 and m_2.

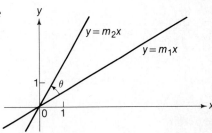

48. If $0 \le u \le \dfrac{\pi}{2}$ and $0 \le v \le \dfrac{\pi}{2}$, show that
$$\sin(u + v) \le \sin u + \sin v.$$

49. If $0 \le u \le \dfrac{\pi}{4}$ and $0 \le v \le \dfrac{\pi}{4}$, show that
$$\tan(u + v) \ge \tan u + \tan v.$$

Special Wave Addition

Energy is transmitted in sound waves, light waves, electrical waves, and water waves in the ocean in such a way that these waves can be described by a periodic function such as

$$y = A \sin \omega t,$$

where y, the displacement, is a function of time t. The number A is called the **amplitude** of the wave, and ω is proportional to the frequency of the wave motion.

Suppose two sound waves are described by

$$y_1 = 0.03 \cos (2000t), \text{ and}$$
$$y_2 = 0.04 \sin (2000t).$$

If the two waves are transmitted through a medium such as air or water, then the displacement of particles in the medium can be described by

$$y = y_1 + y_2.$$

To illustrate this wave motion, graph y_1 and y_2 on their common period $\left[0, \dfrac{\pi}{1000}\right]$, and for several values of t add graphically the ordinates y_1 and y_2.

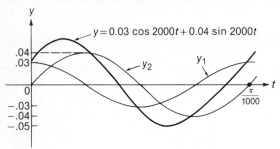

From the graph it appears that the sum function y is itself periodic with a maximum height, or amplitude, of about 0.05 and period $\dfrac{\pi}{1000}$. By using the following theorem you can verify these observations.

Theorem For all $t \in \mathbf{R}$,

$$A \cos \omega t + B \sin \omega t = C \cos (\omega t - D),$$

where $C = \sqrt{A^2 + B^2}$ and

$$D = \arctan \left(\dfrac{B}{A}\right).$$

A proof of this theorem depends upon expanding $\cos (\omega t - D)$ and is left to the exercises.

For the functions given above,

$$y(t) = 0.03 \cos 2000t + 0.04 \sin 2000t,$$
$$A = 0.03 \text{ and } B = 0.04, \text{ so that}$$
$$C = \sqrt{(0.03)^2 + (0.04)^2} = 0.05 \text{ and}$$
$$D = \arctan \dfrac{0.04}{0.03} \doteq 0.93.$$

So by the theorem,

$$y(t) \doteq 0.05 \cos (2000t - 0.93).$$

If two sine or cosine curves with different periods are added, the result is not an exact sine or cosine curve. This is illustrated below for $y_1 = \cos x$, $y_2 = \sin 2x$, and $f(x) = \cos x + \sin 2x$.

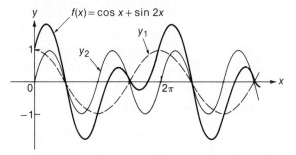

Exercises

1. Prove the theorem given above.

Find C, ω, and D so that $f(t) = C \cos (\omega t - D)$.

2. $f(t) = \sin t + \cos t$
3. $f(t) = \sqrt{3} \cos 2t + \sin 2t$
4. $f(t) = 3 \cos t + 4 \sin t$
5. $f(t) = 2 \cos 3t - \sin 3t$
6. Sketch a graph of $f(x) = \cos x - \sin 2x$.
7. Find the maximum value of the function $g(x) = 4 \cos 2x + 5 \sin 2x$.
8. Sketch a graph of $f(x) = 2 \cos 2x + 2 \sin x$.

11–4 Double-Argument Identities

The following theorems contain the double-argument, or double-angle, identities and all follow directly from the addition identities in section 11–3.

Theorem 11–5 For all $x \in R$, $\sin 2x = 2 \sin x \cos x$.

PROOF In the identity for $\sin (x + y)$, let $y = x$.

$$\sin (x + x) = \sin x \cos x + \cos x \sin x$$
$$\sin 2x = 2 \sin x \cos x$$

Theorem 11–6 For all $x \in R$,

$$\cos 2x = \begin{cases} \cos^2 x - \sin^2 x \\ 2 \cos^2 x - 1 \\ 1 - 2 \sin^2 x. \end{cases}$$

PROOF In the identity $\cos (x + y) = \cos x \cos y - \sin x \sin y$, let $y = x$ to get

$$\cos 2x = \cos^2 x - \sin^2 x.$$

Now use the Pythagorean identity $\sin^2 x + \cos^2 x = 1$ to obtain the last two double-argument identities for cosine.

Theorem 11–7 $\tan 2x = \dfrac{2 \tan x}{1 - \tan^2 x}$ provided $\tan^2 x \neq 1$, and x, $2x \in \text{Dom tan}$.

The proof of Theorem 11–7 follows directly from Theorem 11–4, or from Theorems 11–5 and 11–6, and is left to the exercises.

Example 1 If $\cos x = -\frac{3}{5}$ and $\tan x > 0$, find $\sin 2x$ and $\tan 2x$.

SOLUTION Since $\sin 2x = 2 \sin x \cos x$, the value of $\sin x$ is needed.
From the Pythagorean identity and $\cos x = -\frac{3}{5}$, $\sin^2 x = \frac{16}{25}$.
Since $\cos x < 0$ and $\tan x > 0$, $W(x)$ is in the third quadrant and hence

$$\sin x = -\frac{4}{5}.$$
$$\sin 2x = 2\left(-\frac{4}{5}\right)\left(-\frac{3}{5}\right) = \frac{24}{25}$$
$$\tan x = \frac{\sin x}{\cos x} = \frac{4}{3}$$
$$\tan 2x = \frac{2\left(\frac{4}{3}\right)}{1 - \left(\frac{4}{3}\right)^2} = -\frac{24}{7}$$

Example 2 Prove the identity: $\tan x \tan 2x = \sec 2x - 1$.

SOLUTION First simplify the left side as follows.

$$\tan x \tan 2x = \frac{\sin x}{\cos x} \cdot \frac{\sin 2x}{\cos 2x}$$

$$= \frac{\sin x \,(2 \sin x \cos x)}{\cos x \cos 2x}$$

$$= \frac{2 \sin^2 x}{\cos 2x}$$

Now simplify the right side of the identity.

$$\sec 2x - 1 = \frac{1}{\cos 2x} - 1 = \frac{1 - \cos 2x}{\cos 2x}$$

$$= \frac{1 - (1 - 2 \sin^2 x)}{\cos 2x}$$

$$= \frac{2 \sin^2 x}{\cos 2x}$$

So, by the transitive property, $\tan x \tan 2x = \sec 2x - 1$.

Example 3 Express $\cos 3x$ in terms of $\cos x$.

SOLUTION
$$\cos 3x = \cos(2x + x)$$
$$= \cos 2x \cos x - \sin 2x \sin x$$
$$= (2 \cos^2 x - 1) \cos x - (2 \sin x \cos x) \sin x$$
$$= 2 \cos^3 x - \cos x - 2 \cos x \,(1 - \cos^2 x)$$
$$= 4 \cos^3 x - 3 \cos x$$

Some equations that could not be solved before can now be solved by using the addition, subtraction, and double-argument identities.

Example 4 Solve for x on $[0, 2\pi]$: $\sin 2x = \sin x$.

SOLUTION
$$2 \sin x \cos x = \sin x$$
$$\sin x \,(2 \cos x - 1) = 0$$

$\sin x = 0$ or $2 \cos x - 1 = 0$

$\sin x = 0$ or $\cos x = \frac{1}{2}$

$x = 0, \pi, 2\pi$ or $x = \frac{\pi}{3}, \frac{5\pi}{3}$

Example 5 Solve for x on $[0, 2\pi]$: $\cos 2x + \cos x = 0$.

SOLUTION Replace $\cos 2x$ with $2\cos^2 x - 1$ since there is already a $\cos x$ term in the equation.

$$(2\cos^2 x - 1) + \cos x = 0$$
$$2\cos^2 x + \cos x - 1 = 0$$
$$(2\cos x - 1)(\cos x + 1) = 0$$
$$\cos x = \tfrac{1}{2} \quad \text{or} \quad \cos x = -1$$

So the solution set is $\left\{\dfrac{\pi}{3}, \dfrac{5\pi}{3}, \pi\right\}$.

Example 6 Solve for x on $[0°, 360°]$: $\cos 2x + \sin x + 2 = 0$.

SOLUTION Replace $\cos 2x$ with $1 - 2\sin^2 x$ since there is already a $\sin x$ term in the equation.

$$1 - 2\sin^2 x + \sin x + 2 = 0$$
$$2\sin^2 x - \sin x - 3 = 0$$
$$(2\sin x - 3)(\sin x + 1) = 0$$
$$\sin x = \tfrac{3}{2} \quad \text{or} \quad \sin x = -1$$

There is no solution for $\sin x = \tfrac{3}{2}$, so the only solution is $x = 270°$.

Exercises

A 1. If $\sin x = \tfrac{3}{5}$ and $0 < x < \tfrac{\pi}{2}$, find $\sin 2x$.

2. If $\sin x = \tfrac{3}{5}$ and $0 < x < \tfrac{\pi}{2}$, find $\cos 2x$.

3. If $\cos x = -\tfrac{3}{5}$ and $\pi < x < \tfrac{3\pi}{2}$, find $\tan 2x$.

4. If $\cos x = -\tfrac{3}{5}$ and $\pi < x < \tfrac{3\pi}{2}$, find $\cot 2x$.

5. If $\sin \theta = \tfrac{5}{13}$ and $90° < \theta < 180°$, find $\sin 2\theta$.

6. If $\sec \theta = 3$ and $-\pi < \theta < 0$, find $\cos 2\theta$.

7. If $\cos x = -0.8$ and $\tan x > 0$, find $\cos 2x$.

8. If $\cos x = 0.6$ and $\tan x < 0$, find $\sin 2x$.

Simplify each expression.

9. $2\sin^2 2t + \cos 4t$

10. $2\sin^2 x + \cos 2x$

Simplify each expression.

11. $\dfrac{\cot 2x}{\cot x - \tan x}$

12. $\dfrac{\sin^2 2t}{\sin^2 t} + 4 \sin^2 t$

Solve for all $x \in [0, 2\pi]$.

13. $\sin 2x - \cos x = 0$

14. $\sin 2x + \sin x = 0$

15. $\cos x + \cos 2x = 0$

16. $\cos 2x - \cos x = 0$

B 17. $\cos 2x + 5 \cos(x + \pi) = -3$

18. $2 \sin^2 x = 3 \cos x$

19. $4 = 2 \sin^2 x + 5 \cos x$

20. $\tan 2x = \tan x$

Prove each identity.

21. $\csc 2x + \cot 2x = \cot x$

22. $\dfrac{1 - \cos 2x}{\sin 2x} = \tan x$

23. $\dfrac{2 \tan \theta}{1 + \tan^2 \theta} = \sin 2\theta$

24. $\sin 4t = 2 \sin 2t \cos 2t$

25. $\sec 2x = \dfrac{\csc x}{\csc x - 2 \sin x}$

26. $\csc 2x = \dfrac{1}{2} \sec x \csc x$

27. $\cos^2 3x - \sin^2 3x = \cos 6x$

28. $\cot x = \csc 2x + \cot 2x$

29. $\tan 3x = \dfrac{3 \tan x - \tan^3 x}{1 - 3 \tan^2 x}$

30. $\tan A = \csc 2A - \cot 2A$

31. $2 \csc 2\theta = \tan \theta + \cot \theta$

32. $\sec \theta = \dfrac{\sin 2\theta}{\sin \theta} - \dfrac{\cos 2\theta}{\cos \theta}$

Simplify each expression.

33. $\dfrac{\sin 3\theta}{\sin \theta} - \dfrac{\cos 3\theta}{\cos \theta}$

34. $\dfrac{1 + \sin 2\theta + \cos 2\theta}{1 + \sin 2\theta - \cos 2\theta}$

C 35. Express $\sin 3x$ and $\sin 4x$ in terms of $\sin x$.

36. Express $\cos 4x$ and $\cos 5x$ in terms of $\cos x$.

37. Simplify: $\dfrac{\cos^3 x \sin^2 x}{2 \cos x - \cos 3x - \cos 5x}$.

38. What is the maximum value of the function defined by $f(x) = \sin x \cos x$?

Find approximations of all solutions in $[0, 2\pi]$.

39. $3 \cos^2 x - 5 \sin x - 1 = 0$

40. $2 \cos 2x = 1 - 3 \sin 2x$

41. $\cos 2x + \cos^2 x = 2(\cos x + 1)$

42. $\cos 2x = 2 \cos x$

43. Prove Theorem 11–7.

44. Solve for all $\theta \in [0°, 360°]$: $\cos^4 2\theta - \sin^4 2\theta = 1$.

11-5 Half-Argument Identities

The half-argument, or half-angle, identities have applications in trigonometry and in integral calculus.

Theorem 11-8 For all $x \in \mathbf{R}$:

a. $\sin^2 \frac{x}{2} = \frac{1 - \cos x}{2}.$ **b.** $\cos^2 \frac{x}{2} = \frac{1 + \cos x}{2}.$

c. $\tan^2 \frac{x}{2} = \frac{1 - \cos x}{1 + \cos x}$, $\cos x \neq -1.$

PROOF **a.** Since $\cos 2t = 1 - 2\sin^2 t$ for all $t \in \mathbf{R}$,

$$2 \sin^2 t = 1 - \cos 2t, \text{ or}$$

$$\sin^2 t = \frac{1 - \cos 2t}{2}.$$

Now let $t = \frac{x}{2}$ to obtain $\sin^2 \frac{x}{2} = \frac{1 - \cos x}{2}.$

The proofs of parts **b** and **c** are similar and are left to the exercises.

Example 1 Evaluate: $\sin \frac{\pi}{12}.$

SOLUTION In $\sin^2 \frac{x}{2} = \frac{1 - \cos x}{2}$, let $x = \frac{\pi}{6}$ to obtain

$$\sin^2 \frac{\pi}{12} = \frac{1 - \cos \frac{\pi}{6}}{2} = \frac{1 - \frac{\sqrt{3}}{2}}{2} = \frac{2 - \sqrt{3}}{4}.$$

Since $\sin \frac{\pi}{12} > 0$, $\sin \frac{\pi}{12} = \frac{1}{2}\sqrt{2 - \sqrt{3}}.$

Example 2 Simplify: $\dfrac{2 \cos^2 \frac{t}{2}}{1 + \sec t}.$

SOLUTION $\dfrac{2 \cos^2 \frac{t}{2}}{1 + \sec t} = \dfrac{2\left(\frac{1 + \cos t}{2}\right)}{1 + \frac{1}{\cos t}}$, provided $\sec t \neq -1$ and $\cos t \neq 0$.

$$= \frac{1 + \cos t}{1 + \frac{1}{\cos t}} = \frac{(1 + \cos t) \cos t}{\cos t + 1}$$

$$= \cos t$$

The half-argument identities can be applied to simplify powers of trigonometric functions.

Example 3 Express $\sin^4 t$ with no powers of 2 or more and simplify.

SOLUTION
$$\sin^4 t = (\sin^2 t)^2$$
$$= \left(\frac{1 - \cos 2t}{2}\right)^2$$
$$= \tfrac{1}{4}(1 - 2\cos 2t + \cos^2 2t)$$
$$= \tfrac{1}{4}\left(1 - 2\cos 2t + \frac{1 + \cos 4t}{2}\right)$$
$$= \tfrac{1}{8}(2 - 4\cos 2t + 1 + \cos 4t)$$
$$= \tfrac{1}{8}(3 - 4\cos 2t + \cos 4t)$$

Example 4 Show that the function defined by $f(x) = \sin^2 2x$ is periodic. Find the period of f and sketch its graph.

SOLUTION
$$f(x) = \sin^2 2x = \frac{1 - \cos 4x}{2}$$
$$= \tfrac{1}{2} - \tfrac{1}{2}\cos 4x$$

Since f is basically a cosine function, with appropriate compressions and a vertical shift, it is periodic with period $\frac{2\pi}{4}$, or $\frac{\pi}{2}$.

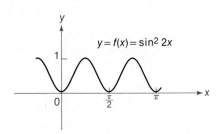

Exercises

A Evaluate each expression, given that α and β are measures of acute angles such that $\csc \alpha = \frac{5}{3}$ and $\cos \beta = \frac{8}{17}$.

1. $\sin(\alpha + \beta)$
2. $\cos(\alpha - \beta)$
3. $\sin 2\alpha$
4. $\cos 2\beta$
5. $\cos \frac{\alpha}{2}$
6. $\sin \frac{\alpha}{2}$
7. $\sin \frac{\beta}{2}$
8. $\cos \frac{\beta}{2}$
9. $\tan \frac{\beta}{2}$
10. $\tan 2\beta$

Use half-argument identities to evaluate each expression exactly.

11. $\sin \frac{5\pi}{12}$
12. $\cos \frac{5\pi}{12}$
13. $\cos 105°$
14. $\sin 15°$
15. $\tan \left(-\frac{\pi}{12}\right)$
16. $\tan \frac{\pi}{8}$
17. $\sin \left(22\tfrac{1}{2}\right)°$
18. $\cos \left(112\tfrac{1}{2}\right)°$

19. $\cos 165°$ **20.** $\sin \dfrac{7\pi}{12}$

Let $P(-3, -4)$ be a point on the terminal side of an angle in standard position with measure α.

21. Find $\sin \dfrac{\alpha}{2}$. **22.** Find $\cos \dfrac{\alpha}{2}$.

B Simplify each expression.

23. $\dfrac{(1 + \cos x) \sec \dfrac{x}{2}}{\cos \dfrac{x}{2}}$ **24.** $\sec x - \tan x \tan \dfrac{x}{2}$

25. Evaluate $\cos \dfrac{\pi}{12}$ by expanding $\cos \left(\dfrac{\pi}{3} - \dfrac{\pi}{4}\right)$ and also by using the half-argument identity to find $\cos \dfrac{1}{2}\left(\dfrac{\pi}{6}\right)$.

26. Evaluate $\sin \dfrac{7\pi}{12}$ by expanding $\sin \left(\dfrac{\pi}{3} + \dfrac{\pi}{4}\right)$ and also by using the half-argument identity to find $\sin \dfrac{1}{2}\left(\dfrac{7\pi}{6}\right)$.

Prove each identity.

27. $4 \sin^2 x \cos^2 x + \cos^2 2x = 1$ **28.** $\csc^2 \dfrac{x}{2} = \dfrac{2 \sec x}{\sec x - 1}$

29. $\tan \dfrac{A}{2} = \csc A - \cot A,\ 0 < A < \dfrac{\pi}{2}$ **30.** $\cos^2 \dfrac{t}{2} = \dfrac{1 + \sec t}{2 \sec t}$

31. $6 \sin \dfrac{x}{2} \cos \dfrac{x}{2} = 3 \sin x$ **32.** $\sec^2 \dfrac{t}{2} = \dfrac{2 + 2 \cos t}{1 + 2 \cos t + \cos^2 t}$

33. $\tan \dfrac{x}{2} = \dfrac{1 - \cos x}{\sin x},\ 0 < x < \dfrac{\pi}{2}$ **34.** $\tan \dfrac{x}{2} = \dfrac{\sin x}{1 + \cos x},\ 0 < x < \dfrac{\pi}{2}$

35. $\tan \dfrac{x}{2} = \dfrac{1}{\csc x + \cot x},\ 0 < x < \dfrac{\pi}{2}$ **36.** $\left|\csc \dfrac{t}{2}\right| = \dfrac{\sqrt{2 + 2 \cos t}}{|\sin t|}$

37. If $\cos t = \dfrac{5}{12}$ and $\dfrac{3\pi}{2} < t < 2\pi$, find $\sin \dfrac{t}{2}$.

38. If $\tan t = -\dfrac{4}{3}$ and $-\dfrac{\pi}{2} < t < 0$, find $\tan \dfrac{t}{2}$.

39. Express $\cos^4 t$ with no powers of 2 or more.

40. Express $\sin^4 2x$ with no powers of 2 or more.

41. Let $f(x) = 2 \sin^2 \dfrac{3x}{2}$. Show that f is periodic and find its period.

42. Let $f(x) = 1 - 3\cos^2\left(\dfrac{3x}{5}\right)$. Show that f is periodic and find its period.

Use half-argument identities to evaluate exactly.

43. $\sin\left(\dfrac{1}{2}\arccos\dfrac{3}{5}\right)$

44. $\cos\left(\dfrac{1}{2}\arcsin\dfrac{4}{5}\right)$

C Solve for all $x \in [0, 2\pi]$.

45. $2\cos^2\dfrac{x}{2} - 3\cos x = 0$

46. $\left|\sin\dfrac{x}{2}\right| = |\cos x|$

47. $\left|\tan\dfrac{x}{2}\right| = |\sin x|$

48. $\cot\dfrac{x}{2} = 2\sin\dfrac{x}{2}$

49. Prove for $0° < \theta < 90°$ that $\tan\dfrac{\theta}{2} = \dfrac{\sin\theta}{1+\cos\theta}$ and explain how the diagram at the right helps demonstrate this geometrically.

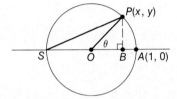

50. Prove Theorem 11–8, parts **b** and **c**.

11–6 Further Equation Solving

For a particular generator the electric current I in amperes is given by $I = \sin\pi t$, where t is in seconds. At what times $t > 0$ is the current 1 ampere? To answer this question, solve the equation

$$\sin\pi t = 1.$$

Since $\sin\left(\dfrac{\pi}{2} + 2\pi k\right) = 1$ for all $k \in J$,

$$\pi t = \dfrac{\pi}{2} + 2\pi k,$$

$$t = \dfrac{1}{2} + 2k,\ k \in J.$$

But the domain of t does not include negative numbers, so the solution set is

$$\left\{t: t = \dfrac{1}{2} + 2k,\ k \in W\right\}.$$

Typical solutions $t = \dfrac{1}{2}, 2\dfrac{1}{2}, 4\dfrac{1}{2}$, etc., are shown on the graph at the right.

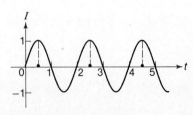

To solve a circular-function equation on a restricted domain, it is usually necessary first to find the general solution and then select from those solutions the ones that are in the domain.

Example 1 Solve exactly for all $x \in [0, 2\pi]$: $\cos 2x = -1$.

SOLUTION $\cos 2x = -1$

$$2x = \pi + 2\pi k, k \in J$$

So $x = \frac{\pi}{2} + \pi k, k \in J$ is the general solution.

If $k = 0$, $x = \frac{\pi}{2}$.

If $k = 1$, $x = \frac{\pi}{2} + \pi = \frac{3\pi}{2}$.

Other integers k lead to solutions that are outside the interval $[0, 2\pi]$. The solution set is $\left\{\frac{\pi}{2}, \frac{3\pi}{2}\right\}$.

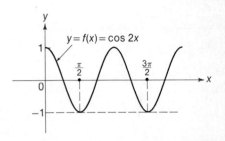

In example 1 it is tempting to say that the only number in $[0, 2\pi]$ whose cosine is -1 is π, so that $2x = \pi$, or $x = \frac{\pi}{2}$, is the only solution. But a sketch of $f(x) = \cos 2x$ shows that the period of f is π and hence there are two cycles in the interval $[0, 2\pi]$ and two solutions of $\cos 2x = -1$.

Example 2 Solve exactly for $x \in [0, 2\pi]$: $2 \cos (2x - \pi) + 1 = 0$.

SOLUTION Subtract 1 from both sides and divide by 2:

$$\cos (2x - \pi) = -\frac{1}{2}.$$

$$2x - \pi = \frac{2\pi}{3} + 2\pi k \quad \text{or} \quad 2x - \pi = \frac{4\pi}{3} + 2\pi k, k \in J$$

$$2x = \frac{5\pi}{3} + 2\pi k \quad \text{or} \quad 2x = \frac{7\pi}{3} + 2\pi k, k \in J$$

Therefore, $x = \frac{5\pi}{6} + \pi k \quad \text{or} \quad x = \frac{7\pi}{6} + \pi k, k \in J,$

is the general solution.

The graph of $y = \cos (2x - \pi)$ shows that there are *four* solutions in $[0, 2\pi]$.

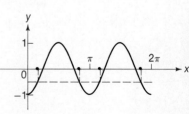

If $k = 0$, $x = \frac{5\pi}{6}$ or $x = \frac{7\pi}{6}$.

If $k = 1$, $x = \frac{11\pi}{6}$ or $x = \frac{13\pi}{6}$, and $\frac{13\pi}{6} \notin [0, 2\pi]$.

If $k > 1$, $x \notin [0, 2\pi]$.

(cont. on p. 414)

If $k = -1$, $x = -\frac{\pi}{6} \notin [0, 2\pi]$ or $x = \frac{\pi}{6}$.

If $k < -1$, $x \notin [0, 2\pi]$.

Therefore the solution set is $\left\{\frac{\pi}{6}, \frac{5\pi}{6}, \frac{7\pi}{6}, \frac{11\pi}{6}\right\}$.

Example 3 Solve exactly for all $x \in [0, 2\pi]$: $\left(2 \sin \frac{x}{5}\right) + \sqrt{3} = 0$.

SOLUTION An algebraic approach to this problem yields

$\sin \frac{x}{5} = -\frac{\sqrt{3}}{2}$ and

$\frac{x}{5} = \frac{4\pi}{3} + 2\pi k$ or $\frac{x}{5} = \frac{5\pi}{3} + 2\pi k$, $k \in J$,

$x = \frac{20\pi}{3} + 10\pi k$ or $x = \frac{25\pi}{3} + 10\pi k$, $k \in J$.

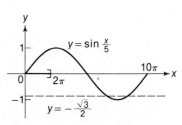

You soon discover there are no integers k that yield an $x \in [0, 2\pi]$.

A strategy for solving circular-function equations is now evident.
1. Use a graph to help determine the number of solutions and to estimate their approximate values.
2. Solve the equation for the indicated argument, $2x$, $3x$, $2x - \pi$, etc.
3. Solve for x in general.
4. Select from the general solution those solutions that are in the specified domain.

The same strategy applies in solving inequalities.

Example 4 Solve for $x \in [0, 2\pi]$: $1 \leq \tan x < 3$.

SOLUTION A sketch of tan on $[0, 2\pi]$ shows that the solution is the union of two half-open intervals. Solve each associated equation on $[0, 2\pi]$.

$\tan x = 1$ $\tan x = 3$

$x = \frac{\pi}{4} + \pi k$ $x \doteq 1.25 + \pi k$

$x = \frac{\pi}{4}, \frac{5\pi}{4}$ $x \doteq 1.25, 4.39$

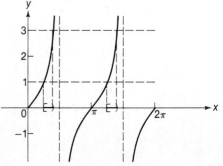

Therefore the solution set is approximately $\left[\frac{\pi}{4}, 1.25\right) \cup \left[\frac{5\pi}{4}, 4.39\right)$.

Exercises

A Find x for $k = -1, 0, 1$.

1. $x = 0.2 + \pi k$
2. $x = \frac{\pi}{6} + 2\pi k$

Find all $x \in [0, 2\pi]$, assuming k is any integer.

3. $x = \frac{\pi}{6} + \pi k$
4. $x = -\frac{\pi}{3} + \pi k$
5. $x = 0.2 + 2k$
6. $x = -0.1 + 2k$
7. $x = -\frac{\pi}{3} + \frac{\pi k}{2}$
8. $x = \frac{\pi}{4} - \frac{\pi k}{2}$

Solve for all $x \in [0, 2\pi]$.

9. $\cos 2x = 0$
10. $\sin 3x = 1$
11. $\sin 2x = \frac{\sqrt{3}}{2}$
12. $\cos \frac{x}{2} = \frac{1}{2}$
13. $\cos 3x = \frac{\sqrt{2}}{2}$
14. $\sin 2\pi x = 0.5$

B
15. $\sin (2x - \pi) = 1$
16. $\cos \frac{x}{2} = \frac{\sqrt{3}}{2}$
17. $\cos (2\pi x - 1) = 0$
18. $2 \sin (2x + 1) + 1 = 0$
19. $\tan \left(x - \frac{\pi}{2}\right) = \sqrt{3}$
20. $\tan 2x = -1$
21. $\cos^2 2x + \sin^2 x = 1$
22. $2 \sin^4 2x = \cos^2 2x$

Solve for all $x \in [0, 2\pi]$; give answers to the nearest hundredth.

23. $\sin 2x = 0.8$
24. $\cos \frac{x}{2} = -0.4$
25. $\cos (2x + 3) = 0.6$
26. $\sin (\pi x - 4) = -0.2$
27. $\sin \frac{x + 1}{2} = 0.3$
28. $\cos (1 - 2x) = 0.7$

C Find approximations graphically of the smallest positive solution of each equation.

29. $\cos \pi x + (x - 1)^2 = 0$
30. $\sin 2x = x$

31. Use a calculator to approximate to the nearest hundredth the solution of the equation in exercise 30.

32. The depth of water at an ocean pier varies with the tide. Suppose the depth can be expressed as a sine function of time with range 10 ft. to 18 ft. and with period 24 hours. What fraction of the day is the depth of the water more than 16 ft.?

33. Solve on $[0, 2\pi]$: $3 \cos 2x + 2 \sin 2x = 1$.

Solve for all $x \in [0, 2\pi]$.

34. $\sin 3x \geq \frac{1}{2}$ **35.** $-1 \leq \tan x < 0$ **36.** $2 \cos 2x > -1$

37. $1 + \sin x \geq \cos 2x$ **38.** $1 < 2 \sin x < 4$ **39.** $2 \sin 2x < \sqrt{2}$

Solve for all $x \in [0, 2\pi]$; give answers to the nearest hundredth.

40. $\cos 2x \leq 0.38$ **41.** $\sin 2x \geq -0.21$

42. $3 \cos^2 x + \cos x > 2$ **43.** $8 \leq 6 \sin x + 9 \cos^2 x$

How many solutions does each equation have?

44. $x^2 = 6 \cos \pi x$ **45.** $\log_3 x = \sin \pi x$

11–7 Complex Numbers in Polar Form

Real numbers may be represented geometrically by points on a coordinate line. Complex numbers may be represented geometrically by points in a coordinate plane. Every complex number $z = a + bi$ determines a unique ordered pair of real numbers (a, b). The point $P(a, b)$ in a coordinate plane is called the **geometric representation**, or the graph, of the complex number $a + bi$. This coordinate plane with a complex number associated with each point is called the **complex plane**. It looks like an ordinary Cartesian plane with the x-axis renamed the **real axis** and the y-axis renamed the **imaginary axis**. The graphs of some complex numbers are shown in the figure. Observe that the graphs of $z = a + bi$ and of its complex conjugate $\bar{z} = a - bi$ are reflections of each other in the real axis.

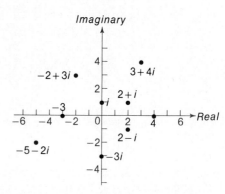

The absolute value of any real number a may be interpreted as the distance from the origin to the graph of a on a coordinate line. It is a natural extension of this idea to define the absolute value of a complex number as a distance also.

Definition The **absolute value** of a complex number $a + bi$, written $|a + bi|$, is the distance from (a, b) to the origin; that is, the real number

$$\sqrt{a^2 + b^2}.$$

416   Chapter 11: Circular Function Identities

The absolute value of a complex number z is sometimes called the **modulus** of z.

Example 1 Graph the complex numbers $z_1 = 2\sqrt{3} - 2i$, $z_2 = -2i$, $z_3 = -1 + i$, and find their absolute values.

SOLUTION
$|z_1| = \sqrt{12 + 4} = 4$
$|z_2| = \sqrt{0 + 4} = 2$
$|z_3| = \sqrt{1 + 1} = \sqrt{2}$

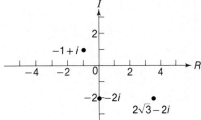

Every complex number can be expressed in polar form. Suppose the point $P(a, b)$ is the geometric representation of the complex number $z = a + bi$.

If polar coordinates of the point P are (r, θ), then

$a = r \cos \theta$ and $b = r \sin \theta$.

Since $z = a + bi = (r \cos \theta) + (r \sin \theta)i$, the complex number z may be written as

$z = r(\cos \theta + i \sin \theta)$.

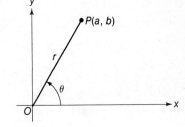

This is called the **polar form**, or the trigonometric form, of the complex number z.

By convention, r is the nonnegative number $\sqrt{a^2 + b^2}$. The angle θ associated with z is called the **argument** of z. For $z = a + bi$, the angle θ is not unique. Recall that $\tan \theta = \frac{b}{a}$ but that $\arctan \frac{b}{a}$ will not always yield the desired θ.

Example 2 Write each complex number in polar form:

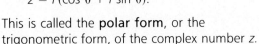
a. $z_1 = -4 + 4i$ **b.** $z_2 = 2\sqrt{3} - 2i$ **c.** $z_3 = 2 + 6i$ **d.** $z_4 = -3$.

SOLUTION The problem is one of converting from Cartesian to polar coordinates.

a. For $z_1 = -4 + 4i$, $a = -4$ and $b = 4$, so that $r = \sqrt{(-4)^2 + 4^2} = 4\sqrt{2}$. Since $\tan \theta = \frac{4}{-4} = -1$ and $(-4, 4)$ is in the second quadrant, use $\theta = \frac{3\pi}{4}$, so that

$$z = 4\sqrt{2}\left(\cos \frac{3\pi}{4} + i \sin \frac{3\pi}{4}\right),$$

(cont. on p. 418)

or, with θ in degrees,

$$z_1 = 4\sqrt{2}(\cos 135° + i \sin 135°).$$

Parts **b**, **c**, and **d** are done in the same manner.

b. $z_2 = 2\sqrt{3} - 2i = 4\left[\cos\left(-\dfrac{\pi}{6}\right) + i \sin\left(-\dfrac{\pi}{6}\right)\right]$

c. $z_3 = 2 + 6i = 2\sqrt{10}[\cos(\arctan 3) + i \sin(\arctan 3)]$
$\doteq 6.32(\cos 71.57° + i \sin 71.57°)$

d. $z_4 = -3 + 0i = 3(\cos \pi + i \sin \pi)$

Graphs of these four complex numbers, with both polar and Cartesian coordinates labeled, are shown below.

a. $z_1 = -4 + 4i$

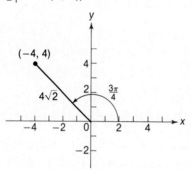

b. $z_2 = 2\sqrt{3} - 2i$

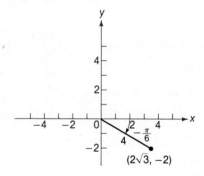

c. $z_3 = 2 + 6i$

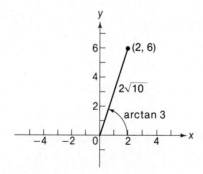

d. $z_4 = -3$

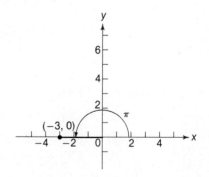

Multiplication and division of complex numbers is particularly easy if the complex numbers are expressed in trigonometric form.

Theorem 11–9 If $z_1 = r_1(\cos \theta_1 + i \sin \theta_1)$ and $z_2 = r_2(\cos \theta_2 + i \sin \theta_2)$, then

a. $z_1 z_2 = r_1 r_2 [\cos (\theta_1 + \theta_2) + i \sin (\theta_1 + \theta_2)]$, and

b. $\dfrac{z_1}{z_2} = \dfrac{r_1}{r_2}[\cos (\theta_1 - \theta_2) + i \sin (\theta_1 - \theta_2)]$, $z_2 \neq 0$.

PROOF **a.** Use the usual rule for multiplying two complex numbers and then simplify.

$$z_1 z_2 = [r_1(\cos \theta_1 + i \sin \theta_1)] \cdot [r_2(\cos \theta_2 + i \sin \theta_2)]$$
$$= r_1 r_2 [(\cos \theta_1 \cos \theta_2 - \sin \theta_1 \sin \theta_2) + i(\sin \theta_1 \cos \theta_2 + \cos \theta_1 \sin \theta_2)]$$
$$= r_1 r_2 [\cos(\theta_1 + \theta_2) + i \sin(\theta_1 + \theta_2)]$$

The proof of **b** is similar and is left to the exercises.

Example 3 If $z_1 = \sqrt{7}(\cos 40° + i \sin 40°)$ and $z_2 = 2(\cos 130° + i \sin 130°)$, find $z_1 z_2$.

SOLUTION By Theorem 11–9, $z_1 z_2 = 2\sqrt{7}(\cos 170° + i \sin 170°)$.

Example 4 Express $z_1 = -3\sqrt{3} - 3i$ and $z_2 = 1 + \sqrt{3}i$ in polar form and find $\dfrac{z_1}{z_2}$.

SOLUTION For z_1, $r = \sqrt{27 + 9} = 6$, and $\arctan \dfrac{-3}{-3\sqrt{3}} = \dfrac{\pi}{6}$; but $\pi < \theta < \dfrac{3\pi}{2}$ is needed, so let

$$\theta = \pi + \dfrac{\pi}{6} = \dfrac{7\pi}{6}.$$

$$z_1 = 6\left(\cos \dfrac{7\pi}{6} + i \sin \dfrac{7\pi}{6}\right)$$

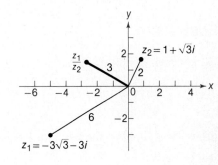

For z_2, $r = \sqrt{1 + 3} = 2$ and $\theta = \dfrac{\pi}{3}$, so that

$$z_2 = 2\left(\cos \dfrac{\pi}{3} + i \sin \dfrac{\pi}{3}\right).$$

$$\dfrac{z_1}{z_2} = \dfrac{6\left(\cos \dfrac{7\pi}{6} + i \sin \dfrac{7\pi}{6}\right)}{2\left(\cos \dfrac{\pi}{3} + i \sin \dfrac{\pi}{3}\right)} = 3\left[\cos \left(\dfrac{7\pi}{6} - \dfrac{\pi}{3}\right) + i \sin \left(\dfrac{7\pi}{6} - \dfrac{\pi}{3}\right)\right]$$

(cont. on p. 420)

$$\frac{z_1}{z_2} = 3\left(\cos\frac{5\pi}{6} + i\sin\frac{5\pi}{6}\right)$$
$$= 3\left(-\frac{\sqrt{3}}{2} + \frac{1}{2}i\right) = -\frac{3\sqrt{3}}{2} + \frac{3}{2}i$$

In this last example, you can verify that $\dfrac{z_1}{z_2} = \dfrac{-3\sqrt{3} - 3i}{1 + \sqrt{3}i}$ equals $-\dfrac{3\sqrt{3}}{2} + \dfrac{3}{2}i$ by multiplying the numerator and denominator by the complex conjugate of z_2.

Raising complex numbers in trigonometric form to powers is easy. For instance, if

$$z = r(\cos\theta + i\sin\theta),$$

then by Theorem 11–9,

$$z^2 = r^2(\cos 2\theta + i\sin 2\theta), \text{ and}$$
$$z^4 = r^4[\cos(2\theta + 2\theta) + i\sin(2\theta + 2\theta)]$$
$$= r^4(\cos 4\theta + i\sin 4\theta).$$

Now you can use this pattern to find $(1 + i)^4$ by proceeding as follows.

$$z = 1 + i = \sqrt{2}\left(\cos\frac{\pi}{4} + i\sin\frac{\pi}{4}\right)$$
$$(1 + i)^4 = 4(\cos\pi + i\sin\pi)$$
$$= 4[(-1) + i(0)]$$
$$= -4$$

Exercises

A Graph each complex number.

1. $1 + 2i$
2. $-3 + 4i$
3. $2i$
4. 7
5. $(1 - i)(1 + 2i)$
6. $(2 - i)(1 - i)$

Graph the complex number z and its conjugate $\bar{z}$ in the complex plane.

7. $z = 2 + 3i$
8. $z = -1 - 4i$

For each complex number, find $|z|$ and $|\bar{z}|$.

9. $z = 1 - i$
10. $z = -1 + 2i$
11. $z = 1 + \sqrt{3}i$
12. $z = 2 - 7i$

Express each complex number in polar form.

13. $1 - i$
14. $1 + \sqrt{3}i$
15. $2i$
16. $-5i$
17. $-3 + 3\sqrt{3}i$
18. $-3 - 3i$
19. $\sqrt{3} - i$
20. $-\dfrac{1}{2} - \dfrac{\sqrt{3}}{2}i$

Express the product $z_1 z_2$ in polar form and in the form $a + bi$, with a and b given to the nearest hundredth.

21. $z_1 = 2(\cos 20° + i \sin 20°)$, $z_2 = 3(\cos 50° + i \sin 50°)$
22. $z_1 = \sqrt{3}\left(\cos \frac{\pi}{5} + i \sin \frac{\pi}{5}\right)$, $z_2 = \cos\left(-\frac{3\pi}{5}\right) + i \sin\left(-\frac{3\pi}{5}\right)$
23. Verify example 4 by showing, without using polar form, that
$$\frac{-3\sqrt{3} - 3i}{1 + \sqrt{3}i} = -\frac{3\sqrt{3}}{2} + \frac{3}{2}i.$$

Use Theorem 11–9 to find $\frac{z_1}{z_2}$.

24. $z_1 = 1 + \sqrt{3}i$, $z_2 = -2\sqrt{3} + 2i$
25. $z_1 = -2 - 2i$, $z_2 = -1 + \sqrt{3}i$

B 26. Prove Theorem 11–9b.

Find z^4.

27. $z = -1 - i$
28. $z = -1 + \sqrt{3}i$
29. $z = 2(\cos 100° + i \sin 100°)$
30. $z = \cos(-20°) + i \sin(-20°)$

31. Graph in the complex plane all the complex numbers z such that $|z| = 3$.

In exercises 32–35, z is a complex number.

32. Prove: $|z| = |\bar{z}|$.
33. Prove: $|-z| = |z|$.
34. Prove: $\bar{\bar{z}} = z$.
35. Prove: $z + \bar{z}$ is a real number.

In exercises 36–39, z and w are complex numbers.

36. Prove: $|zw| = |z| \cdot |w|$.
37. Prove: $\overline{z + w} = \bar{z} + \bar{w}$.
38. Prove: $\overline{zw} = \bar{z} \cdot \bar{w}$.
39. Prove: $\overline{z \div w} = \bar{z} \div \bar{w}$, $w \neq 0$.

C 40. Graph the set of complex numbers z such that $|z - (1 + i)| = 2$.
41. Give a geometric argument that $|z + w| \leq |z| + |w|$ for complex numbers z and w.
42. Express the complex number $(\sqrt{3} + i)^7$ in the form $a + bi$.
43. Express the complex number $(\sqrt{3} - i)^{12}$ in the form $a + bi$.
44. Find 8 different complex numbers z such that $z^8 = 1$.

Chapter Review Exercises

11–1, page 391

1. Prove the identity $\sin x + \cot x \cos x = \csc x$.

2. Simplify: $\dfrac{\cos^2 A - \cot^2 A}{\sin A \cot A \cos A}$.

11–2, page 394

Find exactly all solutions on **R**.

3. $\sin x = -\dfrac{\sqrt{2}}{2}$

4. $\cos x - \sin^2 x = 1$

5. Find all solutions of $\tan x = 2.0$ accurate to the nearest hundredth.

11–3, page 398

6. Simplify: $\cos\left(x - \dfrac{\pi}{2}\right)$.

7. Find the exact value of $\sin \dfrac{5\pi}{12}$.

11–4, page 405

8. Prove the identity: $\dfrac{\cos 2x + \sin^2 x}{1 - \sin x} = 1 + \sin x$.

9. If $\cos x = -\dfrac{4}{5}$ and $\pi < x < \dfrac{3\pi}{2}$, find $\sin 2x$ exactly.

10. Solve for all $x \in [0, 2\pi]$: $\sin 2x + \cos x = 0$.

11–5, page 409

11. Evaluate $\cos \dfrac{\pi}{12}$ exactly.

12. Prove: $\tan^2 \dfrac{x}{2} = \dfrac{\sec x - 1}{\sec x + 1}$.

11–6, page 412

13. Find all $x \in [0, 2\pi]$, assuming k is a positive integer: $x = -\dfrac{\pi}{6} + \pi k$.

14. Solve for all $x \in [0, 2\pi]$: $\cos 2x = \dfrac{\sqrt{3}}{2}$.

11–7, page 416

15. Graph the complex numbers z and $\bar{z}$ in the complex plane if $\bar{z} = 5 - 7i$.

16. If $z_1 = 15(\cos 30° + i \sin 30°)$ and $z_2 = 3(\cos 10° + i \sin 10°)$,

    express $\dfrac{z_1}{z_2}$ in polar form.

Chapter Test

1. Simplify: $\tan\left(\dfrac{\pi}{4} - x\right)$.
2. Approximate all solutions of $\sin x = -0.82$ to the nearest hundredth.
3. Express the complex number $\sqrt{3} + i$ in polar form.

Find all solutions on R exactly.

4. $3\cos x - 1 = \cos x$
5. $\sqrt{3}\tan x + 1 = 0$
6. $\sec^2 x = 2$

7. Find the exact value of $\sin(x+y)$ if $\sin x = -\dfrac{3}{5}$ where $\pi < x < \dfrac{3\pi}{2}$ and $\cos y = -\dfrac{4}{5}$ where $\dfrac{\pi}{2} < y < \pi$.

8. If $z_1 = \sqrt{2} + \sqrt{2}i$ and $z_2 = -\sqrt{3} - i$, use Theorem 11–9 to find $z_1 z_2$ in polar form.

Solve for all $x \in [0, 2\pi]$.

9. $\cos 2x = \cos^2 x$
10. $\sin 2x = -0.5$
11. Prove: $\dfrac{2\cos 2x}{\sin 2x + 2\sin^2 x} = \cot x - 1$.

12. Simplify: $\dfrac{2\sin^2\left(\dfrac{x}{2}\right)}{\csc x - \cot x}$.

13. Solve for all $x \in [0, 2\pi]$ to the nearest hundredth: $\cos(2x + 1) = 0.3$.

Vectors

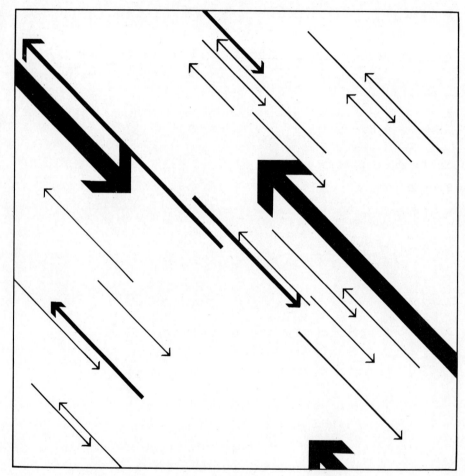

12

12–1 Vectors and Arrows

Suppose two ropes are attached to a crate, as in the figure. One person pulls to the right on a rope with a force of 30 pounds. A second person pulls straight up on the other rope with a force of 40 pounds.

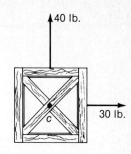

Vectors provide a simple way of representing the effective, or net, force acting on the crate at C. That force is the vector $\langle 30, 40 \rangle$.

Definition A two-dimensional **vector** is an ordered pair of numbers v_1 and v_2, written $\langle v_1, v_2 \rangle$. The numbers v_1 and v_2 are called **components** of the vector. The vectors $\langle a_1, a_2 \rangle$ and $\langle b_1, b_2 \rangle$ are **equal** if and only if $a_1 = b_1$ and $a_2 = b_2$.

A single letter with an arrow above it, as in $\vec{F} = \langle 30, 40 \rangle$, will be used to represent a vector. Some books use boldface letters without the arrow to represent vectors.

In analytic geometry you associate an ordered pair of real numbers with a point in the plane. To keep the ideas of point and vector separate, a vector is represented as an arrow. For example, the vector $\vec{F} = \langle 30, 40 \rangle$ may be represented in the coordinate plane as an arrow from $O(0, 0)$ to $P(30, 40)$. Since the length of arrow $\overrightarrow{OP}$ is $\sqrt{30^2 + 40^2} = 50$, it is reasonable to define the length of the vector $\vec{F}$ as 50 also.

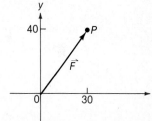

Definition The **length**, or **magnitude**, of the vector $\vec{V} = \langle v_1, v_2 \rangle$, which is written $\|\vec{V}\|$, is $\sqrt{v_1^2 + v_2^2}$.

Example 1 Suppose a rowboat on a lake travels 100 meters east and 50 meters north. Represent the displacement of the rowboat as a vector $\vec{D}$, sketch an arrow to represent $\vec{D}$, and find $\|\vec{D}\|$.

SOLUTION Let $\vec{D} = \langle 100, 50 \rangle$. In the figure the arrow $\overrightarrow{OT}$ represents $\vec{D}$.

$$\|\vec{D}\| = \sqrt{100^2 + 50^2} = \sqrt{12{,}500}$$
$$\doteq 111.8 \text{ m}$$

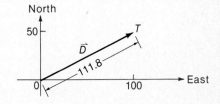

In example 1, the displacement of the rowboat from some initial position to its final position is approximately 111.8 m. As long as the rowboat goes 100 m east and 50 m north, the displacement does not depend upon where the boat started. So it is reasonable to say that each of the arrows in the figure represents the displacement vector $\vec{D} = \langle 100, 50 \rangle$. Each

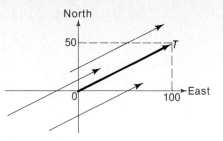

arrow has a starting point and an ending point; each arrow has length 111.8, and each is pointed in the same direction. Any two arrows with the same length pointing in the same direction are called **equivalent** arrows. Both represent the same vector $\vec{V}$. If $\vec{V} = \langle a, b \rangle$, the arrow $\overrightarrow{OP}$ where $P = (a, b)$ is called the **standard representative** of $\vec{V}$.

Example 2 Given points $S(-2, 1)$ and $T(1, 2)$. Find point P such that $\overrightarrow{ST}$ and $\overrightarrow{OP}$ are equivalent arrows.

SOLUTION By subtracting corresponding coordinates, it is clear that arrow $\overrightarrow{ST}$ represents a vector displacement of 3 units to the right and 1 unit up. Therefore from $O(0, 0)$ to P, go 3 units to the right and 1 unit up. Hence P is the point $(3, 1)$.

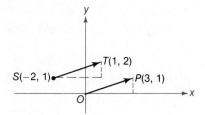

In example 2, the arrow from O to $P(3, 1)$ is the standard representative of the vector $\langle 3, 1 \rangle$. The arrow $\overrightarrow{ST}$ is another representative of the vector $\langle 3, 1 \rangle$. If you are asked to sketch a vector $\vec{V} = \langle v_1, v_2 \rangle$, you may draw any arrow equivalent to $\overrightarrow{OP}$, where $P = (v_1, v_2)$.

Example 3 Sketch on the coordinate plane the standard representative of $\vec{V} = \langle -1, -2 \rangle$. For $S = (3, 2)$, find T such that $\overrightarrow{ST}$ is also a representative of $\vec{V}$.

SOLUTION The standard representative of $\vec{V} = \langle -1, -2 \rangle$ is $\overrightarrow{OP}$, where $P = (-1, -2)$. You can locate point T by observing that $OPTS$ must be a parallelogram, since $\overrightarrow{OP}$ and $\overrightarrow{ST}$ must have the same length and point in the same direction. Find the coordinates of T by subtracting 1 from the first coordinate of S and by subtracting 2 from the second coordinate of S. The result is $T = (2, 0)$.

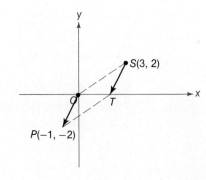

Example 4 Suppose a force of 3 pounds is represented by an arrow of length 3 that makes a 30° angle with the positive x-axis. Express this force as a vector.

SOLUTION Let $\overrightarrow{OP}$ be the arrow, as in the figure. Draw the perpendicular $\overline{BP}$ from P to the x-axis.

$$BP = 3 \sin 30° = \frac{3}{2}$$

$$OB = 3 \cos 30° = \frac{3\sqrt{3}}{2}$$

Therefore $\overrightarrow{OP}$ is the standard representative of $\vec{V} = \left\langle \frac{3\sqrt{3}}{2}, \frac{3}{2} \right\rangle$.

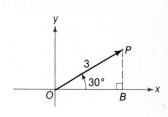

Exercises

A Find the length of each vector.

1. $\vec{V} = \langle 1, 3 \rangle$ **2.** $\vec{V} = \langle -1, 3 \rangle$ **3.** $\vec{V} = \langle 3, -4 \rangle$ **4.** $\vec{V} = \langle -3, -4 \rangle$

Sketch three different arrows that are representatives of $\vec{V}$.

5. $\vec{V} = \langle 1, 3 \rangle$ **6.** $\vec{V} = \langle 2, -1 \rangle$

The arrow $\overrightarrow{ST}$ is a representative of the vector $\vec{P}$. Find $\vec{P}$ and sketch its standard representative.

7. 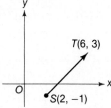   **8.** 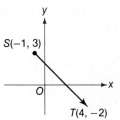   **9.** 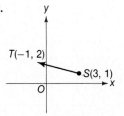   **10.**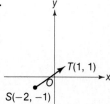

11. $S(1, 4)$, $T(-1, -3)$ **12.** $S(-1, -2)$, $T(-1, 4)$

13. Two forces act on an object. One is a force of 5 pounds in the positive x-direction and the other is a force of 10 pounds in the positive y-direction. Express the effective force on the object as a vector and sketch it.

14. A boat travels on a lake 100 meters east and 60 meters south. Express its displacement as a vector and sketch the vector.

B A force vector $\vec{F} = \langle a_1, a_2 \rangle$ is represented by an arrow of length k that makes an angle θ with the positive x-axis. Find a_1 and a_2.

15. $k = 6$, $\theta = 30°$ **16.** $k = 4$, $\theta = 45°$ **17.** $k = 2$, $\theta = 60°$ **18.** $k = 5$, $\theta = 120°$

Find k so that $\overrightarrow{ST}$ is equivalent to $\overrightarrow{QR}$.

19. $S(0, 3)$, $T(2, 1)$, $Q(7, -2)$, $R(9, k)$
20. $S(-1, 2)$, $T(1, 4)$, $Q(k, 1)$, $R(6, 3)$
21. $S(a, b)$, $T(k, b + 2)$, $Q(a + 2, b - 2)$, $R(a + 3, b)$

Find the coordinates of point T such that $\overrightarrow{ST}$ is a representative of $\langle 3, 1 \rangle$.

22. $S(1, 1)$ 23. $S(-5, 3)$ 24. $S(-1, 1)$ 25. $S(1, -2)$

Find all real values of k such that $\|\vec{V}\| = 1$.

26. $\vec{V} = \langle 1, k \rangle$ 27. $\vec{V} = \langle k, k \rangle$ 28. $\vec{V} = \left\langle k, \dfrac{k}{2} \right\rangle$ 29. $\vec{V} = \left\langle \dfrac{k}{3}, \dfrac{1}{2} \right\rangle$

C Given the three points $A(a_1, a_2)$, $B(b_1, b_2)$, and $C(c_1, c_2)$, find an expression for each of the following.

30. $\|\overrightarrow{AB}\|$
31. Vector $\vec{X}$ such that $\overrightarrow{AC}$ is a representative of $\vec{X}$.
32. The coordinates of point T such that $\overrightarrow{AB}$ and $\overrightarrow{CT}$ are equivalent.

12–2 Vector Addition and Subtraction

Suppose a boat travels east on a lake 500 feet and a cross wind blows the boat north 100 feet. The displacement due to rowing may be represented by $\vec{A} = \langle 500, 0 \rangle$ and the displacement due to the wind by $\vec{B} = \langle 0, 100 \rangle$. The resulting movement, 500 feet east and 100 feet north, may then be represented by $\langle 500, 100 \rangle$ which is the vector sum $\vec{A} + \vec{B}$.

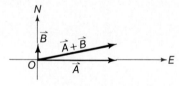

Definition For any two vectors $\vec{A} = \langle a_1, a_2 \rangle$ and $\vec{B} = \langle b_1, b_2 \rangle$, the **vector sum** of $\vec{A}$ and $\vec{B}$ is the vector $\vec{A} + \vec{B} = \langle a_1 + b_1, a_2 + b_2 \rangle$.

Example 1 Sketch $\vec{A} = \langle 5, 2 \rangle$, $\vec{B} = \langle 2, 4 \rangle$, and $\vec{A} + \vec{B}$.

SOLUTION $\vec{A} + \vec{B} = \langle 5, 2 \rangle + \langle 2, 4 \rangle = \langle 7, 6 \rangle$

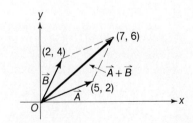

The figure in example 1 shows that sketching the vector sum $\vec{A} + \vec{B}$ is easily done by drawing the diagonal of a parallelogram whose sides are standard representatives of $\vec{A}$ and $\vec{B}$. The figure at the right illustrates that the vector sum can be found by adding graphically the components of $\vec{A}$ and $\vec{B}$.

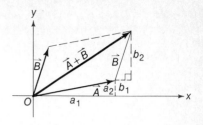

Example 2 An airplane travels east with an airspeed of 400 mph but drifts to the north because of a 120-mph wind blowing 60° north of east. Sketch and find the components of a vector that represents the velocity of the airplane. Find the resulting groundspeed of the plane and its direction of travel.

SOLUTION Let $\vec{A} = \langle 400, 0 \rangle$ represent the plane's velocity in still air. If $\vec{W}$ represents the wind velocity, then

$$\vec{W} = \langle 120 \cos 60°, 120 \sin 60° \rangle$$
$$\doteq \langle 60, 103.9 \rangle.$$

The velocity of the plane is then

$$\vec{V} = \vec{A} + \vec{W} = \langle 460, 103.9 \rangle,$$

as sketched. The plane's ground speed is

$$\|\vec{V}\| \doteq \sqrt{(460)^2 + (103.9)^2} \doteq 471.6 \text{ mph},$$

and its direction of travel is θ degrees north of east where

$$\theta \doteq \arctan\left(\frac{103.9}{460}\right) \doteq 12.7°.$$

Sometimes it is useful to represent two vectors $\vec{A}$ and $\vec{B}$ by arrows without showing coordinate axes. The vector sum $\vec{A} + \vec{B}$ may still be represented as the diagonal of a parallelogram.

The figure illustrates the following:

1. The vector $\vec{A} + \vec{B}$ is the same as the vector $\vec{B} + \vec{A}$.

2. Since the opposite sides of a parallelogram are congruent, you could also find $\vec{A} + \vec{B}$ by drawing a representative of $\vec{A}$ at the end of $\vec{B}$ and then completing the triangle as shown.

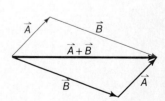

The algebra of vectors is governed by the following theorem.

Theorem 12–1 For any vectors $\vec{A}$, $\vec{B}$, and $\vec{C}$:

a. $\vec{A} + \vec{B} = \vec{B} + \vec{A}$.

b. The zero vector $\vec{O} = \langle 0, 0 \rangle$ satisfies $\vec{A} + \vec{O} = \vec{O} + \vec{A} = \vec{A}$.

c. There is a vector $-\vec{A}$ such that $\vec{A} + (-\vec{A}) = (-\vec{A}) + \vec{A} = \vec{O}$.

d. $\vec{A} + (\vec{B} + \vec{C}) = (\vec{A} + \vec{B}) + \vec{C}$.

PROOF a. By definition $\vec{A} + \vec{B} = \langle a_1 + b_1, a_2 + b_2 \rangle$ and
$$\vec{B} + \vec{A} = \langle b_1 + a_1, b_2 + a_2 \rangle.$$
Since the commutative law of addition holds for real numbers,
$$\vec{A} + \vec{B} = \vec{B} + \vec{A}.$$

The proofs of parts **b**, **c**, and **d** are left to the exercises.

Part **d** of Theorem 12–1 can be verified by sketching, as shown below.

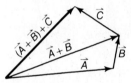 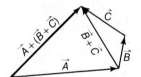

The vector $-\vec{A}$ is represented by an arrow of the same length as $\vec{A}$ but in the opposite direction. If $\vec{A} = \langle a_1, a_2 \rangle$, then $(-\vec{A}) = \langle -a_1, -a_2 \rangle$ so that $\vec{A} + (-\vec{A}) = \langle 0, 0 \rangle$.

The vector sum $\vec{A} + (-\vec{B})$ may then be sketched as in the figure.

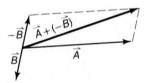

Notice in this last figure that an arrow from the end of $\vec{B}$ to the end of $\vec{A}$ is equivalent to $\vec{A} + (-\vec{B})$. Adding the vector $\vec{A} + (-\vec{B})$ to $\vec{B}$ gives $\vec{A}$. It seems natural then to write $\vec{A} + (-\vec{B})$ as $\vec{A} - \vec{B}$.

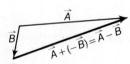

Definition For any two vectors $\vec{A}$ and $\vec{B}$, $\vec{A} - \vec{B} = \vec{A} + (-\vec{B})$.

If $\vec{A} = \langle a_1, a_2 \rangle$ and $\vec{B} = \langle b_1, b_2 \rangle$, then $(-\vec{B}) = \langle -b_1, -b_2 \rangle$ and
$$\vec{A} - \vec{B} = \langle a_1, a_2 \rangle + \langle -b_1, -b_2 \rangle$$
$$= \langle a_1 - b_1, a_2 - b_2 \rangle.$$

Example 3 For $\vec{A} = \langle 3, 2 \rangle$ and $\vec{B} = \langle -1, 1 \rangle$, find $\vec{A} - \vec{B}$.

SOLUTION $\vec{A} - \vec{B} = \langle 3, 2 \rangle - \langle -1, 1 \rangle$
$= \langle 4, 1 \rangle$

Exercises

A Find $\vec{A} + \vec{B}$ and $\vec{A} - \vec{B}$.

1. $\vec{A} = \langle 1, 3 \rangle$
   $\vec{B} = \langle 3, -2 \rangle$
2. $\vec{A} = \langle 5, -1 \rangle$
   $\vec{B} = \langle 2, 3 \rangle$
3. $\vec{A} = \langle -1, 6 \rangle$
   $\vec{B} = \langle 2, -1 \rangle$
4. $\vec{A} = \langle 2, 5 \rangle$
   $\vec{B} = \langle 0, -1 \rangle$

Copy $\vec{A}$ and $\vec{B}$ and sketch $\vec{A} + \vec{B}$ and $\vec{A} - \vec{B}$.

5.

6.

7.

8.

Find real numbers s and t that satisfy the following.

9. $\langle s, t \rangle + \langle 2, 3 \rangle = \langle 4, -6 \rangle$
10. $\langle 1, -5 \rangle + \langle s, 6 \rangle = \langle -3, t \rangle$

For each figure write an expression for $\vec{X}$ in terms of $\vec{A}$ and $\vec{B}$.

11.

12.

13.

14.

B 15. If $\vec{V} = \langle a, b \rangle$, prove that $2\|\vec{V}\| = \|\vec{V} + \vec{V}\|$.
16. Prove parts b, c, and d of Theorem 12–1 algebraically.
17. If P is the point (a, b) and Q the point (c, d), prove that $\vec{PQ} + \vec{QP} = \vec{O}$.
18. Prove algebraically that if $\vec{V} + \vec{W} = \vec{O}$, then $\vec{V} = -\vec{W}$.

Copy $\vec{A}$, $\vec{B}$, and $\vec{C}$ and sketch $\vec{A} + \vec{B} + \vec{C}$ and $\vec{A} + \vec{B} - \vec{C}$.

19. 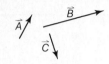   **20.**

21. A motorboat can travel 12 mph in still water. If a river is flowing at 3 mph and the motorboat heads directly across the river, what is its resulting velocity and direction of travel?

22. The force vector resulting from the wind's action on an airplane wing is $\vec{F}$ and it is 10° off the vertical, as in the figure. The drag is the horizontal component of $\vec{F}$ and the lift is the vertical component. If the lift is 2000 pounds, find the drag.

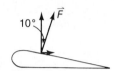

23. Prove algebraically that $\|(-\vec{V})\| = \|\vec{V}\|$.

C An airplane is headed east with an airspeed s mph. The plane drifts to the north because of a wind blowing w mph on a bearing $\theta°$ north of east. For the data given in exercises 24–27, find (a) the components of the plane's velocity vector, (b) the resulting groundspeed of the plane, and (c) the direction of travel of the plane.

24. $s = 300$
$w = 50$
$\theta = 60°$

25. $s = 400$
$w = 100$
$\theta = 45°$

26. $s = 350$
$w = 50$
$\theta = 120°$

27. $s = 400$
$w = 60$
$\theta = 150°$

12–3 Scalar Multiplication

If $\vec{F}$ represents a force vector acting on an object, then $2\vec{F}$ represents a force acting in the same direction with twice the magnitude. Geometrically, $2\vec{F}$ is a vector which is in the same direction and twice as long as $\vec{F}$. If a vector is multiplied by a real number, the product is called a **scalar multiple** of the vector.

Definition For any real number k and vector $\vec{A} = \langle a_1, a_2 \rangle$, a **scalar multiple** of $\vec{A}$ is defined by

$$k\vec{A} = \langle ka_1, ka_2 \rangle.$$

The number k is called a **scalar**.

Notice that $(-1)\vec{A} = -1\langle a_1, a_2 \rangle = \langle -a_1, -a_2 \rangle = -\vec{A}$. If $k > 0$, then $k\vec{A}$ is in the direction of $\vec{A}$. If $k < 0$, then $k\vec{A}$ has the opposite direction of $\vec{A}$.

Example 1 Given $\vec{A}$ and $\vec{B}$ as sketched.
Sketch $2\vec{A} + \vec{B}$ and $\vec{B} - 2\vec{A}$.

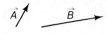

SOLUTION

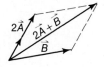

The algebra of scalar multiples of vectors is what you would expect it to be. For example, $2\vec{A} + 3\vec{A} = 5\vec{A}$ and $2(\vec{A} + \vec{B}) = 2\vec{A} + 2\vec{B}$. Furthermore, $k\vec{A}$ is $|k|$ times as long as $\vec{A}$.

Theorem 12–2 For any scalars r and s and vectors $\vec{A}$ and $\vec{B}$:

a. $1\vec{A} = \vec{A}$
b. $(-r)\vec{A} = -(r\vec{A})$
c. $(rs)\vec{A} = r(s\vec{A})$
d. $(r + s)\vec{A} = r\vec{A} + s\vec{A}$
e. $r(\vec{A} + \vec{B}) = r\vec{A} + r\vec{B}$
f. $\|r\vec{A}\| = |r|\|\vec{A}\|$

The proof of Theorem 12–2 is left to the exercises.

Example 2 Given $\vec{A} = \langle 5, 2 \rangle$, find a scalar k such that $\|k\vec{A}\| = 1$.

SOLUTION By Theorem 12–2, $\|k\vec{A}\| = |k|\|\vec{A}\|$. Since $\|k\vec{A}\| = 1$, then $|k| = \dfrac{1}{\|\vec{A}\|}$.

For $\vec{A} = \langle 5, 2 \rangle$, $\|\vec{A}\| = \sqrt{25 + 4} = \sqrt{29}$.

Hence $|k| = \dfrac{1}{\sqrt{29}}$ and $k = \pm \dfrac{1}{\sqrt{29}}$.

A **unit vector** is any vector of length 1.

Theorem 12–3 For any nonzero vector $\vec{A}$, the vector $\dfrac{1}{\|\vec{A}\|}(\vec{A})$ is a unit vector in the direction of $\vec{A}$.

The proof of Theorem 12–3 is left to the exercises.

Example 3 Find a unit vector $\vec{V}$ in the direction of $\vec{A} = \langle 3, 4 \rangle$.

SOLUTION The length of $\vec{A}$ is $\sqrt{3^2 + 4^2}$, or 5. Therefore $\frac{1}{5}\langle 3, 4 \rangle$ is a unit vector in the direction of $\vec{A} = \langle 3, 4 \rangle$.

$$\vec{V} = \langle \tfrac{3}{5}, \tfrac{4}{5} \rangle$$

If a unit vector $\vec{V}$ is represented by the standard arrow $\overrightarrow{OP}$, as in the figure, then $\vec{V} = \langle \cos \theta, \sin \theta \rangle$.

For the unit vector $\vec{V} = \langle \tfrac{3}{5}, \tfrac{4}{5} \rangle$ in example 3,

$$\theta = \arctan \tfrac{4}{3} \doteq 53.1°.$$

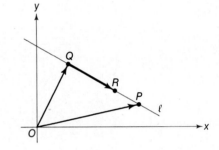

Vectors and scalars may be used together to write vector equations of lines. Let Q and R be two distinct points on a line ℓ. If P is any point on ℓ, then there is a real number t such that

$$\overrightarrow{QP} = t\overrightarrow{QR}.$$

By vector addition,

$$\overrightarrow{OP} = \overrightarrow{OQ} + \overrightarrow{QP}.$$

Hence, $\overrightarrow{OP} = \overrightarrow{OQ} + t\overrightarrow{QR}$.

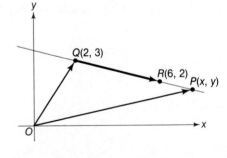

If $P = (x, y)$, $Q = (2, 3)$, and $R = (6, 2)$, as in the figure, then $\overrightarrow{QR} = \langle 4, -1 \rangle$ and

$$\langle x, y \rangle = \langle 2, 3 \rangle + t\langle 4, -1 \rangle,$$

which is called a **vector equation** of ℓ.

The equations $x = 2 + 4t$
$y = 3 + (-1)t$
obtained by equating components are called **parametric equations** of ℓ.

The arrow $\overrightarrow{QR}$ in the diagram above represents the vector $\langle 4, -1 \rangle$, and $\langle 4, -1 \rangle$ determines the direction of line ℓ just as the slope $-\tfrac{1}{4}$ does. This vector is called **a direction vector** of the line. The vector approach to lines has the advantage that it can be generalized to three dimensions as is done in section 12–6.

Example 4 Write a vector equation of the line ℓ through $Q(-2, 1)$ and $R(3, 4)$.

SOLUTION The arrow $\overrightarrow{QR}$ represents $\langle 5, 3 \rangle$, which is a direction vector for ℓ. Therefore a vector equation for ℓ is

$$\langle x, y \rangle = \langle -2, 1 \rangle + t \langle 5, 3 \rangle.$$

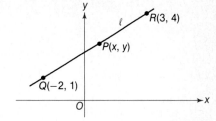

Example 5 The line ℓ has a vector equation $\vec{P} = \langle 1, -2 \rangle + t \langle 3, 3 \rangle$. Find three points on ℓ and sketch the line.

SOLUTION Let $\vec{P} = \langle x, y \rangle$.
For $t = 0$: $\langle x, y \rangle = \langle 1, -2 \rangle$.
For $t = 1$: $\langle x, y \rangle = \langle 4, 1 \rangle$.
For $t = -1$: $\langle x, y \rangle = \langle -2, -5 \rangle$.

Hence the points $(1, -2)$, $(4, 1)$, and $(-2, -5)$ all lie on ℓ.

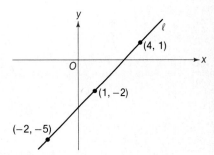

Every direction vector for a line is a scalar multiple of every other direction vector for the line. You can use any one that is convenient. Clearly then, a line does not have a unique vector equation. In example 5,

$$\vec{V} = \langle -2, -5 \rangle + k \langle 6, 6 \rangle \quad \text{and} \quad \vec{S} = \langle 4, 1 \rangle + r \langle -3, -3 \rangle$$

are also vector equations of the same line ℓ.

Exercises

A If $\vec{V_1} = \langle 2, -5 \rangle$ and $\vec{V_2} = \langle 1, 3 \rangle$, find $\vec{V}$.

1. $\vec{V} = 2\vec{V_1} - 3\vec{V_2}$
2. $\vec{V} = \vec{V_2} - \vec{V_1}$
3. $\vec{V} = \vec{V_1} + 3\vec{V_2}$
4. $\vec{V} = \vec{V_2} + (-7)\vec{V_1}$

Simplify each vector and represent $\vec{V}$ as an arrow in the coordinate plane.

5. $\vec{V} = 3\langle 1, -2 \rangle - 1\langle 2, -4 \rangle$
6. $\vec{V} = -\langle -1, 4 \rangle - 3\langle 2, -1 \rangle$

Copy $\vec{A}$ and $\vec{B}$ and sketch $2\vec{A} + \vec{B}$ and $2\vec{A} - \vec{B}$.

7.

8.

Find a unit vector in the direction of $\vec{A}$.

9. $\vec{A} = \langle 3, 4 \rangle$
10. $\vec{A} = \langle -1, 3 \rangle$
11. $\vec{A} = \langle 2, -3 \rangle$
12. $\vec{A} = \langle 12, -5 \rangle$

Find a direction vector for the line through points Q and R.

13. $Q = (1, 4)$
    $R = (-2, 3)$
14. $Q = (2, -1)$
    $R = (4, 5)$
15. $Q = (2, -1)$
    $R = (-1, 2)$
16. $Q = (1, 3)$
    $R = (-4, 0)$

B Find a vector in the direction of $\vec{A}$ that has length 2.

17. $\vec{A} = \langle 1, 1 \rangle$
18. $\vec{A} = \langle 1, -1 \rangle$
19. $\vec{A} = \langle 2, -1 \rangle$
20. $\vec{A} = \langle 3, 4 \rangle$

For what values of the scalar t is $\langle x, y \rangle = \langle 1, 2 \rangle$?

21. $\langle x, y \rangle = \langle 1, 2 \rangle + t \langle -1, 3 \rangle$
22. $\langle x, y \rangle = \langle 0, 0 \rangle + t \langle 1, 2 \rangle$

Write a vector equation of the line ℓ through points Q and R.

23. $Q(1, 2), R(5, -1)$
24. $Q(-1, 4), R(2, 2)$
25. $Q(3, -1), R(-1, 2)$
26. $Q(1, 5), R(2, 0)$
27. $Q(0, 2), R(1, 2)$
28. $Q(2, 0), R(1, -4)$

Find points corresponding to $t = 0$, $t = 1$, and $t = -2$ for the given vector equation of a line. Sketch the line.

29. $\langle x, y \rangle = \langle 3, 1 \rangle + t \langle 11, -8 \rangle$
30. $\langle x, y \rangle = \langle 5, 7 \rangle + t \langle 0, -1 \rangle$

C 31. Prove Theorem 12–2c.
32. Prove Theorem 12–2d.
33. Prove Theorem 12–2e.
34. Prove Theorem 12–2f.
35. Prove Theorem 12–3.

Find the point of intersection of the given lines.

36. $\ell_1: \langle x, y \rangle = \langle 1, 2 \rangle + t \langle -1, 3 \rangle$
    $\ell_2: \langle x, y \rangle = \langle 3, -2 \rangle + s \langle 5, 1 \rangle$
37. $\ell_1: x = 1 + t, y = 3 - 2t$
    $\ell_2: x = -2 + 3s, y = 5 - s$

38. If line ℓ has slope m, find a direction vector for ℓ in terms of m.
39. $\vec{A} = \langle a, b \rangle$ is a direction vector for ℓ_1. Lines ℓ_1 and ℓ_2 are perpendicular. Find a direction vector for line ℓ_2 in terms of a and b.

12-4 Vectors in Physics

Vectors play an important role in many applications. Each of the examples below begins by representing each relevant quantity by a vector, and then determines the components of each vector.

Example 1 A motorcycle rider leaves a point A going straight toward point S, which is 4 miles east and 4 miles north. At the same time a bicycle rider leaves A going toward T, which is 4 miles west and 2 miles north. If the bicycle goes 12 mph and the motorcycle 40 mph, what is the rate at which the two riders are separating?

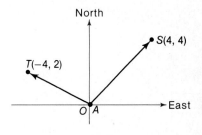

SOLUTION Let $\vec{M}$ represent the velocity of the motorcycle. Then $\vec{M}$ is in the direction of $\vec{AS}$ and $\|\vec{M}\|$ is 40. Since $\langle \cos 45°, \sin 45° \rangle$ is a unit vector in the direction of $\vec{M}$,

$$\vec{M} = 40\langle \cos 45°, \sin 45° \rangle$$
$$\doteq \langle 28.3, 28.3 \rangle.$$

Let $\vec{B}$ represent the velocity of the bicycle. Then $\|\vec{B}\| = 12$ and $\vec{B}$ is in the direction of $\vec{AT} = \langle -4, 2 \rangle$. Since the length of $\langle -4, 2 \rangle$ is $\sqrt{16 + 4} = 2\sqrt{5}$, the unit vector in the direction of $\vec{B}$ is $\dfrac{1}{2\sqrt{5}} \langle -4, 2 \rangle$. Therefore,

$$\vec{B} = 12 \left(\dfrac{1}{2\sqrt{5}} \langle -4, 2 \rangle \right) \doteq \langle -10.7, 5.4 \rangle.$$

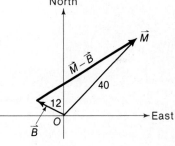

The rate at which the two riders are separating is $\|\vec{M} - \vec{B}\|$.

$$\vec{M} - \vec{B} \doteq \langle 28.3 - (-10.7), 28.3 - 5.4 \rangle$$
$$= \langle 39.0, 22.9 \rangle$$
$$\|\vec{M} - \vec{B}\| \doteq \sqrt{39.0^2 + 22.9^2} \doteq 45.2$$

ANSWER The riders are separating at approximately 45.2 miles per hour.

Example 2 Two forces act on an object W. Let $\vec{F_1}$ represent a force of 2 pounds acting at 60° to the horizontal, and let $\vec{F_2}$ represent a force of 3 pounds acting straight upward. Find a force $\vec{F_3}$ so that the system will be in equilibrium, which means that

$$\vec{F_1} + \vec{F_2} + \vec{F_3} = \vec{0}.$$

(cont. on p. 438)

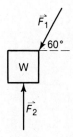

SOLUTION Represent W by a point at the origin, and sketch standard representatives of $\vec{F}_1$ and $\vec{F}_2$. Then the sum $\vec{F}_1 + \vec{F}_2$ is the resultant force acting on W. To achieve equilibrium, $\vec{F}_3$ must equal $-(\vec{F}_1 + \vec{F}_2)$.

Let $\vec{F}_2 = \langle 0, 3 \rangle$.

Let $\vec{F}_1 = 2\langle \cos 240°, \sin 240° \rangle$

$\doteq \langle -1, -1.73 \rangle$.

Then $\vec{F}_1 + \vec{F}_2 \doteq \langle -1, 1.27 \rangle$ and

$\vec{F}_3 \doteq \langle 1, -1.27 \rangle$.

$\|\vec{F}_3\| \doteq 1.61$ pounds

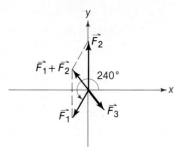

Exercises

A 1. An ocean liner and a tug leave a port at the same time. The liner moves due east at 15 knots and the tug northeast at 5 knots.
   a. Express the velocity of the liner as a vector $\vec{L}$.
   b. Express the velocity of the tug as a vector $\vec{T}$.
   c. Find $\vec{L} - \vec{T}$.
   d. Find $\|\vec{L} - \vec{T}\|$, the rate at which the two vessels are separating.

2. An ocean liner and a tug leave a port at the same time. The liner moves south at 12 knots and the tug travels 30° east of north at 4 knots.
   a. Express the velocity of the liner as a vector $\vec{L}$.
   b. Express the velocity of the tug as a vector $\vec{T}$.
   c. Find $\vec{L} - \vec{T}$.
   d. Find $\|\vec{L} - \vec{T}\|$, the rate at which the two vessels are separating.

Two forces $\vec{F}_1$ and $\vec{F}_2$ act on a weight W as pictured in exercises 3–8. Sketch to correct scale a force $\vec{F}_3$ such that $\vec{F}_1 + \vec{F}_2 + \vec{F}_3 = \vec{0}$.

3.

4.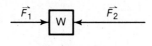

5.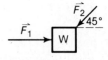

6.

7.

8.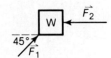

Find $\vec{F}_3$ such that $\vec{F}_1 + \vec{F}_2 + \vec{F}_3 = \vec{O}$ given

B 9. The figure for exercise 3 with $\|\vec{F}_1\| = \|\vec{F}_2\| = 2$.
10. The figure for exercise 4 with $\|\vec{F}_1\| = 1$ and $\|\vec{F}_2\| = 3$.
11. The figure for exercise 5 with $\|\vec{F}_1\| = \|\vec{F}_2\| = 1$.
12. The figure for exercise 6 with $\|\vec{F}_1\| = 1$ and $\|\vec{F}_2\| = 2$.

C 13. A ship sails in a direction 30° north of east at 12 mph. A man running south on the deck of the ship goes 6 mph relative to the deck. How fast is the man moving relative to the water?

14. Two cars set out on straight roads from point A at the same time. The first car travels 40 mph toward point B, which is 20 miles west and 10 miles north of A. The second car travels 60 mph toward point C, which is 10 miles east and 5 miles north of A. How far apart will the cars be after 10 minutes?

12–5 Space

On any plane you can construct a Cartesian coordinate system with two perpendicular axes, commonly called the x- and y-axes, and use the system to locate a point in the plane. Likewise, in three-dimensional space (often called 3-space) you can construct a coordinate system with three mutually perpendicular axes, commonly called the x-, y-, and z-axes. Normally we picture the z-axis pointing straight up, the y-axis pointing to the right, and the x-axis coming out of the paper toward you, as in the figure. The three axes intersect at a point called the origin.

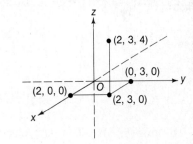

A point in 3-space may then be located by an ordered triple (x, y, z) of real numbers. For example, to locate (2, 3, 4) go 2 units from the origin in the positive x-direction, 3 units to the right to (2, 3, 0), and 4 units up.

The three coordinate axes determine three planes, the xy-plane, the xz-plane, and the yz-plane. For every point in the xz-plane, the y-coordinate is 0. Hence $y = 0$ is an equation of the xz-plane. The equation $y = 4$ describes all points with y-coordinate 4, which is a plane parallel to the xz-plane.

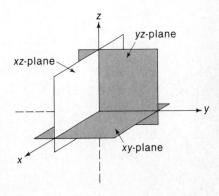

It will be shown in section 12–8 that $Ax + By + Cz = D$ is an equation of a plane as long as A, B, and C are not all zero.

Example 1 Find the point of intersection of the plane $2x + 3y + 4z = 12$ with each of the coordinate axes and sketch the plane.

SOLUTION Along the z-axis x and y are zero. Hence $4z = 12$ and $z = 3$. So $(0, 0, 3)$ is the point of intersection. Along the y-axis x and z are zero, so that $3y = 12$ and $y = 4$. The point of intersection is $(0, 4, 0)$. Similarly, $(6, 0, 0)$ is the point of intersection with the x-axis.

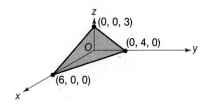

To find the distance between two points in 3-space you can use the Pythagorean theorem and the figure shown below.

Suppose $A = (x_1, y_1, z_1)$ and $B = (x_2, y_2, z_2)$.

Let C be the point (x_2, y_2, z_1) that is directly above or below B at the same height as A. Then $\triangle ABC$ is a right triangle with $\overline{AB}$ as hypotenuse. Clearly,

$$BC = \sqrt{(z_2 - z_1)^2},$$

and by the Pythagorean theorem,

$$AC = \sqrt{(x_2 - x_1)^2 + (y_2 - y_1)^2}.$$

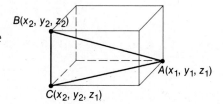

Since $AB^2 = BC^2 + AC^2$, this process establishes the distance formula in the following theorem.

Theorem 12–4 **Distance Formula in 3-Space**

The distance between $A(x_1, y_1, z_1)$ and $B(x_2, y_2, z_2)$ is

$$AB = \sqrt{(x_2 - x_1)^2 + (y_2 - y_1)^2 + (z_2 - z_1)^2}.$$

Example 2 Find the distance d between $A(3, 0, 1)$ and $B(1, 4, 3)$.

SOLUTION Use Theorem 12–4:

$$d = \sqrt{(1 - 3)^2 + (4 - 0)^2 + (3 - 1)^2}$$
$$= \sqrt{24}$$
$$= 2\sqrt{6}.$$

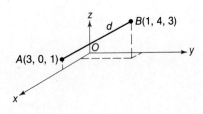

The distance formula can be used to find an equation of a sphere. Since every point on a sphere is the same distance, the radius r, from the center (h, k, ℓ), an equation for a sphere has the form

$$(x - h)^2 + (y - k)^2 + (z - \ell)^2 = r^2.$$

Example 3 Show that $x^2 + y^2 - 2y + z^2 = 3$ is the equation of a sphere. Find its center and its radius.

SOLUTION Complete the square on the y-terms:

$$x^2 + (y^2 - 2y + 1) + z^2 = 3 + 1.$$
$$x^2 + (y - 1)^2 + z^2 = 4$$

So $(0, 1, 0)$ is the center and the radius is 2.

A plane curve such as the ellipse $\frac{(x-1)^2}{1} + \frac{(y-2)^2}{4} = 1$ describes a type of cylinder in 3-space. Sketch the ellipse in the xy-plane. Then any point above, on, or below that ellipse is part of a surface called an **elliptic cylinder**.

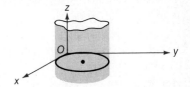

Another 3-space surface is the **paraboloid**, which is generated by rotating a parabola about its axis of symmetry. The graph of the paraboloid $z = x^2 + y^2$ is shown. For any positive value of z, the cross section of the paraboloid is a circle. In the yz-plane, where $x = 0$, the cross section is the parabola $z = y^2$.

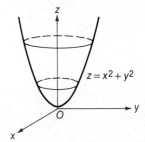

The graph of $z = \frac{x^2}{a^2} + \frac{y^2}{b^2}$ is also a paraboloid, but each cross section parallel to the xy-plane is an ellipse, so the figure is called an **elliptic paraboloid**. These surfaces are examples of **quadric surfaces**.

Exercises

A Sketch the points A and B in a coordinate system and find the distance AB.

1. $A(1, 0, 1)$
   $B(0, 1, 0)$

2. $A(1, 0, 2)$
   $B(1, 1, 0)$

3. $A(1, 2, -1)$
   $B(-1, 1, 1)$

4. $A(1, 2, 3)$
   $B(-1, 3, 0)$

5. $A(2, 1, 1)$
   $B(-1, 2, -1)$

6. $A(0, 1, -2)$
   $B(1, 3, 1)$

Sketch the plane with the given equation.

7. $y = 1$
8. $z = 2$
9. $x = 3$
10. $y = -1$
11. $z = y$
12. $z = x$

Find the points of intersection of the given plane with the coordinate axes and sketch the plane.

13. $2x + y + z = 4$
14. $x + 2y + 3z = 6$
15. $3x + 2y + 4z = 12$
16. $x + y + z = 3$

Find the center and the radius of each sphere.

17. $x^2 + y^2 + z^2 - 2y = 8$
18. $x^2 + y^2 + z^2 + 2y - 4z = -1$
19. $x^2 + y^2 + z^2 - 6x = -8$
20. $x^2 + y^2 + z^2 - 2x + 4z = -1$

B Write an equation of the plane through point P that is parallel to the plane $z = 4$.

21. $P = (1, 1, 1)$
22. $P = (1, 2, -3)$

Determine the coordinates of three different points on the line of intersection of the two given planes and sketch that line on a coordinate system.

23. $x = 2$
    $y = 3$

24. $y = 2$
    $z = 1$

25. $x = y$
    $x = 3$

26. $y = 2x$
    $y = 6$

Find three different points on the axis of symmetry of the right circular cylinder.

27. $(x - 1)^2 + y^2 = 4$
28. $(y + 2)^2 + z^2 = 1$

Sketch each paraboloid and find its intersections with the plane $z = 4$.

29. $z = x^2 + y^2 + 1$
30. $z = 4x^2 + 4y^2$

C Graph each 3-space cylinder.

31. $x^2 - y^2 = 1$
32. $x^2 + 4y^2 = 4$

12–6 Vectors and Lines in 3-Space

Everything you learned about two-dimensional vectors has a three-dimensional analogue. A three-dimensional vector is an ordered triple of real numbers a_1, a_2, a_3, written $\langle a_1, a_2, a_3 \rangle$, that satisfies certain properties of equality, addition, and scalar multiplication.

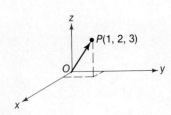

Definition For any vectors $\vec{A} = \langle a_1, a_2, a_3 \rangle$ and $\vec{B} = \langle b_1, b_2, b_3 \rangle$ and for a scalar k:

a. Equality $\vec{A} = \vec{B}$ if and only if $a_1 = b_1$, $a_2 = b_2$, and $a_3 = b_3$

b. Addition $\vec{A} + \vec{B} = \langle a_1 + b_1, a_2 + b_2, a_3 + b_3 \rangle$

c. Scalar Multiplication $k\vec{A} = \langle ka_1, ka_2, ka_3 \rangle$

d. Subtraction $\vec{A} - \vec{B} = \vec{A} + (-1)\vec{B}$

e. Vector Length $\|\vec{A}\| = \sqrt{a_1^2 + a_2^2 + a_3^2}$

If $\vec{A}$ and $\vec{B}$ are nonzero vectors that are not scalar multiples of one another, their standard representatives determine a unique plane. If you consider this page to be such a plane, you can sketch vectors $\vec{A}$ and $\vec{B}$ as in the figure. Then vectors such as $2\vec{A} + \vec{B}$ and $\vec{B} - \vec{A}$ may be sketched just as in 2-space.

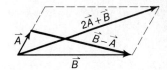

Three-dimensional vectors obey the same algebraic rules as two-dimensional vectors. Theorems 12–1, 12–2, and 12–3 remain valid, and the proofs are the same, with one more dimension.

Example 1 Given $\vec{A} = \langle 1, 2, -4 \rangle$ and $\vec{B} = \langle 3, 0, -7 \rangle$. Find $\|\vec{A} - 5\vec{B}\|$.

SOLUTION $\vec{A} - 5\vec{B} = \langle 1, 2, -4 \rangle - 5\langle 3, 0, -7 \rangle$

$= \langle 1, 2, -4 \rangle - \langle 15, 0, -35 \rangle = \langle -14, 2, 31 \rangle$

$\|\vec{A} - 5\vec{B}\| = \sqrt{(-14)^2 + (2)^2 + (31)^2}$

$= \sqrt{1161} \doteq 34.07$

Example 2 Given points $Q = (1, 3, -5)$ and $R = (0, -1, 2)$, find a unit vector in the direction of $\overrightarrow{QR}$.

SOLUTION The points Q and R and the origin O determine a plane. The arrow $\overrightarrow{QR} = \overrightarrow{OR} - \overrightarrow{OQ}$ represents the vector $\langle -1, -4, 7 \rangle$ whose length is $\sqrt{1 + 16 + 49}$, or $\sqrt{66}$. So $\left\langle \dfrac{-1}{\sqrt{66}}, \dfrac{-4}{\sqrt{66}}, \dfrac{7}{\sqrt{66}} \right\rangle$ is a unit vector in the direction of $\overrightarrow{QR}$.

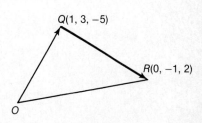

12–6: *Vectors and Lines in 3-Space* **443**

Vectors are particularly useful for writing equations of lines in 3-space. Suppose Q and R are two distinct points on a line ℓ and O is the origin. Just as in section 12–3, let P be any point on ℓ. There is a real number t such that

$$\overrightarrow{QP} = t\overrightarrow{QR}.$$

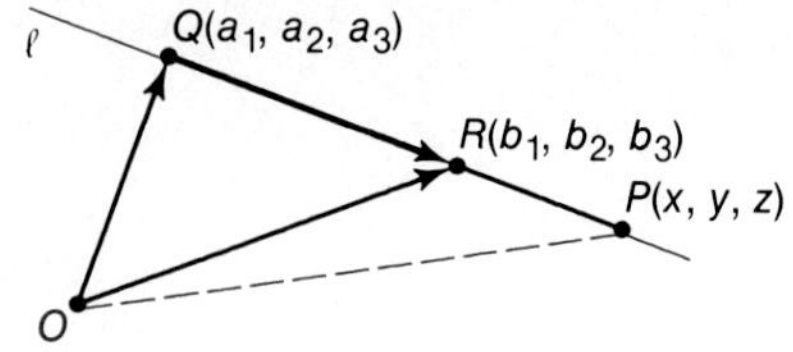

By vector addition,

$$\overrightarrow{OP} = \overrightarrow{OQ} + \overrightarrow{QP} = \overrightarrow{OQ} + t\overrightarrow{QR}.$$

If points P, Q, and R have coordinates as in the figure, then $\overrightarrow{OQ} = \langle a_1, a_2, a_3 \rangle$, and $\overrightarrow{QR} = \langle b_1 - a_1, b_2 - a_2, b_3 - a_3 \rangle$ is a direction vector for ℓ.

Substitution in $\overrightarrow{OP} = \overrightarrow{OQ} + t\overrightarrow{QR}$ establishes the following useful theorem.

Theorem 12–5 A vector equation of the line through the distinct points (a_1, a_2, a_3) and (b_1, b_2, b_3) is

$$\langle x, y, z \rangle = \langle a_1, a_2, a_3 \rangle + t\langle b_1 - a_1, b_2 - a_2, b_3 - a_3 \rangle.$$

The parametric form of the equation in Theorem 12–5 is

$$x = a_1 + t(b_1 - a_1)$$
$$y = a_2 + t(b_2 - a_2)$$
$$z = a_3 + t(b_3 - a_3).$$

The line joining points $Q(1, 3, -5)$ and $R(0, -1, 2)$ in example 2 has $\langle -1, -4, 7 \rangle$ as a direction vector. So an equation of this line is

$$\langle x, y, z \rangle = \langle 1, 3, -5 \rangle + t\langle -1, -4, 7 \rangle,$$

or in parametric form,

$$x = 1 - t$$
$$y = 3 - 4t$$
$$z = -5 + 7t.$$

To write an equation of a line in 3-space you need the coordinates of a point on the line and a direction vector for the line. Since every direction vector for a line is a scalar multiple of every other direction vector, you can use any one of them that is convenient.

Example 3 Determine an equation of the line ℓ that contains $(1, 3, 2)$ and $(2, 1, 0)$. Find the intersection of ℓ with the yz-plane.

SOLUTION Let $Q = (1, 3, 2)$, $R = (2, 1, 0)$, and let (x, y, z) be a point on ℓ. Then $\langle 2 - 1, 1 - 3, 0 - 2 \rangle$, or $\langle 1, -2, -2 \rangle$, is a direction vector of ℓ.

(cont. on p. 445)

A vector equation of line ℓ is $\langle x, y, z \rangle = \langle 1, 3, 2 \rangle + t\langle 1, -2, -2 \rangle$. A parametric form of this equation is

$$x = 1 + t$$
$$y = 3 - 2t$$
$$z = 2 - 2t.$$

The yz-plane has equation $x = 0$, so at the point of intersection $1 + t = 0$. For $t = -1$,

$$x = 0$$
$$y = 3 - 2(-1) = 5$$
$$z = 2 - 2(-1) = 4.$$

The point of intersection is $(0, 5, 4)$.

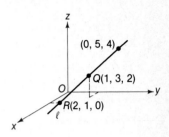

Example 4 Find the point of intersection of the lines

$\ell_1: \langle x, y, z \rangle = \langle 1, 1, 1 \rangle + t\langle 1, -1, 2 \rangle$ and
$\ell_2: \langle x, y, z \rangle = \langle -1, 4, -4 \rangle + s\langle 0, 1, -1 \rangle$.

SOLUTION For ℓ_1: $\langle x, y, z \rangle = \langle 1 + t, 1 - t, 1 + 2t \rangle$.
For ℓ_2: $\langle x, y, z \rangle = \langle -1, 4 + s, -4 - s \rangle$.

At the point of intersection the coordinates must be identical. So you need to solve the following system of three equations in two variables, t and s.

$$1 + t = -1$$
$$1 - t = 4 + s$$
$$1 + 2t = -4 - s$$

From the first equation, $t = -2$, and by substituting in the second equation, $s = -1$. You can verify that these values satisfy the third equation.

By substitution in the equation for ℓ_1 or ℓ_2 you find that the point of intersection is $(-1, 3, -3)$.

Exercises

A Find $2\vec{A} - 3\vec{B}$.

1. $\vec{A} = \langle 1, 3, -2 \rangle$, $\vec{B} = \langle 0, -1, 2 \rangle$
2. $\vec{A} = \langle -1, 1, 0 \rangle$, $\vec{B} = \langle 2, -2, 3 \rangle$

Find a vector in the direction of $\overrightarrow{QR}$.

3. $Q(1, 3, -1)$, $R(2, -1, 4)$
4. $Q(-1, 2, 3)$, $R(1, 2, -3)$

Show that point P lies on line ℓ.

5. $\ell: \langle x, y, z \rangle = \langle 2, -1, 3 \rangle + t\langle 1, 4, -2 \rangle$; $P(4, 7, -1)$
6. $\ell: \langle x, y, z \rangle = \langle -2, 0, 5 \rangle + t\langle -1, 1, 3 \rangle$; $P(1, -3, -4)$

Show that point P does not lie on line ℓ.

7. $\ell: x = 1 + 3t, y = 2, z = 1 - 2t; P(1, 2, 0)$
8. $\ell: x = -1, y = 2k, z = 2 + 3k; P(-1, 2, 8)$

Find a unit vector in the direction of $\overrightarrow{QR}$.

9. $Q(-1, 1, 0); R(2, 1, 3)$
10. $Q(2, -1, 1); R(1, 0, 1)$

For what $k \in \mathbf{R}$ are $\vec{V_1}$ and $\vec{V_2}$ direction vectors for the same line?

11. $\vec{V_1} = \langle 1, 2, 3 \rangle; \vec{V_2} = \langle -3, -6, k \rangle$
12. $\vec{V_1} = \langle 2, 0, k \rangle; \vec{V_2} = \langle 1, 0, -1 \rangle$

Write a vector equation of the line through points Q and R.

13. $Q(1, 2, -5), R(-1, 0, 3)$
14. $Q(-2, 1, 7), R(0, -4, 2)$
15. $Q(-3, 0, 2), R(-3, 1, -4)$
16. $Q(0, 0, -3), R(-1, 2, 1)$
17. $Q(1, 2, -12), R(-5, 6, 1)$
18. $Q(5, 7, 12), R(-5, 1, -6)$

Find the coordinates of the point of intersection of ℓ with the yz-plane.

19. $\ell: \langle x, y, z \rangle = \langle 1, -2, 3 \rangle + t\langle -1, 1, 2 \rangle$
20. $\ell: \langle x, y, z \rangle = \langle 3, 0, 1 \rangle + t\langle 1, -1, 0 \rangle$
21. $\ell: \langle x, y, z \rangle = \langle -1, 1, 2 \rangle + t\langle 2, 0, 1 \rangle$
22. $\ell: \langle x, y, z \rangle = \langle 4, 0, -1 \rangle + t\langle 1, 2, 3 \rangle$

B Find the point of intersection of the two given lines.

23. $\ell_1: \langle x, y, z \rangle = \langle 1, 2, -3 \rangle + t\langle 1, -\frac{1}{2}, 0 \rangle$
    $\ell_2: \langle x, y, z \rangle = \langle -1, 5, -4 \rangle + s\langle 2, 1, -1 \rangle$
24. $\ell_1: \langle x, y, z \rangle = \langle -1, 0, 2 \rangle + s\langle 3, -1, 2 \rangle$
    $\ell_2: \langle x, y, z \rangle = \langle 5, 9, 11 \rangle + t\langle -1, 4, 1 \rangle$

Show that ℓ does not intersect the yz-plane.

25. $\ell: \langle x, y, z \rangle = \langle 1, 0, 2 \rangle + t\langle 0, -1, 3 \rangle$
26. $\ell: \langle x, y, z \rangle = \langle -1, 1, 1 \rangle + t\langle 0, 2, -1 \rangle$

C 27. Prove Theorem 12–1 for three-dimensional vectors $\vec{A}, \vec{B},$ and $\vec{C}$.

28. Prove Theorem 12–2 for three-dimensional vectors $\vec{A}$ and $\vec{B}$ and scalars r and s.

Suppose that an object of mass M is placed at the origin and subjected to forces $\vec{F_1}, \vec{F_2},$ and $\vec{F_3}$. Find a force $\vec{F_4}$ such that the system will be in equilibrium; that is, the vector sum of the forces is zero.

29.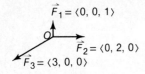
30.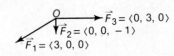

12–7 The Dot Product

In section 12–3 you learned how to find the product of a scalar and a vector. Now we will define a product of two vectors and use this idea to determine the angle between two vectors. In the next section this idea is applied to find equations of planes in 3-space.

Definition For any two vectors $\vec{A} = \langle a_1, a_2, a_3 \rangle$ and $\vec{B} = \langle b_1, b_2, b_3 \rangle$ the **dot product** of $\vec{A}$ and $\vec{B}$, written $\vec{A} \cdot \vec{B}$, is
$$\vec{A} \cdot \vec{B} = a_1 b_1 + a_2 b_2 + a_3 b_3.$$

As you would expect, in 2-space the dot product of $\vec{A}$ and $\vec{B}$ is
$$\vec{A} \cdot \vec{B} = a_1 b_1 + a_2 b_2.$$

Notice that the dot product of two vectors is a scalar (a real number) and not a vector.

Example 1 If $\vec{A} = \langle 1, 2, -1 \rangle$ and $\vec{B} = \langle 2, -1, 3 \rangle$, find $\vec{A} \cdot \vec{B}$.

SOLUTION $\vec{A} \cdot \vec{B} = (1)(2) + (2)(-1) + (-1)(3) = -3$

Definition The angle between two nonzero vectors $\vec{A}$ and $\vec{B}$ is the smallest angle ϕ between their standard representatives. If $\vec{A} = k\vec{B}$, then $\phi = 0$ if $k > 0$ and $\phi = 180°$ if $k < 0$.

Two vectors are **orthogonal** (perpendicular) if the angle between them is 90°.

For points $A(a_1, a_2)$ and $B(b_1, b_2)$, suppose that $\overline{OA} \perp \overline{OB}$ so that by definition the angle between $\vec{A}$ and $\vec{B}$ is 90°. The vectors are orthogonal. Since $\overline{OA} \perp \overline{OB}$, the product of the slopes of the segments is -1. Thus
$$\frac{a_2}{a_1} \cdot \frac{b_2}{b_1} = -1.$$
$$a_2 b_2 = -a_1 b_1$$
$$a_1 b_1 + a_2 b_2 = 0$$

Hence $\vec{A} \cdot \vec{B} = 0$. A similar result holds for three-dimensional vectors, but to prove this it is useful to establish the following theorem first.

Theorem 12–6 If ϕ is the angle between two nonzero vectors $\vec{A}$ and $\vec{B}$, then
$$\vec{A} \cdot \vec{B} = \|\vec{A}\| \, \|\vec{B}\| \cos \phi.$$

(cont. on p. 448)

PROOF Let $\vec{A} = \langle a_1, a_2, a_3 \rangle$ and $\vec{B} = \langle b_1, b_2, b_3 \rangle$.

By the Law of Cosines applied to $\triangle OAB$,
$$\overline{AB}^2 = \overline{OA}^2 + \overline{OB}^2 - 2(\overline{OA})(\overline{OB}) \cos \phi.$$
In vector form, this is
$$\|\vec{B} - \vec{A}\|^2 = \|\vec{A}\|^2 + \|\vec{B}\|^2 - 2\|\vec{A}\|\|\vec{B}\| \cos \phi$$
Since $\vec{B} - \vec{A} = \langle b_1 - a_1, b_2 - a_2, b_3 - a_3 \rangle$,
$$(b_1 - a_1)^2 + (b_2 - a_2)^2 + (b_3 - a_3)^2 =$$
$$(a_1^2 + a_2^2 + a_3^2) + (b_1^2 + b_2^2 + b_3^2) - 2\|\vec{A}\|\|\vec{B}\| \cos \phi.$$
Squaring and subtracting equal terms from both sides yields
$$-2a_1b_1 - 2a_2b_2 - 2a_3b_3 = -2\|\vec{A}\|\|\vec{B}\| \cos \phi.$$
$$a_1b_1 + a_2b_2 + a_3b_3 = \|\vec{A}\|\|\vec{B}\| \cos \phi$$
$$\vec{A} \cdot \vec{B} = \|\vec{A}\|\|\vec{B}\| \cos \phi$$

Theorem 12–6 may be used to determine the angle between two vectors.

Example 2 Find the angle between $\vec{A} = \langle -1, 2, 2 \rangle$ and $\vec{B} = \langle 1, 2, 1 \rangle$.

SOLUTION Substitute in $\vec{A} \cdot \vec{B} = \|\vec{A}\|\|\vec{B}\| \cos \phi$ to get
$$-1 + 4 + 2 = 3\sqrt{6} \cos \phi.$$
$$\cos \phi = \frac{5}{3\sqrt{6}} \doteq 0.6804$$
Therefore $\phi \doteq 47.1°$, or 0.82 radian.

Example 3 If $A = (-1, 2, 1)$ and $B = (2, 0, 0)$ find $\angle AOB$.

SOLUTION Let $\vec{A} = \langle -1, 2, 1 \rangle$ and $\vec{B} = \langle 2, 0, 0 \rangle$.
Then $\vec{A} \cdot \vec{B} = -2$, $\|\vec{A}\| = \sqrt{6}$, $\|\vec{B}\| = 2$.
Substituting into Theorem 12–6 gives
$$-2 = 2\sqrt{6} \cos \phi,$$
$$\cos \phi = -\frac{1}{\sqrt{6}} \doteq -0.4082, \text{ or}$$
$$\phi \doteq 114.1°.$$

The angle between two lines that intersect in 3-space is the angle between two of their respective direction vectors.

Example 4 Given $\ell_1: \langle x, y, z \rangle = \langle 2, 1, 0 \rangle + t\langle -1, 2, 1 \rangle$ and
$\ell_2: \langle x, y, z \rangle = \langle 2, 1, 0 \rangle + s\langle -1, -1, 1 \rangle$.
Find the angle of intersection of lines ℓ_1 and ℓ_2.

SOLUTION $\vec{A} = \langle -1, 2, 1 \rangle$ is a direction vector for ℓ_1 and $\vec{B} = \langle -1, -1, 1 \rangle$ is a direction vector for ℓ_2.

$$\vec{A} \cdot \vec{B} = 1 - 2 + 1 = 0$$

Therefore $\|\vec{A}\| \|\vec{B}\| \cos \phi = 0$.

Since $\vec{A}$ and $\vec{B}$ are nonzero vectors, $\cos \phi = 0$ and $\phi = 90°$.

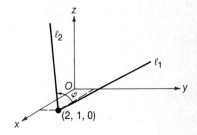

Example 4 suggests the following corollary, whose proof follows immediately from Theorem 12–6 and is left to the exercises.

Corollary 12–6 Two nonzero vectors are orthogonal if and only if their dot product is zero.

Example 5 Show that $\ell_1: \langle x, y, z \rangle = \langle 2 + t, -1 + 2t, -3t \rangle$ and
$\ell_2: \langle x, y, z \rangle = \langle 5 + 3k, -1, 1 + k \rangle$ are perpendicular.

SOLUTION Find the point of intersection of the two lines by equating x-, y-, and z-coordinates and solving for t and k. The result is $t = 0$ and $k = -1$. The point of intersection is $(2, -1, 0)$.

In the vector equation for ℓ_1, the coefficients of t determine a direction vector for ℓ_1 to be $\langle 1, 2, -3 \rangle$. Similarly, for ℓ_2 a direction vector is $\langle 3, 0, 1 \rangle$. Since $\langle 1, 2, -3 \rangle \cdot \langle 3, 0, 1 \rangle = 0$, the vectors are orthogonal and the two lines are perpendicular.

Exercises

A Find $\vec{A} \cdot \vec{B}$.

1. $\vec{A} = \langle 1, 2 \rangle, \vec{B} = \langle -1, 1 \rangle$
2. $\vec{A} = \langle 0, 1 \rangle, \vec{B} = \langle -1, 0 \rangle$
3. $\vec{A} = \langle 3, 1 \rangle, \vec{B} = \langle -2, 4 \rangle$
4. $\vec{A} = \langle 1, -1 \rangle, \vec{B} = \langle 3, -1 \rangle$
5. $\vec{A} = \langle 1, 0, 5 \rangle, \vec{B} = \langle 0, 1, 2 \rangle$
6. $\vec{A} = \langle -1, 1, 0 \rangle, \vec{B} = \langle 0, -1, 1 \rangle$
7. $\vec{A} = \langle 0, 1, -3 \rangle, \vec{B} = \langle 1, -1, 1 \rangle$
8. $\vec{A} = \langle 1, 1, 1 \rangle, \vec{B} = \langle 1, 1, -1 \rangle$

Use $\vec{A} \cdot \vec{B} = \|\vec{A}\| \|\vec{B}\| \cos \phi$ to find the angle ϕ between $\vec{A}$ and $\vec{B}$.

9. $\vec{A} = \langle 1, 0 \rangle, \vec{B} = \langle 0, 1 \rangle$
10. $\vec{A} = \langle 1, 1 \rangle, \vec{B} = \langle 1, 0 \rangle$
11. $\vec{A} = \langle 1, 0, 0 \rangle, \vec{B} = \langle 0, 1, 0 \rangle$
12. $\vec{A} = \langle 0, 0, 1 \rangle, \vec{B} = \langle 0, -1, 0 \rangle$
13. $\vec{A} = \langle 3, -5, 1 \rangle, \vec{B} = \langle 1, -1, 4 \rangle$
14. $\vec{A} = \langle 7, 2, -1 \rangle, \vec{B} = \langle 2, 5, 3 \rangle$

15. $\vec{A} = \langle -1, 0, 4 \rangle$, $\vec{B} = \langle 5, 2, -1 \rangle$
16. $\vec{A} = \langle 3, -2, 4 \rangle$, $\vec{B} = \langle 1, -4, 0 \rangle$

Show that lines ℓ_1 and ℓ_2 are perpendicular.

17. $\ell_1: \langle x, y, z \rangle = \langle 1, 2, 3 \rangle + t\langle 1, 2, 1 \rangle$
    $\ell_2: \langle x, y, z \rangle = \langle 1, 2, 3 \rangle + s\langle -3, 0, 3 \rangle$

18. $\ell_2: \langle x, y, z \rangle = \langle 1, 5, 10 \rangle + t\langle 1, -1, 1 \rangle$
    $\ell_2: \langle x, y, z \rangle = \langle 1, 5, 10 \rangle + s\langle 2, 1, -1 \rangle$

B Determine whether ℓ_1 and ℓ_2 intersect. If they do, find the angle of intersection.

19. $\ell_1: \langle x, y, z \rangle = \langle 1, -1, 1 \rangle + t\langle 2, 1, 0 \rangle$
    $\ell_2: \langle x, y, z \rangle = \langle 4, -2, 3 \rangle + s\langle -1, 2, -2 \rangle$

20. $\ell_1: \langle x, y, z \rangle = \langle 1, -1, 1 \rangle + t\langle 0, 2, -1 \rangle$
    $\ell_2: \langle x, y, z \rangle = \langle -1, 1, 0 \rangle + s\langle 1, -1, 2 \rangle$

21. $\ell_1: x = 1 + t, y = 3 - 2t, z = 2t$
    $\ell_2: x = -1 + k, y = 1, z = 2 - k$

22. $\ell_1: x = -3 + 2t, y = 1 - t, z = 2 + t$
    $\ell_2: x = 6 - s, y = 5 + 2s, z = -3 - 3s$

23. Prove Corollary 12–6.

24. Find three different vectors each of which is orthogonal to $\langle 1, -1, 2 \rangle$.

Prove each of the following for three-dimensional vectors $\vec{A}$, $\vec{B}$, and $\vec{C}$.

25. $\vec{A} \cdot \vec{B} = \vec{B} \cdot \vec{A}$
26. $\vec{A} \cdot \vec{A} = \|\vec{A}\|^2$
27. $(\vec{A} + \vec{B}) \cdot \vec{C} = \vec{A} \cdot \vec{C} + \vec{B} \cdot \vec{C}$

C 28. $|\vec{A} \cdot \vec{B}| \leq \|\vec{A}\| \|\vec{B}\|$

12–8 Planes

In the study of elementary geometry, the word *plane* is often taken as undefined, even though you have a strong intuitive feeling for its meaning. Notice in the figure how a nonzero vector $\vec{N}$ can have an unlimited number of vectors $\overrightarrow{OP}$ orthogonal to it at Q. The set of their endpoints P determines a unique plane. We use this idea to define *plane*.

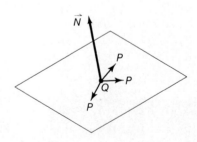

Definition Given a point Q in 3-space and a nonzero vector $\vec{N}$, the set of all points P such that $\vec{QP} \cdot \vec{N} = 0$ is a plane E. The vector $\vec{N}$ is called a **normal vector** to plane E.

Suppose $Q(x_0, y_0, z_0)$ lies in a plane E and $\vec{N}$ is a normal to E at Q. For any point $P(x, y, z)$ in E,

$$\vec{QP} = \langle x - x_0, y - y_0, z - z_0 \rangle.$$

Since $\vec{N}$ and $\vec{QP}$ are orthogonal by the definition of a plane,

$$\vec{QP} \cdot \vec{N} = 0$$

by Corollary 12–6. This proves the following theorem.

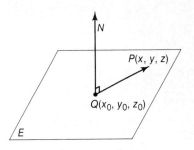

Theorem 12–7 An equation of the plane through (x_0, y_0, z_0) with normal vector $\vec{N} = \langle a, b, c \rangle$ is

$$a(x - x_0) + b(y - y_0) + c(z - z_0) = 0.$$

Example 1 Write an equation of the plane through $Q\left(1, \frac{1}{3}, \frac{1}{2}\right)$ with normal $\vec{N} = \langle 3, 6, 2 \rangle$.

SOLUTION By substitution in Theorem 12–7 or from

$$\left\langle x - 1, y - \tfrac{1}{3}, z - \tfrac{1}{2} \right\rangle \cdot \langle 3, 6, 2 \rangle = 0,$$

you find

$$3(x - 1) + 6\left(y - \tfrac{1}{3}\right) + 2\left(z - \tfrac{1}{2}\right) = 0, \text{ or}$$

$$3x + 6y + 2z = 6.$$

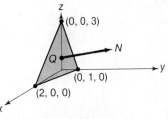

Example 1 shows that an equation of a plane in the form $a(x - x_0) + b(y - y_0) + c(z - z_0) = 0$ can be simplified to $ax + by + cz = d$, where d is the constant $ax_0 + by_0 + cz_0$.

Corollary 12–7 Every plane has an equation of the form $ax + by + cz = d$, where the vector $\langle a, b, c \rangle$ is normal to the plane.

Example 2 Find an equation of the line ℓ through $P(1, 4, 2)$ and perpendicular to plane $E: 2x - y + 3z = 3$.

SOLUTION By Corollary 12–7 the vector $\vec{N} = \langle a, b, c \rangle = \langle 2, -1, 3 \rangle$ is normal to E. Therefore $\langle 2, -1, 3 \rangle$ is a direction vector for ℓ, and an equation of ℓ is

$$\langle x, y, z \rangle = \langle 1, 4, 2 \rangle + t \langle 2, -1, 3 \rangle.$$

Example 3 Find an equation of the plane that contains $A(1, 1, 2)$, $B(2, 3, 1)$, and $C(3, 1, 4)$.

SOLUTION By subtraction of coordinates, $\vec{AB} = \langle 1, 2, -1 \rangle$ and $\vec{AC} = \langle 2, 0, 2 \rangle$. Let $\vec{N} = \langle a, b, c \rangle$ be a normal to the plane of the three points, so that $\vec{N} \cdot \vec{AB} = 0$ and $\vec{N} \cdot \vec{AC} = 0$. Then

$$1a + 2b - c = 0 \text{ and}$$
$$2a + 0b + 2c = 0.$$

From the second equation, $a = -c$. Substituting in the first equation gives $b = c$. So the normal vector $\langle a, b, c \rangle$ must satisfy $a = -c = -b$. Hence any nonzero vector of the form $\langle a, -a, -a \rangle$ is a normal vector. Using $\langle 1, -1, -1 \rangle$ as a convenient normal vector, by Theorem 12–7, an equation of the plane through $(1, 1, 2)$ is

$$1(x - 1) + (-1)(y - 1) + (-1)(z - 2) = 0, \text{ or } x - y - z = -2.$$

Example 4 Find the point of intersection of the plane $E: 3x - 2y + z = 2$ and the line $\ell: \langle x, y, z \rangle = \langle 0, 4, 1 \rangle + t \langle 1, -2, 2 \rangle$.

SOLUTION An equation of ℓ in parametric form is

$$x = t, y = 4 - 2t, z = 1 + 2t.$$

To find the point of intersection, substitute these expressions in the equation of the plane.

$$3(t) - 2(4 - 2t) + (1 + 2t) = 2$$

The solution of this equation is $t = 1$.

Therefore $x = 1$, $y = 4 - 2(1) = 2$, $z = 1 + 2(1) = 3$, and the point of intersection is $(1, 2, 3)$.

Exercises

A Write an equation of the plane through point P with normal $\vec{N}$.

1. $P(1, 2, 3), \vec{N} = \langle 1, -1, 2 \rangle$
2. $P(-1, 2, 3), \vec{N} = \langle 1, -1, 2 \rangle$
3. $P(0, 1, -2), \vec{N} = \langle 2, 1, -1 \rangle$
4. $P(0, -1, 3), \vec{N} = \langle 3, -1, 2 \rangle$
5. $P(1, 2, 5), \vec{N} = \langle 7, 1, 2 \rangle$
6. $P(-1, 2, -3), \vec{N} = \langle 3, -1, 4 \rangle$

Write an equation of the line through P that is perpendicular to plane E.

7. $P(1, 2, 3), E: 3x - 2y + z = 5$
8. $P(-1, 2, 3), E: 3x - y + 2z = -4$
9. $P(2, -5, 4), E: x = 2y - z + 3$
10. $P(-1, 3, -2), E: y + 3z = 2x - 1$

B Write an equation of the plane determined by points A, B, and C.

11. $A(2, 2, 2)$, $B(3, 1, 1)$, $C(6, -4, -6)$
12. $A(1, 3, -2)$, $B(-1, 2, 2)$, $C(3, -1, 4)$
13. $A(-1, 0, 2)$, $B(2, 0, -1)$, $C(0, 3, 4)$
14. $A(0, 0, 5)$, $B(-1, 1, 10)$, $C(2, -1, -2)$

Find the point of intersection of ℓ and E.

15. $\ell: \langle x, y, z \rangle = \langle 1, 2, 3 \rangle + t\langle 1, -1, 2 \rangle$
    $E: 2x + 7y + z = -2$

16. $\ell: \langle x, y, z \rangle = \langle -1, -2, 13 \rangle + t\langle 1, 1, -3 \rangle$
    $E: 3x - y + 2z = 5$

C Two planes are *parallel* if normal vectors to each plane have the same or opposite direction. Find an equation of the plane parallel to E and passing through point P.

17. $P(1, -2, 3)$, $E: 3x - 2y + z = 2$
18. $P(-1, 5, 2)$, $E: x - y + 4z = -7$

Two planes are *perpendicular* if normal vectors to each plane are orthogonal. Show that planes E_1 and E_2 are perpendicular.

19. $E_1: 3x - 2y + z = 5$
    $E_2: 2x - y - 8z = 9$

20. $E_1: x - y + 2z = -1$
    $E_2: 5x + 3y - z = 5$

Two lines ℓ_1 and ℓ_2 that intersect in a single point determine a unique plane. Find an equation of the plane for the given lines.

21. $\ell_1: \langle x, y, z \rangle = \langle 1, 2, 3 \rangle + t\langle 1, 0, 2 \rangle$
    $\ell_2: \langle x, y, z \rangle = \langle -8, -2, 5 \rangle + s\langle 1, 1, -3 \rangle$

The line ℓ and a point P not on ℓ determine a unique plane E in 3-space. Find an equation of plane E.

22. $P(1, 2, 3)$, $\ell: \langle x, y, z \rangle = \langle -1, 1, 2 \rangle + t\langle 1, 2, -1 \rangle$

Special Matrices

A **matrix** is a rectangular array of numbers, with horizontal *rows* and vertical *columns*. Here are some examples.

$$A = \begin{bmatrix} 1 & -2 & 3 & -1 \\ 0 & 1 & -1 & 0 \\ -3 & -2 & 5 & 0 \end{bmatrix} \quad B = \begin{bmatrix} -1 & -5 \\ 2 & 6 \end{bmatrix}$$

$$C = \begin{bmatrix} 3 \\ 1 \\ -4 \end{bmatrix}$$

The numbers of rows and columns, in that order, of a matrix are called the **dimensions** of the matrix. Thus A is a 3×4 (read *three-by-four*) matrix, B is 2×2 and is called a **square matrix of order two**, and C is 3×1. A matrix like C with only one column is called a *column vector*, or column matrix. A matrix with a single row is called a *row vector*, or row matrix. One important use of matrices is to represent a system of equations. Matrix A represents the coefficients and right-side values of the system

$$\begin{aligned} 1x - 2y + 3z &= -1 \\ 0x + 1y - 1z &= 0 \\ -3x - 2y + 5z &= 0. \end{aligned}$$

Two matrices of the same dimensions can be added, with each entry of $M + N$ found by taking the sum of the corresponding entries in M and N.

$$\begin{bmatrix} 2 & -1 \\ -3 & -4 \end{bmatrix} + \begin{bmatrix} -2 & 3 \\ 1 & 5 \end{bmatrix} = \begin{bmatrix} 0 & 2 \\ -2 & 1 \end{bmatrix}$$

Notice that the sum

$$\begin{bmatrix} -1 & 9 \\ 2 & -3 \end{bmatrix} + \begin{bmatrix} 0 & -1 & 1 \\ 5 & 2 & 6 \end{bmatrix}$$

is not defined.

As with vectors, the product of a scalar and a matrix is defined as the product of the scalar with each entry of the matrix. Thus for the matrix B given above,

$$\tfrac{1}{4}B = \begin{bmatrix} -\tfrac{1}{4} & -\tfrac{5}{4} \\ \tfrac{1}{2} & \tfrac{3}{2} \end{bmatrix}.$$

The product of two matrices is defined by using the dot product of vectors. If R is an $m \times n$ matrix and Q is an $n \times p$ matrix, then the product RQ is the $m \times p$ matrix whose entries are the dot products of the rows of R with the columns of Q.

Example 1 If $R = \begin{bmatrix} 1 & 0 & -2 \\ -2 & 3 & 4 \end{bmatrix}$ and

$Q = \begin{bmatrix} -5 & 0 \\ 1 & -2 \\ 1 & 3 \end{bmatrix}$, find RQ.

SOLUTION The entry in the first row and first column of RQ is the dot product of the first row of R with the first column of Q: $(1)(-5) + (0)(1) + (-2)(1) = -7$. The entry in the second row, first column of RQ is the dot product of the second row of R with the first column of Q: $(-2)(-5) + (3)(1) + (4)(1) = 17$. You should evaluate the two remaining dot products to verify that RQ is the 2×2 matrix

$$\begin{bmatrix} -7 & -6 \\ 17 & 6 \end{bmatrix}.$$

Matrix multiplication is not commutative. For the matrices in example 1, QR is a 3×3 matrix that clearly is not the same as RQ. Often a product BA is not even defined, even though AB is defined.

When Q is an $n \times 1$ column vector and R is an $m \times n$ matrix, the product RQ will be an $m \times 1$ column vector. For example,

$$\begin{bmatrix} -1 & -5 \\ 2 & 6 \end{bmatrix} \begin{bmatrix} 3 \\ -1 \end{bmatrix} = \begin{bmatrix} 2 \\ 0 \end{bmatrix}.$$

This leads to a convenient way of expressing a system of equations in matrix form. The system $\begin{aligned} -x - 5y &= 4 \\ 2x + 6y &= 8 \end{aligned}$ can be written as a single matrix equation:

$$\begin{bmatrix} -1 & -5 \\ 2 & 6 \end{bmatrix} \begin{bmatrix} x \\ y \end{bmatrix} = \begin{bmatrix} 4 \\ 8 \end{bmatrix}, \text{ or } BX = C,$$

where $B = \begin{bmatrix} -1 & -5 \\ 2 & 6 \end{bmatrix}$ is the *coefficient matrix* and

$X = \begin{bmatrix} x \\ y \end{bmatrix}$ and $C = \begin{bmatrix} 4 \\ 8 \end{bmatrix}$ are column vectors. A unique solution of this system can be found if there is a matrix B^{-1} such that $B^{-1}B = \begin{bmatrix} 1 & 0 \\ 0 & 1 \end{bmatrix}$. This matrix is called the **identity matrix of order 2**, and B^{-1} is called the (multiplicative) **inverse** of B. You can check that

$$B^{-1} = \begin{bmatrix} \frac{3}{2} & \frac{5}{4} \\ -\frac{1}{2} & -\frac{1}{4} \end{bmatrix}$$

satisfies $B^{-1}B = \begin{bmatrix} 1 & 0 \\ 0 & 1 \end{bmatrix}$. Now multiply both sides of the matrix equation $BX = C$ by B^{-1}.

$$B^{-1}BX = B^{-1}C$$

$$\begin{bmatrix} \frac{3}{2} & \frac{5}{4} \\ -\frac{1}{2} & -\frac{1}{4} \end{bmatrix} \begin{bmatrix} -1 & -5 \\ 2 & 6 \end{bmatrix} \begin{bmatrix} x \\ y \end{bmatrix} = \begin{bmatrix} \frac{3}{2} & \frac{5}{4} \\ -\frac{1}{2} & -\frac{1}{4} \end{bmatrix} \begin{bmatrix} 4 \\ 8 \end{bmatrix}$$

$$\begin{bmatrix} 1 & 0 \\ 0 & 1 \end{bmatrix} \begin{bmatrix} x \\ y \end{bmatrix} = \begin{bmatrix} 16 \\ -4 \end{bmatrix}$$

Solution of the system: $\begin{bmatrix} x \\ y \end{bmatrix} = \begin{bmatrix} 16 \\ -4 \end{bmatrix}$.

If the coefficient matrix has no inverse, the system either has no solutions (it is *inconsistent*) or infinitely many solutions (one equation is *redundant*).

For 2×2 matrices, $A = \begin{bmatrix} a & b \\ c & d \end{bmatrix}$ will have an inverse if and only if $ad - bc \neq 0$. The quantity $ad - bc$ is called the **determinant** of A, written det A. You can show that

if $A^{-1} = \frac{1}{\det A} \begin{bmatrix} d & -b \\ -c & a \end{bmatrix}$, then $A^{-1}A = \begin{bmatrix} 1 & 0 \\ 0 & 1 \end{bmatrix}$.

Example 2   Find the inverse of $A = \begin{bmatrix} 4 & -2 \\ -3 & -1 \end{bmatrix}$.

SOLUTION   det $A = (4)(-1) - (-2)(-3) = -10$

$$A^{-1} = -\frac{1}{10} \begin{bmatrix} -1 & 2 \\ 3 & 4 \end{bmatrix} = \begin{bmatrix} \frac{1}{10} & -\frac{1}{5} \\ -\frac{3}{10} & -\frac{2}{5} \end{bmatrix}.$$

Example 3   Express the system $\begin{array}{r} 4x - 2y = -10 \\ -3x - y = 5 \end{array}$ in matrix form and find its solution.

SOLUTION   With $A = \begin{bmatrix} 4 & -2 \\ -3 & -1 \end{bmatrix}$ as coefficient matrix and $C = \begin{bmatrix} -10 \\ 5 \end{bmatrix}$, the system can be written $AX = C$. Now multiply both sides by A^{-1} from example 2.

$$\begin{bmatrix} \frac{1}{10} & -\frac{1}{5} \\ -\frac{3}{10} & -\frac{2}{5} \end{bmatrix} \begin{bmatrix} 4 & -2 \\ -3 & -1 \end{bmatrix} \begin{bmatrix} x \\ y \end{bmatrix} = \begin{bmatrix} \frac{1}{10} & -\frac{1}{5} \\ -\frac{3}{10} & -\frac{2}{5} \end{bmatrix} \begin{bmatrix} -10 \\ 5 \end{bmatrix}$$

$$\begin{bmatrix} 1 & 0 \\ 0 & 1 \end{bmatrix} \begin{bmatrix} x \\ y \end{bmatrix} = \begin{bmatrix} -2 \\ 1 \end{bmatrix}$$

Solution of the system: $\begin{bmatrix} x \\ y \end{bmatrix} = \begin{bmatrix} -2 \\ 1 \end{bmatrix}$.

Matrices are especially useful when solving larger systems of equations.

Exercises

Given $D = \begin{bmatrix} -7 & -2 & 8 \\ 0 & 1 & -1 \end{bmatrix}$, $E = \begin{bmatrix} -2 & 1 \\ -3 & 3 \end{bmatrix}$, and $F = \begin{bmatrix} 5 & -1 \\ 0 & 2 \end{bmatrix}$.

1. Specify the dimensions of D, E, and F.
2. Find the matrix $E + F$ and show that $E + F = F + E$.
3. Find the matrix $-2F$.
4. Find the matrices EF and FE.
5. Find the matrix FD.
6. Explain why the product DF is not defined.
7. Find the inverse of E.
8. Use matrices to solve $\begin{array}{r} -2x + y = 6 \\ -3x + 3y = -9 \end{array}$.
9. Show that the system $\begin{array}{r} -2x + y = 6 \\ 4x - 2y = -1 \end{array}$ is inconsistent.
10. Show that for the system $\begin{array}{r} -x + 7y = 3 \\ 2x - 14y = -6 \end{array}$ one of the equations is redundant.
11. Show that if $A = \begin{bmatrix} a & b \\ c & d \end{bmatrix}$ and det $A \neq 0$, then
$$A^{-1} = \frac{1}{\det A} \begin{bmatrix} d & -b \\ -c & a \end{bmatrix}.$$

Chapter Review Exercises

12–1, page 425

1. Sketch two different arrows that are representatives of $\vec{V} = \langle -3, 4 \rangle$.
2. Find the vector represented by the arrow from $S(2, -3)$ to $T(-4, 1)$.

12–2, page 428

3. If $\vec{A} = \langle 2, -5 \rangle$ and $\vec{B} = \langle -1, 3 \rangle$, find $\|\vec{A} + \vec{B}\|$.
4. Copy $\vec{A}$ and $\vec{B}$ and sketch $\vec{A} + \vec{B}$ and $\vec{A} - \vec{B}$.

12–3, page 432

5. Find a unit vector in the direction of $\langle -4, 6 \rangle$.
6. Write a vector equation of the line through $Q(-3, 2)$ and $R(6, -7)$.

12–4, page 437

7. An airplane is heading northeast at 500 knots. Express its velocity as a vector.
8. Given forces $\vec{F}_1$ and $\vec{F}_2$ as in the figure, sketch a force $\vec{F}_3$ so that $\vec{F}_1 + \vec{F}_2 + \vec{F}_3 = 0$.

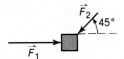

9. In problem 8 let $\|\vec{F}_1\| = 10$ and $\|\vec{F}_2\| = 6$ where $\vec{F}_2$ makes an angle of 30° with the horizontal. Find $\vec{F}_3$ so that the system is in equilibrium.

12–5, page 439

10. Find the distance between points $A(1, -3, 5)$ and $B(-3, 6, 9)$.
11. Find the center and the radius of the sphere $x^2 + y^2 + z^2 + 4x - 6y = 7$.

12–6, page 442

12. Show that $(-1, 9, -2)$ lies on the line $\langle x, y, z \rangle = \langle 3, 7, -10 \rangle + t\langle -2, 1, 4 \rangle$.
13. Write an equation of the line through the points $(7, -2, 5)$ and $(-1, 11, 2)$.

12–7, page 447

14. Find the acute angle ϕ between $\vec{A} = \langle 3, -2, 5 \rangle$ and $\vec{B} = \langle 7, -3, 9 \rangle$.
15. Show that lines $\ell_1: \langle x, y, z \rangle = \langle 4, 2, -7 \rangle + t\langle -3, 2, 4 \rangle$ and $\ell_2: \langle x, y, z \rangle = \langle 1, 4, -3 \rangle + s\langle 2, 5, -1 \rangle$ intersect in a right angle.

12–8, page 448

16. Write an equation of the plane through $(-7, 13, 4)$ with normal $\langle 5, -6, 1 \rangle$.
17. Write an equation of the line through $(4, 0, -9)$ that is perpendicular to the plane $5x - 4y + 12z = 4$.

Chapter Test

1. If $\vec{V_1} = \langle 3, -5, 2 \rangle$ and $\vec{V_2} = \langle 7, 1, -9 \rangle$, find $\|\vec{V_1}\|$ and $\vec{V_1} - 2\vec{V_2}$.

2. Find k so that $\overrightarrow{ST}$ is equivalent to $\overrightarrow{QR}$ for $S(1, 4)$, $T(-2, 3)$, $Q(k, -3)$, and $R(7, -4)$.

3. Copy $\vec{A}$, $\vec{B}$, and $\vec{C}$ and sketch $\vec{B} - 2\vec{A}$ and $\vec{A} + \vec{B} + \vec{C}$.

4. Find real numbers s and t such that $\langle s, 3 \rangle = \langle -1, t \rangle + \langle 5, -7 \rangle$.

5. Find the coordinates of the point of intersection of line ℓ and the plane $y = 3$ if ℓ is $\langle x, y, z \rangle = \langle -3, 7, 4 \rangle + t\langle -1, 2, 5 \rangle$.

6. Find the acute angle of intersection of the lines
    $\ell_1: x = 1 + 4t,\ y = -7 + t,\ z = 11 - 3t$ and
    $\ell_2: x = 1 - 5s,\ y = -7 + 3s,\ z = 11 + s$.

7. If the wind is blowing to the east at 70 knots and the groundspeed of an airplane traveling on a bearing of 60° is 400 knots,
    a. Express the plane's velocity in still air as a vector.
    b. Find the effective velocity of the airplane as a vector.

8. Given point $P(11, -4, 2)$ and the lines
    $\ell_1: \langle x, y, z \rangle = \langle 3, -7, 2 \rangle + t\langle -1, 0, 4 \rangle$ and
    $\ell_2: \langle x, y, z \rangle = \langle 4, -10, -5 \rangle + s\langle 2, 3, -5 \rangle$,
    a. Show that P does not lie on ℓ_1.
    b. Find Q, the point of intersection of the two lines.
    c. Find an equation of the line through P and Q.

9. Given point $P(1, 8, -4)$, line $\ell: \langle x, y, z \rangle = \langle -1, 7, -4 \rangle + t\langle 2, -3, 9 \rangle$, and plane $E: 6x - 5y + z = 27$.
    a. Find the point of intersection of line ℓ and plane E.
    b. Find an equation of the line perpendicular to plane E and passing through P.
    c. Find an equation of the plane through P and perpendicular to line ℓ.

10. The forces $\vec{F_1}$, $\vec{F_2}$, and $\vec{F_3}$ act on an object in the directions indicated in the figure. If $\|\vec{F_1}\| = 10$, $\|\vec{F_2}\| = 6$, and $\|\vec{F_3}\| = 8$, find a force $\vec{F_4}$ so that the system will be in equilibrium.

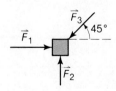

Sequences and Series

13

13-1 Arithmetic Sequences

In the year 1202 A.D., an Italian merchant, Leonardo of Pisa (better known as Fibonacci), upon his return from the Middle East wrote a book presenting new ideas in arithmetic that he had accumulated during his travels. In this book he asked:

"How many pairs of rabbits can be produced from a single pair in a year if
  a. Each pair becomes productive at the end of its first two months, and thereafter begets a new pair each month, and
  b. deaths do not occur?"

Starting from a newborn pair, at the beginning of the second month, there is still one pair of rabbits. At the beginning of the third month, this pair produces its first new pair of rabbits. At the beginning of the fourth month, the original pair produces yet another pair of rabbits, but the second pair is not old enough to have offspring. So at the start of the fourth month there are three pairs. As the fifth month begins, two pairs of rabbits each produce a pair, bringing the total number of rabbit pairs to five.

This process can be described by a function F which assigns to every argument n the number of rabbit pairs at the beginning of the nth month.

$$F(1) = 1 \qquad F(5) = 5$$
$$F(2) = 1 \qquad F(6) = 8$$
$$F(3) = 2 \qquad F(7) = 13$$
$$F(4) = 3 \qquad F(8) = 21$$

Notice that for each $n > 2$, $F(n) = F(n-1) + F(n-2)$. You can verify that at the beginning of the 12th month the rabbit population stands at 144 pairs, that is, $F(12) = 144$.

For this problem, F only needs to have domain $\{n \in \mathbf{N}: n \leq 12\}$. If the domain of F is taken instead as the set of all natural numbers (in effect, postulating that the rabbits are immortal), then F is called the **Fibonacci sequence**. Sequences form an important class of functions.

> **Definition**
>
> A **sequence** is a function which has as its domain the set of natural numbers.

If f is a sequence, then the value $f(n)$ is called the nth *term* of the sequence. To describe a sequence, it is common to give a partial list of the terms:

$$f(1), f(2), f(3), \ldots, f(n), \ldots,$$

or

$$a_1, a_2, a_3, \ldots, a_n, \ldots, \text{ where } a_n = f(n).$$

The Fibonacci sequence is usually written

$$1, 1, 2, 3, 5, 8, 13, \ldots, F(n), \ldots,$$

where

$$F(n) = F(n-1) + F(n-2), n > 2.$$

Similarly, the sequence $g: n \to n^2$ with $n \in \mathbf{N}$ is indicated by

$$1, 4, 9, 16, \ldots, n^2, \ldots.$$

To shorten this notation further, you may refer to the nth term and denote g by $\{n^2\}$. The sequence $a_1, a_2, a_3, \ldots, a_n, \ldots$ is then written simply as $\{a_n\}$.

Example 1 Determine the seventh term of the sequence $h: 4, 7, 10, 13, \ldots, 3n + 1, \ldots$.

SOLUTION $h(7) = 3(7) + 1 = 22$

Example 2 List the first four terms of the sequence $\{a_n\}$, where $a_n = \dfrac{(-2)^{n+1}}{n+3}$.

SOLUTION $a_1 = \dfrac{(-2)^{1+1}}{1+3} = 1 \qquad a_3 = \dfrac{(-2)^{3+1}}{3+3} = \dfrac{8}{3}$

$a_2 = \dfrac{(-2)^{2+1}}{2+3} = -\dfrac{8}{5} \qquad a_4 = \dfrac{(-2)^{4+1}}{4+3} = -\dfrac{32}{7}$

ANSWER The first four terms of $\{a_n\}$ are $1, -\dfrac{8}{5}, \dfrac{8}{3},$ and $-\dfrac{32}{7}$.

Example 3 Determine whether 17 is a term of $\left\{\dfrac{3n+1}{2n}\right\}$.

SOLUTION If 17 is a term of the sequence, then for some $k \in \mathbf{N}$, $17 = \dfrac{3k+1}{2k}$. Solving for k yields $k = \dfrac{1}{31}$. Since $\dfrac{1}{31} \notin \mathbf{N}$, 17 is not a term of the given sequence.

The sequence $\{3n + 1\}$ in example 1 is characterized by the fact that consecutive terms differ by the same amount. Any sequence with this property is called an arithmetic sequence.

Definition An **arithmetic sequence** is a sequence $\{a_n\}$ such that for some constant d

$$a_{n+1} - a_n = d$$

for every $n \in \mathbf{N}$.

For the sequence $\{3n + 1\}$ the constant d, which is called the **common difference**, is 3.

It is easy to determine any term of an arithmetic sequence if the first term and the common difference are known.

Example 4 If $a_1 = 2$ and $d = 11$ for an arithmetic sequence, find a_4.

SOLUTION Since $a_{n+1} - a_n = d$ for all $n \in \mathbf{N}$, $a_2 - a_1 = 11$ and $a_2 = 13$.
Likewise, $a_3 = a_2 + d = 13 + 11 = 24$, and
$$a_4 = a_3 + d = 24 + 11 = 35.$$

If you wanted to find a_{70} instead of a_4, the method in example 4 would be very slow. Instead, observe the following pattern:

$$a_2 = a_1 + d$$
$$a_3 = a_2 + d = (a_1 + d) + d = a_1 + 2d$$
$$a_4 = a_3 + d = (a_1 + 2d) + d = a_1 + 3d.$$

This suggests that for any $n \in \mathbf{N}$,

$$a_n = a_1 + (n - 1)d.$$

So, for the sequence 2, 13, 24, 35, ..., $2 + (n - 1)11$, ...

$$a_{70} = 2 + (69)11 = 761.$$

Example 5 If 2 and 12 are the first and fourth terms of an arithmetic sequence, find the tenth term.

SOLUTION $a_n = a_1 + (n - 1)d$
$12 = 2 + (4 - 1)d$
$d = \frac{10}{3}$ Notice that d does not have to be an integer.
$a_{10} = 2 + (10 - 1)\left(\frac{10}{3}\right) = 32$

A more difficult problem is to find the sum of the first n terms of an arithmetic sequence. Again consider the sequence $\{2 + (n - 1)11\}$ and suppose that you are to find the sum of the first 70 terms. You could compute the 70 terms and add them, but if you look at the terms as suggested in the diagram below, then the task can be greatly simplified.

$$2, 13, 24, \ldots, 728, 739, 750, 761$$

Notice that the pairing shown by the arrows yields

$$2 + 761 = 13 + 750 = 24 + 739 = \ldots = 763.$$

Thus the required sum is actually the sum of 35 terms of the form $a_k + a_{71-k}$, each of which equals 763. Therefore

$$\text{Sum of 70 terms} = 35(2 + 761) = 26{,}705.$$

These results can be generalized as follows.

Theorem 13–1 Let $\{a_n\}$ be an arithmetic sequence with d the common difference and let S_n be the sum of the first n terms. Then

a. $a_n = a_1 + (n - 1)d$, and

b. $S_n = \dfrac{n}{2}(a_1 + a_n)$.

PROOF **b.** $S_n = a_1 + a_2 + a_3 + \ldots + a_{n-2} + a_{n-1} + a_n$

Now use the fact that $a_k = a_1 + (k - 1)d$ and $a_{n-k} = a_n - kd$ to write

$$S_n = a_1 + (a_1 + d) + (a_1 + 2d) + \ldots + (a_n - 2d) + (a_n - d) + a_n.$$

Reversing the order of the terms yields

$$S_n = a_n + (a_n - d) + (a_n - 2d) + \ldots + (a_1 + 2d) + (a_1 + d) + a_1.$$

Now add the two preceding equations term by term.

$$2S_n = \underbrace{(a_1 + a_n) + (a_1 + a_n) + \ldots + (a_1 + a_n) + (a_1 + a_n)}_{n \text{ terms}}.$$

$$2S_n = n(a_1 + a_n)$$

$$S_n = \dfrac{n}{2}(a_1 + a_n)$$

Example 6 Find the sum of the multiples of 7 that lie between 100 and 1000.

SOLUTION First, by division, determine that 105 and 994 are the first and last terms in the sum. Since consecutive multiples of 7 differ by 7, use the arithmetic sequence $\{a_n\}$, with $a_1 = 105$ and $d = 7$. Now find a value of n such that $a_n = 994$.

$$a_n = a_1 + (n - 1)d$$
$$994 = 105 + (n - 1)7, \text{ or } n = 128$$

$$S_{128} = \dfrac{128}{2}(105 + 994) = 70{,}336$$

Exercises

A List the first five terms of each sequence.

1. $\{4n + 3\}$
2. $\{2n - 1\}$
3. $\left\{\dfrac{2n}{3n - 1}\right\}$
4. $\left\{\sin\dfrac{n\pi}{2}\right\}$
5. $a_n = 2^{n-1}$
6. $f(n) = \dfrac{2^{n-1}}{3^n}$
7. $a_n = \dfrac{(-1)^n}{n}$
8. $g(n) = (-1)^{n-1}\left(\dfrac{2}{3}\right)^n$

Determine whether the given number is a term of the given sequence. If it is, specify which term.

9. $\dfrac{1}{33}$; $\left\{\dfrac{1}{5n - 3}\right\}$
10. $\dfrac{1}{32}$; $\left\{\dfrac{(-1)^n}{2^{n+1}}\right\}$
11. $\dfrac{11}{75}$; $\left\{\dfrac{2n + 1}{3n^2}\right\}$
12. 1024; $\left\{4\left(\dfrac{2}{3}\right)^n\right\}$

13. If -2 and 7 are the first and second terms of an arithmetic sequence, find the ninth term and the sum of the first 9 terms.

14. If 4 and 1 are the second and third terms of an arithmetic sequence, find the seventh term and the sum of the first 7 terms.

B 15. For the arithmetic sequence $-1, 3, 7, \ldots$, which term is 47?

16. If $\dfrac{25}{4}$ and $\dfrac{19}{4}$ are the first and third terms of an arithmetic sequence, which term is $-\dfrac{1}{2}$?

17. If 2 and -30 are the first and ninth terms of an arithmetic sequence, find the sum of the first 15 terms.

18. If $\{a_n\}$ is an arithmetic sequence with $a_3 = 17$ and $a_9 = 49$, find S_{20}.

19. Find the sum of the integers n such that $100 < n < 1000$.

20. Find the sum of the multiples of 3 that lie between 50 and 125.

21. Find a formula for the sum of the first n natural numbers.

22. Find a formula for the sum of the first n odd natural numbers.

23. A college is graduating 300 students this year and hopes to increase its graduating class by 25 each year for the next 9 years. If this is done, how many graduates will the college have in the 10-year period?

24. In the first week of his third-year Spanish course, Darryl read 4 pages of a novel in Spanish. Each week he increased his weekly reading rate by the same number of pages. He finished the 432-page novel on the last day of the 36-week school year. How many pages did he read in the last 10 weeks of the year?

25. The sequence $\{b_n\}$ is defined by $b_1 = 1$, $b_2 = 3$, and $b_n = b_{n-1} + 2b_{n-2}$ for all $n > 2$. Prove or disprove that b_n is prime for all $n > 1$.

26. A sequence $\{a_n\}$ has the property that each term after the first is obtained by multiplying the preceding term by $\frac{3}{2}$. If $a_1 = 6$, determine a_5 and find a formula for the nth term.

C 27. Find the sum of the integers between 100 and 200 that are multiples of neither 3 nor 5.

28. Write a computer program so that for any given n the computer will print out the first n terms of the Fibonacci sequence and their sum.

13–2 Geometric Sequences

From 1967 through 1979, inflation in the U.S. averaged 6.7% a year as measured by the cost-of-living index. This means that what $15,000 could buy in 1967 would require 6.7% more, or $16,005, in 1968. You can multiply the amount for one year by 1.067 to obtain the equivalent amount for the next year: $17,077 for 1969, $18,222 for 1970, and so on. The increasing cost of living can be described by a sequence:

$$15{,}000,\ 16{,}005,\ 17{,}077,\ 18{,}222,\ \ldots,\ a_n,\ \ldots,$$

where a_n is the amount $n - 1$ years after 1967. To find a formula for a_n, notice the following.

$$15{,}000 = a_1 = \text{Amount in 1967}$$
$$16{,}005 = a_2 = \text{Amount 1 year later} = 15{,}000(1.067)$$
$$17{,}077 = a_3 = \text{Amount 2 years later} = [15{,}000(1.067)](1.067)$$
$$= 15{,}000(1.067)^2$$
$$18{,}222 = a_4 = \text{Amount 3 years later} = [15{,}000(1{,}067)^2](1.067)$$
$$= 15{,}000(1.067)^3$$
$$\vdots$$
$$a_n = \text{Amount } n - 1 \text{ years later} = 15{,}000(1.067)^{n-1}$$

When the ratio of the consecutive terms of a sequence is a constant, the sequence is called a **geometric sequence**. The cost-of-living sequence above is one example of a geometric sequence, and $-\frac{1}{3}, \frac{1}{9}, -\frac{1}{27}, \ldots, \left(-\frac{1}{3}\right)^n, \ldots$ is another.

Definition A **geometric sequence** is a sequence $\{a_n\}$ such that for some constant r, not equal to 0 or 1,

$$\frac{a_{n+1}}{a_n} = r, \text{ for every } n \in \mathbf{N}.$$

The constant r is called the **common ratio**. For instance, for the sequence $9, 6, 4, \frac{8}{3}, \frac{16}{9}, \ldots$, the common ratio is $\frac{2}{3}$.

Example 1 If $\{a_n\}$ is a geometric sequence with $a_1 = -3$, $a_3 = -6$, and $a_2 > 0$, find the common ratio r.

SOLUTION Since $\dfrac{a_2}{a_1} = \dfrac{a_3}{a_2}$, $a_2{}^2 = a_1 \cdot a_3 = (-3)(-6) = 18$.

Since $a_2 > 0$, $a_2 = 3\sqrt{2}$.

Since $\dfrac{a_2}{a_1} = r$, $r = \dfrac{3\sqrt{2}}{-3} = -\sqrt{2}$.

As the cost-of-living illustration showed, you can find any term of a geometric sequence $\{a_n\}$ if the first term a_1 and the common ratio r are known.

Since $\dfrac{a_{n+1}}{a_n} = r$ for every $n \in \mathbf{N}$,

$$a_2 = a_1 r$$
$$a_3 = a_2 r = (a_1 r)r = a_1 r^2$$
$$a_4 = a_3 r = (a_1 r^2)r = a_1 r^3$$

so that $a_n = a_1 r^{n-1}$.

This result, together with a formula for getting the sum of the first n terms of a geometric sequence, is incorporated in the next theorem.

Theorem 13–2 Let $\{a_n\}$ be a geometric sequence with r the common ratio, and let S_n be the sum of the first n terms; then

a. $a_n = a_1 r^{n-1}$, and

b. $S_n = \dfrac{a_1(1 - r^n)}{1 - r}$, $r \neq 1$.

(cont. on p. 466)

PROOF   **b.** Since $a_n = a_1 r^{n-1}$,
$$S_n = a_1 + a_1 r + a_1 r^2 + \ldots + a_1 r^{n-2} + a_1 r^{n-1}.$$

Multiply both sides of this equation by r.
$$rS_n = a_1 r + a_1 r^2 + \ldots + a_1 r^{n-1} + a_1 r^n$$

Subtract the second equation from the first.
$$S_n - rS_n = a_1 - a_1 r^n$$
$$S_n = \frac{a_1(1 - r^n)}{1 - r}, \; r \neq 1$$

Example 2   If $\frac{9}{16}$ and $\frac{243}{16}$ are the first and fourth terms of a geometric sequence $\{a_n\}$, find a_2 and a_3.

SOLUTION   Since $a_4 = a_1 r^{4-1}$, $\frac{243}{16} = \frac{9}{16} r^3$.
Solve for r: $r^3 = 27$, or $r = 3$.
Hence $a_2 = a_1 r = \frac{9}{16} \cdot 3 = \frac{27}{16}$, and
$a_3 = a_2 r = \frac{27}{16} \cdot 3 = \frac{81}{16}$.

Example 3   A foolish gambler believes he has a system which will enable him always to win at roulette. He will initially bet $3 on even and then double his bet each time he loses until he wins, at which point he will start all over. If the rolling ball settles on an odd number for six consecutive spins of the wheel, how much money will the gambler have lost at that time?

SOLUTION   The problem amounts to finding the sum of the first 6 terms of a geometric sequence in which $a_1 = 3$ and $r = 2$. Substituting in the formula
$$S_n = \frac{a_1(1 - r^n)}{1 - r} \quad \text{gives} \quad S_6 = \frac{3(1 - 2^6)}{1 - 2} = 189.$$

ANSWER   He will have lost $189.

Example 4   A prospective business-school graduate reads that with an MBA degree she can expect an annual salary increase of 15% during the first 7 years of her career. If her goal is to earn at least $250,000 in her first 7 years of working, what starting salary should she seek as she looks for a job?

(cont. on p. 467)

SOLUTION Use the formula for the sum of the first 7 terms of a geometric sequence, setting $S_7 \geq 250{,}000$ and $r = 1.15$, and solve the inequality

$$\frac{a_1(1 - 1.15^7)}{1 - 1.15} \geq 250{,}000.$$

$$a_1 \geq \frac{250{,}000(1 - 1.15)}{1 - 1.15^7} \doteq 22{,}590$$

ANSWER She should seek a starting salary of at least $22,590 per year.

Exercises

A In exercises 1–6, $\{a_n\}$ is a geometric sequence.

1. $a_1 = 2$, $r = 3$; find a_5 and S_5.
2. $a_1 = \frac{3}{2}$, $r = -2$; find a_4 and S_4.
3. $a_3 = 6$, $a_2 = 4$; find a_6.
4. $a_3 = \frac{4}{9}$, $a_5 = \frac{16}{81}$; find a_4.
5. $a_1 = -8$, $a_8 = \frac{1}{16}$; find a_3.
6. $a_4 = \frac{\sqrt{3}}{3}$, $a_{10} = 9\sqrt{3}$; find a_{11}.

7. Which term of the geometric sequence $\frac{1}{32}, \frac{1}{16}, \frac{1}{8}, \ldots$ is 256?
8. Which term of the geometric sequence $\frac{1}{3}, -\frac{2}{9}, \frac{4}{27}, \ldots$ is $-\frac{32}{729}$?

B 9. $2, \frac{2}{3}, \frac{2}{9}, \ldots$ is a geometric sequence. Find k if $S_k = \frac{80}{27}$.
10. $-\frac{3}{4}, \frac{3}{2}, -3, \ldots$ is a geometric sequence. Find k if $S_k = -\frac{33}{4}$.
11. A ball when dropped rebounds to a height which is three-fifths the height from which it is dropped. If the ball is dropped from a height of 20 feet, how high will it go following the sixth bounce?
12. Ten years ago a house was appraised at $30,000 and since that time property values have gone up 12% annually. What should be the appraised value now?
13. The amount of money a person pays for medical insurance has been increasing at a rate of 15% per year. In the course of the next 8 years, how much money should a person expect to spend on medical insurance if the amount paid this year is $400?
14. The first month a novel was on sale, 20,000 copies were sold. Subsequently, sales dropped 30% each month. How many copies were sold in the first 8 months?
15. A company that manufactures solar energy cells reports that this year it budgeted $200,000 for research and development (R & D), and that over the past decade it has increased this budget by one third each year.

(cont. on p. 468)

a. What was the R & D budget 5 years ago?
b. Assuming the same rate of increase, how much will the company plan to spend on R & D over the next 10 years?

C 16. Because of the effects of acid rain, the number of fish caught annually in a certain New England lake has been decreasing at a rate of 8% a year. It was reported that in the 7 years from 1977 through 1983, 7600 fish were caught. At this rate, how many fish would you expect to be caught during the 5 years from 1984 through 1988?

17. A computer manufacturer produced 4000 units in 1976 and in the next 4 years built another 176,000 units. If production was increased by the same percent each year, estimate the annual rate of increase.

13-3 Sigma Notation

A grocer taking inventory wishes to determine the number of oranges in a pyramid-shaped display. He notes that there is one orange on top, 4 in the next level, 9 in the third level, and that there are 18 layers in the pyramid. He assumes that each layer is made up of a square of oranges, and that the number of oranges in each layer is given by the sequence 1, 4, 9, 16, ..., n^2, ... To count the oranges he therefore evaluates the sum

$$1 + 4 + 9 + \ldots + 16^2 + 17^2 + 18^2.$$

As in this example, writing out many terms of a sequence may be laborious and awkward. Such sums are more concisely expressed with **sigma notation**. By using the Greek letter Σ (sigma), the above sum is written

$$\sum_{n=1}^{18} n^2.$$

Here the subscript 1 and the superscript 18 indicate that the first through 18th terms of the sequence $\{n^2\}$ are to be added. In general the sum of the first p terms of a sequence $\{a_n\}$ is written

$$\sum_{n=1}^{p} a_n.$$

Example 1 Evaluate $\sum_{n=1}^{3} (2n + 3)$.

SOLUTION The terms of the sum are from the sequence $\{2n + 3\}$.
$$\sum_{n=1}^{3} (2n + 3) = (2 \cdot 1 + 3) + (2 \cdot 2 + 3) + (2 \cdot 3 + 3)$$
$$= 5 + 7 + 9 = 21$$

Example 2 Use sigma notation to express the sum of the 3rd through 200th terms of the sequence 3, 7, 11, 15, ..., $4n - 1$, ...

SOLUTION $11 + 15 + 19 + \ldots + 799 = \sum_{n=3}^{200} (4n - 1)$.

Example 3 Express the sum $4 + 10 + 28 + 82 + 244$ using sigma notation.

SOLUTION Since $4 = 3^1 + 1$, $10 = 3^2 + 1$, $28 = 3^3 + 1$, $82 = 3^4 + 1$, and $244 = 3^5 + 1$, the given sum is equal to $\sum_{n=1}^{5} (3^n + 1)$.

The letter n in a summation such as $\sum_{n=2}^{4} \frac{(-1)^n}{n}$ is called a **dummy variable** because it can be replaced by any other letter without changing the numerical value of the summation. Thus,
$$\sum_{n=2}^{4} \frac{(-1)^n}{n} = \sum_{i=2}^{4} \frac{(-1)^i}{i} = \sum_{j=2}^{4} \frac{(-1)^j}{j},$$
since each represents the sum $\frac{(-1)^2}{2} + \frac{(-1)^3}{3} + \frac{(-1)^4}{4}$.

To express the sum of the first n terms of the sequence $\left\{\frac{(-1)^n}{n}\right\}$, you should use a letter other than n as the dummy variable. Accordingly, you might write $\sum_{i=1}^{n} \frac{(-1)^i}{i}$.

Example 4 Simplify $\sum_{i=1}^{10} i^3 - \sum_{j=1}^{9} j^3$.

SOLUTION $\sum_{i=1}^{10} i^3 - \sum_{j=1}^{9} j^3 = \left(10^3 + \sum_{i=1}^{9} i^3\right) - \sum_{j=1}^{9} j^3$
$$= 1000$$

The sum $(-2)^3 + (-1)^3 + 0^3 + 1^3 + 2^3$ can be written as $\sum_{i=1}^{5} (i-3)^3$, but it can also be expressed $\sum_{i=-2}^{2} i^3$. In other words p and q in the notation $\sum_{i=p}^{q}$ need not be positive. It is only required that they be integers and that $p \leq q$. Also, sigma notation requires that the terms being added must be defined at each of the integers in the interval $[p, q]$.

Example 5 Evaluate (a) $\sum_{k=-3}^{-1} \cos k\pi$ and (b) $\sum_{j=-1}^{3} \frac{1}{j-2}$.

SOLUTION a. $\sum_{k=-3}^{-1} \cos k\pi = \cos(-3\pi) + \cos(-2\pi) + \cos(-\pi)$

$$= -1 + 1 - 1 = -1$$

b. $\sum_{j=-1}^{3} \frac{1}{j-2}$ is not defined, since $\frac{1}{j-2}$ is undefined for $j = 2$.

Example 6 If $\sum_{j=1}^{5} c_j = 37$, evaluate $\sum_{j=1}^{5} 3c_j$.

SOLUTION $\sum_{j=1}^{5} 3c_j = 3c_1 + 3c_2 + \ldots + 3c_5 = 3(c_1 + c_2 + \ldots + c_5)$

$$= 3 \sum_{j=1}^{5} c_j$$

$$= 3 \cdot 37 = 111$$

Two important properties of Σ notation are described by the next theorem.

Theorem 13–3 Given sequences $\{a_n\}$ and $\{b_n\}$ and a constant c, for all $n \in \mathbb{N}$

a. $\sum_{i=1}^{n} ca_i = c \sum_{i=1}^{n} a_i$, and

b. $\sum_{i=1}^{n} (a_i + b_i) = \sum_{i=1}^{n} a_i + \sum_{i=1}^{n} b_i$.

PROOF b. $\sum_{i=1}^{n} (a_i + b_i) = (a_1 + b_1) + (a_2 + b_2) + \ldots + (a_n + b_n)$

$$= (a_1 + a_2 + \ldots + a_n) + (b_1 + b_2 + \ldots + b_n)$$

$$= \sum_{i=1}^{n} a_i + \sum_{i=1}^{n} b_i$$

Example 7 Evaluate $\sum_{i=1}^{50} 3i$.

SOLUTION $\sum_{i=1}^{50} 3i = 3 \sum_{i=1}^{50} i$

Now find a formula for $\sum_{i=1}^{n} i$ in terms of n.

Since $\{n\}$ is an arithmetic sequence with $a_1 = 1$ and $a_n = n$,

$$\sum_{i=1}^{n} i = \frac{n}{2}(1 + n).$$

Hence $\sum_{i=1}^{50} i = \frac{50}{2}(1 + 50) = 25 \cdot 51$, and therefore

$$\sum_{i=1}^{50} 3i = 3(25 \cdot 51) = 3825.$$

Example 8 Evaluate $\sum_{i=1}^{10} \left[\left(\frac{9}{10}\right)^i - 5i \right]$.

SOLUTION $\sum_{i=1}^{10} \left[\left(\frac{9}{10}\right)^i - 5i \right] = \sum_{i=1}^{10} \left(\frac{9}{10}\right)^i + \sum_{i=1}^{10} (-5i) = \sum_{i=1}^{10} \left(\frac{9}{10}\right)^i - 5 \sum_{i=1}^{10} i$

Since $\left\{ \left(\frac{9}{10}\right)^n \right\}$ is a geometric sequence with $a_1 = \frac{9}{10}$ and $r = \frac{9}{10}$,

$$S_{10} = \frac{\frac{9}{10}\left[1 - \left(\frac{9}{10}\right)^{10}\right]}{1 - \frac{9}{10}} = 9\left[1 - \left(\frac{9}{10}\right)^{10}\right] \doteq 5.86.$$

Now use the formula $\sum_{i=1}^{n} i = \frac{n}{2}(n + 1)$.

$$5 \sum_{i=1}^{10} i = 5\left[\frac{10(10 + 1)}{2}\right] = 275$$

Hence $\sum_{i=1}^{10} \left[\left(\frac{9}{10}\right)^i - 5i\right] \doteq 5.86 - 275 = -269.14.$

Exercises

A Evaluate each expression.

1. $\sum_{i=1}^{3} i^2$
2. $\sum_{i=1}^{4} 2^i$
3. $\sum_{j=2}^{5} 2j$
4. $\sum_{j=0}^{3} (2j - 1)$
5. $\sum_{j=-1}^{2} (j^2 + 1)$
6. $\sum_{k=-4}^{-1} k \cdot 2^k$
7. $\sum_{k=3}^{7} \frac{(-1)^k}{k - 1}$
8. $\sum_{i=1}^{4} 2$

9. $\displaystyle\sum_{k=3}^{6} (-1)$
10. $\displaystyle\sum_{k=0}^{4} \sin\frac{k\pi}{2}$
11. $\displaystyle\sum_{k=-1}^{3} \frac{(-1)^k}{3^k}$
12. $\displaystyle\sum_{k=1}^{4} \frac{(-1)^{k+1}}{2k-1}$

Express each sum using sigma notation.

13. $\frac{1}{3} + \frac{1}{4} + \frac{1}{5} + \frac{1}{6}$

14. $2^3 + 3^3 + 4^3 + \ldots + 90^3$

15. $\sin\pi + \sin\frac{\pi}{2} + \sin\frac{\pi}{3} + \ldots + \sin\frac{\pi}{12}$

16. $1 + 3 + 7 + 15 + 31 + 63$

B 17. $\frac{1}{3} - \frac{1}{5} + \frac{1}{7} - \frac{1}{9} + \frac{1}{11} - \frac{1}{13}$

18. $\frac{1}{2\cdot 3} - \frac{1}{3\cdot 4} + \frac{1}{4\cdot 5} - \frac{1}{5\cdot 6} + \frac{1}{6\cdot 7}$

Evaluate each expression.

19. $\displaystyle\sum_{i=1}^{100} i^3 - \sum_{i=1}^{99} i^3$

20. $\displaystyle\sum_{k=1}^{100} (k^2 - k) - \sum_{k=3}^{100} k(k-1)$

21. $\displaystyle\sum_{i=1}^{50} (2i - 1)$

22. $\displaystyle\sum_{i=0}^{10} \left(\frac{1}{3}\right)^i$

23. $\displaystyle\sum_{i=2}^{11} \left(\frac{3}{2}\right)^i$

24. $\displaystyle\sum_{i=1}^{10} \frac{2^{i+1}}{3^i}$

25. $\displaystyle\sum_{i=1}^{12} (2^i - 2^{i-1})$

26. $\displaystyle\sum_{i=1}^{11} [i^3 - (i-1)^3]$

27. $\displaystyle\sum_{k=1}^{15} \frac{1}{k+3} - \sum_{k=4}^{18} \frac{1}{k}$

28. $\displaystyle\sum_{i=1}^{13} \frac{1}{i^2} - \sum_{k=1}^{10} \frac{1}{(k+2)^2}$

Evaluate the following sums if $\displaystyle\sum_{i=1}^{13} i^2 = 819$.

29. $\displaystyle\sum_{k=2}^{13} 5k^2$

30. $\displaystyle\sum_{i=1}^{13} (3i^2 - 1)$

31. $\displaystyle\sum_{i=0}^{13} (3i - 2i^2)$

32. $\displaystyle\sum_{i=1}^{13} (\sqrt{2}i^2 + 4i)$

33. Evaluate $\displaystyle\sum_{k=1}^{18} \frac{1}{(k+2)(k+3)}$ if $\displaystyle\sum_{i=1}^{20} \frac{1}{i(i+1)} = \frac{20}{21}$.

34. Evaluate $\displaystyle\sum_{i=-10}^{10} i^3$.

35. If $f(n) = \displaystyle\sum_{i=1}^{n} i^3$, evaluate $f(10) - f(9)$.

36. If $f(n) = \displaystyle\sum_{i=1}^{n} 3i$, show that $f(2n) - f(n) = k[n^2] + f(n)$ and evaluate the constant k.

37. If $f(n) = \dfrac{n(n+1)(2n+1)}{6}$ and $g(n) = \displaystyle\sum_{i=1}^{n} i^2$, evaluate $f(n) - g(n)$ for $n = 2$, 3, 4, and 5.

C 38. If $f(x) = \dfrac{x}{x^2 + 1}$, evaluate $\sum\limits_{i=1}^{10} [f(i) - f(i-1)]$.

39. By evaluating $\sum\limits_{i=1}^{n} \left[\dfrac{1}{i} - \dfrac{1}{i+1} \right]$, find a formula for the sum

$$\dfrac{1}{1 \cdot 2} + \dfrac{1}{2 \cdot 3} + \dfrac{1}{3 \cdot 4} + \cdots + \dfrac{1}{n(n+1)}.$$

40. Use the fact that $\sum\limits_{i=1}^{n} [i^3 - (i-1)^3] = \sum\limits_{i=1}^{n} (3i^2 - 3i + 1)$ to find a formula for $\sum\limits_{i=1}^{n} i^2$.

13–4 Mathematical Induction

It was suggested in the example at the beginning of the previous section that the grocer should be able to count the oranges in his 18-layer pyramid display by evaluating the sum $\sum\limits_{i=1}^{18} i^2$. This is a useful shortcut, but adding the resulting 18 terms is still a laborious task.

In exercise 37, section 13–3, you were asked to compare the sum $\sum\limits_{i=1}^{n} i^2$ with the product $\dfrac{n(n+1)(2n+1)}{6}$ for $n = 2$, 3, 4, and 5. In each case you should have found the sum and the product to be equal. For example, for $n = 5$, $\sum\limits_{i=1}^{5} i^2 = 1^2 + 2^2 + 3^2 + 4^2 + 5^2 = 55$, and $\dfrac{5(5+1)(2 \cdot 5 + 1)}{6} = 5 \cdot 11 = 55$.

If you could be sure that for every natural number n it is true that $\sum\limits_{i=1}^{n} i^2 = \dfrac{n(n+1)(2n+1)}{6}$, then it would be easy to calculate the number of oranges in the display; it would be $\dfrac{18(18+1)(2 \cdot 18 + 1)}{6}$, or 2109.

You might check the formula for many values of n but this would never prove the formula for all n, as even one counterexample would disprove it.

We introduce now a method of proving that a formula or a statement is true for every natural number n. This method is called **mathematical induction**. To see how it works, imagine that many dominoes are stood on end and lined up so that whenever one domino falls, it causes the next one in the line to fall as well. If you now push over the first domino, it will cause the second to fall, the second will cause the third to fall, and so on. A chain reaction occurs with the result that every domino falls.

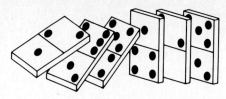

To see how the domino analogy applies as a method of proof, suppose you wish to prove that for every $n \in \mathbf{N}$, $\sum_{i=1}^{n} 2i = n(n + 1)$.

The first part of the proof is to show that the formula holds for $n = 1$. This corresponds to pushing over the first domino.

Part 1: Since $\sum_{i=1}^{1} 2i = 2 \cdot 1 = 2$ and $1(1 + 1) = 2$, it follows that

$$\sum_{i=1}^{1} 2i = 1(1 + 1);$$ the formula holds for $n = 1$.

The second step is to show that if the formula is true for a given natural number k, then it must also be true for $k + 1$. This corresponds to showing that whenever any domino is knocked over, the next one in line is also knocked over. The assumption that the formula is true for $n = k$ is called the **induction hypothesis**.

Part 2: Assume that for a given $k \in \mathbf{N}$, $\sum_{i=1}^{k} 2i = k(k + 1)$. Now prove that

$$\sum_{i=1}^{k+1} 2i = [k + 1][(k + 1) + 1].$$ Observe that

$$\sum_{i=1}^{k+1} 2i = (2 \cdot 1 + 2 \cdot 2 + \cdots + 2k) + 2(k + 1)$$

$$= \sum_{i=1}^{k} 2i + 2(k + 1).$$

Next substitute, using the induction hypothesis, to obtain

$$\sum_{i=1}^{k+1} 2i = k(k + 1) + 2(k + 1) = (k + 1)(k + 2)$$

$$= [k + 1][(k + 1) + 1].$$

This completes the proof that whenever the formula is true for $k \in \mathbf{N}$, it is also true for $k + 1$, the next consecutive integer.

Since the formula holds for $n = 1$, the second part of the proof implies the formula holds for $n = 2$, which in turn implies it is valid for $n = 3$, and so on. Just as pushing over the first domino sets off a chain reaction which flattens all the dominoes, so it is that the combination of the two parts of the proof shown above implies that the formula is true for all natural numbers n.

The intuitive ideas in the domino analogy are placed on a firm mathematical foundation in the principle of mathematical induction.

Principle of Mathematical Induction

Let $P(n)$ be a statement about any natural number n. If
a. the statement is true for $n = 1$, and
b. whenever the statement is true for $n = k$ it is also true for $n = k + 1$,
then the statement $P(n)$ is true for all $n \in \mathbf{N}$.

Example 1 Prove that for every natural number n, $\sum_{i=1}^{n} i^2 = \dfrac{n(n+1)(2n+1)}{6}$.

SOLUTION *Part 1:* Prove that the formula is true for $n = 1$.

$$\sum_{i=1}^{1} i^2 = 1^2 = 1, \text{ and } \frac{1(1+1)(2 \cdot 1 + 1)}{6} = \frac{2 \cdot 3}{6} = 1$$

Thus $\sum_{i=1}^{1} i^2 = \dfrac{1(1+1)(2 \cdot 1 + 1)}{6}$.

Part 2: Given that the formula is true for $n = k$, prove that it is true for $n = k + 1$. That is, assume as the induction hypothesis that for some $k \in \mathbf{N}$

$$\sum_{i=1}^{k} i^2 = \frac{k(k+1)(2k+1)}{6} \text{ and prove that}$$

$$\sum_{i=1}^{k+1} i^2 = \frac{(k+1)([k+1]+1)(2[k+1]+1)}{6}.$$

Since $\sum_{i=1}^{k+1} i^2 = (k+1)^2 + \sum_{i=1}^{k} i^2$, the induction hypothesis implies

$$\sum_{i=1}^{k+1} i^2 = (k+1)^2 + \frac{k(k+1)(2k+1)}{6}.$$

Now simplify the expression on the right.

$$\sum_{i=1}^{k+1} i^2 = [k+1]\left[\frac{6(k+1) + k(2k+1)}{6}\right]$$

$$= [k+1]\left[\frac{2k^2 + 7k + 6}{6}\right] = \frac{(k+1)(k+2)(2k+3)}{6}$$

$$= \frac{[k+1][(k+1)+1][2(k+1)+1]}{6}$$

But this is the formula for $n = k + 1$. This completes the proof that if the formula is true for $n = k$, then it is also true for $n = k + 1$. Therefore

$$\sum_{i=1}^{n} i^2 = \frac{n(n+1)(2n+1)}{6} \text{ for every } n \in \mathbf{N}.$$

The strategy for proving by mathematical induction that some statement about natural numbers is true for every natural number *n* can be summarized as follows.

Part 1: Verify that the statement is true when *n* is 1.

Part 2: a. Write out the induction hypothesis by replacing *n* by *k* everywhere in the statement.
b. By substituting $k + 1$ for *n* in the statement, write out explicitly the conclusion that must be reached.
c. Now use the induction hypothesis to prove the result expressed in b. Often this means expressing the statement for $k + 1$ in such a way that the statement for *k* can be directly substituted.

Conclusion: Having completed parts *1* and *2*, apply the principle of mathematical induction to conclude that the statement is true for every $n \in \mathbf{N}$.

Example 2 Prove that for every $n \in \mathbf{N}$, $\dfrac{1}{1 \cdot 4} + \dfrac{1}{4 \cdot 7} + \cdots + \dfrac{1}{(3n - 2)(3n + 1)} = \dfrac{n}{3n + 1}$.

SOLUTION *Part 1:* Since $\dfrac{1}{1 \cdot 4} = \dfrac{1}{3 \cdot 1 + 1}$, the statement is true for $n = 1$.

Part 2: Given that $\displaystyle\sum_{i=1}^{k} \dfrac{1}{(3i - 2)(3i + 1)} = \dfrac{k}{3k + 1}$,

prove that $\displaystyle\sum_{i=1}^{k+1} \dfrac{1}{(3i - 2)(3i + 1)} = \dfrac{k + 1}{3(k + 1) + 1}$.

First observe that

$$\sum_{i=1}^{k+1} \dfrac{1}{(3i - 2)(3i + 1)} = \dfrac{1}{[3(k + 1) - 2][3(k + 1) + 1]} + \sum_{i=1}^{k} \dfrac{1}{(3i - 2)(3i + 1)}.$$

Use the induction hypothesis to substitute for the last term.

$$\begin{aligned}\sum_{i=1}^{k+1} \dfrac{1}{(3i - 2)(3i + 1)} &= \dfrac{1}{(3k + 1)(3k + 4)} + \dfrac{k}{3k + 1} \\ &= \dfrac{1 + k(3k + 4)}{(3k + 1)(3k + 4)} = \dfrac{3k^2 + 4k + 1}{(3k + 1)(3k + 4)} \\ &= \dfrac{(3k + 1)(k + 1)}{(3k + 1)(3k + 4)} \\ &= \dfrac{k + 1}{3(k + 1) + 1}\end{aligned}$$

Thus if the given statement is true for $n = k$, it must be true for $n = k + 1$.

(cont. on p. 477)

Thus, by the principle of mathematical induction,

$$\sum_{i=1}^{n} \frac{1}{(3i-2)(3i+1)} = \frac{n}{3n+1} \text{ for every } n \in \mathbf{N}.$$

Whenever you wish to prove that a statement or theorem is true for all natural numbers, you should consider using a proof by mathematical induction. If, however, the statement is not restricted to the domain of natural numbers, then this is not a suitable means of proof. Another limitation is that you must know in advance the exact conclusion you are trying to prove. Mathematical induction is of no help in discovering new results.

Exercises

A 1. Is the statement $\sum_{i=1}^{n} 4i = 2n(n+1)$ true for $n = 1$? For $n = 3$?

2. Is the statement $\sum_{i=1}^{n} (3^i - 2) = 6n - 5$ true for $n = 1$? For $n = 2$?

3. Is $n(n^2 + 5)$ divisible by 6 if $n = 1$? If $n = 5$?
4. Is $n^3 + n + 1$ divisible by 3 if $n = 1$? If $n = 4$?
5. Is $2^n \leq n^2$ if $n = 1$? If $n = 2$? If $n = 3$?
6. Is $\sum_{j=1}^{n} \frac{1}{\sqrt{j}} \geq \sqrt{n}$ if $n = 1$? If $n = 2$? If $n = 3$?

In exercises 7–10, suppose you wish to prove the given statement is true for all $n \in \mathbf{N}$ by mathematical induction.
a. Determine whether the statement is true for $n = 1$.
b. State the induction hypothesis.
c. State precisely what it means to say that the statement is true for $n = k + 1$.

7. $\sum_{i=1}^{n} i^3 = \frac{n^2(n+1)^2}{4}$

8. $2n + 1$ is an odd number.

9. $3^n \geq n^3$

10. $\sum_{i=1}^{n} \frac{1}{i(i+1)} = \frac{n}{n+1}$

B Prove by mathematical induction that each of the following statements is true for all natural numbers n.

11. $\sum_{i=1}^{n} i = \dfrac{n(n+1)}{2}$

12. $\sum_{i=1}^{n} (2i-1) = n^2$

13. $\sum_{i=1}^{n} 3^i = \dfrac{3^{n+1}-3}{2}$

14. $\sum_{i=1}^{n} (i^2 + i) = \dfrac{n(n+1)(n+2)}{3}$

15. $\sum_{i=1}^{n} \dfrac{1}{i(i+1)} = \dfrac{n}{n+1}$

16. $\sum_{i=1}^{n} \dfrac{1}{2^i} = 1 - \dfrac{1}{2^n}$

17. $\sin(t + n\pi) = (-1)^n \sin t$

18. $\cos(t + n\pi) = (-1)^n \cos t$

19. $\sum_{i=1}^{n} i^3 = \left[\dfrac{n(n+1)}{2}\right]^2$

20. $\sum_{j=1}^{n} j \cdot 2^{j-1} = 1 + (n-1)2^n$

21. Prove or disprove that $\sum_{i=1}^{n} \dfrac{i^2}{i+1} = \dfrac{8n-5}{6}$ for all $n \in \mathbf{N}$.

22. Consider the statement $\sum_{i=1}^{n} (2i+1) = (n+1)^2$.
   a. If the statement is true for $n = k$, is it true for $n = k+1$?
   b. Is the statement true for all $n \in \mathbf{N}$?

C 23. Prove by mathematical induction that if c is constant and if f is a function defined on $\mathbf{N}$, then $\sum_{i=1}^{n} cf(i) = c \sum_{i=1}^{n} f(i)$.

24. Prove by mathematical induction that if f and g are functions defined on $\mathbf{N}$, then $\sum_{i=1}^{n} [f(i) + g(i)] = \sum_{i=1}^{n} f(i) + \sum_{i=1}^{n} g(i)$.

25. Prove by mathematical induction that for a geometric sequence $\{ar^{n-1}\}$ the sum of the first n terms is given by
$$S_n = \sum_{i=1}^{n} ar^{i-1} = \dfrac{a(1-r^n)}{1-r} \text{ for all } n \in \mathbf{N}.$$

26. Discover a formula for $\sum_{i=1}^{n} \dfrac{1}{4i^2 - 1}$ and prove that it is true for all $n \in \mathbf{N}$.

27. Prove that if $p > -1$, then $(1+p)^n \geq 1 + np$ for all $n \in \mathbf{N}$.

28. Prove that $\sum_{j=1}^{n} \dfrac{1}{\sqrt{j}} \geq \sqrt{n}$ for all $n \in \mathbf{N}$.

29. Prove that $\left(1 - \dfrac{1}{4}\right)\left(1 - \dfrac{1}{9}\right)\left(1 - \dfrac{1}{16}\right) \cdots \left(1 - \dfrac{1}{2500}\right) = \dfrac{51}{100}$.

[HINT: The denominators are consecutive perfect squares.]

13-5 De Moivre's Theorem

The Fundamental Theorem of Algebra, Chapter 6, states that every polynomial equation of degree $n \geq 1$ has exactly n solutions (counted in their multiplicity) in the domain of complex numbers **C**. Thus the equation $x^6 + 64 = 0$ should have six solutions. This is equivalent to saying there should be six 6th roots of -64 in **C**.

It is easy to verify that $2i$ and $-2i$ are 6th roots of 64, but it is not obvious what the other four are. The complex number $\sqrt{3} + i$ happens to be one of them, but to verify that fact you would have to evaluate $(\sqrt{3} + i)^6$, which is a time-consuming task.

In this section we shall apply mathematical induction to prove a famous theorem which makes it easy to find powers and roots of complex numbers. To use this theorem it is necessary to work with complex numbers expressed in polar form.

Consider the complex number $w = \sqrt{3}\left(\cos \frac{\pi}{5} + i \sin \frac{\pi}{5}\right)$. Suppose that it is required to express w^4 in the form $a + bi$. Recall that if

$$z_1 = r_1(\cos \theta_1 + i \sin \theta_1) \text{ and}$$
$$z_2 = r_2(\cos \theta_2 + i \sin \theta_2), \text{ then}$$
$$z_1 z_2 = r_1 r_2 [\cos(\theta_1 + \theta_2) + i \sin(\theta_1 + \theta_2)].$$

Then
$$w^2 = \sqrt{3} \cdot \sqrt{3} \left[\cos\left(\tfrac{\pi}{5} + \tfrac{\pi}{5}\right) + i \sin\left(\tfrac{\pi}{5} + \tfrac{\pi}{5}\right)\right]$$
$$= 3 \left[\cos \tfrac{2\pi}{5} + i \sin \tfrac{2\pi}{5}\right].$$

Now repeat this procedure to obtain w^4.
$$w^4 = (w^2)^2 = 3 \cdot 3 \left[\cos\left(\tfrac{2\pi}{5} + \tfrac{2\pi}{5}\right) + i \sin\left(\tfrac{2\pi}{5} + \tfrac{2\pi}{5}\right)\right]$$
$$w^4 = 9 \left(\cos \tfrac{4\pi}{5} + i \sin \tfrac{4\pi}{5}\right)$$

By calculator, $\quad w^4 \doteq -7.28 + 5.29i.$

The pattern in the expression for w^4 suggests the following theorem.

Theorem 13-4 **De Moivre's Theorem**

If $z = r(\cos \theta + i \sin \theta)$ is any complex number in polar form, then
$$z^n = r^n(\cos n\theta + i \sin n\theta), \text{ for every } n \in \mathbf{N}.$$

PROOF Use mathematical induction on the statement in the theorem.

Part 1: $z^1 = z = r(\cos \theta + i \sin \theta) = r^1[\cos(1 \cdot \theta) + i \sin(1 \cdot \theta)]$,
so the statement is true for $n = 1$.

(cont. on p. 480)

Part 2: Given that $z^k = r^k(\cos k\theta + i \sin k\theta)$ is true (the induction hypothesis), prove that

$$z^{k+1} = r^{k+1}[\cos(k+1)\theta + i\sin(k+1)\theta].$$

First note that $z^{k+1} = z^k \cdot z$. Now use the induction hypothesis and substitute $r^k(\cos k\theta + i \sin k\theta)$ for z^k.

$$\begin{aligned} z^{k+1} &= [r^k(\cos k\theta + i\sin k\theta)] \cdot [r(\cos\theta + i\sin\theta)] \\ &= r^k r[\cos(k\theta + \theta) + i\sin(k\theta + \theta)] \\ &= r^{k+1}[\cos(k+1)\theta + i\sin(k+1)\theta] \end{aligned}$$

Thus the given statement is true for $n = k + 1$ if it is true for $n = k$, and by the principle of mathematical induction,

$$z^n = r^n(\cos n\theta + i\sin n\theta) \text{ for every } n \in \mathbf{N}.$$

Example 1 Express $(-1 + \sqrt{3}i)^5$ in the form $a + bi$.

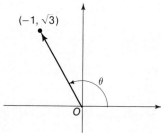

SOLUTION First express $z = -1 + \sqrt{3}i$ in polar form:

$$r = |z| = \sqrt{(-1)^2 + (\sqrt{3})^2} = 2, \quad \tan\theta = \frac{\sqrt{3}}{-1},$$

and as the figure shows, $\frac{\pi}{2} < \theta < \pi$. Hence $\theta = \frac{2\pi}{3}$.

Thus $z = 2\left(\cos\frac{2\pi}{3} + i\sin\frac{2\pi}{3}\right)$, and

$$z^5 = 2^5\left[\cos 5\left(\frac{2\pi}{3}\right) + i\sin 5\left(\frac{2\pi}{3}\right)\right] = 32\left(\cos\frac{10\pi}{3} + i\sin\frac{10\pi}{3}\right)$$

$$= 32\left(-\frac{1}{2} - \frac{\sqrt{3}}{2}i\right) = -16 - 16\sqrt{3}i.$$

In the next example, observe how de Moivre's Theorem can be applied to the problem stated earlier of finding all the solutions of the equation $x^6 + 64 = 0$.

Example 2 Find the six 6th roots of -64.

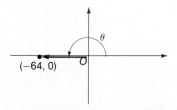

SOLUTION First express -64 in polar form. From the graph, $r = 64$ and $\theta = \pi$. Thus

$$-64 = 64(\cos\pi + i\sin\pi).$$

Now let $z = s(\cos\phi + i\sin\phi)$ be any 6th root of -64.

(cont. on p. 481)

$[s(\cos \phi + i \sin \phi)]^6 = 64(\cos \pi + i \sin \pi)$, and by de Moivre's Theorem,

$s^6(\cos 6\phi + i \sin 6\phi) = 64(\cos \pi + i \sin \pi)$.

These two complex numbers are equal if and only if

$s^6 = 64$ and $6\phi = \pi + 2k\pi$, where $k \in J$.

So $s = 2$ and $\phi = \dfrac{\pi}{6} + \dfrac{k\pi}{3}$.

The required sixth roots are now obtained by letting $k = 0, 1, 2, 3, 4,$ and 5 and computing the resulting values of ϕ.

$\phi = \dfrac{\pi}{6} + \dfrac{0 \cdot \pi}{3} = \dfrac{\pi}{6}; \dfrac{\pi}{6} + \dfrac{1 \cdot \pi}{3} = \dfrac{3\pi}{6}; \dfrac{\pi}{6} + \dfrac{2 \cdot \pi}{3} = \dfrac{5\pi}{6};$ and so on.

Thus the six sixth roots are as follows.

$z_1 = 2\left(\cos \dfrac{\pi}{6} + i \sin \dfrac{\pi}{6}\right) = \sqrt{3} + i$

$z_2 = 2\left(\cos \dfrac{3\pi}{6} + i \sin \dfrac{3\pi}{6}\right) = 0 + 2i$

$z_3 = 2\left(\cos \dfrac{5\pi}{6} + i \sin \dfrac{5\pi}{6}\right) = -\sqrt{3} + i$

$z_4 = 2\left(\cos \dfrac{7\pi}{6} + i \sin \dfrac{7\pi}{6}\right) = -\sqrt{3} - i$

$z_5 = 2\left(\cos \dfrac{9\pi}{6} + i \sin \dfrac{9\pi}{6}\right) = 0 - 2i$

$z_6 = 2\left(\cos \dfrac{11\pi}{6} + i \sin \dfrac{11\pi}{6}\right) = \sqrt{3} - i$

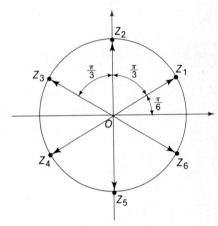

Observe that the graphs of the roots in the last example are equally spaced on the circle centered at the origin with radius 2. Notice also that if values of k other than 0, 1, 2, 3, 4, and 5 are chosen, no additional complex numbers are generated.

The results of this example can be stated as a corollary to de Moivre's theorem. We give the corollary here without proof.

Corollary 13–4 If $n \in N$ and $z = r(\cos \theta + i \sin \theta)$, $r > 0$, is any complex number, then the n distinct nth roots of z are given by

$$\sqrt[n]{r}\left[\cos\left(\dfrac{\theta + 2k\pi}{n}\right) + i \sin\left(\dfrac{\theta + 2k\pi}{n}\right)\right],$$

where $k = 0, 1, 2, \ldots, n - 1$.

Example 3 Find all the 4th roots of $1 - 3i$.

SOLUTION Express $z = 1 - 3i$ in polar form.
$$r = |z| = \sqrt{1^2 + (-3)^2} = \sqrt{10}$$
$$\tan \theta = \frac{-3}{1}$$

The figure shows that $\frac{3\pi}{2} < \theta < 2\pi$.

Since $-\frac{\pi}{2} < \arctan(-3) < 0$, let
$\theta = 2\pi + \arctan(-3)$. From Corollary 13–4, the roots are given by
$$\sqrt[4]{\sqrt{10}} \left[\cos\left(\frac{\theta}{4} + \frac{k\pi}{2}\right) + i \sin\left(\frac{\theta}{4} + \frac{k\pi}{2}\right) \right], \text{ where } \theta = 2\pi + \arctan(-3)$$
and $k = 0, 1, 2,$ and 3.

Using a calculator yields
$\theta \doteq 5.034$, and
$$z_1 \doteq 0.410 + 1.269i$$
$$z_2 \doteq -1.269 + 0.410i$$
$$z_3 \doteq -0.410 - 1.269i$$
$$z_4 \doteq 1.269 - 0.410i.$$

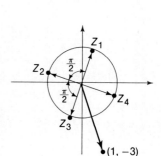

Exercises

A Express each of the following in the form $a + bi$ and simplify.

1. $\left[4\left(\cos\frac{\pi}{3} + i \sin\frac{\pi}{3}\right)\right]^3$
2. $\left[\sqrt{2}\left(\cos\frac{\pi}{4} + i \sin\frac{\pi}{4}\right)\right]^5$
3. $[\sqrt{3}(\cos 15° + i \sin 15°)]^6$
4. $[2(\cos 75° + i \sin 75°)]^4$
5. $(-1 + i)^4$
6. $(\sqrt{3} - i)^5$
7. $(3 - 3\sqrt{3}i)^5$
8. $(2\sqrt{2} + 2\sqrt{2}i)^6$
9. $(3 + 4i)^4$
10. $(-1 + 2i)^3$

Find the required roots and sketch their graphs.

11. The cube roots of $8\left(\cos\frac{2\pi}{3} + i \sin\frac{2\pi}{3}\right)$
12. The 4th roots of $9(\cos 240° + i \sin 240°)$

B 13. The square roots of $3i$
14. The cube roots of -8
15. The 4th roots of $-4 + 4\sqrt{3}i$
16. The 5th roots of $-16 - 16i$
17. The cube roots of $1 + 2i$
18. The 4th roots of $-3 + 4i$

Solve for z on the domain **C**. Express answers in the form $a + bi$.

19. $z^3 + 1 = 0$
20. $z^4 + 16i = 0$
21. $z^2 + 1 - i = 0$
22. $z^4 + 27iz = 0$
23. $z^3 - 2 + 3i = 0$
24. $z^4 + 3 - \sqrt{7}i = 0$

C 25. Use de Moivre's theorem to derive a formula for $\cos 3\theta$ in terms of $\cos \theta$.
[HINT: Consider $(\cos \theta + i \sin \theta)^3$.]

26. Show that if $a + bi$ is a solution of the equation $z^5 + 2a^2 - 1 = 0$, then $a - bi$ is also a solution. [HINT: Let $a + bi = r(\cos \theta + i \sin \theta)$.]

13–6 Limits

Some terms of the sequence $\{a_n\}$, where $a_n = \dfrac{n^2 + 1}{n^2}$, are $2, \dfrac{5}{4}, \dfrac{10}{9}, \dfrac{17}{16}, \dfrac{26}{25}, \dfrac{37}{36}, \ldots$. It is evident that the terms are getting closer and closer to 1 as n increases. The sequence $\{a_n\}$ is said to approach 1 as a **limit** and this is expressed by $\lim a_n = 1$.

Example 1 Write out several terms of each of the following sequences and identify their limits: **a.** $\left\{\dfrac{n}{n+1}\right\}$ **b.** $\left\{\dfrac{(-1)^{n+1}}{n}\right\}$ **c.** $\{1\}$.

SOLUTION **a.** $\left\{\dfrac{n}{n+1}\right\}$: $\dfrac{1}{2}, \dfrac{2}{3}, \dfrac{3}{4}, \dfrac{4}{5}, \dfrac{5}{6}, \ldots;$ $\lim \dfrac{n}{n+1} = 1$

b. $\left\{\dfrac{(-1)^{n+1}}{n}\right\}$: $1, -\dfrac{1}{2}, \dfrac{1}{3}, -\dfrac{1}{4}, \dfrac{1}{5}, -\dfrac{1}{6}, \ldots;$ $\lim \dfrac{(-1)^{n+1}}{n} = 0$

c. $\{1\}$: $1, 1, 1, 1, 1, \ldots;$ $\lim 1 = 1$

Example 2 For the sequence $\left\{\dfrac{n}{n+1}\right\}$, determine which terms lie within 0.1 unit of 1.

SOLUTION On the domain **N** solve the inequality

$$1 - 0.1 < \dfrac{n}{n+1} < 1 + 0.1.$$

$$\dfrac{9}{10} < \dfrac{n}{n+1} < \dfrac{11}{10}$$

$9n + 9 < 10n < 11n + 11$ Since $n + 1 > 0$

$n > 9$ and $-n < 11$

(cont. on p. 484)

Since $-n < 11$ for all $n \in \mathbf{N}$, the solution of the inequality is $\{n \in \mathbf{N}: n > 9\}$. Thus all terms beyond the 9th term will be within 0.1 unit of 1.

The calculations in this example could be repeated to show that for any positive distance ϵ it is possible to find a term a_M in the sequence beyond which all subsequent terms lie within the given distance ϵ of L.

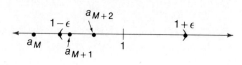

Definition A sequence $\{a_n\}$ has the real number L as a **limit** if and only if for every constant $\epsilon > 0$ there exists a number M such that

$$L - \epsilon < a_n < L + \epsilon$$

whenever $n > M$.

If L is a limit of a sequence $\{a_n\}$, then the sequence is said to **converge** to L, denoted by $\{a_n\} \to L$. If a sequence does not converge (has no limit), it is said to **diverge**.

An inspection of the terms of $\{a_n\}$: $1, \frac{1}{2}, \frac{1}{3}, \frac{1}{4}, \ldots, \frac{1}{n}, \ldots$ suggests that the sequence converges to 0. To prove this conjecture, apply the definition. Finding a suitable M corresponding to a given ϵ means that you must produce a formula which relates M to ϵ so that whenever $n > M$, $0 - \epsilon < a_n < 0 + \epsilon$. Let ϵ be any positive number. First find a formula for M in terms of ϵ by solving the inequality

$$0 - \epsilon < \frac{1}{n} < 0 + \epsilon$$

on the domain $\mathbf{N}$.

$$-n\epsilon < 1 < n\epsilon$$

$$-n\epsilon < 1 \quad \text{and} \quad 1 < n\epsilon$$

$$n > -\frac{1}{\epsilon} \quad \text{and} \quad n > \frac{1}{\epsilon}$$

But since $\epsilon > 0$, $n > -\frac{1}{\epsilon}$ for all $n \in \mathbf{N}$. Thus the solution set of the inequality is $\{n \in \mathbf{N}: n > \frac{1}{\epsilon}\}$. Therefore $0 - \epsilon < \frac{1}{n} < 0 + \epsilon$ whenever $n > M = \frac{1}{\epsilon}$, and the proof that $\{\frac{1}{n}\} \to 0$ is complete.

Theorem 13–5 The sequence $\{\frac{1}{n}\}$ converges to 0.

In working with limits, you will often apply the following theorem concerning the convergence of a sequence of constants.

Theorem 13–6 If c is a constant, the sequence $\{a_n\}$ such that $a_n = c$ for every $n \in \mathbf{N}$ converges to c.

The following important principle is a direct consequence of the definition of a limit.

> If $\{a_n\} \to L$, then every interval of the form $[L - \epsilon, L + \epsilon]$ contains all but a finite number of the terms of the sequence.

This observation implies that a sequence cannot have more than one limit and hence it is appropriate to speak of *the* limit of a sequence.

Not all sequences have limits. For example, it is clear that the terms for $\{n^2\}$: 1, 4, 9, 16, 25, ... grow large without bound. Hence there is no number which the terms approach. A guiding principle is that if a sequence is not *bounded* (does not have both an upper and a lower bound) then it cannot have a limit.

Example 3 Identify by inspection which of the following sequences converge:

a. $\{\sqrt{n}\}$, b. $\left\{\dfrac{1}{2^n}\right\}$, c. $\{(-1)^n\}$.

SOLUTION a. The terms of $\{\sqrt{n}\}$ increase without bound, so the sequence diverges.

b. $\left\{\dfrac{1}{2^n}\right\}$: $\dfrac{1}{2}, \dfrac{1}{4}, \dfrac{1}{8}, \dfrac{1}{16}, \ldots$ converges to 0.

c. $\{(-1)^n\}$: $-1, 1, -1, 1, -1, \ldots$ diverges even though it is bounded. There is no number L for which all but a finite number of terms fall inside the interval $\left[L - \tfrac{1}{2}, L + \tfrac{1}{2}\right]$, for example.

It is often helpful in considering the limit of a sequence to use the fact that sequences can be combined under arithmetic operations. Thus for sequences $\{a_n\}$ and $\{b_n\}$ the sum $\{a_n\} + \{b_n\}$ is defined as the sequence $\{a_n + b_n\}$ obtained by adding corresponding terms.

Example 4 For $\{a_n\}$: 3, 5, 7, ..., $2n + 1$, ... and $\{b_n\}$: 2, 4, 8, ..., 2^n, ..., determine $\{a_n\} + \{b_n\}$ and $\{a_n\} \cdot \{b_n\}$.

SOLUTION $\{a_n\} + \{b_n\} = \{a_n + b_n\}$: 5, 9, 15, ..., $2n + 1 + 2^n$, ..., and

$\{a_n\} \cdot \{b_n\} = \{a_n \cdot b_n\}$: 6, 20, 56, ..., $2^n(2n + 1)$, ...

Example 5 Express $\left\{\dfrac{n+1}{n}\right\}$ as a sum of two sequences.

SOLUTION Since $\dfrac{n+1}{n} = 1 + \dfrac{1}{n}$, $\left\{\dfrac{n+1}{n}\right\} = \{1\} + \left\{\dfrac{1}{n}\right\}$.

Writing some terms of the sequence $\left\{\dfrac{n+1}{n}\right\}$ as $2, \dfrac{3}{2}, \dfrac{4}{3}, \dfrac{5}{4}, \ldots$ suggests that $\lim \dfrac{n+1}{n} = 1$. Also Theorems 13–5 and 13–6 imply that $\lim 1 + \lim \dfrac{1}{n} = 1 + 0 = 1$. Since $\left\{\dfrac{n+1}{n}\right\} = \{1\} + \left\{\dfrac{1}{n}\right\}$, it appears that the limit of a sum of two sequences should equal the sum of their limits. The next theorem, given without proof, confirms that this is true.

Theorem 13–7 Given convergent sequences $\{a_n\}$ and $\{b_n\}$, with $\lim a_n = A$ and $\lim b_n = B$.

a. $\lim (a_n + b_n) = A + B$ **b.** $\lim (a_n \cdot b_n) = AB$

c. $\lim \dfrac{a_n}{b_n} = \dfrac{A}{B}$ provided $B \neq 0$, and for all $n \in \mathbf{N}$, $b_n \neq 0$.

By using Theorem 13–7 you can often carry out convergence proofs without using the M–ϵ definition of a limit.

Example 6 Prove that $\left\{\dfrac{1}{n^2}\right\} \to 0$.

SOLUTION $\left\{\dfrac{1}{n^2}\right\} = \left\{\dfrac{1}{n}\right\} \cdot \left\{\dfrac{1}{n}\right\}$, and $\lim \dfrac{1}{n} = 0$.

By Theorem 13–7b, $\lim \dfrac{1}{n^2} = \left(\lim \dfrac{1}{n}\right)\left(\lim \dfrac{1}{n}\right) = 0 \cdot 0 = 0$.

Example 7 Prove that $\left\{\dfrac{5n^2}{n^2 + 2n}\right\} \to 5$.

SOLUTION $\lim \dfrac{5n^2}{n^2 + 2n} = \lim \dfrac{5}{1 + \dfrac{2}{n}}$ \qquad Divide numerator and denominator by n^2.

$= \dfrac{\lim 5}{\lim \left(1 + \dfrac{2}{n}\right)} = \dfrac{\lim 5}{\lim 1 + \lim 2 \lim \dfrac{1}{n}}$ \qquad Theorem 13–7 c, a, b

$= \dfrac{5}{1 + 2 \cdot 0} = 5$ \qquad Theorems 13–6 and 13–5

Example 8 Show that $\left\{\dfrac{n^3 + 4n}{2n^2 - 1}\right\}$ is a divergent sequence.

SOLUTION
$$\dfrac{n^3 + 4n}{2n^2 - 1} = \dfrac{\left(\dfrac{n^3 + 4n}{n^3}\right)}{\left(\dfrac{2n^2 - 1}{n^3}\right)} = \dfrac{\left(1 + \dfrac{4}{n^2}\right)}{\left(\dfrac{2}{n} - \dfrac{1}{n^3}\right)}$$

For the numerator,
$$\lim\left(1 + \dfrac{4}{n^2}\right) = \lim 1 + (\lim 4)\left(\lim \dfrac{1}{n^2}\right) = 1 + 4 \cdot 0 = 1.$$

For the denominator,
$$\lim\left(\dfrac{2}{n} - \dfrac{1}{n^3}\right) = (\lim 2)\left(\lim \dfrac{1}{n}\right) - \left(\lim \dfrac{1}{n}\right)\left(\lim \dfrac{1}{n^2}\right)$$
$$= 2 \cdot 0 - 0 \cdot 0 = 0.$$

Since the denominator approaches 0 as the numerator approaches 1, the quotient is unbounded and therefore the sequence diverges.

The following theorem generalizes an earlier observation that $\{(\tfrac{1}{2})^n\} \to 0$.

Theorem 13–8 Given a constant r,

a. If $|r| < 1$, then $\{r^n\} \to 0$.

b. If $|r| > 1$, then $\{r^n\}$ diverges.

The proof is left as an exercise. For part **a**, show that when $n > \dfrac{\log \epsilon}{\log |r|}$, then $-\epsilon < r^n < \epsilon$, and for part **b** show that the sequence is unbounded.

Example 9 Prove that $\left\{\dfrac{9^{n+1}}{10^n}\right\} \to 0$.

SOLUTION
$$\left\{\dfrac{9^{n+1}}{10^n}\right\} = \{9\} \cdot \left\{\left(\dfrac{9}{10}\right)^n\right\}$$

Since $\left|\dfrac{9}{10}\right| < 1$, $\lim\left(\dfrac{9}{10}\right)^n = 0$, by Theorem 13–8a.

Hence $\lim \dfrac{9^{n+1}}{10^n} = [\lim 9]\left[\lim \left(\dfrac{9}{10}\right)^n\right] = 9 \cdot 0 = 0$.

Exercises

A Specify whether the given sequence is convergent or divergent. If convergent, identify the limit.

1. $\left\{\dfrac{3}{n}\right\}$
2. $\{n^3\}$
3. $\left\{\left(\dfrac{1}{3}\right)^n\right\}$
4. $\left\{\dfrac{2}{n^2}\right\}$
5. $\left\{\dfrac{1}{n^3+1}\right\}$
6. $\left\{\left(\dfrac{-4}{3}\right)^n\right\}$
7. $\left\{\left(\dfrac{-11}{10}\right)^n\right\}$
8. $\left\{\dfrac{7}{8^n}\right\}$
9. $\left\{\sin\dfrac{n\pi}{3}\right\}$
10. $\left\{\dfrac{2^{n+1}}{5^n}\right\}$
11. $\left\{\dfrac{4n}{\sqrt{n}}\right\}$
12. $\left\{\dfrac{2}{\sqrt{n}}\right\}$
13. $\{\log n\}$
14. $\{2^n\}$
15. $\left\{\left(\dfrac{\pi}{3}\right)^n\right\}$
16. $\{\cos n\pi\}$
17. $\left\{\dfrac{2n+1}{n}\right\}$
18. $\left\{\dfrac{n+2}{n^2}\right\}$
19. $\left\{\dfrac{2^n-1}{2^n}\right\}$
20. $\left\{\dfrac{n^3+7}{3n^2}\right\}$
21. $\left\{\dfrac{2n^2+3n}{5n^2-4}\right\}$
22. $\left\{\dfrac{n+3}{4n^2-2n}\right\}$
23. $\left\{\dfrac{3n^3}{n^4+1}\right\}$
24. $\left\{\dfrac{4n-7n^2}{5+7n}\right\}$

B Consider the given sequence $\{a_n\}$ and the given constants L and ϵ. In each case find the smallest integer M such that whenever $n > M$, $L - \epsilon < a_n < L + \epsilon$.

25. $\left\{\dfrac{1}{n+2}\right\}$; $L = 0$; $\epsilon = \dfrac{1}{8}$
26. $\left\{\dfrac{2n}{n+3}\right\}$; $L = 2$; $\epsilon = \dfrac{1}{20}$
27. $\left\{\dfrac{2n-1}{n+1}\right\}$; $L = 2$; $\epsilon = \dfrac{1}{50}$
28. $\left\{\dfrac{1}{\sqrt{n+3}}\right\}$; $L = 0$; $\epsilon = 0.01$

Determine the first 6 terms and identify the limit of the given sequence.

29. $\{a_n\}$ such that $a_1 = 1$ and for $n \geq 2$, $a_n = 1 + \dfrac{1}{1+a_{n-1}}$.

30. $\{a_n\}$ such that $a_1 =$ the number of students in your class and for $n \geq 2$, $a_n = \dfrac{1}{2}\left(\dfrac{49}{a_{n-1}} + a_{n-1}\right)$.

C Use the M-ϵ definition of a limit to prove that the given sequence converges to the specified limit.

31. $\left\{\dfrac{1}{2n+1}\right\} \to 0$
32. $\left\{\dfrac{n-1}{3n}\right\} \to \dfrac{1}{3}$

33. Define $\{a_n\}$ as follows: a_1 is your age in years; a_2 is your teacher's age; and for $n \geq 3$, $a_n = a_{n-1} + a_{n-2}$. Then form a second sequence $\{b_n\}$ defined by $b_n = \dfrac{a_{n+1}}{a_n}$. Write a computer program that will print the first 20 terms of $\{b_n\}$ and enable you to guess the limit of $\{b_n\}$ (a famous number called the Golden Mean).

Special   The Number π

What is π? A common answer is that π equals $\frac{22}{7}$, but since it was proved in 1761 that π is an irrational number, this answer is clearly false. A better answer is that π is the ratio of the circumference of any circle to its diameter. This of course is true, but if you then ask what is the circumference of a circle, you are apt to be told that it is equal to $2\pi r$. Now you have a definition that essentially defines π in terms of itself.

A different method of defining π approaches the problem in terms of the limit of a sequence of perimeters of regular polygons inscribed in a circle. First note that for every natural number $n \geq 3$, a regular n-sided polygon (sometimes called an *n-gon*) can be inscribed in any circle. From the perimeters of such n-gons in a circle of radius 1, a sequence $\{P_n\}$ can be created as follows.

$P_1 = 1$ } Chosen arbitrarily since 1- and 2-sided
$P_2 = 1$ } polygons don't exist.
P_3 = perimeter of the equilateral triangle
P_4 = perimeter of the square
$\vdots$
P_n = perimeter of the regular n-gon

It is easy to show by elementary geometry that
$P_3 = 3\sqrt{3}$ and $P_4 = 4\sqrt{2}$.

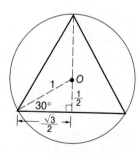 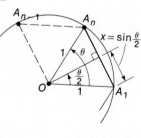

By applying trigonometry you can show that for any regular n-gon the perimeter P_n is given by

$$P_n = n\left(2 \sin \frac{180°}{n}\right)$$
$$= 2n \sin \left(\frac{180}{n}\right)°.$$

From this formula you can compute decimal approximations of terms of the sequence of perimeters: 1, 1, 5.196, 5.657, 5.878, 6.000, ... Observe that the successive terms differ by less and less. This is particularly apparent for the higher-numbered terms.

$P_{100} \doteq 6.2821519$ $P_{200} \doteq 6.2829274$

$P_{101} \doteq 6.2821723$ $P_{201} \doteq 6.2829299$

The sequence $\{P_n\}$ has a limit, which we define to be 2π. It is possible to approximate this limit as closely as you wish by evaluating terms sufficiently far out in the sequence.

A little sketching will make it evident that as the number of sides increases the inscribed polygons more closely approximate the circle in which they are inscribed.

The observation makes it seem reasonable to say that the circumference of a unit circle is equal to 2π and to define the circumference of a circle of radius r as the product $2\pi r$.

Exercises

1. Define a sequence $\{Q_n\}$ of perimeters of regular polygons *circumscribed* about a circle of radius 1 as follows:

   $Q_1 = 1$; $Q_2 = 1$; and for $n \geq 3$
   Q_n = perimeter of the n-gon.

   a. Evaluate Q_3, Q_4, Q_6, and Q_8.
   b. Derive a formula for Q_n.
   c. Using your formula from **b** evaluate Q_{200}.

2. Let $\{R_n\}$ be the sequence defined in terms of $\{P_n\}$ and $\{Q_n\}$ from exercise 1 as follows:

$$R_n = \frac{P_n + Q_n}{2}.$$

   a. Evaluate R_3 and R_6.
   b. Estimate π using R_{200}.

13-7 Infinite Series

In Chapter 8 you were introduced to the exponential function $f: x \to e^x$, which is used extensively in scientific work. The constant e is an irrational number which may be approximated by the sum

$$1 + \frac{1}{1!} + \frac{1}{2!} + \frac{1}{3!} + \frac{1}{4!} + \frac{1}{5!} + \frac{1}{6!},$$

which is an expression whose value is about 2.718056. By adding the next two terms in the pattern above, namely $\frac{1}{7!}$ and $\frac{1}{8!}$, you get a better estimate of e, one which agrees to six figures with the approximation a calculator gives when you press the keys $\boxed{1}$ $\boxed{\text{INV}}$ $\boxed{\ln x}$.

Observe that the denominators in this sum are all expressed in factorial notation.

$$1! = 1 \qquad 2! = 2 \cdot 1 \qquad 3! = 3 \cdot 2 \cdot 1 \qquad 4! = 4 \cdot 3 \cdot 2 \cdot 1$$

For all $n \in \mathbf{N}$,

$$n! = n(n-1)(n-2)(n-3) \cdots 4 \cdot 3 \cdot 2 \cdot 1.$$

By convention, $0! = 1$. The terms of the sum

$$\frac{1}{1!} + \frac{1}{2!} + \frac{1}{3!} + \cdots + \frac{1}{6!}$$

are members of the sequence $\left\{\frac{1}{n!}\right\}$.

Although we cannot add together all the terms of an infinite sequence, we can still write the expression

$$a_1 + a_2 + a_3 + \cdots + a_n + \cdots,$$

or

$$\sum_{n=1}^{\infty} a_n,$$

and call that collection of symbols an **infinite series**. For example, the sequence $\left\{\frac{1}{3^n}\right\}$ produces the infinite series

$$\frac{1}{3} + \frac{1}{9} + \frac{1}{27} + \cdots + \frac{1}{3^n} + \cdots.$$

If, for some natural number n, only the first n terms of an infinite series are added, the result is called the nth **partial sum** and is denoted S_n. Thus

$$S_1 = a_1,$$
$$S_2 = a_1 + a_2,$$
$$S_3 = a_1 + a_2 + a_3,$$

and for any $n \in \mathbf{N}$, $\quad S_n = a_1 + a_2 + \cdots + a_n = \sum_{k=1}^{n} a_k.$

Example 1 Evaluate the first 4 partial sums for the series derived from the sequence $\{n^2\}$.

SOLUTION
$S_1 = 1$
$S_2 = 1 + 4 = 5$
$S_3 = 1 + 4 + 9 = 14$
$S_4 = 1 + 4 + 9 + 16 = 30$

What meaning can be attached to the "sum" of infinitely many terms of a sequence? To answer the question, consider the series

$$\frac{1}{2} + \frac{1}{4} + \frac{1}{8} + \frac{1}{16} + \cdots + \frac{1}{2^n} + \cdots.$$

Create from this series a sequence of partial sums $\{S_n\}$:

$$S_1 = \frac{1}{2}$$

$$S_2 = \frac{1}{2} + \frac{1}{4} = \frac{3}{4}$$

$$S_3 = \frac{1}{2} + \frac{1}{4} + \frac{1}{8} = \frac{7}{8}$$

$$S_n = \frac{1}{2} + \frac{1}{4} + \frac{1}{8} + \cdots + \frac{1}{2^n}.$$

Since the terms of the series are drawn from the geometric sequence $\left\{\left(\frac{1}{2}\right)^n\right\}$, Theorem 13–2, which gives the sum of the first n terms of a geometric sequence, can be used to evaluate S_n.

$$S_n = a_1 \left[\frac{1 - r^n}{1 - r}\right]$$

$$= \frac{1}{2}\left[\frac{1 - \left(\frac{1}{2}\right)^n}{1 - \frac{1}{2}}\right]$$

$$= 1 - \left(\frac{1}{2}\right)^n$$

Thus $\left\{1 - \left(\frac{1}{2}\right)^n\right\}$ is the sequence of partial sums derived from the series

$$\frac{1}{2} + \frac{1}{4} + \frac{1}{8} + \frac{1}{16} + \cdots.$$

Next note that

$$\lim \left(1 - \left[\frac{1}{2}\right]^n\right) = \lim 1 - \lim \left(\frac{1}{2}\right)^n$$

$$= 1 - 0.$$

Thus the sequence of partial sums $\{S_n\}$ has 1 as its limit.

Whenever a series has the property that its sequence of partial sums has a limit S, the series is said to be **convergent** and S is called the **sum of the series**. In the illustration above, 1 is the sum of the series, and we write

$$1 = \sum_{n=1}^{\infty} \frac{1}{2^n}.$$

When a series yields a sequence of partial sums that does not have a limit, the series is said to be **divergent** and there is no number that can be designated as the sum of the series. The series

$$1 + 4 + 9 + 16 + \cdots + n^2 + \cdots$$

is an example of a divergent series. It is not always easy to recognize whether a series converges or diverges, for a series may diverge even though its terms approach 0 as a limit. The so-called harmonic series,

$$1 + \frac{1}{2} + \frac{1}{3} + \frac{1}{4} + \cdots + \frac{1}{n} + \cdots,$$

illustrates this possibility. To show that this series diverges, simply group the terms of the nth partial sum as follows:

$$S_n = \sum_{k=1}^{n} \frac{1}{k} = 1 + \frac{1}{2} + \left(\frac{1}{3} + \frac{1}{4}\right) + \left(\frac{1}{5} + \frac{1}{6} + \frac{1}{7} + \frac{1}{8}\right) + \cdots.$$

Notice that the sum of each group of terms in parentheses exceeds $\frac{1}{2}$. Since the sequence $\left\{\frac{1}{k}\right\}$ contains an infinite number of terms, it is possible to form as many groups of terms, each with sum greater than $\frac{1}{2}$, as needed to produce a partial sum that is greater than any given number. So the sequence of partial sums is unbounded. This implies that the harmonic series is divergent, even though it is created from a sequence that converges to 0.

In the case of a geometric series, it is quite simple to determine whether the series converges, and, if it does, the sum may be found readily.

Theorem 13–9 If $a \neq 0$, the geometric series

$$a + ar + ar^2 + ar^3 + \cdots + ar^{n-1} + \cdots, \quad \text{or} \quad \sum_{n=1}^{\infty} ar^{n-1}$$

a. diverges if $|r| \geq 1$;

b. converges if $|r| < 1$, with $\dfrac{a}{1-r}$ as its sum.

PROOF **b.** First note that $S_n = \dfrac{a(1-r^n)}{1-r} = \dfrac{a}{1-r}(1-r^n)$.

Since $\dfrac{a}{1-r}$ is a constant, $\lim \dfrac{a}{1-r} = \dfrac{a}{1-r}$.

(cont. on p. 493)

Since $|r| < 1$, $\lim r^n = 0$ and thus $\lim(1 - r^n) = 1$.

Hence $\lim \dfrac{a(1 - r^n)}{1 - r} = \lim \dfrac{a}{1 - r} \lim(1 - r^n) = \dfrac{a}{1 - r} \cdot 1$.

Therefore $\lim S_n = \dfrac{a}{1 - r}$ whenever $|r| < 1$.

Example 2 The first 3 terms of a geometric sequence are $\dfrac{7}{3}$, $-\dfrac{21}{8}$, and $\dfrac{189}{64}$. Determine whether the corresponding series, $\dfrac{7}{3} - \dfrac{21}{8} + \dfrac{189}{64} - \cdots$ converges.

SOLUTION Since the series is geometric,
$$r = -\dfrac{21}{8} \div \dfrac{7}{3} = -\dfrac{9}{8}.$$
Since $\left|-\dfrac{9}{8}\right| > 1$, the series diverges.

Example 3 A cider press is initially filled with apples, and each turn of the screw yields three-fifths as much cider as the preceding turn. If 2 liters of cider are squeezed out in the first turn, what is the maximum yield for the batch of apples?

SOLUTION Find the sum of the infinite series that corresponds to the sequence
$$2, \dfrac{6}{5}, \dfrac{18}{25}, \cdots, a_n, \cdots, \text{ where } a_n = 2\left(\dfrac{3}{5}\right)^{n-1}.$$
This is a geometric sequence with $a = 2$ and $r = \dfrac{3}{5}$. Since $|r| < 1$, the series converges, and the sum is
$$S = \dfrac{a}{1 - r} = \dfrac{2}{1 - \dfrac{3}{5}} = 5.$$
The given batch of apples can yield at most 5 liters of cider.

As suggested at the beginning of this section, the infinite series
$$\dfrac{1}{0!} + \dfrac{1}{1!} + \dfrac{1}{2!} + \dfrac{1}{3!} + \cdots = \sum_{k=0}^{\infty} \dfrac{1}{k!}$$
converges to the constant e. This series is a special case of the class of series,
$$\dfrac{1}{0!} + \dfrac{c}{1!} + \dfrac{c^2}{2!} + \dfrac{c^3}{3!} + \cdots = \sum_{k=0}^{\infty} \dfrac{c^k}{k!},$$
where c is a given constant. When $c = 1$, $\sum_{k=0}^{\infty} \dfrac{1^k}{k!} = e$.

Furthermore, it can be shown that no matter what constant c is chosen, the series $\sum_{k=0}^{\infty} \frac{c^k}{k!}$ converges and that e^c is the limit. In other words,

$$1 + \frac{x}{1!} + \frac{x^2}{2!} + \frac{x^3}{3!} + \frac{x^4}{4!} + \cdots = e^x.$$

Example 4 Estimate $e^{1/3}$ using 4 terms of the exponential series, and compare the result with a calculator approximation of $e^{1/3}$.

SOLUTION $e^{1/3} \doteq \sum_{k=0}^{3} \frac{\left(\frac{1}{3}\right)^k}{k!} = \sum_{k=0}^{3} \frac{1}{3^k k!}$

$e^{1/3} \doteq \frac{1}{3^0 \cdot 0!} + \frac{1}{3^1 \cdot 1!} + \frac{1}{3^2 \cdot 2!} + \frac{1}{3^3 \cdot 3!}$

$e^{1/3} \doteq 1 + \frac{1}{3} + \frac{1}{18} + \frac{1}{162} \doteq 1.3950617$

The calculator key sequence $\boxed{3}$ $\boxed{1/X}$ $\boxed{INV}$ $\boxed{\ln x}$ gives $e^{1/3} \doteq 1.3956124$.

How fast the series for e^x converges depends upon how far x is from 0. The smaller $|x|$ is, the more rapidly the corresponding series converges to e^x. In other words, the number of terms required for a given accuracy increases as $|x|$ gets larger. As you saw in example 4, a partial sum of only 4 terms produced an approximation that was within 0.001 of the calculator value of $e^{1/3}$. But for $x = 2$, using a 4-term partial sum of the exponential series to estimate e^2 gives a larger error.

$$\text{Error} = e^2 - \left(1 + \frac{2}{1!} + \frac{2^2}{2!} + \frac{2^3}{3!}\right) \doteq 1.0557$$

Trial and error shows that to get a series approximation that is within 0.001 of e^2 requires not 4 but 10 terms.

Exercises

A For each infinite geometric series, find the sum, or else specify that the series diverges.

1. $\sum_{i=1}^{\infty} 5\left(\frac{3}{4}\right)^{i-1}$

2. $\sum_{i=1}^{\infty} -2(1.02)^{i-1}$

3. $\sum_{i=1}^{\infty} \frac{1}{25}(\sqrt{3})^{i-1}$

4. $\sum_{k=1}^{\infty} 4\left(-\frac{2}{3}\right)^{k-1}$

5. $\sum_{k=1}^{\infty} \frac{-4}{5^{k-1}}$

6. $\sum_{k=1}^{\infty} 3\left(\frac{\sqrt{2}}{2}\right)^{k-1}$

7. $7 + 5 + \frac{25}{7} + \cdots$

8. $-\frac{9}{4} + 3 - 4 + \cdots$

9. $\frac{3}{5} - \frac{9}{25} + \frac{27}{125} + \cdots$

10. $\dfrac{2}{7} + \dfrac{4}{49} + \dfrac{8}{343} + \cdots$ 11. $\sum_{k=1}^{\infty} \dfrac{1}{3^k}$ 12. $\sum_{k=1}^{\infty} \dfrac{-7}{2^k}$

B Evaluate the partial sum $S_k = 1 + \dfrac{p}{1!} + \dfrac{p^2}{2!} + \dfrac{p^3}{3!} + \cdots + \dfrac{p^k}{k!}$ for given values of p and k to obtain a series approximation for e^p. Find the difference between this approximation and the value given by a calculator.

13. a. $p = \dfrac{1}{2};\ k = 3$ b. $p = \dfrac{4}{5};\ k = 3$ 14. a. $p = \dfrac{3}{2};\ k = 3$ b. $p = \dfrac{3}{2};\ k = 6$

15. a. $p = -1;\ k = 3$ b. $p = -1;\ k = 6$ 16. a. $p = -\dfrac{1}{3};\ k = 3$ b. $p = -\dfrac{2}{3};\ k = 3$

17. Suppose that a ball is dropped from a height of 25 feet and that each time it bounces the ball regains three-fourths of its original height. Theoretically, what is the greatest total distance the ball could travel before coming to rest?

18. Each week the amount of oil pumped from a well is 98% of the amount pumped the preceding week. If 800 barrels are produced during the first week of operation, how much oil could the well possibly produce?

If $-1 < x < 1$, then the series $x - \dfrac{x^2}{2} + \dfrac{x^3}{3} - \dfrac{x^4}{4} + \cdots$ converges to $\ln(1 + x)$. Use k terms of this series to approximate the specified logarithm and compare the result with the value given by a calculator.

19. a. $\ln 1.2;\ k = 3$ b. $\ln 1.2;\ k = 5$ [HINT: $\ln 1.2 = \ln(1 + 0.2)$]

20. a. $\ln 1.1;\ k = 4$ b. $\ln 1.5;\ k = 4$

Classify each series as convergent or divergent.

21. $1 - 1 + 1 - 1 + 1 - 1 + \cdots$ 22. $\sum_{k=1}^{\infty} \dfrac{1}{k+2}$

23. $\sum_{k=1}^{\infty} \dfrac{1}{2k}$ 24. $\sum_{k=1}^{\infty} \dfrac{3^{k+3}}{4^k}$

C 25. For the series $\sum_{k=1}^{\infty} \left(\dfrac{2}{3}\right)^k$, find the number of terms required to produce a partial sum S_n which will differ from the infinite sum by less than 0.02.

26. For the series $\sum_{k=1}^{\infty} 1.01^k$, find the number of terms required to produce a partial sum S_n which is greater than 100.

27. Write a computer program that uses the exponential series to approximate e^x for any given argument x and any given number of terms n.

28. The series $x - \dfrac{x^3}{3!} + \dfrac{x^5}{5!} - \dfrac{x^7}{7!} + \cdots$ converges to $\sin x$. Write a computer program to approximate $\sin x$ for any given argument x and any given number of terms n.

Chapter Review Exercises

13–1, page 459

1. Determine whether $\frac{45}{13}$ is a member of $\left\{\frac{3n}{n-1}\right\}$.

2. How many numbers between 100 and 400 are divisible by 13? What is their sum?

13–2, page 464

3. In a geometric sequence $\{a_n\}$, $a_1 = 125$ and $a_6 = \frac{243}{25}$. Evaluate S_5.

4. The Quality Sales Corporation gives each employee an annual bonus and increases that bonus by one-twelfth each year. If an employee receives $200 the first year, how long will it take the employee to accumulate more than $3000 in bonus money?

13–3, page 468

5. Use sigma notation to express the following sum:

$$\frac{1}{17} - \frac{1}{26} + \frac{1}{37} - \frac{1}{50} + \frac{1}{65} - \frac{1}{82}.$$

6. Evaluate: $\sum_{i=0}^{20} \frac{1}{i+1} - \sum_{j=2}^{19} \frac{1}{j+1}$.

13–4, page 473

7. Prove that for every $n \in \mathbb{N}$, $\sum_{i=1}^{n} \frac{1}{(4i-3)(4i+1)} = \frac{n}{4n+1}$.

13–5, page 479

8. Use de Moivre's Theorem to express $(-2 + 2\sqrt{3}i)^5$ in simplest $a + bi$ form.

9. Solve for z on the domain $\mathbb{C}$: $z^3 + 4\sqrt{2} - 4\sqrt{2}i = 0$.

13–6, page 483

10. For each of the following, specify whether the given sequence is convergent or divergent. If convergent, specify the limit.

    a. $\left\{\frac{3n-1}{n+2}\right\}$ b. $\left\{\frac{3^n+1}{3^n}\right\}$ c. $\left\{\sin\frac{n\pi}{2}\right\}$ d. $\left\{\frac{2n+1}{n^2+2n}\right\}$

13–7, page 490

11. For each of the following, determine whether the given series is convergent or divergent. If convergent, find the sum.

    a. $\sum_{k=1}^{\infty} \left(\frac{2}{3}\right)^k$ b. $\sum_{k=1}^{\infty} \frac{1}{k+1}$ c. $\sum_{k=1}^{\infty} \left(\frac{\pi}{3}\right)^k$ d. $\sum_{k=0}^{\infty} \frac{3}{2^k}$

12. Using 3 terms of the exponential series, obtain an approximation for $e^{1/2}$.

Chapter Test

1. Evaluate: $\sum_{i=0}^{4} \frac{(-1)^i}{2^i}$.

2. Find the tenth term of the arithmetic sequence $\{a_n\}$ if $a_1 = -7$ and $a_3 = -4$.

3. Find the sum of the multiples of 3 that lie between 80 and 200.

4. Use sigma notation to express the following sum:
$$3 \sin \frac{\pi}{3} + 9 \sin \frac{\pi}{6} + 27 \sin \frac{\pi}{9} + 81 \sin \frac{\pi}{12}.$$

5. For each of the following, determine whether the given sequence converges or diverges and find the limit if it exists.

    a. $\left\{ \dfrac{4n - 3n^3}{3n^2 + 1} \right\}$

    b. $\left\{ \left(\dfrac{4}{5}\right)^n \right\}$

    c. $\left\{ \dfrac{2n^2}{1 - n^2} \right\}$

6. Express $\left(\frac{1}{2} + \frac{1}{2}i\right)^5$ in simplest $a + bi$ form.

7. Find the cube roots of $-27i$.

8. A publisher introducing a new monthly magazine estimates that during the first year of publication sales should increase by one-eighth each month. To break even in the first year, 200,000 copies must be sold. To accomplish this how many copies must be sold in the first month of publication?

9. For each of the following, determine whether the given series is convergent or divergent. If convergent, find the sum.

    a. $\sum_{i=0}^{\infty} \left(\dfrac{3}{4}\right)^i$

    b. $\sum_{i=1}^{\infty} \dfrac{3}{i}$

    c. $\sum_{i=1}^{\infty} \left(\dfrac{3\sqrt{2}}{2}\right)^i$

10. Prove by mathematical induction that for all $n \in \mathbf{N}$,
$$\sum_{i=1}^{n} (2i - 1)^2 = \frac{4n^3 - n}{3}.$$

Introduction to Calculus
Derivatives

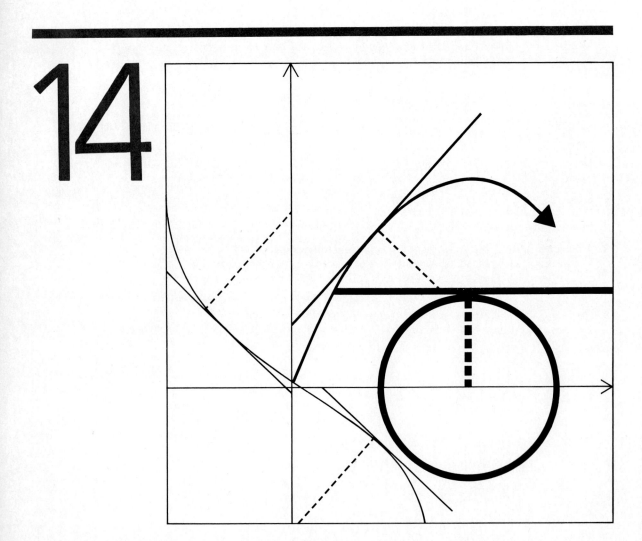

14

14–1 Tangents to Curves

In geometry a tangent to a circle is defined as a line that intersects the circle in exactly one point. In Chapter 5 you saw that this idea could be extended to include tangents to parabolas. Tangency can, in fact, be defined for a much broader class of curves, but it will be necessary to omit the requirement that a tangent must intersect the curve in only one point.

Consider the graph of $y = x^3$, shown twice below, and the lines ℓ_1 and ℓ_2.

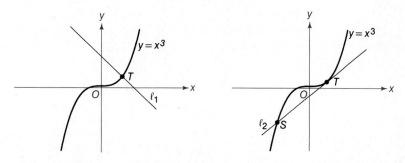

We want ℓ_1 in the figure at the left not to be a tangent despite its intersection with the curve at only one point. We want ℓ_2 in the figure at the right to be tangent to the curve at T even though it intersects the curve at a second point S. That is, we want to consider the tangent at a point on a curve without regard to what the tangent line or the curve does far from the point.

Suppose you want to obtain the tangent to the graph of $y = x^2$ at the point $T(2, 4)$. For every natural number n there corresponds a unique point $P_n(x_n, y_n)$ with $x_n = 2 + \dfrac{1}{n}$ and $y_n = \left(2 + \dfrac{1}{n}\right)^2$. Each of these points when paired with T determines a secant $\overleftrightarrow{TP_n}$ which has a slope

$$m_n = \frac{y_n - 4}{x_n - 2} = \frac{\left(2 + \dfrac{1}{n}\right)^2 - 4}{\left(2 + \dfrac{1}{n}\right) - 2}$$

$$= \frac{\dfrac{4}{n} + \dfrac{1}{n^2}}{\dfrac{1}{n}} = 4 + \dfrac{1}{n}.$$

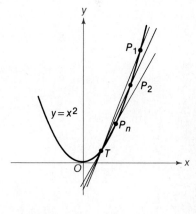

Thus the sequence of points $\{P_n\}$, which approaches T along the curve, generates a corresponding sequence of slopes $\{m_n\} = \left\{4 + \dfrac{1}{n}\right\}$. As the points P_n approach T, the secants more nearly approximate the tangent at T. This suggests that the tangent at T could be defined as the line through T whose slope is the limit of the sequence $\{m_n\}$.

Since $\lim \left(4 + \frac{1}{n}\right) = 4$, an equation for that line is $y - 4 = 4(x - 2)$. By the methods of Chapter 5, you can confirm that this line is in fact the tangent to $y = x^2$ at $(2, 4)$.

This method of finding a tangent at $T(2, 4)$ can be streamlined as follows. For every $h \neq 0$ there corresponds a unique point P on the parabola, distinct from T, which has coordinates $(2 + h, [2 + h]^2)$. Observe that as values of h are chosen closer and closer to 0, the point P gets closer and closer to T. Now create a slope function M which assigns to each nonzero h the slope of the secant $\overleftrightarrow{TP}$.

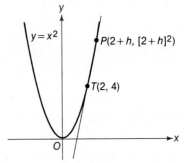

$$M(h) = \frac{(2 + h)^2 - 4}{(2 + h) - 2}$$
$$= \frac{4h + h^2}{h} = 4 + h; \text{ Dom } M = \mathbf{R} - \{0\}$$

As the sequence $\left\{4 + \frac{1}{n}\right\} \to 4$, so the function $M(h) = 4 + h$ is said to approach 4 as a limit as h approaches 0. This is written $\lim\limits_{h \to 0} M(h) = 4$.

Thus the slope of the tangent at T is the limit of the slope function at T. To generalize this method of describing a tangent at any point on a curve, suppose that $T(c, f(c))$ is a given point on the graph of a function f. The slope function at T is then

$$M(h) = \frac{f(c + h) - f(c)}{h}, h \neq 0.$$

Next evaluate $\lim\limits_{h \to 0} M(h)$. Do not be disturbed by the fact that slope functions are not defined at $h = 0$.

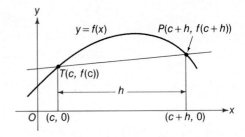

What matters is that values of the slope function will be as close to the limit as you wish if arguments h are chosen sufficiently close to 0. Assuming that $\lim\limits_{h \to 0} M(h)$ exists, let this limit be the slope of a line through T. This is a line that agrees with our extended notion of a tangent to f at T.

Definition If a function f is defined on an open interval containing an argument c and if $\lim\limits_{h \to 0} M(h) = m$, where $M(h) = \dfrac{f(c + h) - f(c)}{h}$, then the line through $T(c, f(c))$ with slope m is the **tangent** to the graph of f at T.

Example 1 Find an equation of the tangent to the graph of $f: x \to \dfrac{1}{x}$ at the point $\left(-2, -\dfrac{1}{2}\right)$.

SOLUTION Let $M(h) = \dfrac{f(-2+h) - f(-2)}{h}$, $h \neq 0$.

To evaluate $\lim\limits_{h \to 0} M(h)$, first simplify $M(h)$.

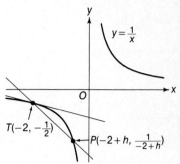

$$M(h) = \dfrac{\dfrac{1}{-2+h} - \dfrac{-1}{2}}{h}$$

$$= \dfrac{2 + (-2+h)}{2(-2+h)} \cdot \dfrac{1}{h} = \dfrac{1}{-4+2h}$$

Now observe that as h approaches 0 the denominator, $-4 + 2h$, approaches -4. Hence $\lim\limits_{h \to 0} M(h) = \lim\limits_{h \to 0} \dfrac{1}{-4+2h} = -\dfrac{1}{4}$. The tangent to the graph of f at $\left(-2, -\dfrac{1}{2}\right)$ is therefore the line $y + \dfrac{1}{2} = -\dfrac{1}{4}(x+2)$.

Example 2 Find the slope of the tangent to the graph of $g(x) = x^3 - 9x$ at $(2, -10)$.

SOLUTION $M(h) = \dfrac{g(2+h) - g(2)}{h}$, $h \neq 0$

$$= \dfrac{[(2+h)^3 - 9(2+h)] - [2^3 - 9 \cdot 2]}{h}$$

$$= \dfrac{1}{h}(h^3 + 6h^2 + 3h)$$

$$= h^2 + 6h + 3$$

Since h^2 and $6h$ each approach 0 as h approaches 0,

$$\lim_{h \to 0} M(h) = \lim_{h \to 0} (h^2 + 6h + 3) = 3.$$

So at $(2, -10)$ the slope of the tangent is 3.

The next example shows that it is not always so easy to determine the limit of the slope function.

Example 3 Find the tangent to $f: x \to \sqrt{x}$ at $(9, 3)$.

SOLUTION $M(h) = \dfrac{f(9+h) - f(9)}{h}$, $h \neq 0$

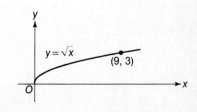

$$= \dfrac{\sqrt{9+h} - 3}{h}$$

In this case, as h approaches 0 the numerator also approaches 0; it is not
(cont. on p. 502)

obvious what is happening to the quotient. Fortunately $\lim_{h \to 0} M(h)$ becomes apparent when the numerator is rationalized. Thus, for $h \neq 0$,

$$M(h) = \frac{\sqrt{9+h} - 3}{h} \cdot \frac{\sqrt{9+h} + 3}{\sqrt{9+h} + 3} = \frac{1}{h}\left(\frac{9+h-9}{\sqrt{9+h}+3}\right) = \frac{1}{\sqrt{9+h}+3}.$$

Since $\sqrt{9+h} + 3$ approaches $\sqrt{9} + 3$ as h approaches 0,

$$\lim_{h \to 0} M(h) = \lim_{h \to 0} \frac{1}{\sqrt{9+h}+3} = \frac{1}{6}.$$

Therefore the tangent at (9, 3) is the line $y - 3 = \frac{1}{6}(x - 9)$.

Exercises

A Given the function f and the point T, evaluate the slope function for the given argument h.

1. $f: x \to x^2 + 1$; $T = (3, 10)$; $h = \frac{1}{2}$

2. $f: x \to \frac{1}{x-1}$; $T = \left(-1, -\frac{1}{2}\right)$; $h = \frac{1}{4}$

Evaluate $\lim_{h \to 0} M(h)$ for each function.

3. $M(h) = 3 + 2h$

4. $M(h) = -4 - 3h$

5. $M(h) = h^2 - 2h + 3$

6. $M(h) = \frac{1}{2 + h^2}$

7. $M(h) = \frac{h^2 + 3h}{h}$

8. $M(h) = \frac{h^2 - 4}{h + 2}$

Find an equation for the tangent to the graph of the given function at the specified point.

9. $f(x) = 2x^2$; $(-2, 8)$

10. $f(x) = 3x^2 - 7$; $(2, 5)$

11. $g(x) = \frac{1}{2x}$; $\left(1, \frac{1}{2}\right)$

12. $g(x) = x^3$; $(2, 8)$

13. $h: x \to 3x^2 - 2x$; $(0, 0)$

14. $h: x \to \frac{1}{x+1}$; $\left(-3, -\frac{1}{2}\right)$

B **15.** $f: x \to 2x^3 - 3x$; $(1, -1)$

16. $f: x \to 2x^2 - x^3$; $(3, -9)$

17. $g(x) = \sqrt{x+1}$; $(3, 2)$

18. $g(x) = \sqrt{2x+1}$; $(4, 3)$

19. $h(x) = \frac{1}{x^2}$; $\left(2, \frac{1}{4}\right)$

20. $h(x) = \frac{1}{x^2 - 4}$; $\left(1, -\frac{1}{3}\right)$

21. If $T(c, c^2)$ is a given point on the graph of $y = x^2$, find the slope of the tangent at T in terms of the constant c.

22. If $T(c, c^3)$ is a given point on the graph of $y = x^3$, find the slope of the tangent at T in terms of the constant c.

C **23.** Write an equation of the tangent to the ellipse $x^2 + 4y^2 = 13$ at (3, 1). [HINT: What function describes the top half of the ellipse?]

24. Find the point (c, d) on the graph of $f(x) = 2x^2 - x$ such that the tangent at (c, d) is parallel to the line $y = 7x$.

Special The Limit of a Function

An extended definition of tangency was given in section 14–1 in terms of the limit of a slope function. It was suggested that the limit of a function is similar to the limit of a sequence, but we did not state precisely what it means to say that $\lim_{h \to 0} M(h) = m$.

The statement $\lim_{x \to c} f(x) = L$ is read, "$f(x)$ approaches L as a limit as x approaches c." Roughly speaking this means that values of f will be as close as you wish to L when arguments are chosen sufficiently close to c. The definition is stated in terms of numbers specified by the Greek letters ϵ (epsilon) and δ (delta).

Definition A function f has a **limit** L at an argument c, denoted as

$$\lim_{x \to c} f(x) = L,$$

if and only if for every $\epsilon > 0$ there exists $\delta > 0$ such that $L - \epsilon < f(x) < L + \epsilon$ whenever $c - \delta < x < c + \delta$ but $x \neq c$.

Thus the statement $\lim_{x \to c} f(x) = L$ implies that for any given distance ϵ, it is possible to find an interval centered at c having width 2δ such that if x, $x \neq c$, is in the interval, then $f(x)$ will lie within the given distance ϵ of L. Note that the definition includes the phrase "but $x \neq c$." This is an important restriction. It means that a function can have a limit at $x = c$ and yet be undefined there. Recall that any slope function $M: h \to \dfrac{f(a + h) - f(a)}{h}$ is undefined at $h = 0$, and yet $\lim_{h \to 0} M(h)$ must exist if there is to be a tangent at $(a, f(a))$.

Example If $M(h) = \dfrac{3(-1 + h)^2 - 3(-1)^2}{h}$, as needed for the tangent to $y = 3x^2$ at $(-1, 3)$, prove that M has a limit at $h = 0$.

SOLUTION First note that Dom $M = \mathbf{R} - \{0\}$. Since $M(h) = \dfrac{-6h + 3h^2}{h}$, it follows that for every $h \in$ Dom M, $M(h) = -6 + 3h$. Even though M is not defined at $h = 0$, it appears that $\lim_{h \to 0} M(h) = -6$.

To prove this, consider any $\epsilon > 0$ and show that there exists $\delta > 0$ such that

$$-6 - \epsilon < M(h) < -6 + \epsilon \text{ whenever}$$
$$0 - \delta < h < 0 + \delta, \text{ but } h \neq 0.$$

To find such a δ, solve the inequality

$$-6 - \epsilon < \quad M(h) \quad < -6 + \epsilon.$$
$$-6 - \epsilon < -6 + 3h < -6 + \epsilon$$
$$-\frac{\epsilon}{3} < \quad h \quad < \frac{\epsilon}{3}$$

Now let $\delta = \dfrac{\epsilon}{3}$, and because the steps above are reversible, it follows that if

$$0 - \frac{\epsilon}{3} < h < 0 + \frac{\epsilon}{3},$$

then for $h \neq 0$,

$$-6 - \epsilon < \frac{-6h + 3h^2}{h} < -6 + \epsilon.$$

Thus for any $\epsilon > 0$, a δ has been found as required by the definition. This completes the proof that $\lim_{h \to 0} M(h) = -6$.

Exercises

1. Given that $f(x) = 5x + 2$, that $c = 3$, and that $\epsilon = 0.02$, identify $L = \lim_{x \to c} f(x)$ and find δ such that if $3 - \delta < x < 3 + \delta$ then $L - 0.02 < f(x) < L + 0.02$.

2. If $M(h) = \dfrac{2h^2 + 7h}{h}$, prove $\lim_{h \to 0} M(h) = 7$.

3. Prove: $\lim_{x \to -1} (3x - 2) = -5$.

4. Prove: $\lim_{x \to 4} \dfrac{x^2 - 16}{x - 4} = 8$.

5. If $f(x) = \dfrac{2x^2 - x - 15}{x - 3}$, identify $L = \lim_{x \to 3} f(x)$ and prove that L is the limit.

14-2 The Derivative Function

The previous section addressed the question, "If T is a point on a curve, what line is tangent to the curve at T?" Now the question is turned around: "If ℓ is a line tangent to a given curve, where on the curve is the point of tangency?" A related question is, "At what point (or points) of a graph does the tangent have a given slope m?" Or more specifically, "Where on the graph of a given function does the tangent have slope 0?"

To see why the last question is of particular interest, suppose that a rectangular beam is to be cut from a log with diameter 1 foot. It turns out that the strength of the beam can be expressed as a function S of the width x of the beam defined by

$$S(x) = K(x - x^3),$$

where K is a positive constant that depends upon the kind of wood. What width will yield the strongest beam?

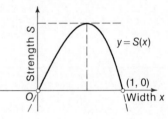

Since the diameter of the log is 1 foot, the domain of S is restricted to $\{x: 0 < x < 1\}$, and the graph of S will be as sketched. The graph suggests that the width that will yield the strongest beam can be found by determining where on the graph of S the tangent is horizontal, that is, where the tangent has slope 0. We shall now develop the mathematical tools for solving this and related problems.

Example 1 Find the point, or points, on the graph of $f: x \to x^3$ where the slope of the tangent is 12.

SOLUTION The graph suggests there are two points on the curve where the slope of the tangent is 12. At (c, c^3) the slope function is

$$M(h) = \frac{f(c + h) - f(c)}{h}, \quad h \neq 0$$

$$= \frac{(c + h)^3 - c^3}{h}$$

$$= \frac{[c^3 + 3c^2h + 3ch^2 + h^3] - c^3}{h}$$

$$= 3c^2 + 3ch + h^2.$$

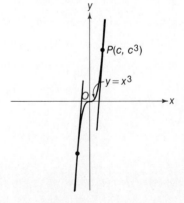

Since c is a constant, it is evident that as h approaches 0, both $3ch$ and h^2 also approach 0. Hence $\lim_{h \to 0} M(h) = \lim_{h \to 0} (3c^2 + 3ch + h^2) = 3c^2$. So at the point (c, c^3) the slope of the tangent is $3c^2$. Therefore to find the point

(cont. on p. 505)

where the slope of the tangent is 12, solve the equation $3c^2 = 12$ to get $c = \pm 2$.

ANSWER The tangents at $(2, 8)$ and $(-2, -8)$ have slope 12.

Example 1 shows that at any point (c, c^3) on the graph of $f: x \to x^3$, the slope of the tangent is $3c^2$. For instance, at $\left(\frac{3}{2}, \frac{27}{8}\right)$ the tangent has slope $3\left(\frac{3}{2}\right)^2$, or $\frac{27}{4}$.

Thus the function $g: x \to 3x^2$ can be seen as the function that assigns to every $x \in \text{Dom } f$ a value that is the slope of the tangent to the graph of $f: x \to x^3$ at the point (x, x^3).

This function g is derived from the given function f and so it is called the derived function, or the derivative, of f and denoted by the symbol f' (f prime). That is, the derivative of $f: x \to x^3$ is $f': x \to 3x^2$.

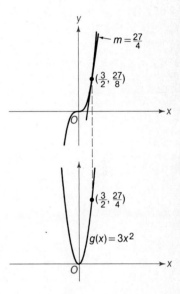

Definition The **derivative** f' of a given function f is the function defined by

$$f'(x) = \lim_{h \to 0} \frac{f(x + h) - f(x)}{h}$$

for every x such that the limit exists.

If $c \in \text{Dom } f'$, the value $f'(c)$ is called the *derivative of f at c*.

Example 2 If $f: x \to x^3$, evaluate $f'(4)$.

SOLUTION From example 1, $f'(x) = 3x^2$. Therefore $f'(4) = 3(4)^2 = 48$.

Example 3 If $f(x) = \dfrac{1}{2x + 1}$, find $f'(x)$.

SOLUTION Assume c is any argument in Dom f.

$$\frac{f(c + h) - f(c)}{h} = \frac{\dfrac{1}{2(c + h) + 1} - \dfrac{1}{2c + 1}}{h}, h \neq 0$$

(cont. on p. 506)

$$\frac{f(c+h) - f(c)}{h} = \frac{2c + 1 - 2c - 2h - 1}{h(2c + 2h + 1)(2c + 1)}$$

$$= \frac{-2}{(2c + 2h + 1)(2c + 1)}$$

Since c is a constant, $2c + 2h + 1$ approaches $2c + 1$ as h approaches 0.

Therefore $\lim_{h \to 0} \frac{-2}{(2c + 2h + 1)(2c + 1)} = \frac{-2}{(2c + 1)(2c + 1)}$, and

$$f'(c) = \frac{-2}{(2c + 1)^2}.$$

But c was assumed to be any argument in Dom f. So

$$f'(x) = \frac{-2}{(2x + 1)^2} \text{ and Dom } f' = \mathbf{R} - \left\{-\frac{1}{2}\right\}.$$

Example 4 Find any points on the graph of $f: x \to x^2$ where the tangent is parallel to the line $y = 3x + 8$.

SOLUTION The tangent will be parallel to the given line when the derivative has the value 3, so first find $f'(x)$.

Let c be any number in the domain of f.

$$\frac{f(c+h) - f(c)}{h} = \frac{(c+h)^2 - c^2}{h}, h \neq 0$$

$$= 2c + h$$

Since c is a constant, $2c + h$ approaches $2c$ as h approaches 0. Hence,

$$\lim_{h \to 0} \frac{f(c+h) - f(c)}{h} = 2c.$$

Thus $f'(x) = 2x$. Now solve

$$f'(x) = 3.$$
$$2x = 3$$
$$x = \frac{3}{2}$$

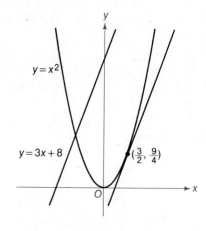

So at $\left(\frac{3}{2}, \frac{9}{4}\right)$ the tangent to the curve is parallel to the line $y = 3x + 8$.

Example 5 Find the width of the strongest beam that can be cut from a log with diameter 1 foot.

SOLUTION As noted at the beginning of this section, the strength of the beam is given by the function $S: x \to K(x - x^3)$, where x is the width in feet.

(cont. on p. 507)

The width of the strongest beam is found by determining where the tangent to the graph of S is horizontal. Since a tangent is horizontal if its slope is 0, the strategy will be to find $S'(x)$ and then determine the zeros of S'.

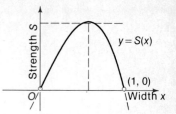

Assume $c \in \text{Dom } S = \{x : 0 < x < 1\}$.

$$\frac{S(c+h) - S(c)}{h} = \frac{K([c+h] - [c+h]^3) - K(c - c^3)}{h}$$

$$= \frac{K(c + h - c^3 - 3c^2h - 3ch^2 - h^3 - c + c^3)}{h}$$

$$= \frac{K(h - 3c^2h - 3ch^2 - h^3)}{h}$$

Hence, $\lim_{h \to 0} \frac{S(c+h) - S(c)}{h} = \lim_{h \to 0} K(1 - 3c^2 - 3ch - h^2)$.

As h approaches 0, $3ch$ and h^2 each approach 0, so that

$$\lim_{h \to 0} K(1 - 3c^2 - 3ch - h^2) = K(1 - 3c^2).$$

Thus $S'(c) = K(1 - 3c^2)$, and for all x, $S'(x) = K(1 - 3x^2)$. Now solve $S'(x) = 0$.

$$K(1 - 3x^2) = 0$$

Since $K > 0$, $1 - 3x^2 = 0$, and $x = \pm\frac{\sqrt{3}}{3}$. But $\frac{-\sqrt{3}}{3} \notin \text{Dom } S$, so the strongest beam is produced when the log is cut to a width of $\frac{\sqrt{3}}{3}$ ft., or about 0.58 ft.

Exercises

A 1. If $f(x) = x^2$, find $f'(7)$.

2. If $g(x) = x^3$, find $g'(-3)$.

3. Find all points on the graph of $y = x^3$ where the tangent has slope 48.

4. Find all points on the graph of $y = x^2$ where the tangent has slope -10.

5. Write an equation of the tangent to the graph of $y = x^3$ at $\left(-\frac{2}{3}, -\frac{8}{27}\right)$.

6. Write an equation of the tangent to $y = \frac{1}{2x + 1}$ at $\left(-2, -\frac{1}{3}\right)$.

7. Write an equation of the tangent to the graph of $y = \frac{1}{2x + 1}$ at the point where the graph intersects the y-axis.

8. Write an equation for the tangent to $y = x^2$ which is perpendicular to the line $x - 6y + 2 = 0$.

In exercises 9–20, find the derivative of each function.

9. $f(x) = x$
10. $h(x) = \dfrac{1}{x}$
11. $g(x) = 3x^2$
12. $F(x) = 5x + 1$
13. $G(x) = \dfrac{1}{x+1}$
14. $H(x) = 4$

B 15. $f(x) = -\dfrac{1}{2}$
16. $g(x) = 4x^2 + 3$
17. $h(x) = \sqrt{x}$
18. $F(x) = \dfrac{x+1}{x}$
19. $G(x) = 2x^3 - x$
20. $H(x) = \sqrt{x+2}$

In exercises 21–24, find all points on the graph of the given function where the slope of the tangent is m.

21. $f: x \to 3x^2 - 2;\ m = \dfrac{2}{3}$
22. $g: x \to x^3 - 1;\ m = 12$
23. $h: x \to \dfrac{1}{x};\ m = -4$
24. $f: x \to \dfrac{1}{x+1};\ m = -9$

In exercises 25 and 26, find all points on the graph of the given function where the tangent is horizontal.

25. $f(x) = x^3 - 4x$
26. $g(x) = x^3 + 3x^2$

27. Use the definition of a derivative to show that if f is a constant function, then $f'(x) = 0$.

28. Use the definition of a derivative to show that if m is a constant and $f(x) = mx$, then $f'(x) = m$.

C 29. Find all points on the graph of $y = \dfrac{x}{x^2 - 1}$ where the tangent is parallel to the tangent to the curve at the origin.

30. Find the tangents to the graph of $y = x^3 - 16$ that pass through the origin.

31. Let $M(h) = \dfrac{\sin(c + h) - \sin(c)}{h}$, where c is a given constant.

  a. For $c = 1$, run a computer program to print values of $M(h)$ for $h = 2^{-1}$, $2^{-2}, 2^{-3}, \ldots, 2^{-10}$ and compare the final value with $\cos c$.
  b. Repeat a for $c = 0.5, 1.5$, and 2.5.
  c. What generalization is suggested by your results?

32. Sketch the graph of $f(x) = \sin x$ on $[-2\pi, 2\pi]$.
  a. From the graph determine the zeros of f'.
  b. If $f'(0) = 1$, what is $f'(\pi)$?
  c. Solve the inequality $f'(x) > 0$ on $[-2\pi, 2\pi]$.
  d. Sketch the graph of $y = f'(x)$ on $[-2\pi, 2\pi]$.

14–3 Theorems for Differentiation

Finding derivatives, which is called *differentiation*, can be laborious. Fortunately there are a number of theorems that provide shortcuts. Before considering these theorems, we point out a few useful properties of limits of functions. Some of these have already been applied intuitively in finding tangents. For example, in evaluating $\lim_{h \to 0} (h^2 + 6h + 3)$, we noted that $\lim_{h \to 0} h^2 = 0$, that $\lim_{h \to 0} 6h = 0$, and that $\lim_{h \to 0} 3 = 3$; we then concluded that

$$\lim_{h \to 0} (h^2 + 6h + 3) = \lim_{h \to 0} h^2 + \lim_{h \to 0} 6h + \lim_{h \to 0} 3$$

$$= 0 + 0 + 3 = 3.$$

In other words, it was assumed implicitly that the limit of a sum of two or more functions equals the sum of their limits. These function-limit properties parallel those for sequences that you have already used. They are stated in the following theorem without proof.

Theorem 14–1 If f and g are functions and each has a limit at c such that

$$\lim_{x \to c} f(x) = L \quad \text{and} \quad \lim_{x \to c} g(x) = K,$$

then $f + g$, $f - g$, $f \cdot g$, and $\dfrac{f}{g}$ each has a limit at c, and

a. $\lim_{x \to c} [f + g](x) = L + K$

b. $\lim_{x \to c} [f - g](x) = L - K$

c. $\lim_{x \to c} [f \cdot g](x) = L \cdot K$

d. $\lim_{x \to c} \left[\dfrac{f}{g}\right](x) = \dfrac{L}{K}$, provided $K \neq 0$.

Following are several useful theorems concerning derivatives.

Theorem 14–2 If $f(x) = k$ is a constant function, then $f'(x) = 0$ for all x.

PROOF Assume c is any real number.

$$\lim_{h \to 0} \frac{f(c + h) - f(c)}{h} = \lim_{h \to 0} \frac{k - k}{h} = \lim_{h \to 0} \frac{0}{h} = \lim_{h \to 0} 0 = 0$$

Hence $f'(c) = 0$, and for all $x \in \mathbf{R}$, $f'(x) = 0$.

In previous examples and exercises, you have probably noticed the following pattern:

If $f(x) = x^1$, then $f'(x) = 1 = 1x^0$.

If $f(x) = x^2$, then $f'(x) = 2x = 2x^1$.

If $f(x) = x^3$, then $f'(x) \quad\quad = 3x^2$.

In exercise 26 at the end of this section you are asked to prove:

If $f(x) = x^4$, then $f'(x) = 4x^3$.

From this pattern you might guess that if $f(x) = x^n$, then $f'(x) = nx^{n-1}$. The following theorem states this result and is given here without proof.

Theorem 14–3 If $f(x) = x^n$, $n \in \mathbb{N}$, then f has a derivative and $f'(x) = nx^{n-1}$.

Example 1 Find all points on the graph of $g(x) = x^5$ where the slope of the tangent is 20.

SOLUTION Since $g(x) = x^5$, $g'(x) = 5x^{5-1} = 5x^4$.

Now solve $5x^4 = 20$.
$$x^4 = 4$$
$$x = \pm\sqrt{2}$$

Thus the required points are $(\sqrt{2}, 4\sqrt{2})$ and $(-\sqrt{2}, -4\sqrt{2})$.

To find the derivative of a monomial function such as $g: x \to 7x^3$, the following theorem is helpful.

Theorem 14–4 If k is a constant and f has a derivative at x, then $g: x \to kf(x)$ has a derivative at x and $g'(x) = kf'(x)$.

PROOF Assume $c \in \text{Dom } f'$. Then

$$g(c + h) - g(c) = kf(c + h) - kf(c) = k[f(c + h) - f(c)], \text{ and}$$

$$\lim_{h \to 0} \frac{g(c + h) - g(c)}{h} = \lim_{h \to 0} k\left[\frac{f(c + h) - f(c)}{h}\right].$$

But f has a derivative at c, so $\lim_{h \to 0} \frac{f(c + h) - f(c)}{h} = f'(c)$. Thus,

$$\lim_{h \to 0} \frac{g(c + h) - g(c)}{h} = \lim_{h \to 0} k \cdot \lim_{h \to 0} \frac{f(c + h) - f(c)}{h} = k \cdot f'(c).$$

Therefore $g'(c) = k \cdot f'(c)$ and for all $x \in \text{Dom } f'$, $g'(x) = k \cdot f'(x)$.

Example 2 Write an equation of the tangent to the graph of $g(x) = \frac{-x^3}{6}$ at $(-1, \frac{1}{6})$.

SOLUTION Use Theorem 14–4 with $k = -\frac{1}{6}$ and $f(x) = x^3$. Then

$$g'(x) = -\frac{1}{6}(3x^2) = \frac{-x^2}{2}.$$

(cont. on p. 511)

$$g'(-1) = \frac{-(-1)^2}{2} = -\frac{1}{2}$$

So the tangent at $(-1, \frac{1}{6})$ is the line $y - \frac{1}{6} = -\frac{1}{2}(x + 1)$.

Theorem 14–5 If f and g both have a derivative at x, then $f + g$ has a derivative at x and
$$[f + g]'(x) = f'(x) + g'(x).$$

PROOF Assume $c \in \text{Dom } f' \cap \text{Dom } g'$. Then
$$[f + g](c + h) - [f + g](c) = [f(c + h) + g(c + h)] - [f(c) + g(c)]$$
$$= [f(c + h) - f(c)] + [g(c + h) - g(c)].$$

$$\lim_{h \to 0} \frac{[f + g](c + h) - [f + g](c)}{h} = \lim_{h \to 0} \left[\frac{f(c + h) - f(c)}{h} + \frac{g(c + h) - g(c)}{h} \right].$$

But f and g each have derivatives at c, so
$$\lim_{h \to 0} \frac{f(c + h) - f(c)}{h} = f'(c) \quad \text{and} \quad \lim_{h \to 0} \frac{g(c + h) - g(c)}{h} = g'(c).$$

Then by Theorem 14–1a,
$$\lim_{h \to 0} \frac{[f + g](c + h) - [f + g](c)}{h} = \lim_{h \to 0} \frac{f(c + h) - f(c)}{h} + \lim_{h \to 0} \frac{g(c + h) - g(c)}{h}$$
$$= f'(c) + g'(c).$$

Therefore $f + g$ has a derivative at c and
$$[f + g]'(x) = f'(x) + g'(x), \text{ for all } x \in \text{Dom } f' \cap \text{Dom } g'.$$

Example 3 Find $h'(-2)$ if $h(x) = \frac{x^4 + 1}{x}$.

SOLUTION Since $\frac{x^4 + 1}{x} = x^3 + \frac{1}{x}$, $h = f + g$, where $f: x \to x^3$ and $g: x \to \frac{1}{x}$.

Thus by Theorem 14–5, $h'(x) = f'(x) + g'(x)$. You know from the exercises that $g'(x) = -\frac{1}{x^2}$. So $h'(x) = 3x^2 - \frac{1}{x^2}$, and

$$h'(-2) = 3(-2)^2 - \frac{1}{(-2)^2} = 12 - \frac{1}{4} = \frac{47}{4}.$$

Using all four of the derivative theorems together, it is now easy to find the derivative of any polynomial function by inspection.

Example 4 If $f(x) = x^5 + 3x^2 + 5x - 4$, find $f'(x)$.

SOLUTION
$$f'(x) = 5x^4 + 3(2x) + 5(1) - 0$$
$$= 5x^4 + 6x + 5$$

Example 5 Find all points on the graph of $g(x) = 2x^3 + 3x^2 - 36x + 2$ where the tangent is horizontal.

SOLUTION The tangent is horizontal when $g'(x) = 0$.
$$g'(x) = 2(3x^2) + 3(2x) - 36(1) + 0$$
$$= 6x^2 + 6x - 36$$

Now find the zeros of g'.
$$6x^2 + 6x - 36 = 0$$
$$6(x + 3)(x - 2) = 0$$
$$x = -3 \quad \text{or} \quad x = 2$$

Since $g(-3) = 83$ and $g(2) = -42$, the graph has horizontal tangents at $(-3, 83)$ and at $(2, -42)$.

Unlike the pattern for the derivative of a sum, the derivative of a product of two functions is *not* equal to the product of their derivatives. For example, if $h(x) = x^5$, $g(x) = x^3$, and $f(x) = x^2$, then $h = f \cdot g$. But notice that $g'(x) \cdot f'(x) = (3x^2)(2x) = 6x^3$, while $h'(x) = 5x^4$. Thus $h'(x) \neq f'(x) \cdot g'(x)$. The correct rule for differentiating a product is given in the following theorem. The proof is omitted.

Theorem 14–6 If f and g are functions that have a derivative at x, then the product $f \cdot g$ has a derivative at x and

$$[f \cdot g]'(x) = f'(x) \cdot g(x) + f(x) \cdot g'(x).$$

Example 6 If $f(x) = \dfrac{x^3}{2x + 1}$, find $f'(x)$.

SOLUTION Express $f(x) = x^3 \cdot \dfrac{1}{2x + 1}$ as a product $g \cdot h$, where $g(x) = x^3$ and $h(x) = \dfrac{1}{2x + 1}$. By Theorem 14–6, $f'(x) = g'(x) \cdot h(x) + g(x) \cdot h'(x)$. Recall from example 3, page 505, that $h'(x) = \dfrac{-2}{(2x + 1)^2}$. Hence,

$$f'(x) = 3x^2 \left(\dfrac{1}{2x + 1} \right) + x^3 \dfrac{-2}{(2x + 1)^2} = \dfrac{4x^3 + 3x^2}{(2x + 1)^2}.$$

Exercises

A Use the appropriate theorems to find $f'(x)$.

1. $f(x) = x^6$
2. $f(x) = 4x^2$
3. $f(x) = x^3 + 7$
4. $f(x) = 6x + 3$
5. $f(x) = -2\sqrt{3}$
6. $f(x) = 3 - x$
7. $f(x) = 7x^3 - 5x + 2$
8. $f(x) = \frac{1}{3}x^3 - 4x^2 - 5$
9. $f(x) = \frac{3}{2x + 1}$
10. $f(x) = 2x^5 - 3x^3 + 2x^2 + 7$

B 11. $f(x) = -3x^4 + 4x^3 - 8x + \frac{1}{x}$
12. $f(x) = \frac{1}{x} + 7x^3 + 2x^2 - 4x$

Write an equation of the tangent to the graph of the given function at the specified point.

13. $f(x) = x^3 + 4x^2 - 13x - 1;\ (2, -3)$
14. $g(x) = 2x^4 - 6x^3 - 3x^2 + 8x + 3;\ (1, 4)$

Find all points on the graph of the given function where the tangent is horizontal.

15. $f(x) = 3x^2 - 2x + 1$
16. $g(x) = x^3 - 4x + 9$
17. $h(x) = 2x^3 + 5x^2 - 3$
18. $F(x) = x^4 - \frac{x^3}{3} - 1$

For the given function, find intervals on which f' is positive.

19. $f(x) = 2x^3 - 3x^2 - 12x + 7$
20. $f(x) = x^4 - 4x^3 + \sqrt{2}$

Use Theorem 14–6 to find $h'(x)$.

21. $h(x) = \frac{x^3 + 1}{x}$
22. $h(x) = \frac{3 - 2x}{2x + 1}$
23. $h(x) = \frac{x^2}{2x + 1}$
24. $h(x) = \frac{1}{x^2}$

C 25. By considering the graphs of $f: x \to x^6$ and $g: x \to (x - 5)^6$, evaluate $g'(6)$, $g'(3)$, and $g'(x)$.

26. If $f(x) = x^4$, prove that $f'(x) = 4x^3$.

27. Assume that f has a derivative at x and that $g(x) = [f(x)]^2$.
   a. Show that $g'(x) = 2f(x) \cdot f'(x)$.
   b. Use the result in **a** to evaluate $g'(-1)$ if $g(x) = \left(x^3 + 2x + \frac{1}{x}\right)^2$.

28. Given that f has a derivative at x and that $g(x) = [f(x)]^n$, $n \in \mathbb{N}$.
   a. Use mathematical induction and Theorem 14–6 to prove that $g'(x) = n[f(x)]^{n-1} f'(x)$.
   b. Use the result in **a** to evaluate $g'(-1)$ if $g(x) = (x^3 + 4x^2 - 2)^4$.

14-4 Curve Sketching Using Derivatives

To sketch the graph of $f: x \to x^3 - 9x$ you would probably first find the zeros of f. Since $f(x) = x(x^2 - 9)$, the zeros are -3, 0, and 3. This information does not, however, tell you which of the graphs sketched at the right best represents f. It would be helpful to be able to locate easily the high and low points of the graph, which are called relative maximum and relative minimum points.

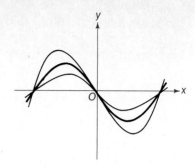

Definition A point $(c, f(c))$ on the graph of a function f is a **relative maximum point** if there exists a positive number p such that $f(x) \leq f(c)$ for every x in the interval $[c - p, c + p]$.

A corresponding definition describes a **relative minimum point**.

For the graph shown, the point $C(3, 8)$ is a relative maximum point even though $f(-8) > f(3)$. Likewise, $B(-3, 2)$ and $D(10, -4)$ are relative minimum points.

In Chapter 6, you encountered the problem of finding relative maximum and minimum points by trial and error. On smooth curves such points can occur only where the tangent is horizontal. This means that the derivative function is an effective tool for locating maximum and minimum points since the tangent is horizontal where the derivative is 0.

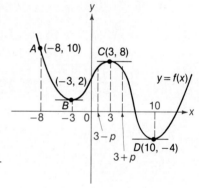

Example 1 Determine the relative maximum and minimum points of the graph of $f: x \to x^3 - 9x$.

SOLUTION $f'(x) = 3x^2 - 9$. Now find the zeros of f'.

$$3x^2 - 9 = 0, \text{ or } x = \pm\sqrt{3}$$

The values corresponding to the zeros are $f(\sqrt{3}) = (\sqrt{3})^3 - 9(\sqrt{3}) = -6\sqrt{3}$, and $f(-\sqrt{3}) = 6\sqrt{3}$. From the basic shape of the graph, you can see that $(\sqrt{3}, -6\sqrt{3})$ is a relative minimum and $(-\sqrt{3}, 6\sqrt{3})$ is a relative maximum point.

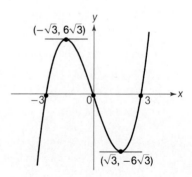

It is possible to have a relative maximum or minimum point occur where the derivative is not defined. For example, the absolute-value function, $f(x) = |x|$, has a relative minimum point at (0, 0); but this is not a smooth curve and the function does not have a derivative at $x = 0$.

Points where the tangent is horizontal or where the derivative is not defined are called **critical points**. Once you have found the critical points you must decide which are relative maximum points, which are relative minimum points, and which are neither. As the figure suggests, it is not necesssarily true that the lowest of the critical points is a relative minimum or that the highest is a relative maximum. In this case, point A, the lowest of the critical points, is a relative maximum, and C, the highest, is neither a relative maximum nor a relative minimum.

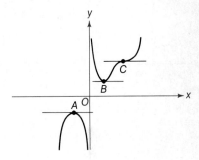

Classifying critical points can usually be done by considering whether the function is increasing or decreasing on either side of the critical point. Thus, if f has a horizontal tangent at $(c, f(c))$ and if f is increasing immediately to the left of c but decreasing immediately to the right of c, then $(c, f(c))$ is a relative maximum point. In like manner, if for some $p > 0$, f is decreasing on $[d - p, d]$ and increasing on $[d, d + p]$, then $(d, f(d))$ is a relative minimum point.

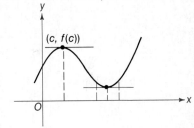

The derivative can also be used to determine where a function is increasing and where it is decreasing. The function pictured is clearly increasing on intervals $(-8, -3)$ and $(4, 11)$. Notice that on these intervals the tangents slope upward to the right; that is, they have positive slope. On the other hand, the slopes of the tangents are negative on $(-3, 4)$ and $(11, 14)$ where the function is decreasing.

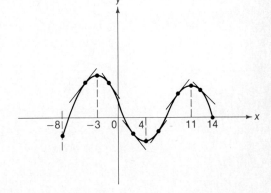

14-4: Curve Sketching Using Derivatives

Example 2 For the function $f: x \to x^2 - 6x$ find the intervals on which f is increasing and $f'(x) \geq 0$.

SOLUTION Since $f(x) = (x^2 - 6x + 9) - 9 = (x - 3)^2 - 9$, the graph is a parabola that opens upward and has its vertex at $(3, -9)$. The graph shows that f is increasing on $[3, \infty)$. Now solve

$$f'(x) \geq 0.$$
$$2x - 6 \geq 0$$
$$x \geq 3$$

Hence $f'(x) \geq 0$ on $[3, \infty)$. Thus f is increasing on the same interval on which f' is greater than or equal to 0.

It appears that a function is increasing on any interval where the derivative is positive and decreasing where the derivative is negative. This principle is stated in the following theorem, which is given without proof.

Theorem 14–7 Given that f is a function defined on an interval (a, b).

a. If $f'(x) > 0$ for all $x \in (a, b)$, then f is increasing on (a, b).
b. If $f'(x) < 0$ for all $x \in (a, b)$, then f is decreasing on (a, b).

Example 3 If $f(x) = \frac{1}{3}x^3 - \frac{3}{2}x^2 - 4x + 1$, find the intervals on which f is increasing.

SOLUTION The function f is increasing when $f'(x) > 0$.

$$f'(x) = x^2 - 3x - 4$$

Solving $(x - 4)(x + 1) > 0$ yields $(-\infty, -1) \cup (4, \infty)$ as the solution set. Therefore by Theorem 14–7, f is increasing on $(-\infty, -1)$ and on $(4, \infty)$.

The critical points are $\left(-1, \frac{19}{6}\right)$ and $\left(4, -\frac{53}{3}\right)$. At $\left(-1, \frac{19}{6}\right)$ the function changes from increasing to decreasing. This means that $\left(-1, \frac{19}{6}\right)$ is a relative maximum point.

Similarly, the change of sign of the derivative at $\left(4, -\frac{53}{3}\right)$ indicates that this point is a relative minimum point.

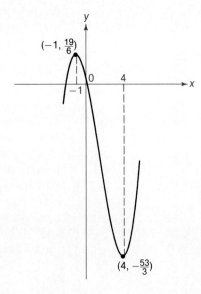

The graph of $f: x \to x^3$ shows that a function may have its derivative equal to 0 at a point P without P being either a relative maximum or minimum. For if $f(x) = x^3$ then $f'(x) = 3x^2$ and therefore $f'(0) = 0$. But the graph shows clearly that (0, 0) is neither a maximum nor a minimum point even though the tangent at the origin is horizontal. The key in this case is that the derivative does not change sign across the critical argument $x = 0$. Note that for a number $p > 0$, $f'(x) > 0$ when $x \in (-p, 0)$ and likewise $f'(x) > 0$ when $x \in (0, p)$. So f is increasing on both sides of $x = 0$. This confirms what the graph shows, namely, that (0, 0) is neither a relative maximum nor a relative minimum point.

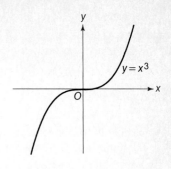

Example 4 For the function $f: x \to \dfrac{x^4}{4} - \dfrac{4x^3}{3} + 18$,

a. find all critical points;
b. find intervals on which f is increasing; and
c. sketch the graph of f.

SOLUTION To locate critical points, find f' and solve $f'(x) = 0$.

$$f'(x) = \tfrac{1}{4}(4x^3) - \tfrac{4}{3}(3x^2) + 0 = x^2(x - 4) = 0$$

Thus $f'(x) = 0$ when $x = 0$ or $x = 4$.

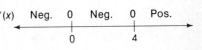

Since $f(0) = 18$ and $f(4) = -\dfrac{10}{3}$, the critical points are (0, 18) and $\left(4, -\dfrac{10}{3}\right)$.

To find where f is increasing, plot a sign diagram for $f'(x)$. From the diagram you can see that f is increasing on $(4, \infty)$.

Now note that f is decreasing on $(-\infty, 0)$ and on (0, 4). In fact, f is decreasing on $(-\infty, 4)$. The sign change of f' from negative to positive at $x = 4$ implies that $\left(4, -\dfrac{10}{3}\right)$ is a relative minimum point. However, since there is no sign change for f' at $x = 0$, (0, 18) is neither a relative maximum nor a relative minimum point. Combining information obtained from the derivative leads to the graph shown at the right.

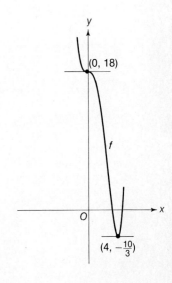

14-4: Curve Sketching Using Derivatives

Exercises

A For each function find all relative maximum and minimum points.

1. $f(x) = x^3 + 3x^2 + 10$
2. $f(x) = 2x^3 - 24x + 1$
3. $g(x) = 4x^3 - 3x^2 - 6x + 2$
4. $g(x) = x^4 - 4x^3 + 2x^2 - 12x + 10$

Consider the given derivative f' and classify the point $(c, f(c))$ as a relative maximum point, a relative minimum point, or neither.

5. $f'(x) = x - 4;\ c = 4$
6. $f'(x) = 3 - 2x;\ c = \frac{3}{2}$
7. $f'(x) = x^2 - x - 2;\ c = -1$
8. $f'(x) = x^3 + 3x^2;\ c = 0$
9. $f'(x) = x^3 - 2x^2 + x;\ c = 1$
10. $f'(x) = \frac{x^2 - 4}{x + 3};\ c = 2$

Determine the intervals on which the given function is increasing.

11. $g(x) = 2x^3 - 9x^2 + 12x - 7$
12. $h(x) = x^3 + 6x^2 - 7$
13. $f(x) = 3x^4 - 16x^3 + 24x^2 + 5$
14. $g(x) = x^4 - 3x^3 + 1$

B 15. $g(x) = \frac{3x^5}{5} - 5x^3 + 12x$
16. $h(x) = 3x^4 - 4x^3 - 24x^2 - 48x - 2$

For each function, (a) determine intervals on which the function is increasing, (b) identify critical points, and (c) sketch the graph.

17. $f(x) = x^3 - 3x + 1$
18. $h(x) = \frac{x^4}{2} - 4x^2 - 3$
19. $g(x) = 2 - x^3 - \frac{x^4}{4}$
20. $F(x) = 7 + 3x - x^3$
21. $G(x) = x^3 + 3x^2 + 3x - 2$
22. $H(x) = x^3 - 6x^2 + 9x + 1$
23. $f(x) = \frac{x^4}{4} - 8x + 3$
24. $g(x) = 1 + \frac{20x^3}{3} - x^5$

C 25. $F(x) = \frac{16 + x^3}{x};\ \text{Dom}\ F = (0, \infty)$

26. Find a function f such that $f'(x) = 4x + 1$ and $f(-1) = 2$.

27. The graph of a polynomial function f is pictured at the right. What is the lowest degree possible for f? Explain your answer.

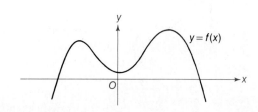

28. If $f(x) = x^3 + bx^2 + 3x + 5$, for what values of b will there be no horizontal tangents to the graph of f?

14–5 Maximum-Minimum Problems

In Chapter 6, the following question was posed: If 4 congruent squares are cut from a 6 in. by 5 in. rectangular sheet of aluminum and the sheet is then folded to form an open box, how large should each square be to obtain the box with largest volume?

The first step in solving the problem was to let x be the width of a side of a corner square and then create a volume function:

$$V(x) = x(6 - 2x)(5 - 2x)$$
$$= 4x^3 - 22x^2 + 30x.$$

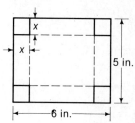

By trial and error it was then estimated that the maximum volume would result if the side of each square removed was 0.91 in.

Derivative functions are of great value in solving problems of finding a maximum or a minimum value. To see how the derivative of V can be applied to answer the question stated above, first consider the graph of V, which has zeros at 0, $\frac{5}{2}$, and 3. The dimensions of the box are such that $x \leq \frac{5}{2}$, so the domain is $\left[0, \frac{5}{2}\right]$. Thus the maximum volume occurs at point A.

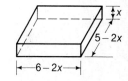

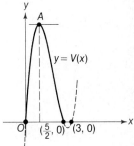

Find the zeros of V'.

$$V'(x) = 12x^2 - 44x + 30$$
$$2(6x^2 - 22x + 15) = 0$$
$$x = \frac{22 \pm \sqrt{484 - 360}}{12} = \frac{11 \pm \sqrt{31}}{6}$$

Since $\dfrac{11 + \sqrt{31}}{6} \doteq 2.761 > \frac{5}{2}$, the x-coordinate of the relative maximum point must be

$$x = \frac{11 - \sqrt{31}}{6} \doteq 0.9054.$$

The box with maximum volume will be obtained if x is approximately 0.9054 in.

Example 1   A cylindrical can is to hold 216 cubic inches of motor oil and is to be mass produced from thin aluminum sheet metal. If the circular ends are to be stamped out of square sheets with the leftover material thrown away, what radius for the can will minimize the aluminum used?

(cont. on p. 520)

SOLUTION The can is to be thin-walled, so the amount of aluminum used may be assumed to be proportional to the lateral surface area plus the area of the squares used in making the ends. So, express the total area T as a function of the radius r and find the minimum amount of aluminum required by minimizing T.

If h is the altitude of the can, then
$$T = 2\pi rh + 2(2r)^2.$$

To express T in terms of r alone, use the fact that the volume is 216 cubic inches.
$$\pi r^2 h = 216$$
$$h = \frac{216}{\pi r^2}$$
$$T(r) = 2\pi r \left(\frac{216}{\pi r^2}\right) + 8r^2$$
$$= \frac{432}{r} + 8r^2; \text{ Dom } T = (0, \infty).$$

Use the derivative to find where T is increasing and where it is decreasing.
$$T'(r) = 432\left(-\frac{1}{r^2}\right) + 16r = \frac{16r^3 - 432}{r^2}$$

Now solve the inequality $T'(r) > 0$.
$$\frac{16(r^3 - 27)}{r^2} > 0$$
$$r > 3$$

$T'(r)$ $\vert$ Neg. 0 Pos.
    0 3

The change of sign of T' at $r = 3$ from negative to positive implies that $(3, T(3))$ is a relative minimum point. Since T is decreasing on $(0, 3)$, $T(3)$ is the minimum value of the function on $(0, 3]$. $T(3)$ is also the minimum value of T on $[3, \infty)$ because T is increasing on that interval. $T(3)$ is therefore the minimum value of T on the entire domain $(0, \infty)$.

Thus a can with radius 3 inches will require the least amount of aluminum in its manufacture.

Exercises

A A box open on top is to be formed by cutting squares from the corners of a rectangular sheet of aluminum. To obtain a box with maximum volume, what size square should be removed from each corner if the sheet is a

1. 1 inch square?
2. 6 inch by 12 inch rectangle?

3. The sum of two positive numbers is 20. If *P* is the product of one number and the square of the other, what is the greatest value *P* can have?

4. Given that *a* and *b* are positive numbers whose sum is 5. What choice of *a* and *b* will make the sum of the cube of *a* and the square of *b* as small as possible?

B 5. Find the area of the largest rectangle that can be inscribed in the region bounded by the graph of $f: x \rightarrow 8 - x^2$ and the *x*-axis.

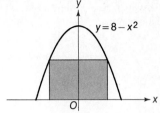

6. Let *P* be any point on the graph of $y = x^2$ between the points (0, 0) and (3, 9), and let *A*(3, 0) be a given point. If *Q* is the foot of the perpendicular from *P* to the *x*-axis, what point *P* will maximize the area of $\triangle APQ$?

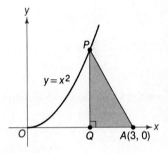

7. Determine the radius and height of the cylinder with maximum volume that can be obtained by rotating a rectangle of perimeter 24 inches about one of its sides.

8. A piece of wire 240 cm in length is cut into 6 pieces, 2 of one length and 4 of another. Each of the 2 pieces that have the same length is bent to form a square and the two squares are then connected by the remaining 4 pieces to make a frame for a box with a square base. What is the largest volume such a box can have?

9. A cylindrical reservoir, open on top, with a capacity of 20,000 cubic feet, is to be constructed using sheet steel one-quarter inch thick. What dimensions for the reservoir should be chosen to minimize the amount of sheet steel required?

10. A poster is to be designed that will have 80 square inches of printed material. The unprinted border will be 2 inches wide at the top and bottom and 1 inch wide on the sides. What should be the overall dimensions of the poster if the amount of paper required is to be a minimum?

11. Find the point on the graph of $y = \sqrt{x}$ that is closest to *A*(3, 0). [HINT: Minimize the square of the distance.]

12. A rectangle *OPQR* is inscribed in the region bounded by the *x*-axis, the *y*-axis, and the graph of $y = 9 - 3x^2$. What is the maximum volume the cylinder formed by rotating *OPQR* about the *y*-axis can have?

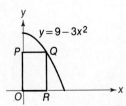

14–5: Maximum-Minimum Problems

C **13.** A closed cylindrical can of volume *K* is to be made of thin sheet metal. Assuming no wasted metal, find the ratio of altitude to radius for the can that requires the least metal.

14. A right circular cylinder is to be inscribed in a right circular cone of radius *a* and altitude *b*. What choice of dimensions *r* and *h* for the cylinder will yield the greatest volume for the cylinder?

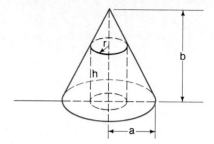

14–6 Velocity

If distance traveled *d* varies directly as the time *t*, then $d = kt$, where the constant *k* is the rate, or velocity. To evaluate velocity at any instant, simply divide the distance *d* by the time *t*.

If distance is not a linear function of time, the notion of velocity at a given instant is not so simple. We illustrate this with the following problem, and show that the difficulties in describing motion when the rate is not constant may be resolved by using derivatives.

A construction worker drops a bolt from a height of 256 feet above ground. How fast is the bolt falling 2 seconds after it is dropped?

The distance in feet traveled by a freely falling body starting from rest is given by the function $f(t) = 16t^2$, where *t* is the time elapsed in seconds from the moment of release. Thus at the end of 2 seconds the bolt has fallen $f(2)$, or 64 feet. If this distance is divided by the time, the result, 32 feet per second, is the *average* velocity during the first 2 seconds of fall. But the bolt is accelerating as it falls and the number 32 is very different from the *instantaneous velocity* after exactly 2 seconds, as you will see.

A rough approximation of the instantaneous velocity *V* at the end of 2 seconds can be found by computing the average velocity during the time interval from $t = 2$ seconds to $t = 3$ seconds. The distance traveled during that second is $f(3) - f(2)$; so if v_1 is the average velocity during the third second of fall, then

$$v_1 = \frac{f(3) - f(2)}{3 - 2} = 16(3)^2 - 16(2)^2 = 80 \text{ ft./sec.}$$

A better approximation of V is the average velocity of the bolt during a shorter time interval, say from $t = 2$ to $t = 2\frac{1}{2}$. If v_2 is this second approximation of V, then

$$v_2 = \frac{f\left(\frac{5}{2}\right) - f(2)}{\frac{5}{2} - 2} = \frac{16\left(\frac{5}{2}\right)^2 - 16(2)^2}{\frac{1}{2}} = 72 \text{ ft./sec.}$$

A still better estimate of V is the average velocity v_3 over the time interval $t = 2$ to $t = 2\frac{1}{4}$, that is, the interval $\left[2, \frac{9}{4}\right]$.

$$v_3 = \frac{f\left(\frac{9}{4}\right) - f(2)}{\frac{9}{4} - 2} = 68 \text{ ft./sec.}$$

For the time interval $\left[2, \frac{17}{8}\right]$,

$$v_4 = 66 \text{ ft./sec.}$$

Let $v(h)$ be the average velocity during an interval of length $h > 0$ from $t = 2$ to $t = 2 + h$. Then

$$v(h) = \frac{f(2 + h) - f(2)}{h}$$

$$= \frac{16(2 + h)^2 - 16(2)^2}{h} = 64 + 16h.$$

Intuition suggests that as shorter time intervals are considered, the average velocity should approach the instantaneous velocity at $t = 2$. Although we have not previously defined velocity at a particular instant, it seems reasonable to say that the velocity at the instant $t = 2$ is the limit of $v(h)$ as h approaches 0.

$$V = \lim_{h \to 0} v(h)$$

$$= \lim_{h \to 0} (64 + 16h) = 64 \text{ ft./sec.}$$

But notice that

$$\lim_{h \to 0} v(h) = \lim_{h \to 0} \frac{f(2 + h) - f(2)}{h}.$$

The expression on the right is simply $f'(2)$, the derivative of f at $t = 2$. Thus we are led to the following definition of velocity.

Definition If an object moves in a straight line so that its position with respect to a fixed point on the line of motion is given by a function of time $f(t)$, then the **velocity** $V(t)$ at time t is

$$V(t) = f'(t).$$

Example 1 For the falling bolt described earlier, find the velocity at the moment the bolt strikes the ground.

SOLUTION Since $f(t) = 16t^2$ describes the position of the bolt, the velocity at any time t is

$$V(t) = f'(t) = 16(2t) = 32t.$$

So to find the velocity of the bolt when it hits the ground, first determine the time it takes to reach the ground. Since the bolt falls from a height of 256 ft., solve the equation $16t^2 = 256$ to find that the required time is 4 seconds. So,

$$V(4) = 32 \cdot 4 = 128.$$

The velocity of the bolt at the instant it hits the ground is 128 ft./sec.

Example 2 The position of a particle moving on the x-axis is given by the function

$$x(t) = t^3 - 6t + 1, \text{ Dom } x(t) = [0, 4].$$

a. Find the velocity when $t = 2$.

b. When is the particle moving to the left?

c. Find the total distance traveled during the 4-second interval.

SOLUTION **a.** $V(t) = x'(t) = 3t^2 - 6$

$V(2) = 3(2)^2 - 6 = 6$

b. Since the positive direction is to the right, the particle is moving to the left when the position function $x(t)$ is decreasing, that is, when

$$x'(t) < 0.$$
$$V(t) < 0$$
$$3t^2 - 6 < 0$$
$$t^2 < 2$$
$$-\sqrt{2} < t < \sqrt{2}$$

Since the domain is the time interval $[0, 4]$, the particle is moving to the left when $0 \leq t < \sqrt{2}$.

c. If the particle moved in the same direction throughout the interval $[0, 4]$, the total distance traveled would simply be $|x(4) - x(0)|$. But the particle changes direction at $t = \sqrt{2}$, so the total distance traveled is found by adding the distance moved to the left and the distance moved to the right.

Total distance $= |x(\sqrt{2}) - x(0)| + |x(4) - x(\sqrt{2})|$

$x(0) = 1;$
$x(\sqrt{2}) = (\sqrt{2})^3 - 6\sqrt{2} + 1 = -4\sqrt{2} + 1;$ and
$x(4) = 4^3 - 6 \cdot 4 + 1 = 41.$

(cont. on p. 525)

Total distance $= |(-4\sqrt{2} + 1) - 1| + |41 - (-4\sqrt{2} + 1)|$
$= 4\sqrt{2} + 40 + 4\sqrt{2} = 40 + 8\sqrt{2}$

Whenever a quantity is expressed as a function of time, the derivative describes the instantaneous *rate of change* of the given quantity. Thus applications requiring consideration of rates of change of quantities such as temperature, population, volume, and energy consumption all require use of derivatives. The next example uses the fact that **acceleration**, which is defined as the rate of change of velocity, is the derivative of velocity.

Example 3 Over the first 3 seconds of a race a sprinter's distance s in yards from the starting line is given by the function

$$s: t \to \frac{-10t^3}{27} + \frac{10t^2}{3},$$

where t is in seconds.

a. Find the runner's acceleration $a(t)$ at the starting gun.

b. Find the runner's velocity when acceleration ceases.

SOLUTION **a.** Since the runner's velocity $v(t)$ is $s'(t)$,

$$v(t) = \frac{-10}{27}(3t^2) + \frac{10}{3}(2t)$$

$$= \frac{-10t^2}{9} + \frac{20}{3}t.$$

By definition, $a(t) = v'(t)$. Hence

$$a(t) = \frac{-20}{9}t + \frac{20}{3}.$$

The runner's initial acceleration is therefore

$$a(0) = \frac{20}{3} \text{ (yd./sec.)/sec.}$$

b. The runner stops accelerating when $a(t) = 0$.

$$\frac{-20}{9}t + \frac{20}{3} = 0, \text{ or } t = 3$$

Thus the velocity when acceleration stops is $v(3)$.

$$v(3) = \frac{-10}{9}(3^2) + \frac{20}{3}(3) = 10 \text{ yd./sec.}$$

Exercises

A Exercises 1–4 refer to the falling-bolt problem in the text.

1. Find the velocity of the bolt after it has fallen for 1 second.
2. Find the velocity when the bolt has fallen 144 feet.
3. Find the acceleration when $t = 3$ sec.
4. What is the height of the bolt at the moment when the acceleration is numerically equal to the velocity?

In exercises 5–12, assume that a particle moves along the x-axis in such a way that during the specified time interval, its x-coordinate is given by $x(t)$. Find the velocity at $t = 2$ and the time intervals during which the particle is moving to the right.

5. $x(t) = 3t^2 - 18t$; $t \in [0, 4]$
6. $x(t) = 6t^2 + 6t - 1$; $t \in [-3, 1]$
7. $x(t) = \dfrac{t^3}{3} - \dfrac{t^2}{2} - 2t + 3$; $t \in [-3, 3]$
8. $x(t) = \dfrac{2t^3}{3} - 3t^2 - 5$; $t \in [-2, 4]$

Find the acceleration at the specified instant.

9. $t = 2$; $x(t) = 2t^3 - 4t^2 - 6t + 10$; $t \in [-1, 3]$
10. When the velocity is 6; $x(t) = t^3 - 6t$; $t \in [0, 5]$

B Find the total distance traveled during the given time interval.

11. $x(t) = \frac{2}{3}t^3 + \frac{3}{2}t^2 - 2t + 1$; $[-3, 3]$
12. $x(t) = \frac{1}{4}t^4 - \frac{4}{3}t^3 + 6$; $[-1, 4]$

For exercises 13–16, assume that a ball is thrown upward from the top of a platform so that its height above ground in feet t seconds after release is given by $h(t) = 24 + 128t - 16t^2$.

13. What is the acceleration when $t = 1$?
14. What is the height of the ball at the moment it starts to descend?
15. How long after release is the ball falling at a velocity of 16 ft./sec.?
16. How fast is the ball moving when it hits the ground?

C 17. The distance a car travels after its brakes have been applied is given by $s: t \to 80t - 8t^2$, where t is in seconds and $s(t)$ is in feet. How far does the car travel before coming to a stop?

18. A dot is programmed to move horizontally across a computer screen for 10 seconds so that after t seconds its distance in inches from the left edge of the screen is given by $f(t) = \dfrac{16}{1 + t}$. Find the velocity when $t = 2$.

19. A plane dives in such a way that its altitude in feet is given by $h(t) = \dfrac{2000}{t^2 + 1}$, where t is the time in seconds after the dive begins. How fast is the plane approaching the ground after 3 seconds of the dive?

20. A ferryboat makes a one-way trip across a lake in such a way that its distance from its starting point is given by
$s(t) = 250t^2 - \dfrac{5}{4}t^4$, where t is in minutes and s is in feet.

  a. What is the distance across the lake?

  b. What is the highest speed reached by the ferryboat?

Chapter Review Exercises

14–1, page 499

1. By finding the limit of the slope function $M(h)$ at $x = -2$, find the slope of the tangent to the graph of $f(x) = \dfrac{1}{x - 2}$ at $\left(-2, -\dfrac{1}{4}\right)$.

14–2, page 504

2. Find all points on the graph of $y = x^3$ where the tangent is perpendicular to the line $y = \dfrac{-x}{12} + 3$.

3. If $f(x) = \dfrac{1}{3x + 1}$, determine $f'(x)$.

14–3, page 509

4. Write an equation of the tangent to the graph of $f: x \to \dfrac{x^3}{3} + 3x^2 - 5x - 31$ at $(-3, 2)$.

5. If $f(x) = \dfrac{2x^3}{3} + \dfrac{3x^2}{2} - 2x + 10$, find the intervals on which f' is positive.

6. If $f(x) = \dfrac{x^3 - 3}{x}$, find $f'(x)$.

14–4, page 514

7. If f is a function such that $f'(x) = x^2(x + 2)$, find arguments of all the critical points and classify each as a relative maximum point, a relative minimum point, or neither.

8. For the function $f: x \to 2 + 3x + x^2 - \dfrac{x^3}{3}$

   a. determine intervals on which f is increasing,
   b. identify all critical points, and
   c. sketch the graph.

14–5, page 519

9. Find the area of the largest rectangle that can be inscribed in the region bounded by $y = x^2$, the x-axis, and the line $x = 3$.

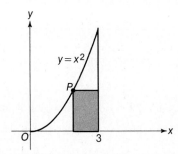

10. A closed rectangular box of volume 128 cu. ft. is to be made of sheet aluminum. The top and bottom are to be square and to have a thickness twice that of the sides. What dimensions should be chosen to minimize the amount of aluminum used?

14–6, page 522

11. A particle moves along the x-axis in such a way that during the time interval [0, 10] its x-coordinate is given by $x(t) = 2t^3 - 15t^2 + 36t - 1$. Find the acceleration when the velocity is 0 for the second time.

12. Suppose that the brakes of a subway train are designed so that when the train is moving at maximum speed the distance in feet traveled in t seconds after the brakes are applied is given by $S(t) = 120t - t^2 - \dfrac{t^3}{3}$. How far will the train travel after the brakes are applied before coming to a stop, assuming it is going full speed?

Chapter Test

1. Write an equation of the tangent to the graph of $y = x^4 + 9x - 2$ at $(0, -2)$.

2. Find all points on the graph of $f(x) = \dfrac{x^3}{3} + 2x - 1$ where the tangent is parallel to the line $3x - y = 1$.

3. Derive a formula for $f'(x)$ if $f(x) = \dfrac{1}{2-x}$.

4. If $g(x) = \dfrac{x^3}{3} - 2x^2 + 3x + 7$, determine any intervals on which g is decreasing.

5. If $h'(x) = 3(1+x)^2(3-x)$, classify each of the points $(-1, h(-1))$ and $(3, h(3))$ as a relative maximum point, a relative minimum, or neither.

6. A particle moves along the x-axis in such a way that its position at time t is given by $x(t) = t^3 - 3t + 3$ for $-5 \le t \le 5$.

   a. When is the particle moving to the left?

   b. What is the velocity when the acceleration is 9?

7. Given that $f'(x) = \dfrac{1}{2\sqrt{x+1}}$ if $f(x) = \sqrt{x+1}$, and that $h(x) = (x^2 - 3)\sqrt{x+1}$. Evaluate $h'(3)$.

8. If $g(x) = 3x^5 - 5x^3 + 3$:

   a. Determine the coordinates of all critical points.

   b. Sketch the graph of g.

9. Let P be any point on the graph of $y = x^3$ between $(0, 0)$ and $(8, 512)$, and let $C(8, 0)$ be a given point. If Q is the foot of the perpendicular from P to the x-axis, what point P will maximize the area of $\triangle CPQ$?

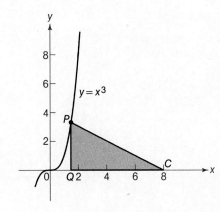

Introduction to Calculus
Integrals

15-1 The Area Under a Curve

In order to estimate the carrying capacity of an irrigation ditch, an engineer needs to compute the area of a cross section. If each cross section is in the shape of a parabola, the area can be computed by using integral calculus. Computing areas is an important application of integral calculus.

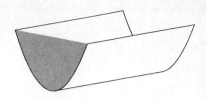

You already know how to find the areas of some familiar figures, such as rectangles and squares, triangles, circles, parallelograms, and trapezoids. Now we turn to computing areas of some less familiar figures.

Suppose that f is a function whose values are all nonnegative on the domain $a \leq x \leq b$. Assume also that f is continuous, that is, that the graph has no breaks in it. The area of the region, shaded in the figure, between the curve and the x-axis is called the *area under the curve*. We denote this area from $x = a$ to $x = b$ by the symbol $S_a^b f(x)$.

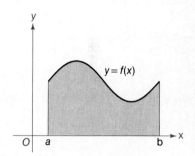

If the graph of f is not a straight line or an arc of a circle, you probably do not have any method for calculating $S_a^b f(x)$ exactly. But you can approximate the area by using areas of appropriate rectangles.

Consider the exponential function $f(x) = \left(\frac{1}{2}\right)^x$ on the domain $0 \leq x \leq 3$. To estimate the area under this curve, note that the shaded region lies inside a large rectangle $ABCD$ whose base is the entire segment $[0, 3]$ on the x-axis and whose height is $f(0) = 1$, the largest value of the function on this interval.

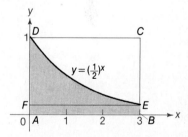

The region also contains a rectangle $ABEF$ of width 3 and height $\frac{1}{8}$, where $f(3) = \frac{1}{8}$ is the smallest function value on $[0, 3]$. Clearly the area under the curve must be between the areas of these two rectangles, that is,

$$\frac{3}{8} \leq S_0^3 \left(\frac{1}{2}\right)^x \leq 3.$$

Since the average of the areas of these two rectangles will be a better estimate than either of the two alone, use the approximation

$$S_0^3 \left(\frac{1}{2}\right)^x \doteq \frac{1}{2}\left(\frac{3}{8} + 3\right) = 1.6875.$$

This approximation can be improved by increasing the number of rectangles and decreasing their widths. You might subdivide the domain interval [0, 3] into 3 smaller intervals, [0, 1], [1, 2], and [2, 3]. For convenience, use subintervals of equal lengths, so that the resulting rectangles will all have equal widths. With these 3 subintervals as bases, first construct rectangles, to be called *upper rectangles*, by using as heights the maximum values of the function on each subinterval. These upper rectangles have heights $1, \frac{1}{2},$ and $\frac{1}{4}$. Their combined area is $(1)(1) + (1)(\frac{1}{2}) + (1)(\frac{1}{4})$, or 1.75. These three upper rectangles cover the region, so that

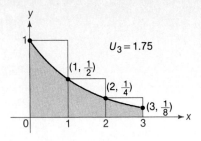

$$S_0^3 \left(\frac{1}{2}\right)^x \leq 1.75.$$

This result is called an **upper estimate** for the area under the curve, and we use the symbol U_3 to denote it (the subscript designates the number of rectangles).

Now use the same 3 subintervals as bases, but construct rectangles that are inside the region, so-called **lower rectangles**, with the heights of these lower rectangles being the minimum values of the function on each subinterval, namely, $\frac{1}{2}, \frac{1}{4},$ and $\frac{1}{8}$. Adding the areas of the 3 lower rectangles gives a number L_3, called a **lower estimate**, that is smaller than the exact area under the curve.

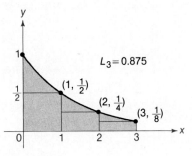

$$L_3 = (1)\left(\frac{1}{2}\right) + (1)\left(\frac{1}{4}\right) + (1)\left(\frac{1}{8}\right) = 0.875 \leq S_0^3\left(\frac{1}{2}\right)^x.$$

The average of L_3 and U_3 yields a better estimate than using either of the two separately.

$$S_0^3 \left(\frac{1}{2}\right)^x \doteq \frac{1}{2}(L_3 + U_3) \doteq 1.3125$$

To improve this new approximation, increase the number of rectangles. You might subdivide the domain interval [0, 3] into 6 subintervals, in which case you will find

$$U_6 = \sum_{i=0}^{5} \frac{1}{2}\left(\frac{1}{2}\right)^{i/2} \quad \text{and} \quad L_6 = \sum_{i=1}^{6} \frac{1}{2}\left(\frac{1}{2}\right)^{i/2}.$$

By calculator, $U_6 \doteq 1.494$ and $L_6 \doteq 1.056$, so that

$$S_0^3 \left(\frac{1}{2}\right)^x \doteq \frac{1}{2}(L_6 + U_6) \doteq 1.275.$$

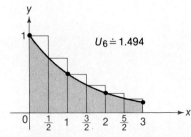

These calculations confirm the expected result that

$$L_1 \leq L_3 \leq L_6 \leq S_0^3 \left(\tfrac{1}{2}\right)^x \leq U_6 \leq U_3 \leq U_1.$$

Exercise 11 on page 535 asks you to improve on these approximations by computing $\tfrac{1}{2}(L_{10} + U_{10})$. It may be of interest to note here that the exact area is

$$S_0^3 \left(\tfrac{1}{2}\right)^x = \frac{7}{8 \ln 2},$$

which is approximately 1.2623582.

A formula for the area under a parabola is not among the reference formulas from geometry or trigonometry, but the area can be approximated by the technique described above.

Example 1 Estimate the area under the parabola $y = x^2$ from $x = 0$ to $x = 1$.

SOLUTION The number to approximate is $S_0^1 x^2$. If you subdivide $[0, 1]$ into 4 subintervals each of width $\tfrac{1}{4}$, you obtain the following upper and lower estimates.

$$U_4 = \frac{1}{4}\left(\frac{1}{4}\right)^2 + \frac{1}{4}\left(\frac{1}{2}\right)^2 + \frac{1}{4}\left(\frac{3}{4}\right)^2 + \frac{1}{4}(1)^2$$

$$= 0.46875$$

$$L_4 = \frac{1}{4}(0)^2 + \frac{1}{4}\left(\frac{1}{4}\right)^2 + \frac{1}{4}\left(\frac{1}{2}\right)^2 + \frac{1}{4}\left(\frac{3}{4}\right)^2$$

$$= 0.21875$$

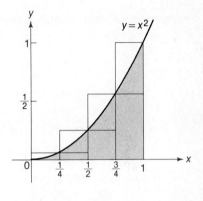

The average of these two estimates is the approximation

$$S_0^1 x^2 \doteq 0.34375.$$

Example 2 Use the average of L_6 and U_6 to find an approximation of $S_2^5 \log x$.

SOLUTION If an interval $[a, b]$ is divided into n subintervals of equal length, the width of each rectangle used in the estimation will be $\dfrac{b - a}{n}$. In this example, the width will be $\dfrac{5 - 2}{6}$, or $\dfrac{1}{2}$.

The heights of the lower rectangles will be log 2, log 2.5, log 3, log 3.5, log 4, and log 4.5.

(cont. on p. 534)

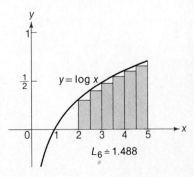

$L_6 \doteq 1.488$

The upper rectangles (not shown) have heights log 2.5, log 3, log 3.5, log 4, log 4.5, and log 5. Thus lower and upper estimates are as follows. Note that we have factored out the common width of the rectangles, $\frac{1}{2}$, from each sum, and used a property of logarithms, namely, log a + log b = log ab.

$$L_6 = \tfrac{1}{2}(\log 2 + \log 2.5 + \log 3 + \log 3.5 + \log 4 + \log 4.5)$$
$$= \tfrac{1}{2} \log 945 \doteq 1.488$$
$$U_6 = \tfrac{1}{2}(\log 2.5 + \log 3 + \log 3.5 + \log 4 + \log 4.5 + \log 5)$$
$$= \tfrac{1}{2} \log 2362.5 \doteq 1.687$$

ANSWER $S_2^5 \log x \doteq \tfrac{1}{2}(L_6 + U_6) \doteq 1.588$

Exercises

A If the area of the shaded region is $S_a^b f(x)$, identify a, b, and $f(x)$.

1.

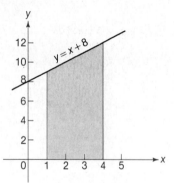

2.

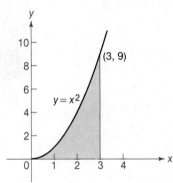

3.

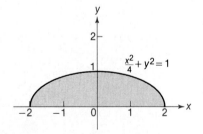

4.

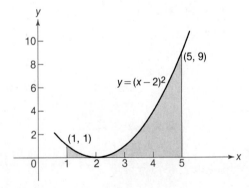

5. Calculate the area of the shaded region in exercise 1.

6. Calculate L_3 and U_3 for the shaded region in exercise 2.

7. Calculate L_4 and U_4 for the shaded region in exercise 3.

8. Calculate L_4 and U_4 for the shaded region in exercise 4.

B 9. Find $\frac{1}{2}(L_2 + U_2)$ for $S_1^3 x^2$.

10. Find $\frac{1}{2}(L_2 + U_2)$ for $S_0^2 x^3$.

11. Find $\frac{1}{2}(L_{10} + U_{10})$ for $S_0^3 \left(\frac{1}{2}\right)^x$ and compare your answer with the results on page 533.

12. Find $\frac{1}{2}(L_8 + U_8)$ for $S_0^1 x^2$ and compare your answer with the results in example I, page 533.

C 13. For $S_0^{\pi/2} \sin x$, find $\frac{1}{2}(L_1 + U_1)$, $\frac{1}{2}(L_2 + U_2)$, and $\frac{1}{2}(L_4 + U_4)$.

14. For $S_1^2 \frac{1}{x+1}$, find $\frac{1}{2}(L_1 + U_1)$ and $\frac{1}{2}(L_2 + U_2)$.

Find an expression for the area of each shaded region, using the notation $S_a^b f(x)$.

15.

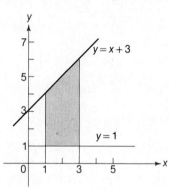

16.

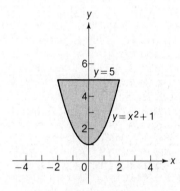

17.

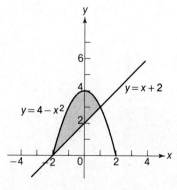

18.

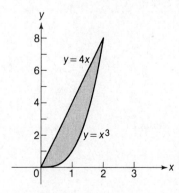

19. Prove that for any linear function f on the interval $[a, b]$, $\frac{1}{2}(L_2 + U_2) = S_a^b f(x)$.

Special Computer Approximations

A computer program in the BASIC language is given below that can be used to speed up the work of approximating areas under curves.

To make an area approximation, the computer must be given certain vital statistics: the right- and left-hand endpoints, the number of rectangles to be used, and, of course, a formula for calculating function values. To use this program you must type in the formula for the function in the first line of the program. The program as written will have the computer ask you WHAT INTERVAL?, to which you respond by typing in the endpoints a and b. The computer then prints HOW MANY RECTANGLES?, whereupon you type in a chosen value of n.

This program uses as heights of the rectangles the values of f at the left end of each subinterval, producing the so-called LEFT APPROXIMATION, L, and then the values of f at the right end of each subinterval, yielding the RIGHT APPROXIMATION, R. So, if the function f is an increasing function on the interval [a, b], the LEFT APPROXIMATION will be the lower estimate L_n, and the RIGHT APPROXIMATION will be the upper estimate U_n. If f is decreasing, the LEFT APPROXIMATION produces U_n and the RIGHT APPROXIMATION produces L_n. The program also has the computer print the AVERAGE approximation, $(L + R)/2$.

The difference $U_n - L_n$ provides a measure of how much error there could be in this average approximation. Since both the true area and the approximation $\frac{1}{2}(U_n + L_n)$ lie between the numbers L_n and U_n, the difference between the approximation and the true area will be at most one-half the distance from L_n to U_n.

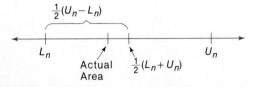

Here is the program, together with a sample RUN for approximating $\int_0^1 x^2$.

LIST

```
5 DEF FNF(X)=X↑2
10 INPUT "WHAT INTERVAL"; A,B
15 INPUT "HOW MANY RECTANGLES"; N
20 LET W=(B−A)/N
25 FOR X=A TO B STEP W
30    IF X > B−W/2 THEN 60
35    LET Y=FNF(X)
40    LET L=L+Y*W
45    LET Y=FNF(X+W)
50    LET R=R+Y*W
55 NEXT X
60 PRINT "LEFT APPROXIMATION = "; L
65 PRINT "RIGHT APPROXIMATION = "; R
70 PRINT "AVERAGE: (L+R)/2 = "; (L+R)/2
75 PRINT "MAXIMUM ERROR IS "; ABS(R−L)/2
80 END
```

RUN

WHAT INTERVAL? 0,1
HOW MANY RECTANGLES? 50
LEFT APPROXIMATION = .3234
RIGHT APPROXIMATION = .3434
AVERAGE: (L+R)/2 = .3334
MAXIMUM ERROR IS .01

Example 1 Using $n = 80$ rectangles, estimate the area $\int_0^{\pi/2} \sin x$.

SOLUTION Each time you want to use the computer program, you must type in a new function definition in line 5. In this example, type SIN(X), which is one of the functions available on the computer.

5 DEF FNF(X) = SIN(X)

RUN

WHAT INTERVAL? 0,1.57
HOW MANY RECTANGLES? 80
LEFT APPROXIMATION = .989359
RIGHT APPROXIMATION = 1.00898
AVERAGE: (L+R)/2 = .999172
MAXIMUM ERROR IS .98125E-2

There are two things you should notice in this example. First, the right-hand end of the interval, $\frac{\pi}{2}$, must be approximated as the decimal 1.57 before it can be used as an input. Second, notice the use of the E-notation in the error statement. The notation .98125E-2 means 0.98125×10^{-2}, that is, the maximum error with 80 rectangles is 0.0098125. Thus

$$S_0^{\pi/2} \sin x \doteq 0.999172,$$

with error less than 0.01.

Example 2 For the function $f: x \to \left(\frac{1}{2}\right)^x$ approximate the area under the curve on the interval [0, 3].

SOLUTION In this example, f is a decreasing function, so the upper estimate will now be the LEFT APPROXIMATION. In the RUN printed below, 1000 rectangles have been used, resulting in an error less than 0.001313.

5 DEF FNF(X)=(1/2)↑X

RUN

WHAT INTERVAL? 0,3
HOW MANY RECTANGLES? 1000
LEFT APPROXIMATION = 1.26367
RIGHT APPROXIMATION = 1.26105
AVERAGE: (L+R)/2 = 1.26236
MAXIMUM ERROR IS .13125E-2

When a function that is neither increasing nor decreasing is used, the error statement in the computer program (line 75) is no longer valid. As the figure below illustrates, for such a function the LEFT APPROXIMATION is neither U_n nor L_n.

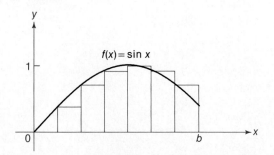

When the program is run for $f(x) = \sin x$ on the interval [0, 3] using 40 rectangles, the results are as follows.

LEFT APPROXIMATION = 1.98377
RIGHT APPROXIMATION = 1.99435
AVERAGE: (L+R)/2 = 1.98906

While these give some idea of the area involved, they are not upper and lower approximations for $S_0^3 \sin x$, and the maximum error statement must be ignored.

Exercises

Use the computer to find an approximation of the given area. In each exercise, use $n = 10$, $n = 50$, and then $n = 200$ rectangles.

1. $S_0^1 x^2$
2. $S_0^2 x^2$
3. $S_1^3 x^2$
4. $S_1^4 x^2$
5. $S_0^1 \sqrt{x}$
6. $S_0^9 \sqrt{x}$
7. $S_0^1 \left(\frac{1}{2}\right)^x$
8. $S_1^2 \left(\frac{1}{2}\right)^x$
9. $S_1^5 \frac{1}{x}$
10. $S_4^{10} \frac{1}{x}$
11. $S_0^{\pi/2} \cos x$
12. $S_{\pi/2}^{\pi} \sin x$

13. Modify the program given above so that each time your program is run, the printout will show information from the use of $n = 10, 50,$ and 200 rectangles, with no input of n required.

Interpret each of the following computer runs of the program given above.

14. 5 DEF FNF(X)=COS(X)

    RUN

    WHAT INTERVAL? 0,1.57
    HOW MANY RECTANGLES? 100
    LEFT APPROXIMATION = 1.00782
    RIGHT APPROXIMATION = .992135
    AVERAGE: (L+R)/2 = .999979
    MAXIMUM ERROR IS .784375E-2

15. Using $n = 50$ and then $n = 100$ rectangles, approximate the area $S_{-\pi/2}^{\pi/2} \cos x$. Comment on the validity of the error statement in the computer printout.

15–2 The Definite Integral $\int_0^1 x^2\, dx$

Suppose you wish to find exactly the area beneath the parabola $f(x) = x^2$ from $x = 0$ to $x = 1$, that is, find $\int_0^1 x^2$. The method for finding this area can be used for a large variety of curves.

Let n be any given natural number. Divide the interval $[0, 1]$ into n subintervals, each of width $\dfrac{1}{n}$.

These subintervals are $\left[0, \dfrac{1}{n}\right]$, $\left[\dfrac{1}{n}, \dfrac{2}{n}\right]$, $\left[\dfrac{2}{n}, \dfrac{3}{n}\right]$, $\ldots$, $\left[\dfrac{n-1}{n}, 1\right]$.

Since $f(x) = x^2$ is an increasing function, the heights of the upper rectangles used to estimate the area are the function values at the right end of each subinterval.

$$f\left(\dfrac{1}{n}\right) = \left(\dfrac{1}{n}\right)^2,\ f\left(\dfrac{2}{n}\right) = \left(\dfrac{2}{n}\right)^2,\ \ldots,\ f(1) = f\left(\dfrac{n}{n}\right) = \left(\dfrac{n}{n}\right)^2$$

Adding the areas of all the upper rectangles gives the upper estimate U_n.

$$U_n = \dfrac{1}{n}\left[\left(\dfrac{1}{n}\right)^2 + \left(\dfrac{2}{n}\right)^2 + \cdots + \left(\dfrac{n}{n}\right)^2\right]$$

$$= \dfrac{1}{n^3}[1^2 + 2^2 + \cdots + n^2] = \dfrac{1}{n^3}\sum_{i=1}^{n} i^2$$

In Chapter 13, it was proved by mathematical induction that, for all $n \in \mathbb{N}$,

$$\sum_{i=1}^{n} i^2 = \dfrac{n(n+1)(2n+1)}{6}.$$

Hence
$$U_n = \dfrac{1}{n^3}\left[\dfrac{n(n+1)(2n+1)}{6}\right]$$

$$= \dfrac{2n^3 + 3n^2 + n}{6n^3}.$$

$$U_n = \dfrac{1}{3} + \dfrac{1}{2n} + \dfrac{1}{6n^2}$$

To find an expression for L_n, you could repeat the algebra used above. However, since $f(x) = x^2$ is an increasing function, you can show that

$$U_n - L_n = \dfrac{b - a}{n}|f(b) - f(a)| = \dfrac{1}{n}|1 - 0| = \dfrac{1}{n},$$

so that
$$L_n = \left(\dfrac{1}{3} + \dfrac{1}{2n} + \dfrac{1}{6n^2}\right) - \dfrac{1}{n} = \dfrac{1}{3} - \dfrac{1}{2n} + \dfrac{1}{6n^2}.$$

538 Chapter 15: Introduction to Calculus—Integrals

As n gets larger, both U_n and L_n approach the area under the parabola. Thus we need to evaluate lim U_n and lim L_n.

We have $\lim \dfrac{1}{2n} = 0$, $\lim \dfrac{1}{6n^2} = 0$, and $\lim \dfrac{1}{3} = \dfrac{1}{3}$.

Hence, $\lim U_n = \lim \dfrac{1}{3} + \lim \dfrac{1}{2n} + \lim \dfrac{1}{6n^2}$

$= \dfrac{1}{3} + 0 + 0 = \dfrac{1}{3}$.

Similarly, $\lim L_n = \dfrac{1}{3}$. So $S_0^1 x^2$, the area under the parabola $f(x) = x^2$, is $\dfrac{1}{3}$.

In integral calculus, we find that the **definite integral**, $\displaystyle\int_a^b f(x)\, dx$, gives the area $S_a^b f(x)$ under the curve, and from now on we will use $\displaystyle\int_a^b f(x)\, dx$ for that area.

With only slight modification of the work above, you can find the area under $y = x^2$ over some other intervals.

Theorem 15–1 If $a > 0$, then $\displaystyle\int_0^a x^2\, dx = \dfrac{a^3}{3}$.

In the proof of this theorem, which is left as an exercise, use n rectangles each of width $\dfrac{a}{n}$.

Example 1 Evaluate $\displaystyle\int_0^2 x^2\, dx$.

SOLUTION $\displaystyle\int_0^2 x^2\, dx = \dfrac{2^3}{3} = \dfrac{8}{3}$

Example 2 Find the area under the parabola $y = x^2$ between $x = 2$ and $x = 5$.

SOLUTION The area of the region from $x = 2$ to $x = 5$ is the difference of two areas, namely, the entire area from $x = 0$ to $x = 5$ and the area from $x = 0$ to $x = 2$.

$\displaystyle\int_2^5 x^2\, dx = \int_0^5 x^2\, dx - \int_0^2 x^2\, dx$

$= \dfrac{5^3}{3} - \dfrac{2^3}{3} = \dfrac{117}{3}$

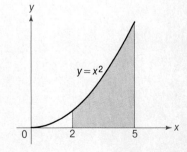

Exercises

A Evaluate each definite integral.

1. $\int_0^3 x^2\,dx$
2. $\int_0^5 x^2\,dx$
3. $\int_0^4 x^2\,dx$
4. $\int_0^{10} x^2\,dx$
5. $\int_2^7 x^2\,dx$
6. $\int_3^5 x^2\,dx$
7. $\int_3^4 x^2\,dx$
8. $\int_9^{10} x^2\,dx$

B 9. When the line $y = x$ is rotated around the x-axis, the resulting solid is a right circular cone. The volume of the cone extending from the origin to $x = b$ can be shown to be the definite integral $\pi \int_0^b x^2\,dx$. Find the volume of such a cone if its circular base has radius 5 units.

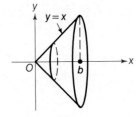

10. A particle moves along a number line in such a way that at time t seconds after it starts its velocity is t^2 ft./sec. It is known that the distance traveled by such a particle during the time interval $[0, a]$ is given by $\int_0^a t^2\,dt$. How far will the particle travel during the first 3 seconds of travel?

11. Show that $\int_0^2 x^2\,dx = \frac{8}{3}$ by duplicating the proof in the text that $\int_0^1 x^2\,dx = \frac{1}{3}$.

12. Prove Theorem 15–1: If $a > 0$, then $\int_0^a x^2\,dx = \frac{a^3}{3}$.

Find the area of each shaded region.

13.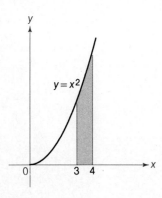

14.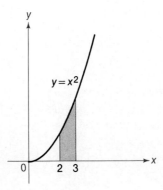

15.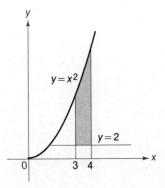

C 16. Use lower and upper estimates to show that $\int_0^1 3x^2\,dx = 1$.

15-3 The definite integral $\int_a^b x^2 \, dx$

In section 15-2 you learned that the area under the parabola $y = x^2$ on the interval $[0, a]$ is $\int_0^a x^2 \, dx = \frac{a^3}{3}$. Now you will see how to find the area under $y = x^2$ on any interval $[a, b]$.

Example 1 Find $\int_{-3}^{-1} x^2 \, dx$, the area under the parabola $y = x^2$ from $x = -3$ to $x = -1$.

SOLUTION The region S whose area is sought (see the figure) is congruent to the region S' between $x = 1$ and $x = 3$. Thus Area $(S) =$ Area (S'), so that

$$\int_{-3}^{-1} x^2 \, dx = \int_1^3 x^2 \, dx$$

$$= \int_0^3 x^2 \, dx - \int_0^1 x^2 \, dx$$

$$= \frac{3^3}{3} - \frac{1^3}{3} = \frac{26}{3}.$$

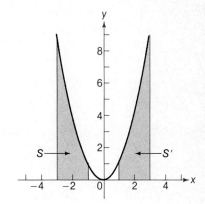

Example 2 Evaluate the area $\int_{-1}^{2} x^2 \, dx$.

SOLUTION The area $\int_{-1}^{2} x^2 \, dx$ is the sum of the areas of regions S_1 and S_2 shown in the figure.

$$\int_{-1}^{2} x^2 \, dx = \text{Area } (S_1) + \text{Area } (S_2)$$

$$= \int_{-1}^{0} x^2 \, dx + \int_0^2 x^2 \, dx$$

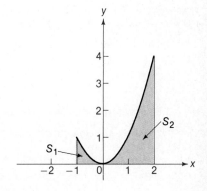

By symmetry, S_1 is congruent to the region under the parabola between $x = 0$ and $x = 1$. Therefore,

$$\int_{-1}^{2} x^2 \, dx = \int_0^1 x^2 \, dx + \int_0^2 x^2 \, dx = \frac{1}{3} + \frac{8}{3} = 3.$$

Notice that $\int_{-1}^{2} x^2 \, dx = \frac{(2)^3}{3} - \frac{(-1)^3}{3}.$

These examples, and example 2 from section 15-2, can be generalized easily to prove the following theorem, which extends the result of Theorem 15-1.

Theorem 15–2 If a and b are any real numbers, with $a < b$, then

$$\int_a^b x^2 \, dx = \frac{b^3 - a^3}{3}.$$

The proof of this theorem is left as an exercise. Note that there are 3 cases to be considered, as illustrated in the following figures.

Case 1: $0 \leq a < b$ Case 2: $a < b \leq 0$ Case 3: $a < 0 < b$

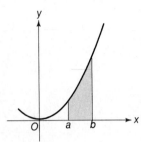
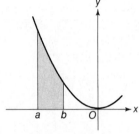
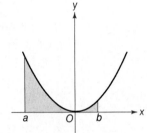

By using Theorem 15–2 you can quickly calculate areas such as the following.

$$\int_{-4}^{1} x^2 \, dx = \frac{1^3 - (-4)^3}{3} = \frac{65}{3} \qquad \int_{-10}^{-5} x^2 \, dx = \frac{(-5)^3 - (-10)^3}{3} = \frac{875}{3}$$

You know enough now to solve the engineer's problem about the irrigation ditch in section 15–1.

Example 3 Find the area of the cross section of an irrigation ditch that has for its shape the parabola $y = x^2$ if its maximum depth is 9 feet.

SOLUTION The required area, the shaded region in the figure, can be found by subtracting the area under the parabola, $\int_{-3}^{3} x^2 \, dx$, from the area of the rectangle of width 6 ft. and height 9 ft. Thus the area A of the cross section is

$$A = (6)(9) - \int_{-3}^{3} x^2 \, dx$$

$$= 54 - \frac{3^3 - (-3)^3}{3}$$

$$= 36 \text{ sq. ft.}$$

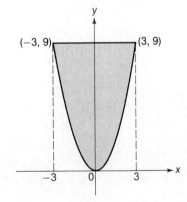

If p is a positive real number (constant), the area under $y = px^2$ will be p times the corresponding area under $y = x^2$. For instance, since each height (y-coordinate) on the graph of $y = 2x^2$ is twice the corresponding height on the graph of $y = x^2$, you would expect the following equality:

$$\int_0^1 2x^2 \, dx = 2 \int_0^1 x^2 \, dx.$$

That this relationship is indeed true can be shown by using upper and lower estimates of the two areas.

The upper estimate for the area under $y = x^2$ is

$$U_n = \frac{1}{n}\left[\left(\frac{1}{n}\right)^2 + \left(\frac{2}{n}\right)^2 + \cdots + \left(\frac{n}{n}\right)^2\right] = \frac{1}{n^3}\sum_{i=1}^{n} i^2.$$

The upper estimate for the area under $y = 2x^2$ is

$$U_n' = \frac{1}{n}\left[2\left(\frac{1}{n}\right)^2 + 2\left(\frac{2}{n}\right)^2 + \cdots + 2\left(\frac{n}{n}\right)^2\right] = \frac{2}{n^3}\sum_{i=1}^{n} i^2.$$

Thus, $U_n' = 2U_n$ and similarly $L_n' = 2L_n$. The factor 2 carries through the rest of the work that was done in showing that $\int_0^1 x^2 \, dx = \frac{1}{3}$, which leads directly to $\int_0^1 2x^2 \, dx = \frac{2}{3}$. (See section 15–2.)

Theorem 15–3 If a, b, and p are real numbers with $a < b$ and $p > 0$, then

$$\int_a^b px^2 \, dx = p\left(\frac{b^3 - a^3}{3}\right).$$

Example 4 Evaluate $\int_{-1}^{5} 4x^2 \, dx$.

SOLUTION $\int_{-1}^{5} 4x^2 \, dx = 4\left(\frac{5^3 - (-1)^3}{3}\right) = 168$

Example 5 Evaluate $\int_2^3 (1 + x^2) \, dx$.

SOLUTION The region S whose area you are to find is divided by the horizontal line $y = 1$ into two regions S_1 and S_2. S_2 is a square whose area is 1; S_1 is congruent to the region under $y = x^2$ on the interval $[2, 3]$.

(cont. on p. 544)

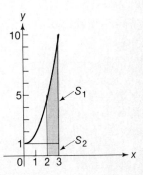

$$\int_2^3 (1 + x^2)\, dx = \text{Area}(S) = \text{Area}(S_1) + \text{Area}(S_2)$$

$$= \int_2^3 x^2\, dx + 1$$

$$= \frac{3^3 - 2^3}{3} + 1 = \frac{22}{3}$$

Exercises

A Evaluate each definite integral.

1. $\int_{-4}^{0} x^2\, dx$
2. $\int_{-3}^{0} x^2\, dx$
3. $\int_{5}^{6} x^2\, dx$
4. $\int_{1}^{3} x^2\, dx$

5. $\int_{-3}^{-2} x^2\, dx$
6. $\int_{-4}^{-1} x^2\, dx$
7. $\int_{-1}^{3} x^2\, dx$
8. $\int_{-2}^{4} x^2\, dx$

9. $\int_{-2}^{1} x^2\, dx$
10. $\int_{-3}^{3} x^2\, dx$
11. $\int_{-1}^{5} 6x^2\, dx$
12. $\int_{-1}^{3} 2x^2\, dx$

13. $\int_{-2}^{1} 4x^2\, dx$
14. $\int_{-4}^{-1} 3x^2\, dx$
15. $\int_{-3}^{-1} 7x^2\, dx$
16. $\int_{-5}^{0} 2x^2\, dx$

17. Find the area of the cross section of an irrigation ditch (as in example 3) if the ditch has for its shape the parabola $y = x^2$ and its maximum depth is 4 ft.

18. Find the area of the cross section of an irrigation ditch if its shape is given by the parabola $y = 2x^2$ and its maximum depth is 2 ft.

B Use the method of example 5 to evaluate each definite integral.

19. $\int_{1}^{2} (1 + x^2)\, dx$
20. $\int_{1}^{3} (2 + x^2)\, dx$
21. $\int_{-1}^{3} (1 + x^2)\, dx$

22. $\int_{-1}^{4} (2 + x^2)\, dx$
23. $\int_{-2}^{-1} (1 + 3x^2)\, dx$
24. $\int_{-5}^{2} (2 + 5x^2)\, dx$

25. Segment PQ has endpoints $P(-1, 1)$ and $Q(2, 4)$. Find the area of the region that lies below $\overline{PQ}$ but above the parabola $y = x^2$.

Find the area of each shaded region.

26.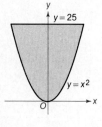
27.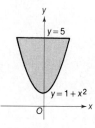
28.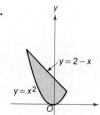
29.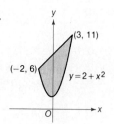

30. Complete the derivation (begun on page 543) that $\int_0^1 2x^2 \, dx = \frac{2}{3}$.

31. Prove Theorem 15–2.

C 32. Prove Theorem 15–3.

33. Find the area under the graph of $y = \sqrt{x}$ on the interval $[0, 1]$; that is, evaluate $\int_0^1 \sqrt{x} \, dx$.

 [HINT: Reflect the graph of $y = \sqrt{x}$ in the identity line.]

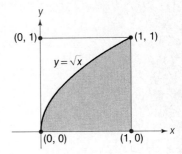

Use the method of exercise 33 to evaluate each definite integral.

34. $\int_0^4 \sqrt{x} \, dx$

35. $\int_0^9 \sqrt{x} \, dx$

36. Find a formula for $\int_0^a \sqrt{x} \, dx$ in terms of a if $a > 0$.

37. Evaluate $\int_1^4 \sqrt{x} \, dx$.

38. Prove: If $0 < a < b$, then $\int_a^b \sqrt{x} \, dx = \frac{2}{3}(b^{3/2} - a^{3/2})$.

39. Use a computer to estimate the area under the upper half of the ellipse $\frac{x^2}{4} + \frac{y^2}{1} = 1$. The true area depends upon π. From your computer results, make a conjecture concerning the true area.

15–4 Areas Under Power Functions

In Theorem 15–1, section 15–2, you learned that areas under the parabola $y = x^2$ can be computed by using the formula

$$\int_0^a x^2 \, dx = \frac{a^3}{3}.$$

The ideas used there can be extended to include areas under $y = x^p$, $p \in \mathbf{N}$. We begin by deriving an expression for the definite integral

$$\int_0^a x^3 \, dx, \quad a > 0.$$

First, divide the domain interval $[0, a]$ into n subintervals of equal width, where n is any fixed natural number. Each subinterval will have width $\frac{a}{n}$. An upper estimate for the integral is

$$U_n = \frac{a}{n}\left(\frac{a}{n}\right)^3 + \frac{a}{n}\left(\frac{2a}{n}\right)^3 + \cdots + \frac{a}{n}\left(\frac{na}{n}\right)^3$$

$$= \sum_{i=1}^{n} \frac{a}{n}\left(\frac{ia}{n}\right)^3$$

$$= \frac{a^4}{n^4}[1^3 + 2^3 + \cdots + n^3] = \frac{a^4}{n^4}\sum_{i=1}^{n} i^3.$$

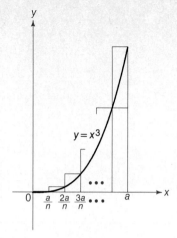

It was proved in exercise 19, section 13–4, that

$$\sum_{i=1}^{n} i^3 = \left[\frac{n(n+1)}{2}\right]^2.$$

Using this formula, you find

$$U_n = \frac{a^4}{n^4}\sum_{i=1}^{n} i^3 = \frac{a^4}{n^4}\left[\frac{n(n+1)}{2}\right]^2$$

$$= \frac{a^4}{4}\left[1 + \frac{2}{n} + \frac{1}{n^2}\right].$$

Since $f(x) = x^3$ is increasing on the interval $[0, a]$, each upper rectangle corresponds to the next lower rectangle at its right. So $U_n - L_n$ is the area of the last upper rectangle at the right. Thus

$$U_n - L_n = \frac{a^4}{n},$$

and by simple algebra,

$$L_n = \frac{a^4}{4}\left[1 - \frac{2}{n} + \frac{1}{n^2}\right].$$

As n gets larger, both U_n and L_n approach $\int_0^a x^3 \, dx$. So we evaluate $\lim U_n$ and $\lim L_n$.

$$\lim U_n = \lim \left(\frac{a^4}{4}\left[1 + \frac{2}{n} + \frac{1}{n^2}\right]\right)$$

$$= \frac{a^4}{4}\left[\lim 1 + \lim \frac{2}{n} + \lim \frac{1}{n^2}\right] = \frac{a^4}{4}[1 + 0 + 0] = \frac{a^4}{4}$$

Similarly, $\lim L_n = \frac{a^4}{4}$. So $\int_0^a x^3 \, dx = \frac{a^4}{4}$.

Example 1 Evaluate $\int_0^5 x^3 \, dx$.

SOLUTION $\int_0^5 x^3 \, dx = \dfrac{5^4}{4} = \dfrac{625}{4}$

By subtracting areas as in sections 15–2 and 15–3, a more general theorem can easily be proved.

Theorem 15–4 If a and b are real numbers, with $0 \le a < b$, then
$$\int_a^b x^3 \, dx = \frac{b^4 - a^4}{4}.$$

PROOF The area of the region beneath $y = x^3$ and between the lines $x = a$ and $x = b$ is

$$\int_a^b x^3 \, dx = \int_0^b x^3 \, dx - \int_0^a x^3 \, dx$$

$$= \frac{b^4}{4} - \frac{a^4}{4}$$

$$= \frac{b^4 - a^4}{4}.$$

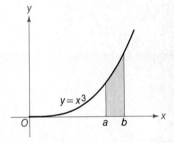

It requires only a slight modification of this work to obtain a result analogous to Theorem 15–3.

Theorem 15–5 If a, b, and p are real numbers, with $0 \le a < b$ and $p > 0$, then
$$\int_a^b px^3 \, dx = p\left(\frac{b^4 - a^4}{4}\right).$$

Example 2 Evaluate $\int_2^5 x^3 \, dx$ and $\int_2^5 8x^3 \, dx$.

SOLUTION By Theorem 15–4, $\int_2^5 x^3 \, dx = \dfrac{5^4 - 2^4}{4} = \dfrac{609}{4}$,

and by Theorem 15–5, $\int_2^5 8x^3 \, dx = 8\left(\dfrac{5^4 - 2^4}{4}\right) = 1218$.

Example 3 Find the area of the shaded region in the figure.

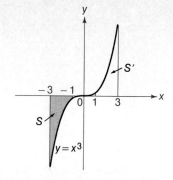

SOLUTION Since the region S is congruent to the region S' under the graph of $y = x^3$ on the interval $[1, 3]$,

Area (S) = Area (S')

$$= \int_1^3 x^3 \, dx = \frac{3^4 - 1^4}{4} = 20.$$

The definition of the area under a curve requires that you work with a function whose values are nonnegative. Since $f(x) = x^3$ is negative when $x < 0$, you must have $a \geq 0$ in both Theorems 15–4 and 15–5.

Notice that the area under $y = x$ from $x = 0$ to $x = a$, $a > 0$, is

$$\int_0^a x \, dx = \frac{a^2}{2},$$

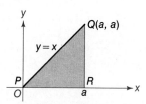

a result which does not require any estimation or mathematical induction. This definite integral is simply the area of a right triangle, $\triangle PQR$.

You may have noticed that there is a pattern developing with these areas. As long as $a > 0$, to ensure nonnegative function values throughout,

$$\int_0^a x \, dx = \frac{a^2}{2}, \quad \int_0^a x^2 \, dx = \frac{a^3}{3}, \quad \text{and} \quad \int_0^a x^3 \, dx = \frac{a^4}{4}.$$

Surely it is reasonable to surmise that

$$\int_0^a x^4 \, dx = \frac{a^5}{5},$$

and that, in general, for every integer $p > 0$,

$$\int_0^a x^p \, dx = \frac{a^{p+1}}{p+1}.$$

With the same method used in proving that $\int_0^a x^3 \, dx = \frac{a^4}{4}$, any one of the formulas above can be proved, provided you know a formula for the appropriate sum: $\sum_{i=1}^n i^4, \sum_{i=1}^n i^5, \cdots, \sum_{i=1}^n i^p$. An easy proof depends upon more advanced calculus methods. The general results are given here as theorems without proofs.

Theorem 15–6 If $p \in \mathbf{N}$ and $a > 0$, then $\int_0^a x^p \, dx = \dfrac{a^{p+1}}{p+1}$.

Corollary 15–6 If $0 \leq a < b$, $p \in \mathbf{N}$, and $k > 0$, then $\int_a^b kx^p \, dx = k\left(\dfrac{b^{p+1} - a^{p+1}}{p+1}\right)$.

Example 4 Evaluate $\int_1^2 3x^5 \, dx$.

SOLUTION $\int_1^2 3x^5 \, dx = 3\left(\dfrac{2^6 - 1^6}{6}\right) = \dfrac{63}{2}$

If p is an even number, then $f(x) = x^p$ has no negative values. In that case it is easy to show that Corollary 15–6 also holds when $a < b \leq 0$ and when $a < 0 < b$. (See Theorem 15–2, page 542.)

Example 5 Evaluate $\int_{-2}^3 2x^4 \, dx$.

SOLUTION $\int_{-2}^3 2x^4 \, dx = 2\left(\dfrac{3^5 - (-2)^5}{5}\right) = 110$

Example 6 Find the area of the region between the graphs of $f(x) = x^2$ and $g(x) = x^3$.

SOLUTION Find the intersection points of the graphs by solving $x^2 = x^3$, obtaining $x = 0$ or $x = 1$. The area of the region between the curves can now be found by subtraction.

Area $(S) = \int_0^1 x^2 \, dx - \int_0^1 x^3 \, dx$

$= \dfrac{1}{3} - \dfrac{1}{4} = \dfrac{1}{12}$

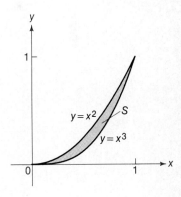

Exercises

A Evaluate each definite integral.

1. $\int_0^4 x^3 \, dx$
2. $\int_2^5 x^3 \, dx$
3. $\int_0^3 2x^3 \, dx$
4. $\int_0^2 4x^3 \, dx$
5. $\int_0^2 x^5 \, dx$
6. $\int_1^2 x^6 \, dx$
7. $\int_{-1}^2 x^4 \, dx$
8. $\int_{-2}^3 x^6 \, dx$

9. $\int_{-1}^{2} 5x^4 \, dx$ 10. $\int_{-2}^{2} 10x^4 \, dx$ 11. $\int_{1}^{5} (2 + x^3) \, dx$ 12. $\int_{-2}^{1} (4 + x^6) \, dx$

B 13. Find the area between the graphs of $y = x^3$ and $y = x^5$ on the interval $[0, 1]$.

14. Find the area of the region that lies above $y = x^4$ but below the segment that joins $(0, 0)$ and $(2, 16)$.

15. If $\int_{0}^{t} x^3 \, dx = 8$, find t.

16. Find k so that $\int_{0}^{k} x^4 \, dx = \int_{k}^{3} x^4 \, dx$.

C Find the area of each shaded region.

17.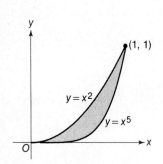

18.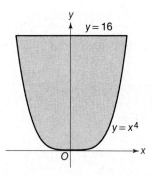

19.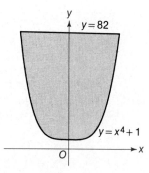

20.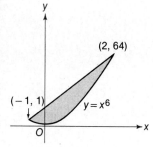

21.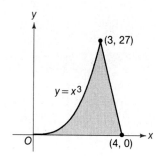

22.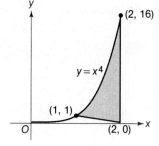

23. Find the area $\int_{0}^{1} \sqrt[3]{x} \, dx$ (compare exercise 33, section 15–3).

24. Use mathematical induction to prove that for all $n \in \mathbb{N}$,

$$\sum_{i=1}^{n} i^4 = \frac{1}{30}[6n^5 + 15n^4 + 10n^3 - n].$$

25. Use the result in exercise 24 to prove: If $a > 0$, $\int_{0}^{a} x^4 \, dx = \frac{a^5}{5}$.

15–5 Area Functions

You now know formulas for the exact value of areas for the power functions:

$$\int_a^b x^p \, dx = \frac{b^{p+1} - a^{p+1}}{p + 1}.$$

For other functions, if the exact area cannot be found you can approximate it by using the average of lower and upper estimates, $\frac{1}{2}(L_n + U_n)$.

Now we consider a new kind of function, called an **area function** A. For each real number $t \geq 0$, define

$$A(t) = \int_0^t x^2 \, dx.$$

$$A(1) = \int_0^1 x^2 \, dx = \frac{1}{3}, \quad A(2) = \int_0^2 x^2 \, dx = \frac{8}{3}, \quad A(\pi) = \int_0^\pi x^2 \, dx = \frac{\pi^3}{3}.$$

In general, $\quad A(t) = \int_0^t x^2 \, dx = \frac{t^3}{3}$, for $t \geq 0$.

You can plot the graph of this area function, keeping in mind that each value $A(t)$ represents an area under $f(x) = x^2$ from $x = 0$ to $x = t$.

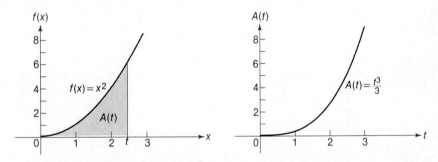

You can see from the left-hand figure that as t increases, the area $A(t)$ does, too; the area function is an increasing function, as in the figure at the right.

Example 1 If $f(x) = 1 - x$ and A is the area function defined by $A(t) = \int_0^t (1 - x) \, dx$, $0 \leq t \leq 1$, compute $A(1)$, $A(\frac{1}{2})$, and $A(0)$.

SOLUTION $A(1) = \int_0^1 (1 - x) \, dx$ is the area of the large triangle in the figure; $A(1) = \frac{1}{2}$.

$A(\frac{1}{2}) = \int_0^{1/2} (1 - x) \, dx$ is the area of the shaded region to the left of $x = \frac{1}{2}$. This is a trapezoid, and $A(\frac{1}{2}) = \frac{3}{8}$. Finally,

$A(0) = \int_0^0 (1 - x) \, dx = 0$, since the

(cont. on p. 552)

interval [0, 0] leads to a "region" that is a single vertical segment.

Example 1 shows that as long as f is defined at $x = a$, it is natural to define $\int_a^a f(x)\, dx = 0$. Note, however, that $\int_0^0 \frac{1}{x}\, dx$ is not defined, since $f(x) = \frac{1}{x}$ is not defined at $x = 0$.

There is a special area function that is especially important in science and mathematics. Begin with the graph of $f: x \to \frac{1}{x}$ on the domain $x > 0$. For each real number $t \geq 1$, define $L(t)$ by

$$L(t) = \int_1^t \frac{1}{x}\, dx.$$

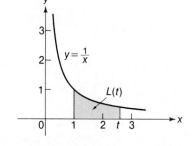

Thus $L(t)$ represents the area beneath the curve $y = \frac{1}{x}$ between the vertical lines $x = 1$ and $x = t$. Clearly $L(1) = 0$, and if $t > 1$, then $L(t) > 0$. Moreover, as with any area function, L is increasing: If $t_1 < t_2$, then $L(t_1) < L(t_2)$.

Individual values, such as $L(2)$ or $L(3)$, must be approximated, since you do not have a formula for $\int_1^2 \frac{1}{x}\, dx$ or $\int_1^3 \frac{1}{x}\, dx$. You can use a computer to find the following values of L, each accurate to the nearest tenth.

t	L(t)
1	0
2	0.7
3	1.1
4	1.4
5	1.6
6	1.8
7	1.9
8	2.1
9	2.2
10	2.3

There is an interesting pattern to be noted:

$L(2) + L(2) = L(4)$

$L(2) + L(3) = L(6)$

$L(2) + L(4) = L(8)$

$L(2) + L(5) = L(10)$

$L(3) + L(3) = L(9)$

and, of course, since $L(1) = 0$,

$L(1) + L(p) = L(p)$.

Each of these equations is an instance of a general property (proved in any calculus course) satisfied by the L function:

$$L(a) + L(b) = L(ab), \quad a, b \geq 1.$$

If you accept this product property for now, you can prove a power property for the L function.

Since $a^2 = a \cdot a$, $\quad L(a^2) = L(a) + L(a) = 2L(a)$.

Then, $\quad\quad\quad L(a^3) = L(a^2 \cdot a) = L(a^2) + L(a) = 3L(a)$.

Since $\quad\quad\quad L(a^{k+1}) = L(a^k \cdot a) = L(a^k) + L(a)$,

it is easy to prove by mathematical induction that, in general,

$$L(a^n) = nL(a), \; n \in \mathbf{N} \text{ and } a \geq 1.$$

Another consequence of the product property is the way the L function behaves with quotients. From the table of approximate values,

$$L\left(\tfrac{8}{2}\right) = L(4) \doteq 1.4 = 2.1 - 0.7 \doteq L(8) - L(2).$$

More generally, if a, b, and $\dfrac{a}{b}$ are all in the domain of L, that is, if $a \geq b \geq 1$, then the product property implies that

$$L\left(\frac{a}{b}\right) + L(b) = L(a)$$

so that

$$L\left(\frac{a}{b}\right) = L(a) - L(b).$$

This is the quotient property of L. Thus L, defined by $L(t) = \displaystyle\int_1^t \frac{1}{x}\,dx$, $t \geq 1$, has the following properties.

a. $L(1) = 0$

b. $L(t) \geq 0 \quad\quad\quad\quad\quad t \geq 1$

c. L is an increasing function.

d. $L(ab) = L(a) + L(b) \quad\quad a, b \geq 1$

e. $L(a^n) = nL(a) \quad\quad\quad\quad n \in \mathbf{N}, a \geq 1$

f. $L\left(\dfrac{a}{b}\right) = L(a) - L(b) \quad\quad a \geq b \geq 1$

By now you have probably guessed why we chose the letter L to represent this function; these properties are also valid for any logarithm function $\log_k$. In fact, this function $L(t)$ is the natural logarithm of t, $\ln t$, first mentioned in Chapter 8. Its values can be found on a calculator by pressing the key $\boxed{\text{LN X}}$. For example, $\ln 1 = 0$, $\ln 2 \doteq 0.693$, $\ln 3 \doteq 1.099$, and $\ln 100 \doteq 4.605$.

You would not expect sudden large changes in the values of $L(t)$ to result from small changes in the argument t. In other words, L is a continuous function. Since L is continuous, with $L(2) < 1$ and $L(3) > 1$, there must be a number between 2 and 3 to which L assigns the value 1, exactly.

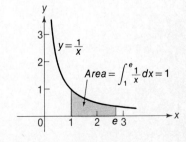

15–5: Area Functions

This is how the number e is defined:

$$\ln e = L(e) = 1,$$

and e is the number such that the area under $y = \frac{1}{x}$ between $x = 1$ and $x = e$ is exactly 1.

$$e \doteq 2.7182818 \quad \text{(by calculator)}.$$

The fact that $L(e) = 1$ reveals that e must be the base of this logarithmic function, that is,

$$\ln t = L(t) = \log_e t, \quad t \geq 1.$$

Example 2 Evaluate $\int_5^9 \frac{1}{x} dx$. Give your answer to the nearest tenth.

SOLUTION The area under $y = \frac{1}{x}$ from $x = 5$ to $x = 9$, namely, $\int_5^9 \frac{1}{x} dx$, can be found as the difference of two areas.

$$\int_5^9 \frac{1}{x} dx = \int_1^9 \frac{1}{x} dx - \int_1^5 \frac{1}{x} dx$$

$$= L(9) - L(5) = \ln 9 - \ln 5$$

$$\doteq 0.6$$

Exercises

A In exercises 1–6, evaluate the given area function $A(t) = \int_a^t f(x) \, dx$ for at least 3 numbers t and use these values to sketch a graph of $A(t)$ on the given interval.

1. $A(t) = \int_0^t x^3 \, dx$; [0, 3]

2. $A(t) = \int_0^t x^4 \, dx$; [0, 3]

3. $A(t) = \int_1^t x^3 \, dx$; [1, 4]

4. $A(t) = \int_1^t x^4 \, dx$; [1, 3]

5. $A(t) = \int_0^t 3x^2 \, dx$; [0, 4]

6. $A(t) = \int_{-1}^t x^2 \, dx$; [−1, 2]

7. Evaluate $\int_2^3 \frac{1}{x} dx$. Use the table on page 552 or a calculator, and give your answer to the nearest tenth.

8. Evaluate $\int_3^7 \frac{1}{x} dx$, accurate to the nearest tenth.

9. The "work" done in moving a particle a distance t micrometers is given by $W(t) = \int_0^t (3 - x)\, dx$. Find the work done in moving the particle $t = 1, 2,$ and 3 micrometers. (HINT: First graph $f(x) = 3 - x$ on the interval $[0, 3]$.)

B If the function f has a graph as shown, sketch the graph of $A(t) = \int_0^t f(x)\, dx$ on the interval $[0, 2]$.

10.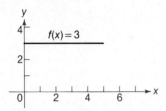

11.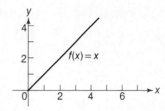

12. The volume obtained by rotating about the x-axis the graph of $y = f(x)$ between $x = 0$ and $x = t$ is given by the function $F(t) = \pi \int_0^t [f(x)]^2\, dx$. Find the volume obtained when $f(x) = x$ and $t = 1, 2,$ and 3.

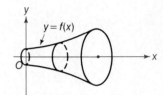

13. An object moves with velocity given by the formula $v(x) = 50 - x$ ft./sec. at time x. The distance traveled between time 0 and time t is given, as a function of t, by the function $D(t) = \int_0^t (50 - x)\, dx$. Find $D(0), D(1),$ and $D(50)$.

C 14. Use $\frac{1}{2}(L_n + U_n)$ for a suitable number of rectangles to approximate $L(12)$, $L(14)$, and $L(15)$ to the nearest tenth.

15. Use your answers to exercise 14, together with values of $L(t)$ in the table on page 552, to verify the product property $L(a) + L(b) = L(ab)$ for all products of positive integers such that $ab \leq 15$.

16. Use mathematical induction to prove the power property:

$L(a^n) = nL(a), \quad n \in N$ and $a \geq 1$.

17. In this section, L has been defined with domain $[1, \infty)$. How must $L(t)$ be defined when $0 < t < 1$ to extend the domain to $(0, \infty)$, as with all logarithm functions?

$\left[\text{HINT: If } 0 < t < 1,\text{ then } L\left(t \cdot \frac{1}{t}\right) = L(1) = 0.\right]$

18. Use a computer for this exercise. The voltage across the terminals of an electrical circuit is given as a function of time t by $V(t) = \int_0^t 5^x\, dx$. Estimate, to the nearest tenth, the voltage at times $t = 0, 1,$ and 2.

15–6 The Fundamental Theorem of Calculus

So far in this chapter, the definite integral has been used to find areas bounded by certain curves and lines. The definite integral was defined as the limit of a sequence of sums, but that same definition may be used even if the function f is not always nonnegative (as was the case in area problems). This broader definition of the definite integral has many applications, including finding centroids, centers of mass, moments of inertia, pressure, and so on, which involve functions that may at times be negative. In this section we will state and discuss the Fundamental Theorem of Calculus without the assumption that the function f in the definite integral $\int_a^b f(x)\,dx$ is nonnegative.

Consider the function $f(x) = x^2$ and the corresponding area function

$$A(t) = \int_0^t f(x)\,dx = \int_0^t x^2\,dx = \frac{t^3}{3}.$$

You know from Theorem 14–3 that

$$A'(t) = \tfrac{1}{3}(3t^2) = t^2,$$

so that $\qquad A'(t) = f(t).$

Thus in this case, the derivative of the area function is the original function f. You can check that the same pattern holds for other power functions $f(x) = x^n$, as summarized in the following table.

$f(x)$	$A(t) = \int_0^t f(x)\,dx$	$A'(t)$
$f(x) = x^0 = 1$	$A(t) = \int_0^t (1)\,dx = t$	$A'(t) = 1$
$f(x) = x^1$	$A(t) = \int_0^t x\,dx = \frac{1}{2}t^2$	$A'(t) = \frac{1}{2}(2t) = t$
$f(x) = x^2$	$A(t) = \int_0^t x^2\,dx = \frac{1}{3}t^3$	$A'(t) = \frac{1}{3}(3t^2) = t^2$
$f(x) = x^3$	$A(t) = \int_0^t x^3\,dx = \frac{1}{4}t^4$	$A'(t) = \frac{1}{4}(4t^3) = t^3$

In general, for all $n \in \mathbf{N}$, if $f(x) = x^n$ and

$$A(t) = \int_0^t x^n\,dx = \frac{1}{n+1}t^{n+1},$$

then $\qquad A'(t) = \dfrac{1}{n+1}[(n+1)t^n] = t^n = f(t).$

That is, $A' = f$. It can be proved using more advanced calculus that this connection between integrals and derivatives holds so long as f is a continuous function. The result is known as the Fundamental Theorem of Calculus.

Theorem 15–7 **The Fundamental Theorem of Calculus**

If f is continuous on the interval $[a, b]$, and if $A(t) = \int_a^t f(x)\, dx$, $a \leq t \leq b$, then $A'(t) = f(t)$.

Example 1 If $A(t) = \int_0^t \sin x\, dx$, find $A'(t)$.

SOLUTION Since sine is continuous, the Fundamental Theorem implies that $A'(t) = \sin t$.

Notice that the Fundamental Theorem applies even though the function $f(x) = \sin x$ is not always nonnegative; that is, the integral

$$A(t) = \int_0^t \sin x\, dx$$

is not always an area function.

The Fundamental Theorem enables you to evaluate certain definite integrals without resorting to the use of upper and lower approximations.

Corollary 15–7 If f is continuous on $[a, b]$, and if G is a function such that $G' = f$, then

$$\int_a^b f(x)\, dx = G(b) - G(a).$$

A formal proof of this corollary depends upon the fact that two functions that have the same derivative differ at most by a constant. Since $G' = f$, and since by the Fundamental Theorem $A' = f$, where

$$A(t) = \int_a^t f(x)\, dx,$$

it follows that for some constant C and for all t,

$$A(t) = G(t) + C.$$

Thus $A(b) = G(b) + C$ and $A(a) = G(a) + C$, so that by subtraction,

$$A(b) - A(a) = [G(b) + C] - [G(a) + C].$$

But $A(a) = \int_a^a f(x)\, dx = 0$ and $A(b) = \int_a^b f(x)\, dx$, so that

$$\int_a^b f(x)\, dx = G(b) - G(a).$$

Example 2 Evaluate $\int_1^4 \frac{1}{2\sqrt{x}}\, dx$.

SOLUTION In Chapter 14 you learned that if $G(x) = \sqrt{x}$, then $G'(x) = \frac{1}{2\sqrt{x}}$. So, by Corollary 15-7,

$$\int_1^4 \frac{1}{2\sqrt{x}}\, dx = G(4) - G(1)$$
$$= \sqrt{4} - \sqrt{1}$$
$$= 1.$$

Example 3 Evaluate $\int_0^1 (3x^2 + 4x + 1)\, dx$.

SOLUTION Let $f(x) = 3x^2 + 4x + 1$. From the rule for finding the derivative of a polynomial function, you can verify that $G(x) = x^3 + 2x^2 + x$ is a function such that $G' = f$. So by Corollary 15-7,

$$\int_0^1 f(x)\, dx = G(1) - G(0),$$

and $\int_0^1 (3x^2 + 4x + 1)\, dx = (1^3 + 2 \cdot 1^2 + 1) - (0^3 + 2 \cdot 0^2 + 0)$
$$= 4.$$

The importance of the Fundamental Theorem and its Corollary is hard to overstate. On the surface, it seems unlikely that there would be any useful connection between derivatives and integrals. Indeed, derivatives have their origin in finding slopes of tangent lines to curves, while integrals were defined to find areas under curves. Integrals can be tedious to calculate, often requiring a computer to obtain the needed accuracy. Yet the Fundamental Theorem and its Corollary enable you to calculate many integrals quickly and simply. Also, the Fundamental Theorem enables you to differentiate important functions like the natural logarithm function.

The two branches of calculus, known as **differential calculus** and **integral calculus**, each have important uses. The connection between them means that the two branches reinforce each other, forming a rich, efficient body of ideas with a great many applications. Newton and Leibnitz, the two men who are credited with inventing the calculus, received this acclaim for proving, independently of each other, the Fundamental Theorem and its Corollary. The separate ideas of derivative and integral had been discovered by others many years earlier.

Exercises

A For the given function f, find $A(t) = \int_0^t f(x)\,dx$ and verify that $A'(t) = f(t)$.

1. $f(x) = x^4$
2. $f(x) = x^5$

For $A(t)$ as given, use the Fundamental Theorem of Calculus to find $A'(t)$.

3. $A(t) = \int_0^t \cos x \, dx$
4. $A(t) = \int_0^t 2^x \, dx$
5. $A(t) = \int_0^t \log(x+1)\,dx$
6. $A(t) = \int_0^t \sqrt{x}\,dx$
7. $A(t) = \int_1^t x^3\,dx$
8. $A(t) = \int_5^t \sqrt{x^2 - 1}\,dx$

Use the Corollary to the Fundamental Theorem to evaluate the following integrals. Use the given fact about derivatives when needed.

9. $\int_0^1 (x + 2)\,dx$
10. $\int_0^1 (x + 3)\,dx$
11. $\int_0^2 (2x^2 + x + 1)\,dx$
12. $\int_1^3 (4x^2 + x^3)\,dx$
13. $\int_1^2 \frac{1}{2\sqrt{x}}\,dx$

B 14. $\int_0^2 e^x \, dx$. If $G(x) = e^x$, then $G'(x) = e^x$.

15. $\int_1^e \frac{1}{x}\,dx$. If $G(x) = \ln x$, then $G'(x) = \frac{1}{x}$.

16. $\int_0^{\pi/3} \cos x\,dx$. If $G(x) = \sin x$, then $G'(x) = \cos x$.

C 17. Use the Fundamental Theorem to prove that the derivative of the natural logarithm function is the reciprocal function. That is, prove that if $f(x) = \ln x$, then $f'(x) = \frac{1}{x}$.

Use this result, together with the properties of ln from the previous section, to graph the function ln.

18. Let the function F be defined for $t \geq -1$ by
$$F(t) = \int_{-1}^t |x^2 - 3x + 2|\,dx.$$

Find the first coordinate of any points on the graph of F where the tangent line is horizontal.

19. a. Find the derivative of the function $h(x) = \sqrt{x^2 + 1}$.

b. Evaluate $\int_0^4 \frac{x}{\sqrt{x^2 + 1}}\,dx$.

Chapter Review Exercises

15–1, page 531

1. Find an expression for the area of the region shaded in the figure, using the notation $S_a^b \, f(x)$.

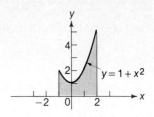

2. Use the average of L_3 and U_3 to find an approximation of $S_1^4 \, (x-2)^2$.

15–2, page 538

Evaluate each definite integral.

3. $\displaystyle\int_0^7 x^2 \, dx$ 4. $\displaystyle\int_4^6 x^2 \, dx$

5. Find the area under the parabola $y = x^2$ on the interval $x = 1$ to $x = 4$.

15–3, page 541

Evaluate each definite integral.

6. $\displaystyle\int_{-2}^0 x^2 \, dx$ 7. $\displaystyle\int_{-3}^1 x^2 \, dx$ 8. $\displaystyle\int_{-1}^2 3x^2 \, dx$

9. Find the area of the region bounded by the graphs of $y = x^2$ and the line $y = 4$.

15–4, page 545

Evaluate each definite integral.

10. $\displaystyle\int_2^4 x^3 \, dx$ 11. $\displaystyle\int_1^4 3x^3 \, dx$ 12. $\displaystyle\int_{-2}^1 5x^4 \, dx$

13. Find the area of the region between the graphs of $y = x^3$ and $y = x^4$.

15–5, page 551

14. Evaluate the area function $A(t) = \displaystyle\int_0^t 2x^3 \, dx$ for four numbers t in the interval $0 \le t \le 3$. Use these values to sketch a graph of $A(t)$ on $[0, 3]$.

15. Evaluate $\displaystyle\int_1^5 \frac{1}{x} \, dx$, accurate to the nearest tenth.

15–6, page 556

16. If $A(t) = \displaystyle\int_0^t \cos x \, dx$, find $A'(t)$.

17. An object moves with velocity given by the formula $v(x) = x^2 + 3x + 2$ ft./sec. at time x. The distance traveled between time 0 and time t is given by $D(t) = \displaystyle\int_0^t v(x) \, dx$. Use the Corollary to the Fundamental Theorem to find $D(6)$.

Chapter Test

Evaluate each definite integral.

1. $\int_0^3 4x^2\, dx$
2. $\int_{-2}^1 5x^2\, dx$
3. $\int_1^3 2x^3\, dx$

4. Find the area under the parabola $y = x^2$ between $x = 4$ and $x = 9$.

Find the area of each shaded region.

5.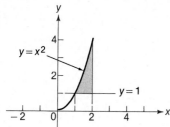

6.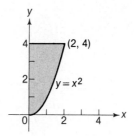

7. Find the area of the region between the graphs of $y = 2x^2$ and $y = 2x + 4$. [HINT: Subtract the area under the parabola from the area of an appropriate trapezoid.]

8. Use the average of L_4 and U_4 to find an approximation of $\int_1^3 \frac{1}{x}$. Compare your answer with ln 3, found from the table on page 552 or from a calculator.

9. If $A(t) = \int_1^t \sqrt{x-1}\, dx$, $t \geq 1$, use the Fundamental Theorem of Calculus to find $A'(t)$.

10. If $F(x) = \sqrt{x^2 + 1}$, then $F'(x) = \dfrac{x}{\sqrt{x^2+1}}$. Use this fact and the Corollary to the Fundamental Theorem to evaluate $\int_0^1 \dfrac{x}{\sqrt{x^2+1}}\, dx$.

11. When the region under the graph of $y = f(x)$ on the interval $[0, 1]$ is rotated around the y-axis, the volume of the resulting solid can be shown to be

$$2\pi \int_0^1 x \cdot f(x)\, dx.$$

Find the volume that results when $f(x) = x^2$.

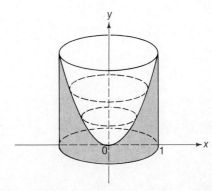

Acknowledgements

For permission to reproduce photographs on the pages indicated, acknowledgment is made to the following:

Cover: Ray F. Hillstrom/Hillstrom Stock Photo

Page ii: Ray F. Hillstrom/Hillstrom Stock Photo
Page 281: © Robert Drea
Page 304: Artstreet
Page 343: National Monuments Record (England)

All photographs not credited are the property of Scott, Foresman and Company.

Appendix

	Page
Tables	564
List of Symbols	571
Glossary	572
Selected Answers to Odd-Numbered Exercises	578
Index	593

Common Logarithms

N	0	1	2	3	4	5	6	7	8	9
10	0000	0043	0086	0128	0170	0212	0253	0294	0334	0374
11	0414	0453	0492	0531	0569	0607	0645	0682	0719	0755
12	0792	0828	0864	0899	0934	0969	1004	1038	1072	1106
13	1139	1173	1206	1239	1271	1303	1335	1367	1399	1430
14	1461	1492	1523	1553	1584	1614	1644	1673	1703	1732
15	1761	1790	1818	1847	1875	1903	1931	1959	1987	2014
16	2041	2068	2095	2122	2148	2175	2201	2227	2253	2279
17	2304	2330	2355	2380	2405	2430	2455	2480	2504	2529
18	2553	2577	2601	2625	2648	2672	2695	2718	2742	2765
19	2788	2810	2833	2856	2878	2900	2923	2945	2967	2989
20	3010	3032	3054	3075	3096	3118	3139	3160	3181	3201
21	3222	3243	3263	3284	3304	3324	3345	3365	3385	3404
22	3424	3444	3464	3483	3502	3522	3541	3560	3579	3598
23	3617	3636	3655	3674	3692	3711	3729	3747	3766	3784
24	3802	3820	3838	3856	3874	3892	3909	3927	3945	3962
25	3979	3997	4014	4031	4048	4065	4082	4099	4116	4133
26	4150	4166	4183	4200	4216	4232	4249	4265	4281	4298
27	4314	4330	4346	4362	4378	4393	4409	4425	4440	4456
28	4472	4487	4502	4518	4533	4548	4564	4579	4594	4609
29	4624	4639	4654	4669	4683	4698	4713	4728	4742	4757
30	4771	4786	4800	4814	4829	4843	4857	4871	4886	4900
31	4914	4928	4942	4955	4969	4983	4997	5011	5024	5038
32	5051	5065	5079	5092	5105	5119	5132	5145	5159	5172
33	5185	5198	5211	5224	5237	5250	5263	5276	5289	5302
34	5315	5328	5340	5353	5366	5378	5391	5403	5416	5428
35	5441	5453	5465	5478	5490	5502	5514	5527	5539	5551
36	5563	5575	5587	5599	5611	5623	5635	5647	5658	5670
37	5682	5694	5705	5717	5729	5740	5752	5763	5775	5786
38	5798	5809	5821	5832	5843	5855	5866	5877	5888	5899
39	5911	5922	5933	5944	5955	5966	5977	5988	5999	6010
40	6021	6031	6042	6053	6064	6075	6085	6096	6107	6117
41	6128	6138	6149	6160	6170	6180	6191	6201	6212	6222
42	6232	6243	6253	6263	6274	6284	6294	6304	6314	6325
43	6335	6345	6355	6365	6375	6385	6395	6405	6415	6425
44	6435	6444	6454	6464	6474	6484	6493	6503	6513	6522
45	6532	6542	6551	6561	6571	6580	6590	6599	6609	6618
46	6628	6637	6646	6656	6665	6675	6684	6693	6702	6712
47	6721	6730	6739	6749	6758	6767	6776	6785	6794	6803
48	6812	6821	6830	6839	6848	6857	6866	6875	6884	6893
49	6902	6911	6920	6928	6937	6946	6955	6964	6972	6981
50	6990	6998	7007	7016	7024	7033	7042	7050	7059	7067
51	7076	7084	7093	7101	7110	7118	7126	7135	7143	7152
52	7160	7168	7177	7185	7193	7202	7210	7218	7226	7235
53	7243	7251	7259	7267	7275	7284	7292	7300	7308	7316
54	7324	7332	7340	7348	7356	7364	7372	7380	7388	7396
N	0	1	2	3	4	5	6	7	8	9

Common Logarithms

N	0	1	2	3	4	5	6	7	8	9
55	7404	7412	7419	7427	7435	7443	7451	7459	7466	7474
56	7482	7490	7497	7505	7513	7520	7528	7536	7543	7551
57	7559	7566	7574	7582	7589	7597	7604	7612	7619	7627
58	7634	7642	7649	7657	7664	7672	7679	7686	7694	7701
59	7709	7716	7723	7731	7738	7745	7752	7760	7767	7774
60	7782	7789	7796	7803	7810	7818	7825	7832	7839	7846
61	7853	7860	7868	7875	7882	7889	7896	7903	7910	7917
62	7924	7931	7938	7945	7952	7959	7966	7973	7980	7987
63	7993	8000	8007	8014	8021	8028	8035	8041	8048	8055
64	8062	8069	8075	8082	8089	8096	8102	8109	8116	8122
65	8129	8136	8142	8149	8156	8162	8169	8176	8182	8189
66	8195	8202	8209	8215	8222	8228	8235	8241	8248	8254
67	8261	8267	8274	8280	8287	8293	8299	8306	8312	8319
68	8325	8331	8338	8344	8351	8357	8363	8370	8376	8382
69	8388	8395	8401	8407	8414	8420	8426	8432	8439	8445
70	8451	8457	8463	8470	8476	8482	8488	8494	8500	8506
71	8513	8519	8525	8531	8537	8543	8549	8555	8561	8567
72	8573	8579	8585	8591	8597	8603	8609	8615	8621	8627
73	8633	8639	8645	8651	8657	8663	8669	8675	8681	8686
74	8692	8698	8704	8710	8716	8722	8727	8733	8739	8745
75	8751	8756	8762	8768	8774	8779	8785	8791	8797	8802
76	8808	8814	8820	8825	8831	8837	8842	8848	8854	8859
77	8865	8871	8876	8882	8887	8893	8899	8904	8910	8915
78	8921	8927	8932	8938	8943	8949	8954	8960	8965	8971
79	8976	8982	8987	8993	8998	9004	9009	9015	9020	9025
80	9031	9036	9042	9047	9053	9058	9063	9069	9074	9079
81	9085	9090	9096	9101	9106	9112	9117	9122	9128	9133
82	9138	9143	9149	9154	9159	9165	9170	9175	9180	9186
83	9191	9196	9201	9206	9212	9217	9222	9227	9232	9238
84	9243	9248	9253	9258	9263	9269	9274	9279	9284	9289
85	9294	9299	9304	9309	9315	9320	9325	9330	9335	9340
86	9345	9350	9355	9360	9365	9370	9375	9380	9385	9390
87	9395	9400	9405	9410	9415	9420	9425	9430	9435	9440
88	9445	9450	9455	9460	9465	9469	9474	9479	9484	9489
89	9494	9499	9504	9509	9513	9518	9523	9528	9533	9538
90	9542	9547	9552	9557	9562	9566	9571	9576	9581	9586
91	9590	9595	9600	9605	9609	9614	9619	9624	9628	9633
92	9638	9643	9647	9652	9657	9661	9666	9671	9675	9680
93	9685	9689	9694	9699	9703	9708	9713	9717	9722	9727
94	9731	9736	9741	9745	9750	9754	9759	9763	9768	9773
95	9777	9782	9786	9791	9795	9800	9805	9809	9814	9818
96	9823	9827	9832	9836	9841	9845	9850	9854	9859	9863
97	9868	9872	9877	9881	9886	9890	9894	9899	9903	9908
98	9912	9917	9921	9926	9930	9934	9939	9943	9948	9952
99	9956	9961	9965	9969	9974	9978	9983	9987	9991	9996
N	0	1	2	3	4	5	6	7	8	9

Values of Trigonometric Functions

Deg.	Rad.	sin	cos	tan	cot	sec	csc
0	0.00	0.0000	1.0000	0.0000	—	1.0000	—
1	0.02	0.0175	0.9998	0.0175	57.2900	1.0002	57.2987
2	0.03	0.0349	0.9994	0.0349	28.6363	1.0006	28.6537
3	0.05	0.0523	0.9986	0.0524	19.0812	1.0014	19.1073
4	0.07	0.0698	0.9976	0.0699	14.3007	1.0024	14.3356
5	0.09	0.0872	0.9962	0.0875	11.4301	1.0038	11.4737
6	0.10	0.1045	0.9945	0.1051	9.5144	1.0055	9.5668
7	0.12	0.1219	0.9925	0.1228	8.1444	1.0075	8.2055
8	0.14	0.1392	0.9903	0.1405	7.1154	1.0098	7.1853
9	0.16	0.1564	0.9877	0.1584	6.3138	1.0125	6.3925
10	0.17	0.1736	0.9848	0.1763	5.6713	1.0154	5.7588
11	0.19	0.1908	0.9816	0.1944	5.1446	1.0187	5.2408
12	0.21	0.2079	0.9781	0.2126	4.7046	1.0223	4.8097
13	0.23	0.2250	0.9744	0.2309	4.3315	1.0263	4.4454
14	0.24	0.2419	0.9703	0.2493	4.0108	1.0306	4.1336
15	0.26	0.2588	0.9659	0.2679	3.7321	1.0353	3.8637
16	0.28	0.2756	0.9613	0.2867	3.4874	1.0403	3.6280
17	0.30	0.2924	0.9563	0.3057	3.2709	1.0457	3.4203
18	0.31	0.3090	0.9511	0.3249	3.0777	1.0515	3.2361
19	0.33	0.3256	0.9455	0.3443	2.9042	1.0576	3.0716
20	0.35	0.3420	0.9397	0.3640	2.7475	1.0642	2.9328
21	0.37	0.3584	0.9336	0.3839	2.6051	1.0711	2.7904
22	0.38	0.3746	0.9272	0.4040	2.4751	1.0785	2.6695
23	0.40	0.3907	0.9205	0.4245	2.3559	1.0864	2.5593
24	0.42	0.4067	0.9135	0.4452	2.2460	1.0946	2.4586
25	0.44	0.4226	0.9063	0.4663	2.1445	1.1034	2.3662
26	0.45	0.4384	0.8988	0.4877	2.0503	1.1126	2.2812
27	0.47	0.4540	0.8910	0.5095	1.9626	1.1223	2.2027
28	0.49	0.4695	0.8829	0.5317	1.8807	1.1326	2.1301
29	0.51	0.4848	0.8746	0.5543	1.8040	1.1434	2.0627
30	0.52	0.5000	0.8660	0.5773	1.7321	1.1547	2.0000
31	0.54	0.5150	0.8572	0.6009	1.6643	1.1666	1.9416
32	0.56	0.5299	0.8480	0.6249	1.6003	1.1792	1.8871
33	0.58	0.5446	0.8387	0.6494	1.5399	1.1924	1.8361
34	0.59	0.5592	0.8290	0.6745	1.4826	1.2062	1.7883
35	0.61	0.5736	0.8192	0.7002	1.4281	1.2208	1.7434
36	0.63	0.5878	0.8090	0.7265	1.3764	1.2361	1.7013
37	0.65	0.6018	0.7986	0.7536	1.3270	1.2521	1.6616
38	0.66	0.6157	0.7880	0.7813	1.2799	1.2690	1.6243
39	0.68	0.6293	0.7771	0.8098	1.2349	1.2868	1.5890
40	0.70	0.6428	0.7660	0.8391	1.1918	1.3054	1.5557
41	0.72	0.6561	0.7547	0.8693	1.1504	1.3250	1.5243
42	0.73	0.6691	0.7431	0.9004	1.1106	1.3456	1.4945
43	0.75	0.6820	0.7314	0.9325	1.0724	1.3673	1.4663
44	0.77	0.6947	0.7193	0.9657	1.0355	1.3902	1.4396

Values of Trigonometric Functions

Deg.	Rad.	sin	cos	tan	cot	sec	csc
45	0.79	0.7071	0.7071	1.0000	1.0000	1.4142	1.4142
46	0.80	0.7193	0.6947	1.0355	0.9657	1.4396	1.3902
47	0.82	0.7314	0.6820	1.0724	0.9325	1.4663	1.3673
48	0.84	0.7431	0.6691	1.1106	0.9004	1.4945	1.3456
49	0.86	0.7547	0.6561	1.1504	0.8693	1.5243	1.3250
50	0.87	0.7660	0.6428	1.1918	0.8391	1.5557	1.3054
51	0.89	0.7771	0.6293	1.2349	0.8098	1.5890	1.2868
52	0.91	0.7880	0.6157	1.2799	0.7813	1.6243	1.2690
53	0.93	0.7986	0.6018	1.3270	0.7536	1.6616	1.2521
54	0.94	0.8090	0.5878	1.3764	0.7265	1.7013	1.2361
55	0.96	0.8192	0.5736	1.4281	0.7002	1.7434	1.2208
56	0.98	0.8290	0.5592	1.4826	0.6745	1.7883	1.2062
57	0.99	0.8387	0.5446	1.5399	0.6494	1.8361	1.1924
58	1.01	0.8480	0.5299	1.6003	0.6249	1.8871	1.1792
59	1.03	0.8572	0.5150	1.6643	0.6009	1.9416	1.1666
60	1.05	0.8660	0.5000	1.7320	0.5774	2.0000	1.1547
61	1.06	0.8746	0.4848	1.8040	0.5543	2.0627	1.1434
62	1.08	0.8829	0.4695	1.8807	0.5317	2.1301	1.1326
63	1.10	0.8910	0.4540	1.9626	0.5095	2.2027	1.1223
64	1.12	0.8988	0.4384	2.0503	0.4877	2.2812	1.1126
65	1.13	0.9063	0.4226	2.1445	0.4663	2.3662	1.1034
66	1.15	0.9135	0.4067	2.2460	0.4452	2.4586	1.0946
67	1.17	0.9205	0.3907	2.3558	0.4245	2.5593	1.0864
68	1.19	0.9272	0.3746	2.4751	0.4040	2.6695	1.0785
69	1.20	0.9336	0.3584	2.6051	0.3839	2.7904	1.0711
70	1.22	0.9397	0.3420	2.7475	0.3640	2.9238	1.0642
71	1.24	0.9455	0.3256	2.9042	0.3443	3.0715	1.0576
72	1.26	0.9511	0.3090	3.0777	0.3249	3.2361	1.0515
73	1.27	0.9563	0.2924	3.2708	0.3057	3.4203	1.0457
74	1.29	0.9613	0.2756	3.4874	0.2867	3.6279	1.0403
75	1.31	0.9659	0.2588	3.7320	0.2680	3.8637	1.0353
76	1.33	0.9703	0.2419	4.0108	0.2493	4.1335	1.0306
77	1.34	0.9744	0.2250	4.3315	0.2309	4.4454	1.0263
78	1.36	0.9781	0.2079	4.7046	0.2126	4.8097	1.0223
79	1.38	0.9816	0.1908	5.1445	0.1944	5.2408	1.0187
80	1.40	0.9848	0.1736	5.6712	0.1763	5.7587	1.0154
81	1.41	0.9877	0.1564	6.3137	0.1584	6.3924	1.0125
82	1.43	0.9903	0.1392	7.1153	0.1405	7.1852	1.0098
83	1.45	0.9925	0.1219	8.1443	0.1228	8.2054	1.0075
84	1.47	0.9945	0.1045	9.5143	0.1051	9.5667	1.0055
85	1.48	0.9962	0.0872	11.4299	0.0875	11.4735	1.0038
86	1.50	0.9976	0.0698	14.3004	0.0699	14.3353	1.0024
87	1.52	0.9986	0.0523	19.0807	0.0524	19.1069	1.0014
88	1.54	0.9994	0.0349	28.6352	0.0349	28.6526	1.0006
89	1.55	0.9998	0.0175	57.2857	0.0175	57.2944	1.0002
90	1.57	1.0000	0.0000	—	0.0000	—	1.0000

Values of Trigonometric Functions

Rad.	Deg.	sin	cos	tan	cot	sec	csc
0.00	0.00	0.0000	1.0000	0.0000	—	1.0000	—
0.05	2.86	0.0500	0.9988	0.0500	19.9833	1.0013	20.0083
0.10	5.73	0.0998	0.9950	0.1003	9.9666	1.0050	10.0167
0.15	8.59	0.1494	0.9888	0.1511	6.6166	1.0114	6.6917
0.20	11.46	0.1987	0.9801	0.2027	4.9332	1.0203	5.0335
0.25	14.32	0.2474	0.9689	0.2553	3.9163	1.0321	4.0420
0.30	17.19	0.2955	0.9553	0.3093	3.2327	1.0468	3.3839
0.35	20.05	0.3429	0.9394	0.3650	2.7395	1.0645	2.9163
0.40	22.92	0.3894	0.9211	0.4228	2.3652	1.0857	2.5679
0.45	25.78	0.4350	0.9004	0.4831	2.0702	1.1106	2.2990
0.50	28.65	0.4794	0.8776	0.5463	1.8305	1.1395	2.0858
0.55	31.51	0.5227	0.8525	0.6131	1.6310	1.1730	1.9132
0.60	34.38	0.5646	0.8253	0.6841	1.4617	1.2116	1.7710
0.65	37.24	0.6052	0.7961	0.7602	1.3154	1.2561	1.6524
0.70	40.11	0.6442	0.7648	0.8423	1.1872	1.3075	1.5523
0.75	42.97	0.6816	0.7317	0.9316	1.0734	1.3667	1.4671
0.80	45.84	0.7174	0.6967	1.0296	0.9712	1.4353	1.3940
0.85	48.70	0.7513	0.6600	1.1383	0.8785	1.5152	1.3311
0.90	51.57	0.7833	0.6216	1.2602	0.7936	1.6087	1.2766
0.95	54.43	0.8134	0.5817	1.3984	0.7151	1.7191	1.2294
1.00	57.30	0.8415	0.5403	1.5574	0.6421	1.8508	1.1884
1.05	60.16	0.8674	0.4976	1.7433	0.5736	2.0098	1.1528
1.10	63.03	0.8912	0.4536	1.9648	0.5090	2.2046	1.1221
1.15	65.89	0.9128	0.4085	2.2345	0.4475	2.4481	1.0956
1.20	68.75	0.9320	0.3624	2.5722	0.3888	2.7597	1.0729
1.25	71.62	0.9490	0.3153	3.0096	0.3323	3.1714	1.0538
1.30	74.48	0.9636	0.2675	3.6021	0.2776	3.7383	1.0378
1.35	77.35	0.9757	0.2190	4.4552	0.2245	4.5661	1.0249
1.40	80.21	0.9854	0.1700	5.7979	0.1725	5.8835	1.0148
1.45	83.08	0.9927	0.1205	8.2381	0.1214	8.2986	1.0073
1.50	85.94	0.9975	0.0707	14.1014	0.0709	14.1368	1.0025
1.55	88.81	0.9998	0.0208	48.0785	0.0208	48.0889	1.0002
$\pi/2$	90.00	1.0000	0.0000	—	0.0000	—	1.0000

Values of Exponential Functions

x	e^x	e^{-x}	x	e^x	e^{-x}
0.0	1.0000	1.0000	4.0	54.5982	0.0183
0.1	1.1052	0.9048	4.1	60.3403	0.0166
0.2	1.2214	0.8187	4.2	66.6863	0.0150
0.3	1.3499	0.7408	4.3	73.6998	0.0136
0.4	1.4918	0.6703	4.4	81.4509	0.0123
0.5	1.6487	0.6065	4.5	90.0171	0.0111
0.6	1.8221	0.5488	4.6	99.4843	0.0101
0.7	2.0138	0.4966	4.7	109.9472	0.0091
0.8	2.2255	0.4493	4.8	121.5104	0.0082
0.9	2.4596	0.4066	4.9	134.2898	0.0074
1.0	2.7183	0.3679	5.0	148.4132	0.0067
1.1	3.0042	0.3329	5.1	164.0219	0.0061
1.2	3.3201	0.3012	5.2	181.2722	0.0055
1.3	3.6693	0.2725	5.3	200.3368	0.0050
1.4	4.0552	0.2466	5.4	221.4064	0.0045
1.5	4.4817	0.2231	5.5	244.6919	0.0041
1.6	4.9530	0.2019	5.6	270.4264	0.0037
1.7	5.4739	0.1827	5.7	298.8674	0.0033
1.8	6.0496	0.1653	5.8	330.2996	0.0030
1.9	6.6859	0.1496	5.9	365.0375	0.0027
2.0	7.3891	0.1353	6.0	403.4288	0.0025
2.1	8.1662	0.1225	6.1	445.8578	0.0022
2.2	9.0250	0.1108	6.2	492.7490	0.0020
2.3	9.9742	0.1003	6.3	544.5719	0.0018
2.4	11.0232	0.0907	6.4	601.8450	0.0017
2.5	12.1825	0.0821	6.5	665.1416	0.0015
2.6	13.4637	0.0743	6.6	735.0952	0.0014
2.7	14.8797	0.0672	6.7	812.4058	0.0012
2.8	16.4446	0.0608	6.8	897.8473	0.0011
2.9	18.1741	0.0550	6.9	992.2747	0.0010
3.0	20.0855	0.0498	7.0	1096.6332	0.0009
3.1	22.1980	0.0450	7.1	1211.9671	0.0008
3.2	24.5325	0.0408	7.2	1339.4308	0.0007
3.3	27.1126	0.0369	7.3	1480.2999	0.0007
3.4	29.9641	0.0334	7.4	1635.9844	0.0006
3.5	33.1155	0.0302	7.5	1808.0424	0.0006
3.6	36.5982	0.0273	7.6	1998.1959	0.0005
3.7	40.4473	0.0247	7.7	2208.3480	0.0005
3.8	44.7012	0.0224	7.8	2440.6020	0.0004
3.9	49.4024	0.0202	7.9	2697.2823	0.0004

Natural Logarithms

x	ln x	x	ln x	x	ln x
0.1	−2.3026	3.6	1.2809	7.1	1.9601
0.2	−1.6094	3.7	1.3083	7.2	1.9741
0.3	−1.2040	3.8	1.3350	7.3	1.9879
0.4	−0.9163	3.9	1.3610	7.4	2.0015
0.5	−0.6931	4.0	1.3863	7.5	2.0149
0.6	−0.5108	4.1	1.4110	7.6	2.0281
0.7	−0.3567	4.2	1.4351	7.7	2.0412
0.8	−0.2231	4.3	1.4586	7.8	2.0541
0.9	−0.1054	4.4	1.4816	7.9	2.0669
1.0	0.0000	4.5	1.5041	8.0	2.0794
1.1	0.0953	4.6	1.5261	8.1	2.0919
1.2	0.1823	4.7	1.5476	8.2	2.1041
1.3	0.2624	4.8	1.5686	8.3	2.1163
1.4	0.3365	4.9	1.5892	8.4	2.1282
1.5	0.4055	5.0	1.6094	8.5	2.1401
1.6	0.4700	5.1	1.6292	8.6	2.1518
1.7	0.5306	5.2	1.6487	8.7	2.1633
1.8	0.5878	5.3	1.6677	8.8	2.1748
1.9	0.6419	5.4	1.6864	8.9	2.1861
2.0	0.6931	5.5	1.7047	9.0	2.1972
2.1	0.7419	5.6	1.7228	9.1	2.2083
2.2	0.7885	5.7	1.7405	9.2	2.2192
2.3	0.8329	5.8	1.7579	9.3	2.2300
2.4	0.8755	5.9	1.7750	9.4	2.2407
2.5	0.9163	6.0	1.7918	9.5	2.2513
2.6	0.9555	6.1	1.8083	9.6	2.2618
2.7	0.9933	6.2	1.8245	9.7	2.2721
2.8	1.0296	6.3	1.8405	9.8	2.2824
2.9	1.0647	6.4	1.8563	9.9	2.2925
3.0	1.0986	6.5	1.8718	10.0	2.3026
3.1	1.1314	6.6	1.8871		
3.2	1.1632	6.7	1.9021		
3.3	1.1939	6.8	1.9169		
3.4	1.2238	6.9	1.9315		
3.5	1.2528	7.0	1.9459		

List of Symbols

N	set of natural numbers, page 1
W	set of whole numbers, page 1
J	set of integers, page 1
Q	set of rational numbers, page 1
R	set of real numbers, page 2
$\subset$	is a subset of, page 2
$\in$	is an element of, page 3
$\neq$	is not equal to, page 6
$>$	is greater than, page 9
$<$	is less than, page 9
$\leq$	is less than or equal to, page 12
$\geq$	is greater than or equal to, page 17
$\notin$	is not an element of, page 18
∞	infinity, page 22
$\cup$	union, page 23
$\cap$	intersection, page 23
$|x|$	absolute value of x, page 28
$\doteq$	is approximately equal to, page 31
C	set of complex numbers, page 40
i	the imaginary unit, page 40
PQ	distance between points P and Q, page 49
$\overline{PQ}$	the line segment with endpoints P and Q, page 50
Dom	domain of a function, page 95
Rng	range of a function, page 95
$f(a)$	the value f assigns to a, page 100
I	identity function, page 107
abs	absolute-value function, page 108
sgn	signum function, page 108
int	greatest-integer function, page 108
$f + g$	sum of functions f and g, page 100
$f \cdot g$	product of functions f and g, page 114
I^2	squaring function, page 116
$f \circ g$	composition of f and g, page 122
$\emptyset$	empty set, page 147
f^{-1}	inverse of a function f, page 248
$\exp_b$	exponential function with base b, page 283
$\log_b$	logarithm function with base b, page 286
log	common logarithm function, base 10, page 287
e	base of the natural logarithm function, page 295
ln	natural logarithm function, page 295
W	wrapping function, page 309
cos	cosine function, page 313
sin	sine function, page 313
sec	secant function, page 328
csc	cosecant function, page 328
tan	tangent function, page 330
cot	cotangent function, page 330
arcsin	inverse sine function, page 344
arccos	inverse cosine function, page 346
arctan	inverse tangent function, page 347
$\vec{V}$	a vector, page 425
$\langle a, b \rangle$	a two-dimensional vector, page 425
$\|\vec{V}\|$	length of a vector $\vec{V}$, page 425
$\overrightarrow{ST}$	arrow from point S to point T, page 426
$\langle a, b, c \rangle$	a three-dimensional vector, page 442
$\vec{A} \cdot \vec{B}$	dot product of vectors $\vec{A}$ and $\vec{B}$, page 447
$\begin{bmatrix} a & b \\ c & d \end{bmatrix}$	a matrix, page 454
$\{a_n\}$	the sequence $a_1, a_2, a_3, \ldots, a_n, \ldots$ page 459
$\sum_{i=1}^{n} a_i$	the sum $a_1 + a_2 + \cdots + a_n$, page 468
$\lim a_n$	limit of the sequence $\{a_n\}$, page 483
$n!$	n factorial, page 490
$\lim_{x \to c} f(x)$	the limit of f as x approaches c, page 503
f'	the derivative function of f, page 505
$\int_a^b f(x)\, dx$	the definite integral, page 539

Glossary

absolute value For any given real number x, the absolute value of x is x if $x \geq 0$; otherwise the absolute value of x is $-x$. page 28

absolute-value function The function abs that assigns to every argument x the value $|x|$. page 108

addition of functions The sum of two functions f and g is the function $f + g$ defined by $[f + g](x) = f(x) + g(x)$. page 110

additive inverse For any given real number a the additive inverse of a is the unique number x such that $a + x = x + a = 0$. page 6

additive inverse of a function The function $-f$ defined by $[-f](x) = -f(x)$ for any given function f. page 111

angular velocity A rate of rotation. page 355

arccosine function The inverse of the cosine function restricted to the domain $[0, \pi]$. page 346

arcsine function The inverse of the sine function restricted to the domain $\left[-\frac{\pi}{2}, \frac{\pi}{2}\right]$. page 344

arctangent function The inverse of the tangent function restricted to the domain $\left(-\frac{\pi}{2}, \frac{\pi}{2}\right)$. page 347.

area function A function of the form $A(t) = \int_c^t f(x)\,dx$. page 551

area under a curve The area of a region bounded by the graph of a function f, the x-axis, and vertical lines $x = a$ and $x = b$, where $f(x) \geq 0$ for all x between a and b. page 531

argument The first coordinate of an ordered pair in a function or a relation. page 95

argument of a complex number See **polar form of a complex number.**

arithmetic sequence A sequence in which the successive terms differ by a constant. page 460

asymptote If a graph approaches a line arbitrarily closely, then the line is an asymptote of the graph. page 107

axis of symmetry A line about which the graph of a function is symmetric. page 141

binary operation on a set S A rule that assigns to every ordered pair of elements of S exactly one element of S. page 3

bounded above A set S is bounded above if there exists a number M such that for every x in S, $x \leq M$. page 23

bounded set A set that is bounded above and bounded below. page 24

Cartesian product The Cartesian product of the sets A and B is the set $A \times B$ of all possible ordered pairs with first element in A and second element in B. page 95

circular function Any of the functions sine, cosine, tangent, cosecant, secant, or cotangent. page 313

common chord The segment joining the intersection points of two circles that intersect in two points. page 70

complex conjugates A pair of complex numbers of the form $a + bi$ and $a - bi$. page 43

complex number A number expressible in the form $a + bi$ where a and b are real numbers and i has the property $i^2 = -1$; a is called the *real part* and b is called the *imaginary part* of the complex number. page 40

complex-number system The set of complex numbers together with the operations of addition and multiplication defined for any two complex numbers $a + bi$ and $c + di$ by $(a + bi) + (c + di) = (a + c) + (b + d)i$ and $(a + bi) \times (c + di) = (ac - bd) + (ad + bc)i$. page 40

complex plane A plane in which each point is associated with exactly one complex number. The x-axis is called the *real axis* and the y-axis is called the *imaginary axis*. page 416

composite function The composite of two functions f and g is the function $f \circ g$ defined by $[f \circ g](x) = f(g(x))$. *page 122*

conic section A curve determined by the intersection of a plane and a double cone; a circle, an ellipse, a hyperbola, or a parabola. *page 170*

constant function For any constant c, the function defined by $f(x) = c$. *page 107*

constant of variation If y is directly proportional to x, with $y = kx$, then k is the constant of variation. *page 119*

continuous function A function whose graph has no breaks. *page 228*

convergent sequence A sequence that has a limit. *page 484*

convergent series A series for which the sequence of partial sums converges to a limit. *page 492*

cosecant function (csc) The reciprocal of the sine function. *page 328*

cosine function (cos) The cosine of a real number t, $\cos t$, is the x-coordinate of the point $W(t)$ determined by the wrapping function W. *page 313*

cotangent function (cot) The reciprocal of the tangent function. *page 330*

critical point A point on a graph where the tangent is horizontal or where the derivative is not defined. *page 515*

decreasing function A function f is decreasing if $f(a) > f(b)$ whenever a and b are in the domain of f and $a < b$. *page 256*

definite integral The quantity $\int_a^b f(x)\, dx$ which for nonnegative-valued functions gives the area under the curve between a and b. *page 539*

degree of a polynomial The highest power to which the variable is raised. *page 209*

derivative The derivative f' of a function f is the function defined by
$$f'(x) = \lim_{h \to 0} \frac{f(x+h) - f(x)}{h}$$
for every x such that the limit exists. *page 505*

direct variation A variable y varies directly as x (or y is directly proportional to x) if $y = kx$, where k is a nonzero constant. *page 119*

direction vector For the line given by the vector equation $\vec{X} = \vec{P} + t\vec{Q}$ the vector $\vec{Q}$ is called a direction vector of the line. *page 434*

directrix See **parabola**.

divergent sequence A sequence that does not have a limit. *page 484*

divergent series A series for which the sequence of partial sums is divergent. *page 492*

domain The set of arguments for a function or a relation. *page 95*

dot product Given two vectors $\vec{A} = \langle a_1, a_2, a_3 \rangle$ and $\vec{B} = \langle b_1, b_2, b_3 \rangle$, the dot product $\vec{A} \cdot \vec{B}$ is the real number $a_1 b_1 + a_2 b_2 + a_3 b_3$. *page 447*

eccentricity For any conic section, the distance from any point of the conic to the focus divided by the distance from that point to the corresponding directrix. *page 202*

ellipse A set of all the points in a plane the sum of whose distances from two fixed points is a constant. *page 186*

equivalent equations or inequalities Equations or inequalities that have the same domain and the same solution set. *page 19*

even function A function whose graph is symmetric about the y-axis, that is, a function f such that $f(-x) = f(x)$ for all x in the domain. *page 117*

exponential function A function of the form $f(x) = b^x$, where $b > 0$ and $b \neq 1$. *page 277*

exponential growth Growth described by a function that is the product of a positive constant and an exponential function. *page 298*

exponential growth principle Exponential growth occurs whenever the rate of change of a quantity is directly proportional to the existing amount of the given quantity. *page 300*

feasible region The set of points that satisfy the system of constraints for a linear-programming problem. *page 80*

focus See **parabola**.

function A relation that assigns exactly one value to each argument. *page 99*

geometric sequence A sequence in which the ratio of consecutive terms is constant. *page 464*

greatest-integer function The function int such that $int(x)$ is the greatest integer less than or equal to x. *page 108*

half-life The time required for a given quantity of radioactive material to disintegrate to one-half the original amount. *page 300*

harmonic sequence A sequence in which the reciprocals of the consecutive terms form an arithmetic sequence. *page 492*

horizontal-line test If every horizontal line intersects the graph of a function in at most one point, then the function is one-to-one. *page 253*

hyperbola A set of all the points in a plane the difference of whose distances from two fixed points is a constant. *page 191*

identity An equation which is true for every member of the domain. *page 391*

identity function The function I such that $I(x) = x$ for all x. *page 107*

increasing function A function f is increasing if $f(a) < f(b)$ whenever a and b are in the domain of f and $a < b$. *page 256*

increasing on an interval A function f is increasing on an interval if $f(a) < f(b)$ whenever a and b are in that interval and $a < b$. *page 257*

infinite series An expression which represents the sum of all the terms of a sequence is called an infinite series. *page 490*

inner function The function g in a composite function $f \circ g$. *page 122*

inverse function A function g is an inverse of f if $f(g(x)) = x$ for every x in the domain of g and $g(f(x)) = x$ for every x in the domain of f. *page 248*

leading coefficient of a polynomial The coefficient of the term with the highest power of the variable. *page 209*

least upper bound of a set The smallest of the upper bounds of a set. *page 24*

length of a vector For a given vector $\langle a, b \rangle$, the length is $\sqrt{a^2 + b^2}$. *page 425*

limit of a function A function f has a limit L at an argument c if and only if for every $\epsilon > 0$ there exists a $\delta > 0$ such that $L - \epsilon < f(x) < L + \epsilon$ whenever $c - \delta < x < c + \delta$ but $x \neq c$. *page 503*

limit of a sequence A number L is the limit of a sequence if for every $p > 0$ there exists an integer M such that all terms beyond the Mth term lie between $L - p$ and $L + p$. *page 484*

line of constant value In a linear-programming problem, if R is a quantity related to two variables x and y by $R = ax + by + c$, then each line obtained by replacing R by a constant is called a line of constant value. *page 83*

linear equation Any equation whose graph is a straight line; that is, an equation of the form $ax + by + c = 0$, where a and b are not both zero. *page 53*

linear function A function of the form $f(x) = mx + b$, where $m \neq 0$. *page 119*

linear programming A decision-making process subject to a system of constraints consisting entirely of linear inequalities. *page 82*

locus The set of points, and only those points, that satisfy a given condition. *page 167*

logarithmic function A function that is the inverse of some exponential function. *page 286*

lower estimate An approximation L_n of the area under a curve determined by n lower rectangles. *page 532*

lower rectangles Nonoverlapping, adjacent rectangles inscribed in a region under a curve. *page 532*

major axis The longest chord of an ellipse. *page 187*

method of least squares A process for determining the best-fitting line for a set of data points. *page 159*

minor axis The chord perpendicular to the major axis of an ellipse at its midpoint. *page 187*

modulus of a complex number The absolute value of the complex number $z = a + bi$, that is, $\sqrt{a^2 + b^2}$. *page 417*

multiplication of functions The product of two functions f and g is the function fg defined by $[fg](x) = f(x)\,g(x)$. *page 114*

multiplicative inverse For any nonzero real number a the multiplicative inverse of a is the unique number x such that $ax = xa = 1$. *page 6*

multiplicity of zeros If P is a polynomial function and $P(x) = (x - a)^r \cdot Q(x)$, then a is a zero of multiplicity r. *page 220*

normal vector to a plane A vector that is perpendicular to the plane. *page 451*

odd function A function whose graph is symmetric about the origin; that is, a function f such that $f(-x) = -f(x)$ for all x in the domain. *page 117*

one-to-one function A function f is one-to-one if and only if no two arguments have the same value. *page 252*

orthogonal vectors Two nonzero vectors whose dot product is 0. *page 449*

outer function The function f in a composite $f \circ g$. *page 122*

parabola A set of all the points in a plane equidistant from a fixed line, called the *directrix*, and a fixed point, called the *focus*. *page 171*

parametric equations A locus of points $P(x, y)$ is described parametrically if there exist two functions f and g such that $x = f(t)$ and $y = g(t)$. The variable t is called the *parameter*. *page 178*

partial sum The nth partial sum of an infinite series is the sum of the first n terms of the series. *page 490*

periodic function A function that repeats itself at regular intervals. The shortest such interval is called the *period*. *page 317*

point-slope form An equation of a line in the form $y - k = m(x - h)$, where m is the slope and (h, k) is a point on the line. *page 54*

polar coordinate system A coordinate system for locating the points of a plane determined by reference to a fixed point called the *pole* and a ray emanating from it called the *polar axis*. *page 380*

polar form of a complex number A complex number expressed in the form $z = r[\cos \theta + i \sin \theta]$, where r is the *modulus* and θ is the *argument* of z. *page 417*

polynomial An expression of the form $a_n x^n + a_{n-1} x^{n-1} + \cdots + a_1 x + a_0$. *page 210*

polynomial function A function f of the form $f(x) = a_n x^n + a_{n-1} x^{n-1} + \cdots + a_1 x + a_0$. *page 209*

power function A function of the form $f(x) = x^n$ for some natural number n. *page 265*

principal nth root If n is a natural number and $c > 0$, then the principal nth root of c is that positive number b such that $b^n = c$. *page 25*

principle of mathematical induction If $P(n)$ is a statement about a natural number n which is true for $n = 1$, and if $P(n)$ is true for $n = k + 1$ whenever $P(n)$ is true for $n = k$, then the statement is true for every natural number n. *page 475*

quadratic equation An equation of the form $ax^2 + bx + c = 0$, where a, b, and c are constants with $a \neq 0$. *page 33*

quadratic function A function f of the form $f(x) = ax^2 + bx + c$, where $a \neq 0$. *page 137*

radian A unit of angle measure. The central angle subtended by an arc of length t on a unit circle has a measure of t radians. *page 353*

range The set of values for a function or a relation. *page 95*

rate of change The instantaneous rate of change of a quantity which is expressed as a function of time is the derivative of the function. *page 525*

rational function A quotient of two polynomial functions. *page 238*

reciprocal function The function f such that $f(x) = \frac{1}{x}$ for all nonzero x. *page 107*

reduced polynomial If $P(x) = (x - a)Q(x)$, where P and Q are polynomial functions, then Q is called a reduced polynomial of P. *page 217*

relation A set of ordered pairs. *page 95*

relative maximum point A point $(c, f[c])$ on the graph of a function f is a relative maximum point if there is an open interval containing c such that $f(x) \leq f(c)$ for every x in the interval. *page 514*

scalar multiple The vector $k\vec{A}$, k a real number, is called a scalar multiple of $\vec{A}$; k is called a *scalar*. *page 432*

secant function (sec) The reciprocal of the cosine function. *page 328*

sequence A function with the set of all natural numbers as its domain. *page 459*

signum function The function sgn which assigns the value -1 to negative arguments, $+1$ to positive arguments, and 0 to 0. *page 108*

sine function (sin) The sine of a real number t, sin t, is the y-coordinate of the point $W(t)$ determined by the wrapping function W. *page 313*

slope The slope of a line or a segment is the number obtained by dividing the difference of the y-coordinates of two points on the line by the difference of the corresponding x-coordinates. *page 53*

slope function A function M defined at a given point $(c, f[c])$ on the graph of f by $M(h) = \dfrac{f(c + h) - f(c)}{h}$, for $h \neq 0$ and $c + h$ in the domain of f. *page 500*

slope-intercept form An equation of a line in the form $y = mx + b$, where m is the slope and b is the y-intercept. *page 56*

solution set of an equation or inequality The set of all members of the domain that yield true statements when substituted for the variable in an equation or inequality. *page 18*

squaring function The function I^2 such that $I^2(x) = x^2$ for all x. *page 116*

standard position of an angle An angle in standard position has its vertex at the origin and one side, called the *initial side*, coincides with the positive x-axis; the other side is called the *terminal side*. *page 353*

standard representative of a vector For a given vector $\vec{A} = \langle a_1, a_2 \rangle$, the arrow joining the origin to the point (a, b) is the standard representative of $\vec{A}$. *page 426*

sum of an infinite series If the sequence of partial sums of an infinite series has a limit, then that limit is the sum of the series. *page 492*

synthetic division The use of synthetic substitution to divide a polynomial $P(x)$ by $x - a$. *page 217*

synthetic substitution A process for calculating polynomial values. *page 211*

tangent function (tan) The function determined by dividing the sine function by the cosine function. *page 330*

tangent to a graph The tangent to the graph of a function f at a given point $T(c, f[c])$ is the line through T with slope equal to $f'(c)$. *page 500*

tangent to a parabola A line that intersects the parabola in exactly one point but is not parallel to the axis of symmetry. *page 182*

trigonometric function A circular function for which the arguments are angle measures rather than real numbers. *page 357*

unit vector A vector of length 1. *page 433*

upper bound of a set A number that is at least as large as any member of the set. *page 23*

upper estimate An approximation U_n for the area under a curve determined by n upper rectangles. *page 532*

upper rectangles Nonoverlapping, adjacent rectangles circumscribed about a region under a curve. *page 532*

variation A quantity y varies directly as x if $y = kx$, where k is a nonzero constant; y varies inversely as x if $y = \frac{k}{x}$. *page 120*

vectors Ordered pairs or triples of real numbers on which the operations of addition and scalar multiplication are defined. *page 425*

vertex of a quadratic function The intersection point of the graph of the function with its axis of symmetry. *page 141*

vertical-line test If every vertical line intersects the graph of a relation in at most one point, then the relation is a function. *page 100*

wrapping function A function used to define the sine and cosine functions. *page 309*

x-intercept The x-coordinate of a point of intersection of a graph with the x-axis. *page 56*

y-intercept The y-coordinate of a point of intersection of a graph with the y-axis. *page 56*

zero of a function A number a is a zero of a function f if $f(a) = 0$. *page 105*

Selected Answers to Odd-Numbered Exercises

Chapter 1

pages 3–4 **1.** Q, R **3.** J, Q, R **5.** Q, R **7.** N, W, J, Q, R **9.** R **11.** Q, R **13.** Q, R **15.** R **17.** Closed **19.** Closed **21.** Not closed (Example: $1 + 1 = 2$, $2 \notin A$) **23.** Not closed (Example: $1 \div 0 \notin Q$) **25.** $\frac{8}{33}$ **27.** $\frac{1088}{333}$ **29.** $b = 2$ or 5, and $c = 3$ **33. a.** 16

pages 8–9 **1.** 868 **3.** 899,000 **5.** Substitution Property, Associative Axiom, Inverse Axiom, Identity Axiom. **7.** Inverse Axiom, Theorem 1–1, Substitution Axiom, Inverse Axiom, Identity Axiom. **9.** $3x(2x^2 + 3)$ **11.** $(y + 6)^2$ **13.** $(b + 2)(b + 7)$ **15.** No, No **17.** Yes **23. a.** $7 * 9 = 9$ **b.** $*$ is commutative and associative **c.** No, Yes

pages 13–15 **1.** $-2x - 2y$ **3.** Corollary 1–7, Substitution, Theorem 1–2a, Theorem 1–6, Corollary 1–6, Theorem 1–2b **5.** $(2a - 3b)(4a^2 + 6ab + 9b^2)$ **7.** $2(4x^2 + 9)(2x + 3)(2x - 3)$ **9.** $(a^2 + 1)(1 + 2b)$ **11.** Definition of subtraction; Distributive Axiom; Theorem 1–6; Definition of Subtraction **13.** $(ac)(bc)^{-1} = (ac)(b^{-1}c^{-1}) = a[c(b^{-1}c^{-1})] = a[(b^{-1}c^{-1})c] = a[b^{-1}(c^{-1}c)] = a[b^{-1} \cdot 1] = a[b^{-1}] = \frac{a}{b}$ **15.** $-\frac{x+1}{2}$, if $x \neq 2$ and $y \neq 2$ **17.** $(4a^2 - 6a + 9)(1 - a)$, if $a \neq -\frac{3}{4}, -\frac{3}{2}$ **19.** $\frac{xy - x^2}{x^3 + y^3}$, if $x \neq -y$ **21.** $\frac{y^2 - 4}{y + 3}$, if $y \neq -4, -3$ **23.** $\frac{a}{b} \cdot \frac{c}{d} = (ab^{-1})(cd^{-1}) = (ac)(b^{-1}d^{-1}) = ac(bd)^{-1} = \frac{ac}{bd}$ **25.** $\frac{a}{b} - \frac{c}{b} = ab^{-1} - cb^{-1} = (a - c)b^{-1} = \frac{a-c}{b}$ **27.** $(a - 1)(a^2 + a + 4)$ **29.** $(x + 2)(x^2 - 2x + 4 - 3y)$ **31.** $(-b)[-(b^{-1})] = b[b^{-1}] = 1$, so by Theorem 1–3b, $-(b^{-1}) = (-b)^{-1}$. $-\left(\frac{a}{b}\right) = -(a \cdot b^{-1}) = a[-(b^{-1})] = a[(-b)^{-1}]$, just proved; thus $-\left(\frac{a}{b}\right) = a\left[\frac{1}{-b}\right] = \frac{a}{-b}$.

Pages 20–21 **1.** $x > \frac{4}{3}$ **3.** $x < 58$ **5.** $x < \frac{6}{11}$ **7.** $x \leq -37$ **9.** $x < \frac{7}{5}$ **11.** $R - \{3\}$ **13.** R **15.** $x < -\frac{1}{5}$ **17.** $x < \frac{-4}{3(c-2)}$ **19.** $x < 3$, $x \neq 0$ **21.** (i) $a \cdot b > 0 \cdot b$, by the Multiplication Axiom (ii) $b \cdot a < 0 \cdot a$, by Theorem 1–11 **23.** If $a > 0$, then $a \cdot a > a \cdot 0 = 0$; if $a = 0$, then $a \cdot a = a \cdot 0 = 0$; if $a < 0$, then $a \cdot a > a \cdot 0 = 0$. **25.** $x < 0$ and $x > -7$ **27.** $x > 2$ or $x < -5$ **29.** $a + c > b + c$; $c + b > d + b$, or $b + c > b + d$. Thus, $a + c > b + d$.

pages 27–28 **1.** $[-2, 3]$ **3.** $(-\infty, -4)$ **5.** $(-\infty, 2) \cup (5, \infty)$ **7.** $[3, \infty)$ **9.** $[2, 5]$ **11.** $(-2, \infty)$ **13.** $[-1, 6)$ **15.** $(-2, 6]$ **17.** $8\sqrt{3}$ **19.** $2\sqrt[4]{3}$ **21.** $5\sqrt{3}$ **23.** $-2\sqrt[3]{7}$ **25.** $2\sqrt{2}$ **27.** $\frac{\sqrt{70}}{7}$ **29.** $\frac{\sqrt[3]{6}}{3}$ **31.** $79\sqrt{2}$ **33.** $4 - \sqrt{6}$ **35.** -37 **37.** $3\sqrt{3} - 9\sqrt{2}$ **39.** $\frac{-5\sqrt{14}}{14}$ **41.** $\frac{11 - 3\sqrt{5}}{12}$ **43.** $\frac{7 + 2\sqrt{6}}{50}$ **45.** $[-2, 3) \cup (3, \infty)$ **47.** Not bounded **49.** Bounded; least upper bound = 1 **51.** Bounded; least upper bound = 4 **53.** $a > 0$ and $b > 0$ imply $(\sqrt{a} - \sqrt{b}) \in R$, so $(\sqrt{a} - \sqrt{b})^2 \geq 0$; $a - 2\sqrt{ab} + b \geq 0$; $\frac{a+b}{2} \geq \sqrt{ab}$. **55.** $\frac{2 + \sqrt[3]{3}}{11}$ **57.** If $a + b\sqrt{2} = c + d\sqrt{2}$, then $a - c = (d - b)\sqrt{2}$. Thus, like its equal, $(d - b)\sqrt{2} \in Q$, which can be only if $(d - b) = 0$. Then $a - c = 0$, also. Therefore, $a = c$ and $b = d$.

pages 31–33 **1.** 8 **3.** -4 **5.** $\frac{32}{243}$ **7.** $\frac{27}{64}$ **9.** $\frac{19}{2}$ **11.** $x = 4$ or -10 **13.** No solution **15.** $x = 3$ or -1 **17.** $-1 < x < 5$ **19.** No solution **21.** $x < -1$ or $x > 6$ **23.** $2p^{11/6}$ **25.** $\frac{3}{2b^{7/3}}$ **27.** $\frac{4}{x^5}$ **29.** $3 + 2p$ **31.** $4m^{2/3} - 12m^{1/3}$ **33.** $-\frac{7}{3} < x < 7$ **35.** $x < -2$ or $x > 2$ **37.** $x > \frac{1}{6}$ or $x < -\frac{7}{6}$ **39. c.** $\sqrt{a^2} = |a|$ for all $a \in R$, so $|-a| = \sqrt{(-a)^2} = \sqrt{a^2} = |a|$ **41.** 224.5 days **43. a.** True **b.** False; $\sqrt[3]{(-1)^3} = -1 \neq |-1|$ **c.** False; $|1 + (-2)| = 1 \neq |1| + |-2|$ **d.** False; $\sqrt{3^2 + 4^2} = 5 \neq |3| + |4|$ **45.** $(-\infty, -\sqrt{5}) \cup (-1, 1) \cup (\sqrt{5}, \infty)$

pages 37–38 **1.** $D = -8$; no real solutions **3.** $D = 0$; one solution **5.** $\frac{1}{2}$, -3 **7.** $-\frac{3}{2}$, 7 **9.** $x \doteq 1.3028$ or -2.3028 **11.** $\frac{1}{2}$, $-\frac{2}{3}$ **13.** $-\frac{m}{2}$, m **15.** $\frac{m \pm 2}{2}$ **17.** ± 2 **19. a.** $r_1 = \frac{-b}{2a} + \frac{\sqrt{D}}{2a}$ and $r_2 = \frac{-b}{2a} - \frac{\sqrt{D}}{2a}$, so $r_1 + r_2 = -\frac{b}{a}$ **b.** $r_1 \cdot r_2 = \left(\frac{b}{2a}\right)^2 - \left(\frac{\sqrt{D}}{2a}\right)^2 = \frac{b^2 - (b^2 - 4ac)}{4a^2} = \frac{c}{a}$ **21.** 8 only **23.** 5 only **25.** $R - \{1, -1\}$; 3 only **27. a.** About 6 seconds **b.** About 0.7 or 4 seconds **c.** No. **29.** 3, ± 2 **31.** ± 1 only **33.** ± 6; $k > 6$ or $k < -6$; $-6 < k < 6$ **35.** $a = \frac{33}{2}$, $b = \frac{1}{2}$

page 39 **1.** 1 8 28 56 70 56 28 8 1 **3.** $32x^5 + 240x^4 + 720x^3 + 1080x^2 + 810x + 243$ **5.** 0.8344

578 Answers

pages 44–45 **1.** $6 + 4i$ **3.** $13 + 6i$ **5.** $21 + 12i$ **7.** $-4i$ **9.** $x = 3, y = 4$ **11.** $(-2 - 3i)$ and $\left(\frac{2}{13} - \frac{3}{13}i\right)$; $(2 + 3i) + (-2 - 3i) = 0 + 0i$ and $(2 + 3i)\left(\frac{2}{13} - \frac{3}{13}i\right) = 1 + 0i$ **13.** $(-1 + 3i)$ and $\left(\frac{1}{10} + \frac{3}{10}i\right)$ **15.** -12
17. $2 + i\sqrt{6}$ **19.** $\frac{-1 \pm i\sqrt{3}}{2}$ **21.** $\frac{1 \pm 3i}{5}$ **23.** $\pm 3i, \frac{1}{2}$ **25.** $(a + bi)(c + di) = (ac - bd) + (ad + bc)i = (ca - db) + (cb + da)i = (c + di)(a + bi)$ **27.** $(a + bi)\left[\left(\frac{a}{a^2 + b^2}\right) + \left(\frac{-b}{a^2 + b^2}\right)i\right] = \frac{a^2 + b^2}{a^2 + b^2} + \left(\frac{-ab}{a^2 + b^2} + \frac{ab}{a^2 + b^2}\right)i = 1 + 0i$.
29. $-\frac{1}{5} + \frac{3}{5}i$ **31.** $b = -6, c = 13$

pages 45–46 **1. a.** J, Q, R **b.** Q, R **c.** R **d.** Q, R **3.** No, $2 + (3 \cdot 4) = 2 + 12 = 14$; $(2 + 3)(2 + 4) = 5 \cdot 6 = 30$
5. a. $3(3x - 1)(9x^2 + 3x + 1)$ **b.** $(3a^2 + 1)(a - 2b)$ **7.** $\{x : x > -\frac{9}{2}\}$ **9.** $[2, 4)$ **11. a.** $6 - 9a^2$
b. $\frac{(-27)^{2/3} x^{4/3}}{b^{-1}} = 9bx^{4/3}$ **13.** $m < -4$ **15.** $\frac{2}{29} - \frac{5}{29}i$

Chapter 2

pages 51–52 **1.** 5, $\left(\frac{3}{2}, 0\right)$ **3.** 8, $(2, -4)$ **5.** 10, $(3, -4)$ **7.** $2\sqrt{2}$, $(15, 16)$ **9.** 9.98, $(3.03, -4)$
11. $\sqrt{(1 - 0)^2 + (0 - 3)^2} = \sqrt{10} = \sqrt{(2 - 1)^2 + (3 - 0)^2}$, Yes **13.** No **15.** $(-2, 7)$ **17.** $(-10, 15)$ **19.** Yes
21. No **23.** Exterior **25.** Interior **27. a.** $\sqrt{109} + \sqrt{26} + \sqrt{41}$ **b.** $\frac{1}{2}(\sqrt{109} + \sqrt{26} + \sqrt{41})$ **29.** Not a right triangle
31. 28 **35.** -2 or 4

page 57 **1.** $\frac{1}{2}$ **3.** 0 **5.** $y - 5 = 3(x - 2)$ **7.** $y + 7 = \frac{1}{2}(x - 2)$ **9.** $y - 17 = -\frac{5}{17}(x + 12)$ **11.** $y = 2x + 9$
13. $y = -2x + 10$ **15.** $y = 3x - 5$ **17.** $y = 3x - 1$ **19.** $y = 6$ **21.** $y + 5 = \frac{1}{3}(x + 2)$ **23.** $y = -\frac{1}{2}x + 3$
25. $m = \frac{b - 0}{0 - a} = -\frac{b}{a}, a \neq 0; y = -\frac{b}{a}x + b; ay + bx = ab$ **27.** $m = 1, b = 2$ **29.** $m = -1, b = 1$ **31.** $m = -\frac{3}{4}$, $b = -\frac{1}{4}$ **33.** 47 **37.** By Theorem 2–4, ℓ has equation $y - b = m(x - 0)$, which simplifies to $y = mx + b$.

pages 60–62 **1.** Slope of $\overline{AB} = \frac{1}{4}$; slope of $\overline{CD} = \frac{1}{4}$; slope of $\overline{BC} = -4$; slope of $\overline{AD} = -4$. Two pairs of parallel sides imply a parallelogram. **3.** $y = 5x + 1$ **5.** $y + 7 = \frac{2}{5}(x - 3)$ **7.** $y = \frac{1}{2}x + 4$ **9.** $y + 2 = -\frac{6}{5}(x - 3)$
11. Yes **13.** $y - 12 = -\frac{1}{3}(x + 3)$ **15.** $\frac{7}{4}$ **17.** $-\frac{4}{3}$ **19.** -17 **21.** $y - 4 = -\frac{1}{6}(x - 6)$ **23.** Slopes of the sides are $\frac{1}{4}, -4, \frac{1}{4}, -4$, so the sides of each angle are perpendicular. **25.** 10 **27.** $(6, 6), (4, -2)$, or $(-4, 2)$

page 65 **1.** $(x + 1)^2 + (y - 5)^2 = 4$ **3.** $(x - 3)^2 + (y - 4)^2 = 25$ **5.** $(x + 1)^2 + (y + 2)^2 = 2$ **7.** $(1, 3); 5$
9. $(-3, -2); \sqrt{5}$ **11.** $(-1, 0); 3$ **13.** $(-3, 5); 6$ **15.** $(0, 0); 2$ **17.** $(-4, -2); 5$ **19.** $(1, 3); 2$ **21.** Point $(3, -2)$
23. Empty set **25.** $y = 5$ **27.** $y - 4 = \frac{3}{4}(x + 3)$ **29.** Distance between (x, y) and $(h, k) = $ radius; $\sqrt{(x - h)^2 + (y - k)^2} = r; (x - h)^2 + (y - k)^2 = r^2$ **31.** $(x + 1)^2 + \left(y + \frac{5}{2}\right)^2 = \frac{25}{4}$ **33.** $p < -65$

pages 70–71 **1.** $\left(\frac{7}{4}, \frac{41}{4}\right)$ **3.** $\left(\frac{3}{5}, \frac{1}{5}\right)$ **5.** $(0, 1)$ **7.** $(1, 2)$ and $(2, 1)$ **9.** $(-6, 8)$ and $(-6, -8)$ **11.** $\left(\frac{2}{5}, \frac{21}{5}\right)$ and $(2, 1)$
13. $(0, 5)$ only **15.** $(-1, 2)$ and $(-2, 1)$ **17.** $(5, 0)$ only **19.** $\pm\sqrt{2}$ **21.** $-\frac{\sqrt{15}}{15} < m < \frac{\sqrt{15}}{15}$ **23.** $(0, 0), (2, 4)$, $(4, 1)$ **25.** The four intersections are $(5, 3), (1, 0), (6, 0)$, and $(10, 3)$. By the distance formula, each side is 5. **27.** $\left(-\frac{18}{7}, \frac{11}{7}\right)$ **29.** $\frac{4\sqrt{5}}{5}$

pages 74–75 **1.** $\frac{8\sqrt{13}}{13}$ **3.** $\frac{11\sqrt{10}}{10}$ **5.** $\sqrt{5}$ **7.** $\frac{19\sqrt{5}}{5}$ **9.** $\frac{29}{2}$ **11.** $\frac{8\sqrt{17}}{17}$ **13.** $y = \frac{\sqrt{6}}{12}(x - 5)$ and $y = -\frac{\sqrt{6}}{12}(x - 5)$
15. $y = -3x + 2 + 2\sqrt{10}$ and $y = -3x + 2 - 2\sqrt{10}$ **17.** $(x + 5)^2 + (y - 5)^2 = 49$ **19. b.** $y = -\frac{3}{4}x$ and $x = 0$ **c.** The vertical tangent line has no slope. **21.** $\frac{\sqrt{17} - 1}{4}$

page 76 **1.** 1000 ft. **3.** 1300 ft. of pipe; 240 ft. from D toward E

pages 80–81 **19.** $y \leq -\frac{4}{5}x + \frac{9}{5}$ **21.** $(x - 3)^2 + (y - 1)^2 > 5$ **23. a.** $x + y \leq 300; x \geq 60; 0 \leq y \leq 175$;
$2x + 3y \leq 750$ **b.** $(60, 0), (60, 175), \left(112\frac{1}{2}, 175\right), (150, 150)$, and $(300, 0)$ **25.** $\left(\frac{-2 - \sqrt{2}}{2}, \frac{6 + \sqrt{2}}{2}\right)$, $\left(\frac{-2 + \sqrt{2}}{2}, \frac{6 - \sqrt{2}}{2}\right)$ **27.** $(5, 0), (3, -4)$

page 85 1. Maximum 56; minimum 7 **3.** Maximum 19; minimum -19 **5.** Maximum 130; minimum 10 **7.** Maximum 93; minimum 3 **9.** $k = 2$

pages 89–90 1. a. $R = 260x + 100y$ **b.** $10,000; $20,666.67; $20,800; 0 **c.** 80 acres corn; no alfalfa

3.

From \ To	Boston	Trenton	Rutland
Hartford	18	0	10
New York	6	16	0

5. (1, 1), (2, 0), (6, 0), (6, 2), (3, 5), (1,5) **7.**

From \ To	X	Y	Z
A	0	6	6
B	8	0	0

9. a. Sell 500 to each school system. **b.** Sell 600 to Northampton, 400 to Easthampton

pages 91–92 1. $\sqrt{26}$; $\left(-\frac{1}{2}, \frac{1}{2}\right)$ **3.** The triangle is isosceles **5.** $y = 10x + 16$ **7.** Not collinear **9.** $y = 7x + 10$ **11.** $y = -\frac{1}{2}x + 2$ **13.** $y = -\frac{3}{2}x - \frac{21}{4}$ **15.** $(-3, 6)$; 7 **17.** $(-7, -23)$ **19.** $(3, -4)$ and $(5, 0)$ **21.** $\frac{6\sqrt{13}}{13}$ **23.** $\frac{13}{2}$ square units **29.** Maximum profit $84,000 for 12,000 gal. Grade A, 8000 gal. Grade B.

Chapter 3

pages 97–98 1. Dom = {5, 173, -1}; Rng = {2, 3} **3.** Dom = {Saco, Charles}; Rng = {Maine, New Hampshire, Massachusetts} **5.** Dom = {7, 6, 0}; Rng = {-7, 7, -6, 0} **7.** Dom = {Kalamazoo, Wheaton, Champaign}; Rng = {49008, 60187, 61820} **13.** Dom = J; Rng = J **15.** Dom = $\{x \in R : x \geq 4\}$; Rng = $\{y \in R: -2 \leq y \leq 2\}$ **23.** $y = \pm 3$ **25.** $k \geq 15$ **27.** $\{(x, y): -4 \leq x \leq 0 \text{ and } y \leq -x\}$

pages 102–103 1. No **3.** Yes **5.** Yes **7.** Yes **9.** No **11.** No **13.** No **15.** Yes **17.** $g(0) = 1$; $g(1) = 3$; $g(2) = 7$ **19.** $f(-3) = 8$; $f\left(\frac{1}{3}\right) = -\frac{8}{9}$; $f(1 + \sqrt{2}) = 2 + 2\sqrt{2}$ **21.** $\frac{1}{2}, \frac{2}{3}, \frac{3}{4}, \frac{4}{5}$ **27.** 5 **29.** 3 **31.** $\frac{-1}{x(x + h)}$ **33.** $-\frac{5}{2}$ or -1 **35. a.** -3 or 1 **b.** $-3 \leq x \leq 1$ **c.** -2 or 3

pages 105–106 1. R **3.** R $-$ {0} **5.** $(-5, \infty)$ **7.** ± 1 **9.** $\frac{1}{3}$, -1 **11.** Yes **13.** Dom = R; Rng = R **15.** Dom = R; Rng = R **17.** Dom = $[4, \infty)$; Rng = $[0, \infty)$ **19.** Dom = R; Rng = $[2, \infty)$ **21.** Dom = R; -2 **23.** Dom = $(-\infty, 4]$; none **25. a.** Dom = $[-3, 3]$: Rng = $[-2, 2]$ **b.** $-3, -1, 2, 3$ **27.** Dom = R $-$ {-1}; Rng = R $-$ {2} **31.** Dom = $\left(\frac{2}{3}, 4\right]$

pages 109–110 1. 5 **3.** -1 **5.** 2 **7.** 1 **9.** -3 **11.** 0 **13.** -1 **15.** 3 **17.** $x \geq 0$ **19.** $x = 1$ **21.** $x = 0$ or 1 **23.** $x = \frac{1}{2}$ **25.** $x = 1$ **27.** $(-\infty, 0) \cup \left(\frac{1}{4}, \infty\right)$ **29.** $x \in W$ **31.** $0 \in$ Dom sgn, but not to Dom f.

pages 113–114 1. $[f + g](x) = x^2 + 3x + 5$; $-g(x) = -3x - 5$; $[f - g](x) = x^2 - 3x - 5$ **3.** $[f + g](x) = x^3 + x^2$; $-g(x) = -x^2$; $[f - g](x) = x^3 - x^2$ **5.** 14; -11; 8 **7.** -18; 18; -18 **21.** Dom $f = $ R $-$ {0}, which differs from Dom I. **23.** $f(x) = -1$; $g(x) = -|x|$ **25.** $f(x) = \frac{1}{x}$; $g(x) = 2$ **27.** 2 **29.** $[f + g](x) = f(x) + g(x) = g(x) + f(x) = [g + f](x)$; similarly, $f(x) + g(x) + h(x)$ is associative.

page 118 1. $[fg](x) = 6x$; 30 **3.** $[fg](x) = x^2 - 2x - 15$; 0 **5.** $[fg](x) = -5|x|$; -25 **9.** Even **11.** Even **13.** Even **15.** Even **17.** Odd **19.** Odd **21.** Even **29.** $[fg](-x) = f(-x) \cdot g(-x) = [-f(x)] \cdot [-g(x)] = f(x) \cdot g(x) = [fg](x)$

page 121 3. -3; 5 **5.** $\frac{1}{2}$; $\frac{1}{2}$ **7.** $f(x) = -2x + 2$ **9.** $f(x) = 2x + 1$ **11.** $f(x) = \frac{r}{2}x + \frac{r}{2}$ **13.** 54 **19.** 16, 17 and 18 are not linear **21.** $-\frac{5}{4}$ **23.** $5,672 **25.** About 9.707 **27.** Both are false.

pages 124–125 1. 1; 1 **3.** 16; 19 **5.** (0, 3), (1, 3), (2, 0) **7.** (0, 0), (1, 1), (4, 4) **9.** $4x - 3$ **11.** $-4x + 3$ **13.** $|x^2 - 4|$; $x^2 - 4$ **15.** $(x - 1)^3$; $x^3 - 1$ **17.** $\frac{5x + 1}{x - 1}$; Dom $f \circ g = $ R $-$ {1} **19.** $4x - 3$; Dom $f \circ f = $ R **21.** $\frac{5 - x}{x + 1}$; R $-$ {-1} **23.** $g(x) = |x|$; $h(x) = x + 1$ **25.** $g(x) = 2x^2$; $h(x) = (x - 1)$ **27.** $\frac{1}{2|1 - x| - 3} + 2$ **29.** One example: $f(x) = 2x - 3$; $g(x) = |x|$; $h(x) = x + 1$ **31.** One example: $f(x) = 5x - 7$; $g(x) = x^2$; $h(x) = x + 6$ **33.** $g(x) = |x|$ **35.** ± 2 **37.** -1 **39.** One example: $g(x) = x^3$; $f(x) = x^2$ **41.** $[f \circ g](-x) = f[g(-x)] = f[-g(x)] = f[g(x)] = [f \circ g](x)$

pages 131–132 **15.** One example: $f: x \to x + 1$; $g: x \to |x|$ **17.** One example: $f: x \to x^2$; $g: x \to x - 1$ **19.** One example; $f: x \to \frac{1}{x}$; $g: x \to x - 1$ **21.** $[f \circ g](x) = x$ **23.** $[f \circ g](x) = |x + 1|$ **37.** $(-\infty, 0]$ **39.** $(-\infty, -3) \cup (0, \infty)$ **41.** One

pages 133–134 **1.** Dom = $\{-1, 0, 1\}$; Rng = $\{0, 3\}$ **3.** Exercise 1 **5.** Yes **7.** Dom = $R - \{2\}$; -2 **9.** 2 **11.** $(-\infty, 0) \cup (1, \infty)$ **13.** Hyperbola, asymptotes $y = 1$ and $x = 0$ **15.** $\frac{1}{2}$ **17.** 0; 0, 2 **19.** $\frac{3}{2}$ **21.** $[f \circ g](x) = \frac{4-x}{x}$; Dom = $R - \{0\}$

Chapter 4

pages 139–140 **1.** $a = 3, b = 7, c = -5$ **3.** $a = 1, b = 0, c = -3$ **5.** $a = -1, b = 4, c = 1$ **7.** $f(x) = -x^2 - x + (3 + \sqrt{3})$; $a = -1, b = -1, c = 3 + \sqrt{3}$ **9.** $1 \pm 2\sqrt{2}$ **11.** $\frac{2 \pm \sqrt{46}}{3}$ **13.** $A(x) = \frac{3\pi x^2}{4}$ **15.** $A(x) = x^2\left(1 + \frac{\pi}{2}\right)$ **17.** $A(x) = (9 + 2x)(12 + 2x)$ **19. a.** 230 **b.** No **c.** 11.86 or 8.65 **d.** About 20 **21.** $\frac{10}{3}$ **23.** $f(x) = \frac{7}{6}x^2 + \frac{3}{2}x - \frac{2}{3}$ **25.** 9 **27.** $f(x) = x(x - k)$

pages 144–145 **1.** $x = 0$ **3.** $x = 0$ **5.** $x = 0$ **7.** Even **9.** Not even **11.** Even **13.** $(1, -3)$; $x = 1$ **15.** $(-1, 1)$; $x = -1$ **17.** $(2, -3)$; $x = 2$; $[-3, \infty)$ **19.** $(1, 1)$; $x = 1$; $(-\infty, 1]$ **21.** $(-3, 1)$; $x = -3$; $(-\infty, 1]$ **23.** $\left(\frac{1}{2}, \frac{3}{2}\right)$; $x = \frac{1}{2}$; $\left(-\infty, \frac{3}{2}\right]$ **25.** 10 m **27.** $(1, 1)$; $(0, 3)$ **29.** $(-1, 8)$; $(0, 5)$ **31.** $(2, 4)$; $(4, 0)$ **33.** -12 **35.** $f(x) = \frac{1}{2}(x - 2)^2$ **37.** $r = -3$; $b = -2$ **39.** 4 **41.** $\left(200, 70\frac{2}{3}\right)$; yes

pages 148–149 **1.** $-5 < x < 5$ **3.** $x \leq -2$ or $x \geq 2$ **5.** $0 < x < 2$ **7.** $x \leq -3$ or $x \geq 1$ **9.** $-3 < x < 3$ **11.** R **13.** $\emptyset$ **15.** $-4 < x < 0$ **17.** $x \leq 3 - \sqrt{10}$ or $x \geq 3 + \sqrt{10}$ **19.** $1 \leq t \leq 4$ **21.** $\frac{-1 - \sqrt{11}}{2} \leq x \leq \frac{-1 + \sqrt{11}}{2}$ **23.** $\emptyset$ **25.** $\frac{3 - \sqrt{177}}{4} \leq x \leq \frac{3 + \sqrt{177}}{4}$ **27.** $(-\infty, 0] \cup [3, \infty)$ **29.** $(-\infty, 0) \cup \left(\frac{1}{2}, \infty\right)$ **31.** $-2\sqrt{6} < k < 2\sqrt{6}$ **33.** One example: $a = \frac{1}{2}$ **35.** $-\frac{3}{2} - \sqrt{5} < p < -\frac{3}{2} + \sqrt{5}$ **37.** $b^2 - b + 1 = b^2 - b + \frac{1}{4} + \frac{3}{4} = \left(b - \frac{1}{2}\right)^2 + \frac{3}{4}$, a positive number **39. a.** $\frac{a+b}{2} \geq \sqrt{ab}$ **b.** $\frac{a+b}{2} = r$, $\sqrt{ab} = BD$ and $BD < r$, the longest side of a right triangle **c.** $(\sqrt{a} - \sqrt{b})^2 \geq 0$, etc. **41.** $1 < t < 3$ **43.** $2 \leq t \leq 3$

pages 151–152 **3.** $x < -3$ or $x > 1$ **5.** $-\frac{1}{3} \leq x \leq \frac{1}{2}$ **7.** $x \leq 0$ or $x \geq 2$ **9.** R **11.** R **13.** $V = \left(\frac{1}{4}, \frac{7}{8}\right)$; $\emptyset$ **15.** $V = \left(\frac{1}{6}, -\frac{11}{12}\right)$; $\emptyset$ **17.** $\emptyset$ **19.** $-1 - \sqrt{3} < x < -1 + \sqrt{3}$ **21.** $\frac{1}{2} - \frac{\sqrt{21}}{2} \leq x \leq \frac{1}{2} + \frac{\sqrt{21}}{2}$ **23.** $x < -3$ or $x > \frac{1}{2}$ **25.** $x < -3$ or $x > -1$ **27.** $t < -3$ or $t > 3$ **29.** $(0, 5), (5, 10)$ **31.** $2 - 2\sqrt{2} < k < 2 + 2\sqrt{2}$ **33.** $-2 \leq x \leq \frac{1}{2}$ **35.** $3 < x < 7$

pages 156–157 **1.** V at $\left(\frac{3}{4}, \frac{1}{8}\right)$, y-intercept = -1 **3.** 4, maximum; $(-\infty, 4]$ **5.** -2, minimum; $[-2, \infty)$ **7.** $\frac{73}{8}$, maximum; $\left(-\infty, \frac{73}{8}\right]$ **9.** $4 \notin$ Rng **11.** About 0.18 ft. **13.** 2, minimum; $[2, \infty)$ **15. a.** 5000 **b.** 3000 **17.** $\frac{65}{2}$ and $\frac{65}{2}$ **19.** 4,050 ft. **21. a.** $A(h) = -\frac{1}{2}h^2 + 6h$ **b.** 18 sq. ft. **23.** 3 **27.** $\frac{3}{4}$

page 158 **1.** $\frac{\sqrt{13}}{25}$

pages 162–163 **1.** $|m - 2|$ **3.** $|-m - 4|$ **5.** $|(2m)|$ **7.** $-\frac{18}{23}$ **9.** $y = \frac{23}{14}x$ **11.** $y = \frac{17}{30}x + 2$ **13.** $\left(\frac{11}{3}, \frac{5}{3}\right)$; $y = \frac{5}{14}x + \frac{5}{14}$ **15.** $y = \frac{9}{10}x + \frac{7}{5}$

pages 163–164 **1.** $f(x) = x^2 - 4x + 5$ **3.** Not quadratic **5.** $A(x) = x^2 + x$; Dom is all positive reals **7.** 6 sec.; about 5.7 sec. **9.** $\left(\frac{1}{6}, \frac{1}{12}\right)$; $t = \frac{1}{6}$ **11.** Parabola: $V = \left(\frac{1}{4}, \frac{7}{8}\right)$; $\left[\frac{7}{8}, \infty\right)$ **13.** $\emptyset$ **15.** $-2 < x < 1$ **17.** $x \leq 1 - \sqrt{2}$ or $x \geq 1 + \sqrt{2}$ **19.** About -0.5667 **21.** $y = \frac{9}{28}t$; $2\frac{25}{28}$ kg.

Chapter 5

pages 168–169 1. $\sqrt{(x+3)^2 + (y-4)^2}$ 3. $|y-4|$ 5. $|x+1|$ 7. $(x-4)^2 + y^2 = 4$, (2, 0) and (6, 0) deleted 9. $y = -2$ 11. $x^2 + y^2 - 2x + 8y + 15 = 0$ 13. $3x^2 + 3y^2 - 12x - 38y + 99 = 0$ 15. $9x^2 + 25y^2 - 225 = 0$ 17. $x^2 + y^2 - 2x + y - 26 = 0$, $(-4, 1)$ and $(6, -2)$ deleted 19. $\frac{y-1}{x+1} = |y+2|$, $x \neq -1$ 21. $x^2 - 2xy - y^2 - 24y - 72 = 0$ 23. $x^2 - 8x - 16y - 16 = 0$

page 170 1. A generating line of the double cone 3. A point, the vertex of the double cone.

pages 175–176 3. (0, 0) 5. (3, 0) 7. $(x-2)^2 = 16(y-3)$ 9. $(x+1)^2 = 12(y-3)$ 11. $(3, -1)$; $y = 4$ 13. $(0, -2)$; $y = -1$ 15. $(1, 0)$; $y = -1$ 17. $(-2, 1)$; $x = -3$ 19. $(2, -1)$; $x = 4$ 21. 4 23. $\frac{a^2+8}{4}$ 25. $(x-5)^2 = -4(y+3)$ 27. $r > 2$ 29. $V = \left(\frac{1}{2}, 0\right)$, opens downward 31. $V = \left(-\frac{1}{2}, 1\right)$, opens to the right 33. $x^2 = 20(y+1)$ 35. $(x-2)^2 = -\frac{1}{5}(y-5)$ 37. $(x-2)^2 = 4y$ 39. $x^2 - 4xy + 4y^2 - 22x - 26y + 16 = 0$

pages 179–180 1. Yes 3. $3\frac{1}{8}$ cm. 5. A portion of a parabola from its vertex (0, 0) to (4, 32) 7. $2\sqrt{2} + 3$ 9. 8 13. $f(y) = y^2 - y + 1$; $\frac{1}{2}$ 15. About 30.08 ft.

pages 185–186 1. $-2\sqrt{2} \leq k \leq 2\sqrt{2}$ 3. $k \geq -\frac{1}{4}$ 5. $y = 8x - 8$ or $y = 0$ 7. $y = 2x - 3$ 9. $y = -x - 1$ or $y = -\frac{1}{4}x - \frac{5}{8}$ 11. None 13. In the "interior" 15. $y = 6x - 6$ 17. $y = -x - 2$ or $y = -9x + 6$; the points of contact are $(-1, -1)$ and $\left(\frac{1}{3}, 3\right)$ 19. $m = 2r \pm 2\sqrt{r^2 - s}$, but $r^2 - s < 0$ 21. Intersections at $(-4, 2)$ and $(4, 2)$; at each point the slopes of the tangents are 1 and -1. 23. $x^2 = 10y + 5$ or $x^2 = 2y + 1$

pages 189–191 1. $3x^2 + 4y^2 - 12 = 0$ 3. x-axis 5. y-axis 7. Vertices: major $(\pm 5, 0)$, minor $(0, \pm 3)$; foci $(\pm 4, 0)$ 9. Vertices: major $(0, \pm 3)$, minor $(\pm 2, 0)$; foci $(0, \pm\sqrt{5})$ 11. Vertices: major $(0, \pm 2)$, minor $(\pm\sqrt{3}, 0)$; foci $(0, \pm 1)$ 13. $\frac{x^2}{9} + \frac{y^2}{4} = 1$ 15. $\frac{y^2}{1} + \frac{y^2}{16} = 1$ 17. $\frac{x^2}{16} + \frac{y^2}{4} = 1$ 19. $\left(\pm\frac{\sqrt{2}}{2}, 0\right)$ 21. (0, 1) and $\left(-\frac{4}{3}, -\frac{1}{3}\right)$ 23. Yes 25. Ellipse, vertical major axis, if $0 < m < 1$; unit circle, if $m = 1$; ellipse, horizontal major axis, if $m > 1$. 27. $y = \frac{1}{3}x \pm \frac{4}{3}$ 29. $A(x) = \frac{2}{3}x\sqrt{9-x^2}$

pages 195–196 1. $15x^2 - y^2 = 15$ 3. x-axis 5. x-axis 7. foci $(\pm\sqrt{13}, 0)$, vertices $(\pm 2, 0)$; $y = \pm\frac{3}{2}x$ 9. foci $(0, \pm\sqrt{5})$, vertices $(0, \pm 2)$; $y = \pm 2x$ 11. foci $(\pm\sqrt{7}, 0)$, vertices $(\pm\sqrt{3}, 0)$; $y = \pm\frac{2\sqrt{3}}{3}x$ 13. foci $\left(0, \pm\frac{\sqrt{6}}{2}\right)$, vertices $\left(0, \pm\frac{\sqrt{2}}{2}\right)$; $y = \pm\frac{\sqrt{2}}{2}x$ 15. $\frac{x^2}{9} - \frac{y^2}{7} = 1$ 17. $\frac{x^2}{9} - \frac{y^2}{9} = 1$ 19. $\frac{y^2}{12} - \frac{x^2}{3} = 1$ 23. $(0, \pm 3)$; $y = \pm\sqrt{2}x$ 25. $\left(\pm\frac{\sqrt{3}}{2}, 0\right)$; $y = \pm\frac{\sqrt{2}}{2}x$ 27. $m < -2$ or $m > 2$ 29. Each hyperbola has asymptotes $y = \pm\frac{5}{3}x$.

pages 200–201 1. $5x^2 - 4y^2 - 10x + 24y - 61 = 0$; $AC = -20$; hyperbola 3. $x^2 - 2x + 4y + 5 = 0$; $AC = 0$; parabola 5. Ellipse: $C(1, -2)$, $V_1(4, -2)$, $V_2(-2, -2)$, $F_1(1 + \sqrt{5}, -2)$, $F_2(1 - \sqrt{5}, -2)$; minor axis endpoints: (1, 0), (1, −4) 7. Hyperbola: $C(1, -2)$, $V_1(4, -2)$, $V_2(-2, 2)$, $F_1(1 + \sqrt{13}, -2)$, $F_2(1 - \sqrt{13}, -2)$; asymptotes $y + 2 = \pm\frac{2}{3}(x - 1)$ 9. Hyperbola: $C(1, -2)$, $V(1, -2 \pm 3)$, $F(1, -2 \pm \sqrt{13})$; asymptotes $y + 2 = \pm\frac{3}{2}(x - 1)$ 11. Ellipse: $C(-2, 1)$, $V(-2 \pm 2, 1)$, $F(-2 \pm \sqrt{3}, 1)$; minor axis endpoints: $(-2, 2)$, $(-2, 0)$ 13. (1, 1) 15. (0, 1) and $(-2, 1)$ 17. $m \neq 0$ 19. $m \neq 0$ and $m \neq -1$ 21. $\frac{(x-1)^2}{16} + \frac{y^2}{7} = 1$ 23. $\frac{(x-1)^2}{25} + \frac{y^2}{9} = 1$ 25. $\frac{(x-4)^2}{1} - \frac{(y+2)^2}{3} = 1$ 27. $m \neq 0$ 29. $0 < m < 1$ 31. $m < 0$ 33. $AC < 0$ and $F \neq 0$ 37. $y = \frac{\sqrt{6}}{3}x + \sqrt{6}$

pages 204–205 1. 0; parabola 3. -32; ellipse 5. $-3x^2 + y^2 + 8x - 4 = 0$; an hyperbola with $C\left(\frac{4}{3}, 0\right)$, $V\left(\frac{4}{3} \pm \frac{2}{3}, 0\right)$ 7. $x^2 - 2x + 2y - 2 = 0$; parabola, $V\left(1, \frac{3}{2}\right)$, opening downward 9. $7x^2 - 2xy + 7y^2 - 14x - 30y + 39 = 0$; ellipse, major axis on $y = x + 1$, $V_1\left(\frac{2}{3}, \frac{5}{3}\right)$, $V_2(2, 3)$ 11. $-3x^2 + 8xy + 3y^2 -$

$18x + 14y + 8 = 0$; hyperbola, transverse axis on $y = -\frac{1}{2}x - \frac{1}{2}$ **13.** Square both sides of $\sqrt{(x+2)^2 + y^2} = \sqrt{2}|x+1|$; rearrange.

pages 205–206 1. $3x^2 + 3y^2 - 22x + 52y + 175 = 0$ **3.** $V(2, -1)$, $F(2, -3)$; $y = 1$ **5.** $V\left(-2, \frac{1}{2}\right)$, $F(-2, 1)$; $y = 0$ **7.** $x^2 - 6x + 9 = 4y$ **9.** $y = 4x^2 - 8x + 5$; about 283 **11.** $k > \frac{5}{4}$ **13.** $36x^2 - 72x - 189 + 100y^2 = 0$; if $y = 2$, $x \notin R$ **15.** $\frac{x^2}{9} + \frac{y^2}{5} = 1$ **17.** $\frac{y^2}{9} - \frac{x^2}{7} = 1$ **19.** $V(-1 \pm 1, 2)$, $F(-1 \pm \sqrt{5}, 2)$ **21.** $V(-1,1)$, $F\left(-1, \frac{1}{2}\right)$ **23.** -32; ellipse **25.** $x^2 + 2y^2 - 4y - 6 = 0$

Chapter 6

pages 213–214 1. No **3.** Yes **5.** Yes **7.** Yes **9.** Yes **11.** 0, 12, -4 **13.** 0, 0, 0 **15.** 0, $\sqrt{3} + \sqrt{2} + 1$, $-\sqrt{3} + \sqrt{2} - 1$ **19.** $f(2) = 33$; $f(-2) = -39$ **21.** $f(2) = 15$; $f(-2) = -13$ **23.** $f(2) = 175$; $f(-2) = -149$ **25.** $f(x) = [(2x + 3)x - 1]x - 4$; $f(1) = 0$; $f(-1) = -2$; $f(3) = 74$; $f(-3) = -28$ **27.** $f(1.207) \doteq 2.680$; $f(-2.23) \doteq -9.03$; $f(0.01) \doteq -4.01$ **29.** $f(3.15) \doteq -76.0$; $f(-1.27) \doteq -39.5$; $f(\sqrt{7}) \doteq -62.0$ **31.** $f(3.15) \doteq 197.2$; $f(-1.27) \doteq 14.7$; $f(\sqrt{7}) \doteq 104.9$ **33.** $(-3, -21)$, $(-2, -4)$, $(-1, 1)$, $(0, 0)$, $(1, -1)$, $(2, 4)$, $(3, 21)$ **35.** $(-3, -486)$, $(-2, -88)$, $(-1, -6)$, $(0, 0)$, $(1, 2)$, $(2, 24)$, $(3, 162)$ **37.** Dom: $0 \le x \le 5$ **39.** 2 **41.** 4 **43.** $g(r^2) = (r^2)^4 - (r^2)^3 - 2(r^2)^2 + 1 = r^8 - r^6 - 2r^4 + 1 = (r^4 + r^3 - 1)(r^4 - r^3 - 1) = f(r) \cdot (r^4 - r^3 - 1) = 0$

pages 218–220 1. 2; -4 **3.** 2; 0 **5.** 48; -16 **7.** 80; -80 **9.** $f(2) = 0$ **11.** $f(2) = 0$ **13.** $f(2) = 0$ **15.** $f\left(-\frac{1}{3}\right) = 0$; $(9x - 3)$ **17.** $f(x) = x^3 - 7x + 6$ **19.** $f(x) = x^5 - 5x^3 + 4x$ **21.** $Q(x) = x^2 + x + 1$, $r = 9$; $Q(x) = x^2 - 2x + 4$, $r = 0$ **23.** $Q(x) = x$, $r = -2$; $Q(x) = x - 3$, $r = 4$ **25.** One example: $f(x) = \frac{3}{2}x^2 - \frac{3}{2}x - 3$ **27.** $f(x) = 6x^2 - 36x + 48$ **29.** $(x + 2)(x - 1)$ **31.** $(x + 3)(x - 2)$ **33.** $(2x^4 - x^2 + 3x) \div (x + 1) = 2x^3 - 2x^2 + x + 2$, $r = -2$ **35.** $(x - 2)(x + 3)(x - 7)$ **37.** $(x - 2)(x + 3)(x + 1)(x + 6)$ **39.** 60 **41.** $k = -8$ **43.** $A = 36$, $B = -108$

page 224 1. ± 1, ± 2 **3.** ± 1, $\pm \frac{1}{3}$ **5.** ± 1, ± 2, $\pm \frac{1}{2}$, $\pm \frac{1}{4}$ **7.** ± 1, ± 3, $\pm \frac{1}{2}$, $\pm \frac{3}{2}$, $\pm \frac{1}{4}$, $\pm \frac{3}{4}$ **9.** ± 1, ± 5, $\pm \frac{1}{2}$, $\pm \frac{5}{2}$, $\pm \frac{1}{3}$, $\pm \frac{5}{3}$, $\pm \frac{1}{6}$, $\pm \frac{5}{6}$ **11.** ± 1, ± 2, ± 3, ± 6, $\pm \frac{1}{2}$, $\pm \frac{3}{2}$, $\pm \frac{1}{3}$, $\pm \frac{2}{3}$, $\pm \frac{1}{6}$ **13.** Theorem does not apply **15.** ± 1, ± 2, ± 3, ± 6, $\pm \frac{1}{2}$, $\pm \frac{3}{2}$, $\pm \frac{1}{4}$, $\pm \frac{3}{4}$, $\pm \frac{1}{5}$, $\pm \frac{2}{5}$, $\pm \frac{3}{5}$, $\pm \frac{6}{5}$, $\pm \frac{1}{10}$, $\pm \frac{3}{10}$, $\pm \frac{1}{20}$, $\pm \frac{3}{20}$ **17.** 0, ± 1 **19.** 1, 2, 3 **21.** -2, $\frac{1}{2}$, 3 **23.** -1 **25.** $\sqrt{3}$ is a zero of $f(x) = x^2 - 3$. The possible rational zeros are ± 1, ± 3, none of which is a zero. Thus $\sqrt{3}$ is irrational.

pages 227–228 1. No positive zeros **3.** No negative zeros **5.** No positive zeros **7.** Neither theorem applies. **9.** Neither theorem applies. **11.** No negative zeros **13.** -1, $\frac{3}{2}$ **15.** 1, $-\frac{1}{2}$, $-\frac{1}{2}$, $\frac{-3 \pm \sqrt{5}}{2}$ **17.** 2 and possibly -1, -2, -3, -6

pages 229–230 1. $f(4) = -13$, $f(5) = 3$ **3.** $f(0) = -5$, $f(1) = 6$; $f(2) = 9$, $f(3) = -20$ **5.** About 2.1 **7.** About 1.7 **9.** About 1.6 **11.** $f(1) = 4$; $f(1.5) = -1.25$; $f(2) = 9$ **13.** $\sqrt[3]{7} \doteq 1.91$ **15.** One of the zeros could have multiplicity two, $f(x)$ touching but not crossing the x-axis, as in $f(x) = x^2(x - 1)$.

pages 232–234 1. -1.00 **3.** -1.73 **5.** $f(x) = x(9 - 2x)(7 - 2x)$, Dom: $0 < x < 3.5$, maximum at 1.30; 1.30 cm. $\times$ 6.40 cm. $\times$ 4.40 cm. **7.** $p = 1.52$, $q = 3.48$ **9.** 10.33 ft. $\times$ 10.33 ft. $\times$ 10.34 ft. **11.** 465,421.13 cu. ft. **13.** $\left(2, \frac{3}{2}\right)$ **15.** 0.59 sq. units **17.** 4.93 **19.** 1.15 ft. $\times$ 1.64 ft.

pages 237–238 1. $(-1, 5)$ **3.** $[-2, -1] \cup [2, \infty)$ **5.** $(-\infty, 0) \cup \left(\frac{5}{2}, \infty\right)$ **7.** $\left[-2, -\frac{1}{2}\right] \cup [3, \infty)$ **9.** $(-4, -2)$ **11.** $(-\infty, -1) \cup [1, \infty)$ **13.** $[0, 1)$ **15.** $[-1, 2) \cup \left[\frac{5}{2}, \infty\right)$ **17.** $\left(-\infty, \frac{1}{3}\right]$ **19.** $\left(-\frac{3}{2}, 0\right) \cup (4, \infty)$ **21.** $(-\sqrt{3}, \sqrt{3}) \cup \left(\frac{7}{2}, \infty\right)$ **23.** $(-\infty, 1)$ **25.** $[-4, -3] \cup [1, 4]$ **27.** $(-\infty, -2) \cup \left(2, \frac{5}{2}\right)$ **29.** $\left(1, \frac{4}{3}\right)$ **31.** $\left(-\infty, \frac{2}{5}\right] \cup (2, \infty)$ **33.** $(-\infty, -3) \cup (3, \infty)$ **35.** $(-\infty, -1) \cup \left[\frac{13}{12}, \frac{3}{2}\right)$ **37.** $\left(-2, -\frac{4}{5}\right) \cup (0, 2)$ **39.** $(-1, 0) \cup \left(\frac{1}{2}, \infty\right)$

pages 241–242 1. Dom = R − {−3}; $x = -3$, $y = 1$; Rng = R − {1} **3.** Dom = R − {1}; $x = 1$, $y = 1$; Rng = R − {1} **5.** Dom = R − {2}; $x = 2$, $y = 2$; Rng = R − {2} **7.** −2, 1; R − {−3, 2}; 2; $y = 2$ **9.** 2, 3; R − {−1}; ∞; none **11.** None; R − {−1, 4}; 0; $y = 0$ **13.** $(-\infty, -1] \cup (0, \infty)$ **15.** $(-\infty, -4] \cup (0, \infty)$ **17.** $(-\infty, 1) \cup [4, \infty)$ **19.** $\left(-\infty, -\frac{1}{2}\right] \cup (0, \infty)$ **21.** $(0.5, 2) \cup (2.2, \infty)$ **23.** $(-\infty, -2] \cup (-1, 1) \cup [2, \infty)$ **25.** 4

page 243 1. About 0.2541

page 244 1. $f(-2) = -18$; $f(3) = 22$ **3.** $f(-3) = 0$, so $(x + 3)$ is a factor of $f(x) = x^5 + 243$ **5.** ± 1, ± 3, $\pm \frac{1}{3}$, $\pm \frac{1}{9}$ **7.** $\frac{1}{2}, \frac{1}{2}, -2$ **9.** $\frac{1}{2}, \frac{1}{2}, \pm\sqrt{2}$ **11.** 3; 2.1 **13.** 6.7 in. × 6.7 in. × 6.6 in. **15.** $(-\infty, -3) \cup \left(3, \frac{7}{2}\right]$ **17.** Dom = R − {−5}; zero at $\frac{1}{2}$; asymptotes: $x = -5$, $y = 2$ **19.** Dom = R − {4, −4}; no zeros; asymptotes: $x = -4$, $x = 4$, $y = 0$

Chapter 7

pages 250–251 1. $f^{-1}(x) = x + 7$ **3.** $h^{-1}(x) = \frac{3}{2}x$ **5.** $f^{-1}(x) = x$ **7.** $h^{-1}(x) = 3(x + 4)$ **9.** $f^{-1}(x) = \frac{x - 7}{2}$ **11.** $f^{-1}(x) = \frac{1 + x}{3x}$ **13.** $h^{-1}(x) = \frac{2x}{3 - x}$ **15.** $f^{-1}(x) = \frac{5x + 3}{3x + 4}$ **17.** Dom f = R − {−2}; Rng f = R − {0}; Dom f^{-1} = R − {0}; Rng f^{-1} = R − {−2} **19.** $\frac{3 \pm \sqrt{17}}{4}$ **21.** $f(x) = \frac{1}{5 - 2x}$ **23. a.** $h(x) = 0$; each f has inverse $-f$ **b.** $g(x) = 1$; each f has inverse $\frac{1}{g}$, excepting $f(x) = 0$

pages 254–255 1. Definition; horizontal line test; composite of two one-to-one functions; apply Theorem 7–1 **3.** $\frac{1}{a} = \frac{1}{b}$ implies $a = b$ **5.** $5 - 2a = 5 - 2b$ implies $a = b$ **7.** $\frac{3}{4 - 3a} = \frac{3}{4 - 3b}$ implies $b = a$ **9.** $\frac{a}{4a - 1} = \frac{b}{4b - 1}$ implies $a = b$ **11.** $\sqrt{a + 3} = \sqrt{b + 3}$ implies $a = b$ **13.** Not one-to-one **15.** One-to-one **17.** One-to-one **19.** No; $f(1) = -2$ and $f(-1) = -2$ **21.** Yes **23.** No; $f(1) = \frac{1}{3}$ and $f(-1) = \frac{1}{3}$ **25.** No; $h(0) = 0$ and $h(\sqrt{3}) = 0$ **27.** Yes **31.** $\frac{4}{3}$ **33.** $h(x) = x^5$; $f(x) = x - 3$; $g = h \circ f$ **35.** $f(x) = \frac{1}{x - 4}$; $g(x) = x^3$; $h = f \circ g$ **37.** Yes

pages 259–260 1. Increasing **3.** Neither **5.** Neither **7.** Neither **11.** $(-\infty, 0), (0, \infty)$ **13.** $(-\infty, -1), (1, \infty)$ **15.** $[-3, \infty)$ **17.** $(-\infty, -1]$ **19.** $[2, \infty)$ **21.** $[2, \infty)$ **23.** $[0, 1)$ and $(1, \infty)$ **25.** $(-\infty, -1]$ **27.** $f(-2) = 1$ and $f(-1) = \frac{1}{2}$ imply not increasing; $f(-5) = \frac{1}{2}$ and $f(-4) = 1$ imply not decreasing. **29.** $a < b$; $4 - 3a > 4 - 3b$; decreasing **31.** $a < c$; $ma + b < mc + b$; increasing **33.** $a < b$; $\frac{2}{4 - a} < \frac{2}{4 - b}$; increasing **35.** $a < b$, $g(a) > g(b)$; $g(b) < g(a)$, $f \circ g(b) > f \circ g(a)$; $f \circ g(a) < f \circ g(b)$; $f \circ g$ is increasing **37.** $h(x) = 2 - 7x$, decreasing; $f(x) = x^3$, increasing; $f \circ h = g$, decreasing **39.** $f(x) = x^3 + 1$, increasing; $g(x) = x^7$, increasing; $g \circ f = h$, increasing **43. b.** $f(1) = 2$; $f(-x) = -x - \frac{1}{x} = -f(x)$ implies symmetry about the origin; for $x > 0$, $f(x) > x$; $f(x) = x$ and $x = 0$ are asymptotes.

pages 264–265 1. {(−1, 2), (7, 3), (−8, 5)} **3.** $x = 5$, $y = -2$ **5.** $[-2, \infty)$ **7.** R − {3} **9.** Dom = Rng = R **11.** Dom = Rng = R **13.** Dom = Rng = R **15.** Dom = Rng = R **21.** R − $\left\{\frac{1}{3}\right\}$ **23.** $\left(\frac{1}{2}, \frac{1}{2}\right)$ **25.** (0, 0), (2, 2) **27.** Dom f = R − {−2}; Rng f = R − {0}; Dom f^{-1} = R − {0}; Rng f^{-1} = R − {−2} **29.** Dom f = R − {0}; Rng f = R − {1}; Dom f^{-1} = R − {1}; Rng f^{-1} = R − {0} **31.** Dom f = $[-2, \infty)$; Rng f = $[0, \infty)$; Dom f^{-1} = $[0, \infty)$; Rng f^{-1} = $[-2, \infty)$ **33.** Slope of $\overleftrightarrow{PQ} = -1$ implies perpendicular to $y = x$; the lines intersect at $\left(\frac{a + b}{2}, \frac{a + b}{2}\right)$. **35.** $m = -1$ **37.** One-to-one means $f(a) = b$, $f^{-1}(b) = a$; odd implies $f(-a) = -b$; thus $f^{-1}(-b) = -a = -f^{-1}(b)$ and f^{-1} is odd.

pages 268–269 1. R, R **3.** $(-\infty, 0], [0, \infty)$ **5.** $[0, \infty), (-\infty, 1]$ **7.** R, R **9.** $[3, \infty), [1, \infty)$ **11.** R, R **13.** $f: x \to 1 + \sqrt{x}$ **15.** $f: x \to \sqrt[3]{3 - x}$ **17.** R − {0}, R − {0} **19.** $(9, \infty), (0, \infty)$ **21.** $(-3, \infty), (-\infty, 0)$ **23.** $[0, \infty), [-3, \infty)$; $f^{-1}(x) = x^2 - 3$ **25.** R, R; $f^{-1}(x) = \sqrt[3]{x - 1}$ **27.** $(0, \infty), (4, \infty)$; $f^{-1}(x) = \frac{1}{x^2} + 4$ **29.** $(-1, 0) \cup (1, \infty)$ **31.** $(1, \infty)$

33. $-2 < r_1 < -1$, $1 < r_2 < 2$ **35.** $-2 < r_1 < -1$, $1 < r_2 < 2$ **37.** $1 < r_1 < 2$, $4 < r_2 < 5$
39. $\left[-5, -\frac{2}{3}\right]$ **41.** $[-4, 1]$ **43.** -1.2

page 271 1. $g(x) = x^3$, $h(x) = x - 1$; $f^{-1} = g^{-1} \circ h^{-1} = \sqrt[3]{x + 1}$ **3.** $f^{-1}(x) = \frac{1}{x} - 4$ **5.** $f^{-1}(x) = x^3 - 3$
7. $f^{-1}(x) = \sqrt[5]{x} + 2$ **9.** $f^{-1}(x) = \sqrt[3]{x + 3} + 2$ **11.** $f^{-1}(x) = \frac{1}{\sqrt[3]{x - 1}} - 2$ **13.** $f^{-1}(x) = \left(\frac{1 - 2x}{x}\right)^2 + 5$
15. $[0, \infty)$; $[1, \infty)$, $[0, \infty)$; $f^{-1}(x) = \sqrt{x - 1}$ **17.** $[0, 3) \cup (3, \infty)$; $(-\infty, 0) \cup \left[\frac{1}{9}, \infty\right)$, $[0, 3) \cup (3, \infty)$;
$f^{-1}(x) = \sqrt{9 - \frac{1}{x}}$ **19.** $[2, \infty)$; $[0, \infty)$, $[2, \infty)$; $f^{-1}(x) = \sqrt{x} + 2$

page 273 1. Not a group **3.** Not a group **5. b.** $7 \oplus (3 \oplus 10) = 8$; $(7 \oplus 3) \oplus 10 = 8$ **c.** 6 and 12 **d.** Yes

page 274 1. $g^{-1}(x) = \frac{x}{2 - x}$ **3.** Assume $f(a) = f(b)$; then $\frac{1}{2a - 7} = \frac{1}{2b - 7}$ implies $b = a$. **5.** Assume $a < b$; then
$5a - 9 < 5b - 9$, so $f(a) < f(b)$. **7.** Assume $a < b$; then $f(a) > f(b)$ and $g(a) > g(b)$; $f(a) + g(a) > f(b) + g(b)$;
$[f + g](a) > [f + g](b)$ **9.** $-\frac{2}{5}$ **11.** $\left[-1, \frac{1}{2}\right]$ **13.** $g^{-1}(x) = \frac{1}{x^3} - 8$

Chapter 8

pages 280–281 1. a. 20.583 **b.** 0.030 **c.** 1.260 **d.** 2.162 **3.** 4 **5.** $\frac{3}{2}$ **7.** 0 **9.** $\frac{1}{2}$ **11.** -2 **13.** -5 **15.** 7 **17.** $\frac{5}{2}$
19. -3 **21.** $x < 3$ **23.** $u \leq -\frac{7}{2}$ **25.** $v < 2$ **27.** $58.6 \cdot 10^6$ **29.** \$179.08, \$236.74, \$310.58 **31.** \$68.90
33. 6 throws **35.** $(-1, 0)$

pages 285–286 1. Asymptotes: $y = 0$; one passes through $(0, 1)$ and $(1, 3)$; other through $(0, 1)$ and $(-1, 3)$
3. Asymptotes: $y = 0$; one through $(0, 1)$ and $(2, 4)$; other through $(0, 1)$ and $(2, 6.25)$ **5.** 4 **7.** $\frac{\pi}{2}$ **9.** $x \leq \frac{1}{2}$
11. $x < -\sqrt{2}$ **13.** Asymptote: $y = 0$; $(-1, 3)$, $(0, 1)$ **15.** Asymptote: $y = 0$; $\left(0, \frac{3}{2}\right)$, $\left(2, \frac{2}{3}\right)$ **17.** Asymptote: $y = 2$;
$(-2, 3)$, $(-1, 5)$ **19.** $\frac{\sqrt{2}}{32}$ **21.** Dom $= (0, \infty)$, Rng $=$ R **23.** 4.1 **25.** \$1,808.65 **27.** Asymptotes: $y = 1$ and $y = 0$;
$\left(0, \frac{1}{2}\right)$, $\left(1, \frac{1}{4}\right)$ **29.** $f(x) = x^3 + 2$, $g(x) = \left(\frac{1}{3}\right)^x$; at $f \cap g$, $f(x) - g(x) = 0$; 1 solution, in $(-1, 0)$

pages 288–289 1. 1 **3.** 3 **5.** 0 **7.** -3 **9.** -2 **11.** 5 **13.** Asymptotes: $x = 0$; intersection at $(1, 0)$; $\log_3 x$
through $\left(\frac{1}{3}, -1\right)$, $(3, 1)$; $\log_{1/3} x$ through $(3, -1)$, $\left(\frac{1}{3}, 1\right)$ **15.** Symmetrical about $(0, 0)$; both asymptotic to
$x = 0$ on opposite sides; $(-2, 1)$, $(-1, 0)$, $\left(-\frac{1}{2}, -1\right)$; $\left(\frac{1}{2}, 1\right)$, $(1, 0)$, $(2, -1)$ **17.** Asymptote: $x = 3$; $(3.5, -1)$,
$(4, 0)$, $(5, 1)$ **19.** Asymptote: $x = 0$; $(1, 3)$, $(3, 2)$ **21.** 64 **23.** 3 **25.** -2 **27.** 0.068 **29.** 1.334 **31.** 0, 0.100 **33.** $\frac{3}{2}$
35. $(-1, 2)$, $\left(-\frac{1}{4}, 0\right)$ **37.** 36 **39.** Dom $f^{-1} =$ R; Rng $f^{-1} = (-2, \infty)$; $f^{-1}(x) = \exp_3(x) - 2$
41. Dom $f^{-1} = (-3, \infty)$; Rng $f^{-1} =$ R; $f^{-1}(x) = \log_2(x + 3) - 1$ **43.** $(0, 1] \cup [3, \infty)$ **45.** Asymptotes: $x = 2$ and
$x = -2$; $(\sqrt{5}, 0)$, $(\sqrt{6}, 1)$, $(\sqrt{8}, 2)$; symmetric about y-axis.

pages 292–293 1. 2 **3.** 2 **5.** 0 **7.** 1 **9.** $\frac{9}{4}$ **11.** $\left(-\frac{1}{2}, \infty\right)$; 1 **13.** $(2, \infty)$, 4 **15.** $(2, \infty)$, $\frac{7}{3}$ **17.** $(2, \infty)$, 4 **19.** $(2, \infty)$, 4
21. Asymptote: $x = 3$; $(4, 0)$, $(7, 1)$ **23.** Asymptote: $x = 2$; $(3, 0)$, $(5, -1)$ **25.** $\left(\frac{2}{3}, \infty\right)$, $\emptyset$ **27.** Dom $= (4, \infty)$;
$(4, 85)$ **29.** Dom $= (-\infty, 3)$; $(-\infty, 1)$ **31.** Dom $= \left(-6, \frac{3}{2}\right)$; $(-6, -1)$ **33. a.** $1 + a$ **b.** $a + 2b$ **c.** $b - a$ **d.** $1 + \frac{2}{3}a$
35. Asymptote: $x = 0$; $(1, 3)$, $(3, 2)$, $(9, 1)$ **37.** 1, 2 **39.** $\frac{\sqrt{2}}{2}$, 16

pages 296–298 1. 1.609 **3.** 0.631 **5.** -3.969 **7.** 0.263 **9.** 3 **11.** 8 **13.** $\log 5$ **15.** 1.765 **17.** -5.679 **19.** 2.565
21. 0.819 **23.** ± 1.352 **25.** $(1.842, 6)$ **27.** $x \leq 1.465$ **29.** $x < -2.893$ **31.** $x \geq -5.229$ **33.** 256 hertz
35. $\log_b a = \frac{\log_a a}{\log_a b} = \frac{1}{\log_a b}$ **37.** $f(x) = \log_b x$, $g(x) = \log_c x$; $f(x) = \frac{\log_c x}{\log_c b} = \frac{1}{\log_c b} \cdot g(x)$ **39.** 42 **41.** 11:41 A.M.
43. 2.096

pages 302–304 1. $1806.11; $1819.40; $1822.03 **3.** $5\frac{3}{4}$ yr. **5. a.** $\frac{15}{4}$ g **b.** $30\sqrt{2}$ g **7.** 40.4 hr. **9.** 99.1 g
11. $1702.55 **13.** 164 **15.** 1985 **17.** 14.6 min.

pages 304–305 1. 320.108 **3.** 5.696 **5.** $\frac{3}{2}$ **7.** 49,000 **9.** Asymptote: $y = -3$; $(1, -2), (2, -1), (3, 1)$ **11.** $x \geq -3\sqrt{2}$
13. $x_1 < x_2$; $g(x_1) > g(x_2)$; $\exp_2[g(x_1)] > \exp_2[g(x_2)]$; decreasing **15.** 0 **17.** 7 **19.** Asymptote: $x = 0$;
$(-3, 2), (-1, 1)$ **21.** $(0, 243]$ **23.** Asymptote: $x = -1$; $(0, 0), \left(2, \frac{1}{2}\right)$ **25.** $(-1, \infty)$; 1 **27.** 2.666 **29.** 1.465
31. $(-1.465, 5)$ **33.** 8.82 min.

page 306 1. 2100 years

Chapter 9

pages 311–312 1. $(0, 1)$ **3.** $\left(\frac{1}{2}\sqrt{3}, \frac{1}{2}\right)$ **5.** $\left(-\frac{1}{2}, -\frac{1}{2}\sqrt{3}\right)$ **7.** $\left(\frac{1}{2}\sqrt{2}, \frac{1}{2}\sqrt{2}\right)$ **9.** $(1, 0)$ **11.** $\left(-\frac{1}{2}\sqrt{3}, -\frac{1}{2}\right)$
13. $\left(\frac{1}{2}\sqrt{3}, -\frac{1}{2}\right)$ **15.** $\left(-\frac{1}{2}\sqrt{2}, -\frac{1}{2}\sqrt{2}\right)$ **17.** $\left(\frac{1}{2}\sqrt{3}, \frac{1}{2}\right)$ **19.** $\left(-\frac{1}{2}\sqrt{2}, \frac{1}{2}\sqrt{2}\right)$ **21.** II **23.** IV **25.** $(0.2, 1.0)$
27. $(0.5, -0.8)$ **29.** $\pi, -\pi, 3\pi, -3\pi, 5\pi$ **31.** $-0.6 \leq x \leq 0.6$; $-0.6 \leq y \leq 0.6$ **33.** $(-a, b), (-a, -b), (a, -b)$
35. All integers

pages 315–316 1. $\frac{1}{2}\sqrt{3}$ **3.** $-\frac{1}{2}\sqrt{3}$ **5.** $\frac{1}{2}$ **7.** $\frac{1}{2}\sqrt{3}$ **9.** $\frac{1}{2}\sqrt{2}$ **11.** 0.04 **13.** 0.5 **15.** -0.84 **17.** $-\frac{1}{2}\sqrt{2}$ **19.** 1 **21.** 1
23. $-\frac{4}{5}$ **25.** $-\frac{1}{3}\sqrt{5}$ **27.** $0, \frac{1}{2}\sqrt{2}, 1, \frac{1}{2}\sqrt{2}, 0$ **29.** 12 sec.

pages 318–319 1. R **3.** $-\frac{\sqrt{2}}{2}$ **5.** $\frac{1}{2}$ **7.** $-\frac{1}{2}$ **9.** $\frac{1}{2}\sqrt{3}$ **11.** $-\frac{1}{2}\sqrt{3}$ **13.** 0 **15.** $-\frac{1}{2}\sqrt{3}$ **17.** $\frac{1}{2}$ **19.** 1 **21. a.** 2 **b.** 1
c. -1 **25.** Cosine: symmetric about y-axis; sine: symmetric about origin **27. a.** 10 FOR N = 0 TO 4;
20 X = 1/10↑N; 30 PRINT (COS(X) − 1)/X,; 40 NEXT N **b.** $-.45970, -.04996, -.00500, -.00050, -.00005$
c. It approaches 0 **29.** $[0, \pi]$ **31.** $\frac{1}{2}\pi, \frac{3}{2}\pi$ **33.** $\frac{1}{3}\pi, \frac{2}{3}\pi$

pages 322–323 1. $0, 2\pi$ **3.** $\frac{1}{6}\pi, \frac{5}{6}\pi$ **5.** $\frac{5}{6}\pi, \frac{7}{6}\pi$ **7.** $\frac{3}{2}\pi$ **9.** $\frac{5}{4}\pi, \frac{7}{4}\pi$ **11.** $[0, 2\pi]$ **13.** $\frac{1}{6}\pi, \frac{5}{6}\pi$ **15.** $\left[0, \frac{1}{3}\pi\right) \cup \left(\frac{5}{3}\pi, 2\pi\right]$
17. $\left[0, \frac{1}{4}\pi\right) \cup \left(\frac{3}{4}\pi, 2\pi\right]$ **19.** $\left[\frac{1}{3}\pi, \frac{2}{3}\pi\right]$ **21.** One; 0.45 **23.** None; 4.26 **25.** Two; 0.64 **27.** $\left[0, \frac{1}{3}\pi\right) \cup \left(\frac{5}{3}\pi, 6\right]$
29. $\pi - r, 2\pi + r, 3\pi - r$ **31. a.** $\frac{1}{3}\pi, \frac{5}{3}\pi$ **b.** $\frac{1}{4}\pi, \frac{7}{4}\pi$ **33.** $[0, 1.88] \cup (4.41, 2\pi]$ **35. a.** $\frac{1}{2}, \frac{3}{2}$ **b.** 2 sec.

pages 325–326 1. sin x shifted down 1 unit; $(2\pi, -1)$ **3.** cos x stretched vertically by 2 and shifted up 1 unit;
$(2\pi, 3)$ **5.** sin x reflected in the x-axis, then shifted up 2 units; $(4\pi, 2)$ **7.** cos x shifted right $\frac{1}{2}\pi$ units; $\left(\frac{1}{2}\pi, 1\right)$
9. sin x stretched vertically by 2, then shifted left $\frac{1}{4}\pi$ units; $\left(\frac{9}{4}\pi, 2\right)$ **11.** sin x stretched vertically by 2, then
shifted right π units; $\left(\frac{3}{2}\pi, 2\right)$ **13.** cos x reflected in the x-axis, then shifted up 1 unit; $(\pi, 2)$ **15.** sin x reflected
in the x-axis, then shifted up 3 units; $\left(\frac{3}{2}\pi, 4\right)$ **17.** sin x shifted right $\frac{1}{2}\pi$ units; $(\pi, 1)$ **19.** cos x stretched
vertically by 2, shifted right $\frac{3}{2}\pi$ units, then shifted up 1 unit; $\left(\frac{\pi}{2}, -1\right)$ **21.** $\left[0, \frac{\pi}{3}\right) \cup \left(\frac{5}{3}\pi, \frac{7}{3}\pi\right)$ **23.** sin x shifted
right 1 unit; $\left(\frac{1}{2}\pi + 1, 1\right)$; $[-1, 1]$ **25.** sin x compressed vertically by $\frac{1}{2}$, then reflected in the x-axis; $\left(\frac{1}{2}\pi, -\frac{1}{2}\right)$;
$\left[-\frac{1}{2}, \frac{1}{2}\right]$ **27.** sin x is stretched vertically by 2, then shifted up 1 unit. The part below $y = 0$ is then reflected
in the x-axis; $(\pi, 1)$; $[0, 3]$ **29.** cos x compressed vertically by $\frac{1}{2}$, then shifted up 1 unit; $\left(\pi, \frac{1}{2}\right)$; $\left[\frac{1}{2}, \frac{3}{2}\right]$
31. $\frac{1}{2}\pi < t < \frac{3}{2}\pi$ **33.** π **35.** sin x is stretched by 2, then shifted right $\frac{1}{2}\pi$ units. The part below $y = 0$ is then reflected
in $y = 0$, and the graph is shifted up 1. **37.** Asymptotes at $\frac{1}{2}\pi + \pi k$, $k \in J$ **39.** sin x compressed horizontally by $\frac{1}{2}$

page 329 1. $\sqrt{2}$ **3.** 1 **5.** 2 **7.** -1 **9.** $-\frac{2}{3}\sqrt{3}$ **11.** $\sqrt{2}$ **13.** -1.00 **15.** -1.19 **17.** sec x stretched vertically by 2
19. sec x reflected in $y = 0$, then shifted up 1 unit **21.** csc x is reflected in $y = 0$, then shifted up 1 unit. The part
below is then reflected in $y = 0$. **23.** 0.8 **25.** $\frac{5}{4}$ **27.** 2.5 **29** $(0, \pi)$ **31.** $(0, \pi) \cup \left[\frac{7}{6}\pi, \frac{11}{6}\pi\right]$ **33.** csc $(-x) = \frac{1}{\sin(-x)} = \frac{1}{-\sin x} = -\csc x$

pages 333–334 **1.** Not possible **3.** $-\sqrt{3}$ **5.** $\sqrt{3}$ **7.** $-\frac{1}{3}\sqrt{3}$ **9.** $\frac{1}{3}\sqrt{3}$ **11.** 1 **13.** 0.727 **15.** -0.071 **17.** $-3 + \sqrt{2}$ **19.** tan x shifted down 1 unit; $\left(\frac{3}{4}\pi, -2\right)$ **21.** tan x shifted left $\frac{1}{2}\pi$ units; $\left(\frac{1}{4}\pi, -1\right)$ **23.** tan x reflected in $y = 0$ **25.** tan x is reflected in $y = 0$, then shifted up 1 unit. The parts below are then reflected in $y = 0$. **29.** $R - \{\pi k, k \in J\}$; R; π; $\left\{\frac{1}{2}\pi + \pi k, k \in J\right\}$ **31.** $\sqrt{5}$ **33.** $\frac{24}{25}, \frac{7}{24}, \frac{25}{24}, \frac{25}{7}, \frac{24}{7}$ **35.** 1.0 min. **37.** 0.98 **39.** 2 **41.** 0.45 **43.** 4.49

pages 337–339 **1.** $\frac{2}{5}\pi$ **3.** $\frac{2}{3}\pi$ **5.** 4 **7.** 4π **9.** sin x stretched vertically by 2, period $\frac{2}{3}\pi$; $\left(\frac{\pi}{2}, -2\right)$ **11.** sin x reflected in $y = 0$, shifted up 1 unit, period $\frac{2}{3}\pi$; $\left(\frac{2}{3}\pi, 1\right)$ **13.** tan x reflected in $y = 0$, period 3π; $\left(-\frac{3}{4}\pi, 1\right)$ **15.** period 2; $\left(\frac{1}{2}, 1\right)$ **17.** sin x reflected in $y = 0$, period π; $\left(\frac{1}{4}\pi, -1\right)$ **19.** cos x shifted up 1 unit, period 2; (2, 2) **21.** sec x reflected in $y = 0$, shifted up 1 unit, period 4π; $(2\pi, 2)$ **23.** 6 **25.** $A = 3$; $k = \pi$ **27.** 60 cycles per sec. **29.** $f(x - 2) = \sin(\pi x - 2\pi) = \sin 2\pi = f(x)$

pages 341–342 **1.** $\left(\frac{1}{2}\pi, -1\right)$ **3.** tan x, period $\frac{\pi}{2}$; $\left(\frac{\pi}{8}, 1\right)$ **5.** sin x shifted right $\frac{1}{2}\pi$ units, period 4π; $(2\pi, 1)$ **7.** cos x shifted left $\frac{1}{2}\pi$ units, period 4π; $(\pi, -1)$ **9.** sin x shifted right $\frac{1}{6}\pi$ units, reflected in $y = 0$, stretched vertically by 5, period $\frac{2}{3}\pi$; $\left(\frac{1}{3}\pi, -5\right)$ **11.** tan x shifted left $\frac{1}{4}$ units, reflected in $y = 0$, shifted up 1 unit; period $\frac{1}{2}$; no y-intercept; Rng $= R$ **13.** cos x shifted right $\frac{3}{2}$ units, stretched vertically by 2, shifted up 1 unit; period π; $(0, -0.98)$; $[-1, 3]$ **15.** sin x reflected in $y = 0$, shifted left 1 unit and down 1 unit, reflected in $y = 0$; period 2π; $(0, 0.16)$; $[0, 2]$ **17.** $[0, \pi - 1]$ **19.** -1 **21.** Infinitely far to the right **23.** Left **25.** $\frac{1}{160}$ sec. **27.** 0.7 **29.** Yes

page 343 **1.** $\left(\frac{e^x + e^{-x}}{2}\right)^2 - \left(\frac{e^x - e^{-x}}{2}\right)^2 = \frac{e^{2x} + 2 + e^{-2x}}{4} - \frac{e^{2x} - 2 + e^{-2x}}{4} = 1$ **3.** $\tanh x = \frac{e^x - e^{-x}}{e^x + e^{-x}}$

pages 348–349 **1.** $-\frac{1}{6}\pi$ **3.** $\frac{1}{2}\pi$ **5.** $\frac{1}{2}\pi$ **7.** $-\frac{1}{3}\pi$ **9.** $-\frac{1}{4}\pi$ **11.** $\frac{1}{4}\pi$ **13.** -0.243 **15.** 0.412 **17.** 14.101 **19.** 1.911 **21.** 3.553, 5.872 **23.** 1.107, 4.249 **25.** 2.034, 5.176 **27.** 0.209 sec. **29.** 0.268 sec **31.** $-1 \leq x < \frac{1}{2}\sqrt{2}$ **33.** $\left[-\frac{\pi}{2}, \frac{\pi}{2}\right]$ **35.** $\left(1, -\frac{1}{4}\pi\right)$ **37.** $\left(-1, \frac{1}{2}\pi\right)$ **39.** $\left(1, \frac{1}{2}\pi\right)$ **41.** $(1, \pi)$ **43.** arcsin $(1 - t)$ **45.** $[-1, -0.416)$

pages 349–350 **1.** $\left(-\frac{1}{2}\sqrt{3}, \frac{1}{2}\right)$ **3.** $\frac{2}{3}\pi, -\frac{4}{3}\pi, \frac{8}{3}\pi$ **5.** $\frac{1}{2}\sqrt{2}$ **7.** $-\frac{1}{2}\sqrt{2}$ **9.** $\frac{2}{3}\pi, \frac{4}{3}\pi$ **11.** $\frac{1}{6}\pi, \frac{11}{6}\pi$ **13.** sin x shifted left $\frac{1}{4}\pi$ units, stretched vertically by 2, period 4π **15.** $\frac{2}{3}\sqrt{3}$ **17.** -1 **19.** $-\sqrt{101}$ **21.** sin x stretched vertically by 3, period π **23.** tan x stretched vertically by 2, shifted left $\frac{1}{2}\pi$ units, period $\frac{1}{3}\pi$ **25.** $-\frac{1}{3}\pi$

Chapter 10

pages 355–356 **1.** $\frac{\pi}{3}$ **3.** $-\frac{3\pi}{4}$ **5.** 0.8639 **7.** 0.3811 **9.** 0.6167 **11.** 10.472 **21.** 97.40° **23.** 47.18° **25.** 2346 $\frac{2}{3}$ rad./min. **27.** 1092.4 rad./min. **29.** 18.8 ft./sec.

pages 358–359 **1.** $\frac{\sqrt{2}}{2}$ **3.** $\frac{\sqrt{3}}{2}$ **5.** 0 **7.** 1 **9.** $-\frac{2\sqrt{3}}{3}$ **11.** $-\frac{\sqrt{3}}{2}$ **13.** 0.2924 **15.** -5.7588 **17.** 0.7986 **19.** 60°, 300° **21.** 45°, 225° **23.** 0°, 90°, 180°, 360° **25.** 27.03° **27.** 7.87° **29.** 78.46°, 281.54° **31.** 41.81°, 138.19°, 228.59° 311.41° **33.** 0°, 60°, 120°, 180°, 240° 300°, 360° **35.** 202.46°, 337.54°

pages 363–364 **1.** $\left(\frac{5}{2}, \frac{5\sqrt{3}}{2}\right)$ **3.** 135° **5.** 56.3° **7.** $y = x - 3$ **9.** $y \doteq -5.67x - 3.67$ **13.** Circle centered at origin, radius 7 **15.** Ellipse, semimajor axis 7, semiminor axis 3 **17.** $y = 3x + 5$ **19.** Part of parabola $y = 3x^2$; $-1 \leq x \leq 1$ and $0 \leq y \leq 3$ **21.** Segment of $x = 2$; endpoints $(2, -3)$, $(2, 3)$ **23.** $\frac{(x - h)^2}{a^2} + \frac{(y - k)^2}{b^2} = 1$; $a = b$, a circle; $a \neq b$, an ellipse with center (h, k), semiaxes a, b **25.** $P = (4 \cot \theta, 4 \sin \theta)$

pages 367–368 **1.** $\frac{\sqrt{33}}{7}$ **3.** $\frac{\sqrt{2}}{2}$ **5.** 1 **7.** $\frac{28}{3}$ **9.** $\frac{4\sqrt{3}}{3}$ **11.** 35.54° **13.** $2\sqrt{3}$ **15.** $\sqrt{2}$ **17.** 60° **19.** 6.21 **21.** 58.78 in. **23.** 35.70 m **25.** 133.3 ft. **27.** $V(h) = \frac{343\pi \cot \theta}{3}$ **29.** $\theta = \arctan \frac{31}{x} - \arctan \frac{20}{x}$

pages 371–372 **1.** $3\sqrt{2}$ **3.** No solution **5.** 75° or 15° **7.** $c \doteq 9.06$, $\angle B \doteq 112°$, $b \doteq 17.89$ **9.** $\angle B \doteq 47.0°$, $\angle C \doteq 67.0°$, $c \doteq 15.12$ **11.** $\angle B \doteq 66.7°$, $\angle A \doteq 79.3°$, $a \doteq 7.38$; or $\angle B \doteq 113.3°$, $\angle A \doteq 32.7°$, $a \doteq 4.06$ **13.** 4348.6 ft. **15.** $BC = 20\sqrt{2} \sin A$ **17.** 1379 ft. or 369 ft. **19.** 224.9 ft. or 82.3 ft. **21.** 68.6 ft.

pages 375–376 **1.** $\frac{11}{16}$ **3.** $\sqrt{34 - 15\sqrt{2}}$ **5.** 2.57 **7.** 101.5° **9.** 10.50, 26.8° **11.** 36.7 in. **13.** 94.3°, 29.3°, 56.4° **15.** $\theta = f(AC) = \arccos\frac{544 - (AC)^2}{480}$ **17.** 14.13 **19.** 120° **21.** 6781 ft. **23.** $\angle C$, opposite longest side, is the only possible obtuse angle.

pages 378–380 **1.** $\frac{21}{2}$ **3.** 9.00 **5.** 9.80 **7.** 7.55 **9.** 0.29 cm² **11.** $xy \sin \theta$ **13.** 93.08 **15.** $A(\theta) = L^2 \sin \theta$ **17.** $\frac{\sqrt{15}}{4}$ **19.** 13,095.9 ft.² **21.** $\frac{5\pi}{36}$ arccos $\frac{x}{5} - x\sqrt{25 - x^2}$ **23.** 9702 ft.² **25.** 1,237.4 ft.²

pages 382–383 **19.** $(1, 30° + n(360°))$, $(1, 150° + n(360°))$, $n \in J$ **21.** $\sqrt{r_1^2 + r_2^2 - 2r_1 r_2 \cos(\theta_2 - \theta_1)}$

page 385 **1.** $(\sqrt{2}, \sqrt{2})$ **3.** $\left(-\frac{1}{2}, -\frac{\sqrt{3}}{2}\right)$ **5.** $\left(4\sqrt{2}, \frac{\pi}{4}\right)$ **7.** $\left(2, \frac{7\pi}{6}\right)$ **9.** $r = 4 \cos \theta$ **11.** $r(2 \cos \theta + \sin \theta) = 6$ **13.** $r = 3$ **15.** $x^2 + y^2 = 2y$ **17.** $x^2 + y^2 = \sqrt{x^2 + y^2} - y$ **19.** $2y = x\sqrt{x^2 + y^2}$ **21.** $x = 2$ **23.** $x^2 = -8(y - 2)$ **25.** $r = 2\sqrt{\sin \theta}$

page 386 **1.** Parabola, vertex (2, 0°), opening left **3.** A hyperbola through (1, 90°), $\left(\frac{1}{3}, 0°\right)$, (1, 270°) and (−1.4, 210°), (−1, 180°), (−1.4, 150°)

pages 387–388 **1.** $\frac{20\pi}{9}$ radians **5.** 565 in./min. **7.** −2 **9.** 0°, 180°, 360°, 30°, 150° **11.** 210° **13.** 0.8 **15.** 4 **17.** 6.12 **19.** No solution **21.** −0.65 **23.** 58.8 cm **25.** 238 cm² **27.** Circle: center (1.5, 0°), radius 1.5 **29.** $(2\sqrt{2}, 2\sqrt{2})$ **31.** $(5\sqrt{2}, 135°)$ **33.** Circle: center (3, 90°), radius 3

Chapter 11

pages 393–394 **19.** 1 **21.** 1 **23.** 2 **25.** $\tan \theta$ **27.** $\sin 2x$ **29.** $\frac{7\pi}{6}, \frac{11\pi}{6}, \frac{\pi}{2}$ **31.** $\frac{\pi}{4}, \frac{3\pi}{4}, \frac{5\pi}{4}, \frac{7\pi}{4}$

pages 396–397 **1.** $\pi k, k \in J$ **3.** $\frac{3\pi}{4} + 2\pi k, \frac{5\pi}{4} + 2\pi k$ **5.** $\pi + 2\pi k$ **7.** $\frac{7\pi}{6} + 2\pi k, \frac{11\pi}{6} + 2\pi k$ **9.** $-\frac{\pi}{4} + \pi k$ **11.** $2\pi - r$ **13.** $-0.6 + 2\pi k, \pi - 0.6 + 2\pi k$ **15.** $0.43 + 2\pi k, 5.86 + 2\pi k$ **17.** $1.25 + \pi k$ **19.** $5.39 + 2\pi k, 4.04 + 2\pi k$ **21.** $1.77 + 2\pi k, 4.51 + 2\pi k$ **23.** $5.55 + 2\pi k, 3.87 + 2\pi k$ **25.** $5.94 + 2\pi k, 3.48 + 2\pi k$ **27.** $\frac{\pi}{2} + \pi k$ **29.** $\frac{\pi}{3} + \pi k, \frac{2\pi}{3} + \pi k$ **31.** $\pi + 2\pi k, \pm\frac{\pi}{3} + 2\pi k$ **33.** $\frac{7\pi}{6} + 2\pi k, \frac{11\pi}{6} + 2\pi k$ **35.** $\frac{2\pi}{3} + 2\pi k, \frac{4\pi}{3} + 2\pi k, 2\pi k$ **37.** 1.23, 5.05, 7.51 **39.** $-0.37 + 2\pi k, \pi + 0.37 + 2\pi k$ **41.** $0.64 + 2\pi k, 5.65 + 2\pi k, 1.71 + 2\pi k, 4.57 + 2\pi k$ **43.** $\frac{\pi}{3} + 2\pi k \leq x \leq \frac{5\pi}{3} + 2\pi k$ **45.** $0.407 + 2\pi k < x < 2.73 + 2\pi k$ **47.** False

pages 402–403 **1.** $-\cos x$ **3.** $\frac{1 + \tan x}{1 - \tan x}$ **5.** $-\sin x$ **7.** $\sin x$ **9.** $\cos x$ **11.** $\sin x$ **13.** $\frac{\sqrt{6} + \sqrt{2}}{4}$ **15.** $\frac{\sqrt{6} - \sqrt{2}}{4}$ **17.** $\frac{\sqrt{6} - \sqrt{2}}{4}$ **19.** $\frac{\sqrt{6} - \sqrt{2}}{4}$ **21.** $\frac{\sqrt{2} - \sqrt{6}}{4}$ **23.** $\frac{140}{221}$ **25.** $\frac{304}{425}$ **27.** $\frac{7\pi}{6}, \frac{11\pi}{6}$ **29.** $\cot A$ **43.** $-\frac{7}{24}$ **47.** $\theta = \arctan \frac{m_2 - m_1}{1 + m_2 m_1}$

page 404 **3.** $\omega = 2, C = 2, D = \frac{\pi}{6}$ **5.** $\omega = 3, C = \sqrt{5}, D \doteq -0.464$ **7.** $\sqrt{41}$

pages 407–408 **1.** $\frac{24}{25}$ **3.** $-\frac{24}{7}$ **5.** $-\frac{120}{169}$ **7.** $\frac{7}{25}$ **9.** 1 **11.** $\frac{1}{2}$ **13.** $\frac{\pi}{2}, \frac{3\pi}{2}, \frac{\pi}{6}, \frac{5\pi}{6}$ **15.** $\frac{\pi}{3}, \frac{5\pi}{3}, \pi$ **17.** $\frac{\pi}{3}, \frac{5\pi}{3}$ **19.** $\frac{\pi}{3}, \frac{5\pi}{3}$ **33.** 2 **35.** $3 \sin x - 4 \sin^3 x$; $(4 \sin x - 8 \sin^3 x)\sqrt{1 - \sin^2 x}$ **37.** $\frac{1}{16}$ **39.** 0.30, 2.84 **41.** 2.38, 3.91

pages 410–412 **1.** $\frac{84}{85}$ **3.** $\frac{24}{25}$ **5.** $\frac{3\sqrt{10}}{10}$ **7.** $\frac{3\sqrt{34}}{34}$ **9.** $\frac{3}{5}$ **11.** $\frac{\sqrt{2+\sqrt{3}}}{2}$ **13.** $\frac{-\sqrt{2-\sqrt{3}}}{2}$ **15.** $-\sqrt{\frac{2-\sqrt{3}}{2+\sqrt{3}}}$ **17.** $\frac{\sqrt{2-\sqrt{2}}}{2}$ **19.** $-\frac{\sqrt{2+\sqrt{3}}}{2}$ **21.** $\frac{2\sqrt{5}}{5}$ **23.** 2 **25.** $\frac{\sqrt{2+\sqrt{6}}}{4}; \frac{\sqrt{2+\sqrt{3}}}{2}$ **37.** $\frac{\sqrt{42}}{12}$ **39.** $\frac{3}{8} + \frac{1}{2}\cos 2t + \frac{1}{8}\cos 4t$ **41.** $f(x) = 1 - \cos 3x$, period $\frac{2\pi}{3}$ **43.** $\frac{\sqrt{10}}{5}$ **45.** $\frac{\pi}{3}, \frac{5\pi}{3}$ **47.** $0, 2\pi, \frac{\pi}{2}, \frac{3\pi}{2}$

pages 415–416 1. $0.2 - \pi$, 0.2, $0.2 + \pi$ **3.** $\frac{\pi}{6}, \frac{7\pi}{6}$ **5.** $0.2, 2.2, 4.2$ **7.** $\frac{\pi}{6}, \frac{2\pi}{3}, \frac{7\pi}{6}, \frac{5\pi}{3}$ **9.** $\frac{\pi}{4}, \frac{3\pi}{4}, \frac{5\pi}{4}, \frac{7\pi}{4}$
11. $\frac{\pi}{6}, \frac{7\pi}{6}, \frac{\pi}{3}, \frac{4\pi}{3}$ **13.** $\frac{\pi}{12}, \frac{3\pi}{4}, \frac{17\pi}{12}, \frac{7\pi}{12}, \frac{5\pi}{4}, \frac{23\pi}{12}$ **15.** $\frac{3\pi}{4}, \frac{7\pi}{4}$ **17.** $\frac{2n+1}{4} + \frac{1}{2\pi}$, for $n = 0, 1, \ldots, 11$ **19.** $\frac{5\pi}{6}, \frac{11\pi}{6}$
21. $0, \frac{\pi}{3}, \frac{2\pi}{3}, \pi, \frac{4\pi}{3}, \frac{5\pi}{3}, 2\pi$ **23.** $0.46, 3.61, 1.11, 4.25$ **25.** $2.11, 5.25, 1.18, 4.32$ **27.** 4.67 **29.** 0.55 **31.** 0.95
33. $0.94, 4.08, 2.79, 5.93$ **35.** $\left[\frac{3\pi}{4}, \pi\right) \cup \left[\frac{7\pi}{4}, 2\pi\right)$ **37.** $[0, \pi] \cup \left[\frac{7\pi}{6}, \frac{11\pi}{6}\right] \cup \{2\pi\}$ **39.** $\left[0, \frac{\pi}{8}\right) \cup \left[\pi, \frac{9\pi}{8}\right) \cup$
$\left(\frac{3\pi}{8}, \pi\right] \cup \left(\frac{11\pi}{8}, 2\pi\right]$ **41.** $[0, 1.68] \cup [3.04, 4.82] \cup [6.18, 2\pi]$ **43.** $[0, 0.94] \cup [2.21, 3.28] \cup [6.15, 2\pi]$
45. Sine curve, amplitude 1 and period 2, intersects $y = \log_3 x$ at 3 points

pages 420–421 1. Point $(1, 2)$ in complex plane **3.** Point $(0, 2)$ **5.** Point $(3, 1)$ **7.** Points $(2, 3)$ and $(2, -3)$
9. $\sqrt{2}$; $\sqrt{2}$ **11.** $1 - \sqrt{3}i$; 2 **13.** $\sqrt{2}\left(\cos\frac{7\pi}{4} + i\sin\frac{7\pi}{4}\right)$ **15.** $2\left(\cos\frac{\pi}{2} + i\sin\frac{\pi}{2}\right)$ **17.** $6\left(\cos\frac{2\pi}{3} + i\sin\frac{2\pi}{3}\right)$
19. $2\left(\cos\frac{11\pi}{6} + i\sin\frac{11\pi}{6}\right)$ **21.** $2.05 + 5.64i$ **25.** $\sqrt{2}\left(\cos\frac{7\pi}{12} + i\sin\frac{7\pi}{12}\right)$ **27.** -4 **29.** $16(\cos 40° + i\sin 40°)$
31. A circle, center $(0, 0)$, radius 3 **43.** $4096 + 0i$

page 422 3. $\frac{5\pi}{4} + 2\pi k, \frac{7\pi}{4} + 2\pi k, k \in J$ **5.** $1.11 + \pi k$ **7.** $\frac{\sqrt{2} + \sqrt{6}}{4}$ **9.** $\frac{24}{25}$ **11.** $\frac{1}{2}\sqrt{2 + \sqrt{3}}$ **13.** $\frac{5\pi}{6}, \frac{11\pi}{6}$

Chapter 12

pages 427–428 1. $\sqrt{10}$ **3.** 5 **7.** $\langle 4, 4 \rangle$ **9.** $\langle -4, 1 \rangle$ **11.** $\langle -2, -7 \rangle$ **13.** $\langle 5, 10 \rangle$ **15.** $3\sqrt{3}, 3$ **17.** $1, \sqrt{3}$ **19.** -4
21. $a + 1$ **23.** $(-2, 4)$ **25.** $(4, -1)$ **27.** $\pm\frac{\sqrt{2}}{2}$ **29.** $\pm\frac{3\sqrt{3}}{2}$ **31.** $\langle c_1 - a_1, c_2 - a_2 \rangle$

pages 431–432 1. $\langle 4, 1 \rangle, \langle -2, 5 \rangle$ **3.** $\langle 1, 5 \rangle, \langle -3, 7 \rangle$ **9.** $s = 2, t = -9$ **11.** $\vec{B} - \vec{A}$ **13.** $-\vec{A} - \vec{B}$ **21.** 12.4 mph, 76°
25. a. $\langle 400 + 50\sqrt{2}, 50\sqrt{2} \rangle$ **b.** 476 mph **c.** 8.5° **27. a.** $\langle 400 - 30\sqrt{3}, 30 \rangle$ **b.** 350 mph **c.** 4.9°

pages 435–436 1. $\langle 1, -19 \rangle$ **3.** $\langle 5, 4 \rangle$ **5.** $\langle 1, -2 \rangle$ **9.** $\langle \frac{3}{5}, \frac{4}{5} \rangle$ **11.** $\langle \frac{2}{\sqrt{13}}, \frac{-3}{\sqrt{13}} \rangle$ **13.** $\langle -3, -1 \rangle$ **15.** $\langle -3, 3 \rangle$
17. $\langle \sqrt{2}, \sqrt{2} \rangle$ **19.** $\langle \frac{4\sqrt{5}}{5}, \frac{-2\sqrt{5}}{5} \rangle$ **21.** 0 **23.** $\langle x, y \rangle = \langle 1, 2 \rangle + t\langle 4, -3 \rangle$ **25.** $\langle x, y \rangle = \langle 3, -1 \rangle + t\langle -4, 3 \rangle$ **27.** $\langle x, y \rangle =$
$\langle 0, 2 \rangle + t\langle 1, 0 \rangle$ **29.** $t = 0$: $(3, 1)$; $t = 1$: $(14, -7)$; $t = -2$: $(-19, 17)$ **37.** $\left(\frac{2}{5}, \frac{21}{5}\right)$ **39.** $\langle -b, a \rangle$

pages 438–439 1. a. $\langle 15, 0 \rangle$ **b.** $\langle 5 \cos 45°, 5 \sin 45° \rangle$ **c.** $\langle 15 - \frac{5\sqrt{2}}{2}, -5\frac{\sqrt{2}}{2} \rangle$ **d.** 12.0 **9.** $\langle -2, -2 \rangle$
11. $\langle \frac{\sqrt{2}}{2} - 1, \frac{\sqrt{2}}{2} \rangle$ **13.** 10.4 mph

pages 441–442 1. $\sqrt{3}$ **3.** 3 **5.** $\sqrt{14}$ **13.** $(0, 0, 4), (2, 0, 0), (0, 4, 0)$ **15.** $(0, 0, 3), (0, 6, 0), (4, 0, 0)$
17. Center $(0, 1, 0)$, radius 3 **19.** Center $(3, 0, 0)$, radius 1 **21.** $z = 1$ **23.** Sample: $(2, 3, 0), (2, 3, 1), (2, 3, -5)$
25. Sample: $(3, 3, 0), (3, 3, 5), (3, 3, 1)$ **27.** Sample: $(1, 0, 0), (1, 0, 5), (1, 0, -7)$ **29.** Intersection: $x^2 + y^2 = 3$

pages 445–446 1. $\langle 2, 9, -10 \rangle$ **3.** $\langle 1, -4, 5 \rangle$ **5.** 2 is solution of each equation: $4 = 2 + t, 7 = -1 + 4t$,
$-1 = 3 - 2t$ **9.** $\langle \frac{\sqrt{2}}{2}, 0, \frac{\sqrt{2}}{2} \rangle$ **11.** -9 **13.** $\langle x, y, z \rangle = \langle 1, 2, -5 \rangle + t\langle -2, -2, 8 \rangle$ **15.** $\langle x, y, z \rangle = \langle -3, 0, 2 \rangle +$
$t\langle 0, 1, -6 \rangle$ **17.** $\langle x, y, z \rangle = \langle 1, 2, -12 \rangle + t\langle -6, 4, 13 \rangle$ **19.** $(0, -1, 5)$ **21.** $\left(0, 1, \frac{5}{2}\right)$ **23.** No point of intersection
25. On ℓ, $x = 1 + t \cdot 0 = 1$, but in yz-plane, $x = 0$ **29.** $\langle -3, -2, -1 \rangle$

pages 449–450 1. 1 **3.** -2 **5.** 10 **7.** -4 **9.** $90°$ **11.** $90°$ **13.** $61.4°$ **15.** $114°$ **19.** $90°$ **21.** No intersection

pages 452–453 1. $x - y + 2z = 5$ **3.** $2x + y - z = 3$ **5.** $7x + y + 2z = 19$ **7.** $\langle x, y, z \rangle = \langle 1, 2, 3 \rangle + t\langle 3, -2, 1 \rangle$
9. $\langle x, y, z \rangle = \langle 2, -5, 4 \rangle + t\langle 1, -2, 1 \rangle$ **11.** $x + 2y - z = 4$ **13.** $x - y + z = 1$ **15.** $(8, -5, 17)$ **17.** $3x - 2y +$
$z = 10$ **21.** $-2x + 5y + z = 11$

page 455 1. 2×3; 2×2; 2×2 **3.** $\begin{bmatrix} -10 & 2 \\ 0 & -4 \end{bmatrix}$ **5.** $\begin{bmatrix} -35 & -11 & 41 \\ 0 & 2 & -2 \end{bmatrix}$ **7.** $\begin{bmatrix} -1 & \frac{1}{3} \\ -1 & \frac{2}{3} \end{bmatrix}$ **9.** Let $\begin{bmatrix} -2 & 1 \\ 4 & -2 \end{bmatrix} = 0$

page 456 **3.** $\sqrt{5}$ **5.** $\langle -\frac{2\sqrt{13}}{13}, \frac{3\sqrt{13}}{13}\rangle$ **7.** $\langle 250\sqrt{2}, 250\sqrt{2}\rangle$ **9.** $\langle 3\sqrt{3} - 10, 3\rangle$ **11.** $(-2, 3, 0)$; $2\sqrt{5}$
13. $\langle x, y, z\rangle = \langle 7, -2, 5\rangle + t\langle -8, 13, -3\rangle$ **17.** $\langle x, y, z\rangle = \langle 4, 0, -9\rangle + t\langle 5, -4, 12\rangle$

Chapter 13

pages 463–464 **1.** 7, 11, 15, 19, 23 **3.** 1, $\frac{4}{5}$, $\frac{3}{4}$, $\frac{8}{11}$, $\frac{5}{7}$ **5.** 1, 2, 4, 8, 16 **7.** $-1, \frac{1}{2}, -\frac{1}{3}, \frac{1}{4}, -\frac{1}{5}$ **9.** No **11.** 5th term **13.** 70; 306 **15.** $n = 13$ **17.** -390 **19.** 494,450 **21.** $\frac{n(n+1)}{2}$ **23.** 4125 **25.** Not prime; $b_5 = 21$ **27.** 8100

pages 467–468 **1.** 242 **3.** $\frac{81}{4}$ **5.** -2 **7.** $n = 14$ **9.** 4 **11.** 0.93 ft. **13.** $5491 **15. a.** $47,461 **b.** $10,054,636
17. 314%

pages 471–473 **1.** 14 **3.** 28 **5.** 10 **7.** $-\frac{23}{60}$ **9.** -4 **11.** $-\frac{61}{27}$ **13.** $\sum_{i=3}^{6} \frac{1}{i}$ **15.** $\sum_{j=1}^{12} \sin\frac{\pi}{j}$ **17.** $\sum_{j=2}^{7} \frac{(-1)^j}{2j-1}$
19. 1,000,000 **21.** 2500 **23.** 254.99 **25.** 4095 **27.** 0 **29.** 4090 **31.** -1365 **33.** $\frac{2}{7}$ **35.** 1000 **37.** 0 for $n = 2, 3, 4,$ and 5 **39.** $\frac{n}{n+1}$

pages 477–478 **1.** Yes; yes **3.** Yes, yes **5.** No; yes; yes **7. a.** True **b.** Assume for a given $k \in N$, $\sum_{i=1}^{k} i^3 = \frac{k^2(k+1)^2}{4}$ **c.** $\sum_{i=1}^{k+1} i^3 = \frac{(k+1)^2([k+1]+1)^2}{4}$ **9. a.** True **b.** Assume that for a given $k \in N$, $3^k \geq k^3$ **c.** $3^{k+1} \geq (k+1)^3$

In exercises 11–19, only parts 1 and 2c of proof by induction are shown. **11.** $\sum_{i=1}^{1} i = 1$ and $\frac{1(1+1)}{2} = 1$; True.

$\sum_{i=1}^{k+1} i = (k+1) + \sum_{i=1}^{k} i = k + 1 + \frac{k(k+1)}{2} = \frac{(k+1)(k+2)}{2} = \frac{(k+1)([k+1]+1)}{2}$ **13.** $\sum_{i=1}^{1} 3^i = 3^1 = 3$; $\sum_{i=1}^{k+1} 3^i = 3^{k+1} + \frac{3^{k+1}-3}{2} = \frac{3 \cdot 3^{k+1} - 3}{2} = \frac{3^{(k+1)+1}-3}{2}$ **15.** $\sum_{i=1}^{1} \frac{1}{i(i+1)} = \frac{1}{1(1+1)} = \frac{1}{2}, \frac{1}{1+1} = \frac{1}{2}$; $\sum_{i=1}^{k+1} \frac{1}{i(i+1)} = \frac{1}{(k+1)(k+2)} + \frac{k}{k+1} = \frac{1}{k+1}\left(\frac{1+k^2+2k}{k+2}\right) = \frac{k+1}{k+2}$ **19.** $\sum_{i=1}^{1} i^3 = 1^3 = 1, \left[\frac{1(1+1)}{2}\right]^2 = 1$; $\sum_{i=1}^{k+1} i^3 = (k+1)^3 + \left[\frac{k(k+1)}{2}\right]^2 = (k+1)^2 \cdot \left[\frac{4(k+1)+k^2}{4}\right] = \frac{(k+1)^2(k+2)^2}{4} = \left[\frac{(k+1)(k+2)}{2}\right]^2$ **21.** False for $n = 3$ **23.** $\sum_{i=1}^{1} c f(i) = c f(1), c \sum_{i=1}^{1} f(i) = c f(1)$;

$\sum_{i=1}^{k+1} c f(i) = c f(k+1) + c \sum_{i=1}^{k} f(i) = c\left[f(k+1) + \sum_{i=1}^{k} f(i)\right] = c \sum_{i=1}^{k+1} f(i)$ **25.** $\sum_{i=1}^{1} ar^{i-1} = ar^{1-1} = a$,

$\frac{a(1-r^1)}{1-r} = a$; $\sum_{i=1}^{k+1} ar^{i-1} = ar^{k+1-1} + \frac{a(1-r^k)}{1-r} = \frac{ar^k - ar^{k+1} + a - ar^k}{1-r} = \frac{a - ar^{k+1}}{1-r} = \frac{a(1-r^{k+1})}{1-r}$ **27.** $(1+p)^1 \geq 1 + 1 \cdot p$;
Since $p > -1$, $1 + p > 0$, and $(1+p)(1+p)^k \geq (1+p)(1+kp)$; $(1+p)^{k+1} \geq 1 + p + kp + kp^2 > 1 + p + kp$;
$(1+p)^{k+1} > 1 + (k+1)p$ **29.** $1 - \frac{1}{(1+1)^2} = \frac{3}{4}, \frac{1+2}{2(1+1)} = \frac{3}{4}$; $\left(1 - \frac{1}{(1+1)^2}\right) \cdots \left(1 - \frac{1}{[k+2]^2}\right) = \left(\frac{k+2}{2[k+1]}\right) \cdot$
$\left(1 - \frac{1}{[k+2]^2}\right) = \frac{(k+2)(k^2+4k+4-1)}{2(k+1)(k+2)^2} = \frac{(k+3)(k+1)}{2(k+1)(k+2)} = \frac{k+3}{2(k+2)}$

pages 482–483 **1.** $-64 + 0i$ **3.** $0 + 27i$ **5.** $-4 + 0i$ **7.** $3888 - 3888\sqrt{3}i$ **9.** $-237.0 - 3116i$ **11.** $1.532 + 1.286i$; $-1.879 + 0.684i$; $0.347 - 1.970i$ **13.** $\frac{\sqrt{6}}{2} + i\frac{\sqrt{6}}{2}; \frac{-\sqrt{6}}{2} - i\frac{\sqrt{6}}{2}$ **15.** $1.456 + 0.841i$; $-0.841 + 1.456i$;
$-1.456 - 0.841i$; $0.841 - 1.456i$ **17.** $1.220 + 0.472i$; $-1.018 + 0.821i$; $-0.201 - 1.292i$ **19.** $\frac{1}{2} + \frac{\sqrt{3}}{2}i$;
$-1 + 0i; \frac{1}{2} - \frac{\sqrt{3}}{2}i$ **21.** $0.455 + 1.099i$; $-0.455 - 1.099i$ **23.** $-0.299 + 1.504i$; $-1.153 - 1.010i$; $1.452 - 0.493i$ **25.** $4\cos^3\theta - 3\cos\theta$

page 488 **1.** Convergent; 0 **3.** Convergent; 0 **5.** Convergent; 0 **7.** Divergent **9.** Divergent **11.** Divergent
13. Divergent **15.** Divergent **17.** Convergent; 2 **19.** Convergent; 1 **21.** Convergent; $\frac{2}{5}$ **23.** Convergent; 0
25. 6 **27.** 149 **29.** 1.5; 1.4; 1.416; 1.414; 1.414; $\lim a_n = \sqrt{2}$

page 489 **1. a.** $6\sqrt{3}$; 8; $4\sqrt{3}$; 6.6274 **b.** $2n \tan\left(\frac{180}{n}\right)$ **c.** 6.2837012

pages 494–495 **1.** 20 **3.** Divergent **5.** −5 **7.** $\frac{49}{2}$ **9.** $\frac{3}{8}$ **11.** $\frac{1}{2}$ **13. a.** 1.6458333; by calculator, 1.6487213; difference, 0.00289 **b.** 2.205333; by calculator, 2.2255409; difference, 0.02021 **15. a.** $\frac{1}{3}$; by calculator, 0.3679; difference, 0.0345 **b.** 0.3681; by calculator, 0.3679; difference, 0.00018 **17.** 175 ft. **19. a.** 0.182667; by calculator, 0.18232 **b.** 0.18233 **21.** Divergent **23.** Divergent **25.** 12

page 496 **1.** Not a member **3.** $288\frac{1}{5}$ **5.** $\sum_{n=4}^{9} \frac{(-1)^n}{n^2+1}$ **9.** $\sqrt{2} + \sqrt{2}i$; $-1.93 + 0.52i$; $0.52 - 1.93i$ **11. a.** Convergent; 2 **b.** Divergent **c.** Divergent **d.** Convergent; 6

Chapter 14

page 502 **1.** $\frac{13}{2}$ **3.** 3 **5.** 3 **7.** 3 **9.** $y = -8x - 8$ **11.** $y = -\frac{1}{2}x + \frac{5}{4}$ **13.** $y = -2x$ **15.** $y = 3x - 4$ **17.** $y = \frac{1}{4}x + \frac{5}{4}$ **19.** $y = -\frac{1}{4}x + \frac{3}{4}$ **21.** 2c **23.** $y = -\frac{3}{4}x + \frac{13}{4}$

page 503 **1.** $L = 17$, $\delta = \frac{1}{250}$ **3.** Let $\delta = \frac{\epsilon}{3}$, $\epsilon > 0$; if $-1 - \frac{\epsilon}{3} < x < -1 + \frac{\epsilon}{3}$, then $-5 - \epsilon < 3x - 2 < -5 + \epsilon$ **5.** Let $\delta = \frac{\epsilon}{2}$, $\epsilon > 0$; if $x \neq 3$, but $3 - \frac{\epsilon}{2} < x < 3 + \frac{\epsilon}{2}$, then $11 - \epsilon < \frac{(2x+5)(x-3)}{x-3} < 11 + \epsilon$, and $L = 11$.

pages 507–508 **1.** 14 **3.** (4, 64), (−4, −64) **5.** $y = \frac{4}{3}x + \frac{16}{27}$ **7.** $y = -2x + 1$ **9.** 1 **11.** 6x **13.** $-\frac{1}{(x+1)^2}$ **15.** 0 **17.** $\frac{1}{2\sqrt{x}}$ **19.** $6x^2 - 1$ **21.** $\left(\frac{1}{9}, -\frac{53}{27}\right)$ **23.** $\left(\frac{1}{2}, 2\right), \left(-\frac{1}{2}, -2\right)$ **25.** $\left(\frac{2}{3}\sqrt{3}, -\frac{16}{9}\sqrt{3}\right), \left(-\frac{2}{3}\sqrt{3}, \frac{16}{9}\sqrt{3}\right)$ **29.** $\left(\sqrt{3}, \frac{1}{2}\sqrt{3}\right), \left(-\sqrt{3}, -\frac{1}{2}\sqrt{3}\right)$ **31. c.** $f'(x) = \cos x$

page 513 **1.** $6x^5$ **3.** $3x^2$ **5.** 0 **7.** $21x^2 - 5$ **9.** $-\frac{6}{(2x+1)^2}$ **11.** $-12x^3 + 12x^2 - 8 - \frac{1}{x^2}$ **13.** $y = 15x - 33$ **15.** $\left(\frac{1}{3}, \frac{2}{3}\right)$ **17.** $(0, -3), \left(-\frac{5}{3}, \frac{44}{27}\right)$ **19.** $(-\infty, -1) \cup (2, \infty)$ **21.** $\frac{2x^3-1}{x^2}$ **23.** $\frac{2x^2+2x}{(2x+1)^2}$ **25.** 6; −192; $f'(x - 5)$ **27. b.** −32

page 518 **1.** Relative maximum at (−2, 14); relative minimum at (0, 10) **3.** Relative maximum at $\left(-\frac{1}{2}, \frac{15}{4}\right)$; relative minimum at (1, −3) **5.** Relative minimum **7.** Relative maximum **9.** Neither **11.** $(-\infty, 1)$ and $(2, \infty)$ **13.** $(0, \infty)$ **15.** $(-\infty, -2), (-1, 1)$ and $(2, \infty)$ **17. a.** $(-\infty, -1)$ and $(1, \infty)$ **b.** Relative maximum at (−1, 3); relative minimum at (1, −1) **19. a.** $(-\infty, -3)$ **b.** $\left(-3, \frac{35}{4}\right)$ is a relative maximum; (0, 2) is a critical point, neither a maximum nor a minimum **21. a.** $(-\infty, -1)$ and $(-1, \infty)$ **b.** (−1, −3) is a critical point, neither a maximum nor a minimum **23. a.** $(2, \infty)$ **b.** (2, −9) is a relative minimum **25. a.** $(2, \infty)$ **b.** (2, 12) is a relative minimum **27.** 4

pages 520–522 **1.** $\frac{1}{6}$ **3.** $\frac{32,000}{27}$ **5.** $12\sqrt{2}$ **7.** Radius 8 in.; height 4 in. **9.** $r = \sqrt[3]{\frac{20,000}{\pi}}$ ft.; $h = \sqrt[3]{\frac{20,000}{\pi}}$ ft. **11.** $\left(\frac{5}{2}, \frac{\sqrt{10}}{2}\right)$ **13.** $\frac{1}{2}$

pages 526–527 **1.** 32 ft./sec. **3.** 32 ft./sec.² **5.** −6; (3, 4] **7.** 0; $[-3, -1) \cup (2, 3]$ **9.** 16 **11.** $\frac{413}{12}$ **13.** −32 ft./sec.² **15.** $\frac{7}{2}$ sec **17.** 200 ft. **19.** 120 ft./sec.²

pages 527–528 **1.** $-\frac{1}{16}$ **3.** $\frac{-3}{(3x+1)^2}$ **5.** $(-\infty, -2)$ and $\left(\frac{1}{2}, \infty\right)$ **7.** (−2, f(−2)) is a relative minimum; (0, f(0)) is neither **9.** 4 **11.** 6

Chapter 15

pages 534–535 **1.** $a = 1$, $b = 4$, $f(x) = x + 8$ **3.** $a = -2$, $b = 2$, $f(x) = \sqrt{1 - \frac{x^2}{4}}$ **5.** $\frac{63}{2}$ **7.** $\sqrt{3}$; $2 + \sqrt{3}$ **9.** 9
11. 1.267 **13.** 0.785; 0.948; 0.987 **15.** $S_1^3(x + 3) - 2$ **17.** $S_{-2}^1(4 - x^2) - \frac{9}{2}$ **19.** Let $f(x) = mx + c$, $m > 0$.
For $\frac{1}{2}(L_2 + U_2)$, each width $= \frac{b-a}{2}$, $L_2 = \frac{b-a}{2}\left[f(a) + f\left(\frac{a+b}{2}\right)\right] = \frac{b-a}{2}\left[ma + c + m\left(\frac{a+b}{2}\right) + c\right]$;
$U_2 = \frac{b-a}{2}\left[m\left(\frac{a+b}{2}\right) + c + mb + c\right]$; $\frac{1}{2}(L_2 + U_2) = \frac{b-a}{2}[ma + mb + 2c]$. Actual area is $S_a^b(mx + c) = $
$\frac{1}{2}$(height)(base + base) $= \frac{b-a}{2}[ma + mb + 2c]$. Proof is similar for $m < 0$.

page 537 **1.** 0.335; 0.3334; 0.333338 **3.** 8.68; 8.6672; 8.6667 **5.** 0.660509; 0.666095; 0.666594
7. 0.721636; 0.721359; 0.721348 **9.** 1.62204; 1.60995; 1.60947 **11.** 0.997945; 0.999918; 0.999995

page 540 **1.** 9 **3.** $\frac{64}{3}$ **5.** $\frac{335}{3}$ **7.** $\frac{37}{3}$ **9.** $\frac{125\pi}{3}$ cubic units **13.** $\frac{37}{3}$ **15.** $\frac{31}{3}$

pages 544–545 **1.** $\frac{64}{3}$ **3.** $\frac{91}{3}$ **5.** $\frac{19}{3}$ **7.** $\frac{28}{3}$ **9.** 3 **11.** 252 **13.** 12 **15.** $\frac{182}{3}$ **17.** $\frac{32}{3}$ **19.** $\frac{10}{3}$ **21.** $\frac{40}{3}$ **23.** 8 **25.** $\frac{9}{2}$ **27.** $\frac{32}{3}$
29. $\frac{125}{6}$ **33.** $\frac{2}{3}$ **35.** $\frac{54}{3}$ **37.** $\frac{14}{3}$ **39.** Upper half of the ellipse has area π.

pages 549–550 **1.** 64 **3.** $\frac{81}{2}$ **5.** $\frac{32}{3}$ **7.** $\frac{33}{5}$ **9.** 33 **11.** 164 **13.** $\frac{1}{12}$ **15.** $2^{5/4}$ **17.** $\frac{1}{6}$ **19.** 388.8 **21.** $\frac{135}{4}$ **23.** $\frac{3}{4}$
25. Use n rectangles, each of width $\frac{a}{n}$. $U_n = \frac{a}{n}\left[\left(\frac{a}{n}\right)^4 + \left(\frac{2a}{n}\right)^4 + \cdots + \left(\frac{na}{n}\right)^4\right] = \frac{a^5}{n^5}[1^4 + 2^4 + \cdots + n^4] = $
$\frac{a^5}{n^5}\left[\frac{1}{30}(6n^5 + 15n^4 + 10n^3 - n)\right] = a^5\left[\frac{1}{5} + \frac{1}{2n} + \frac{1}{3n^2} - \frac{1}{30n^4}\right]$; $\lim U_n = \frac{a^5}{5}$; similarly, $\lim L_n = \frac{a^5}{5}$. Area $= $
$\lim U_n = \lim L_n$; so $\int_0^a x^4\, dx = \frac{a^5}{5}$

pages 554–555 **1.**

t	0	1	2	3
$A(t)$	0	$\frac{1}{4}$	4	$\frac{81}{4}$

3.

t	1	2	3	4
$A(t)$	0	$\frac{15}{4}$	20	$63\frac{3}{4}$

5.

t	0	1	2	3
$A(t)$	0	1	8	27

7. 0.4 **9.** $\frac{5}{2}$; 4; $\frac{9}{2}$
11.

t	0	1	2
$A(t)$	0	$\frac{1}{2}$	2

13. 0; $\frac{99}{2}$; 1250 **15.** $L(2) + L(6) \doteq 2.5 \doteq L(12)$; $L(2) + L(7) \doteq 2.6 \doteq L(14)$; $L(3) + L(4) \doteq $
$2.5 \doteq L(12)$; $L(3) + L(5) \doteq 2.7 \doteq L(15)$ **17.** For t such that $0 < t < 1$, $L(t) = -L\left(\frac{1}{t}\right)$

page 559 **1.** $A(t) = \frac{t^5}{5}$; $A'(t) = t^4$ **3.** $\cos t$ **5.** $\log(t + 1)$ **7.** t^3 **9.** $2\frac{1}{2}$ **11.** $9\frac{1}{3}$ **13.** $\sqrt{2} - 1$ **15.** 1
17. $f(x) = \ln x = \int_1^x \frac{1}{t}\, dt$, so $f'(x) = \frac{1}{x}$ by the Fundamental Theorem of Calculus, $x \geq 1$. Since $f'(x) > 0$ for
all $x \geq 1$, f is an increasing function. **19. a.** $h'(x) = \frac{x}{\sqrt{x^2 + 1}}$ **b.** $\frac{4}{\sqrt{17}}$

page 560 **1.** $S_{-1}^2(1 + x^2)$ **3.** $\frac{343}{3}$ **5.** 21 **7.** $\frac{28}{3}$ **9.** $\frac{32}{3}$ **11.** $\frac{765}{4}$ **13.** $\frac{1}{20}$ **15.** 1.6 **17.** 138

Index

Absolute value, *28–30*
  of complex number, *416–417*
  of real number, *416*
Absolute-value function, *108, 515*
Addition, *3*
  of functions, *110–113*
  of sequences, *485–486*
  of vectors, *428–430*
Addition and subtraction identities, *398–401*
Addition axiom, *15*
Additive inverse, *6*
  of function, *111–112*
Algebra, Fundamental Theorem of, *213, 479*
Amplitude (of wave), *404*
Analytic geometry defined, *170*
Angle(s)
  between two vectors, *447–448*
  degree measure of, *353–354*
  initial side of, *353*
  radian measure of, *353–355*
  standard position of, *353*
  terminal side of, *353*
Angular velocity, *355*
Aphelion, *188*
Applications to real world
  carbon dating, *306*
  compound interest, *298–300*
  distance formulas, *76*
  exponential functions, *298–302, 306*
  inequalities, *17–18, 79–80, 86–88*
  linear programming, *79–80, 86–88*
  maximum and minimum points, *519–520*
  maximum and minimum values, *153–155, 158*
  method of least squares, *159–162*
  parabola, *177–179*
  vectors, *437–438*
  velocity formulas, *522–525*
Appolonius, *170*
Area
  of triangle, *377–378*
  under a curve, *531–534, 536–537, 538–539, 541–544, 545–549, 551–554*
  under parabola, *533–534, 538–539, 545*
  under power function, *545–549*
Area function, *551–554, 556*
Argument (in ordered pair), *95, 309–311*
Arithmetic mean, *149*
Arithmetic sequence, *460–462. See also* **Sequence**
Arrow (for vector), *425–427*
Associative axiom, *5*
Asymptote(s), *107*
  of hyperbola, *193*
Auxiliary rectangle, *193*
Axioms
  addition, *15*
  associative, *5*
  closure, *5*
  commutative, *5*
  completeness, *24*
  distributive, *5*
  equality, *2*
  identity, *5*
  inverse, *5*
  multiplication, *15*
  order, *15–17, 42*
  transitive, *15*
  trichotomy, *15*
Axis (axes)
  of cone, *170*
  of ellipse, *187*
  translation of, *197–200*
Axis of symmetry, *141–144, 186–187*
  as line of reflection, *141*
  of parabola, *171*

Bearing (navigation), *360*
Best fit, line of, *159–162*
Binary operation, *3*
Binomial coefficients, *39*
Binomial Theorem, *39*
Bounded sequence, *485*
Bounded set, *24*
Bounds, *23–24*

Calculus, Fundamental Theorem of, *556–558*
Cancellation properties, *6*
Carbon dating, *306*
Cardioid, *381*
Cartesian product, *95*
Catenary, *343*

Center
  of ellipse, *186*
  of hyperbola, *192*
Change-of-base Theorem, *294–295*
Chapter Review Exercises, *45–46, 91–92, 133–134, 163–164, 205–206, 244, 274, 304–305, 349–350, 387–388, 422, 456, 496, 527–528, 560*
Chapter Tests, *47, 93, 135, 165, 207, 245, 275, 307, 351, 389, 423, 457, 497, 529, 561*
Circle(s), *63–64, 167–168*
  ellipse as, *197, 202–203*
  equation of, *63–64, 167–168*
  unit, *309*
Circular functions, *313. See also* **Cosine function; Sine function; Tangent function**
  cofunctions, *398*
  equations of, *412–414*
  identities for, *391–393, 398–401*
  inequalities involving, *321–322, 414*
  inverse, *344–347*
Closed interval, *22*
Closed set, *3*
Closure axiom, *5*
Coefficient(s) of function, *209*
  leading, *209*
Cofunctions, *398*
Column vector, *454*
Common logarithms, *287*
Commutative axiom, *5*
Completeness axiom, *24*
Complex conjugates, *43*
Complex numbers, *40–44*
  absolute value of, *416–417*
  de Moivre's Theorem and, *479*
  equal, *41*
  field axioms and, *42*
  imaginary part of, *41*
  modulus of, *417*
  multiplication and division of, *418–420*
  order axioms and, *42*
  in polar form, *416–420*
  powers and roots of, *479–482*
  system of, *40*
Complex plane, *416*

Components (of vector), *425*
Composite function, *122–124, 269–271*
  graph of, *127–131, 271*
  inner function, *122*
  outer function, *122*
Compound inequality, *22*
Compound interest, *298–300*
Congruence (of triangles), *374*
Conic sections, *170, 197–200.* See also **Ellipse; Hyperbola; Parabola**
  algebraic definition of, *202–204*
  eccentricity and, *202–204*
  equations of, *202–204, 386*
Conjugate axis (hyperbola), *192*
Conjugates, *26*
Constant (of variation), *119*
Constant function, *107, 509*
Constant term, *209*
Continuous function, *228, 556–557*
Convergent sequence, *484–487*
Convergent series, *492*
Coordinate plane, *49–51, 53–56, 63–64, 66–70, 72–74, 77–80, 81–84, 95–97, 425, 439*
  region in, *96*
  vectors in *425–427*
Coordinate system
  Cartesian. See **Coordinate plane**
  polar, *380–382, 383–385, 386*
  in 3-space, *439–441*
Cosecant function, *327–329*
  graph of, *327–329*
Cosine function, *313–314, 316–318, 320–322, 324–325*
  graph of, *316–317, 320–322, 324–325*
Cosines, Law of, *373–375*
Cotangent function, *330–333*
  graph of, *332*
Critical points (on graph), *515*
Cubic polynomial, *220*
Curve(s)
  area under, *531–534, 536–537, 538–539, 541–544, 545–549, 551–554*
  tangents to, *499–502, 504–507*
  derivatives and, *514–517*

Curve sketching, see also **Graph**
  derivatives and, *514–517*

Decreasing function, *256–259, 515–517*
Definite integral(s), *538–539, 541–544, 545–549, 556–558*
Degree (of function), *209*
Degree measure (of angle), *353–354*
De Moivre's Theorem, *479–482*
Derivatives, *504–507, 556–558*
  acceleration, *525*
  curve sketching and, *514-517*
  differentiation, *509–512*
  problem solving and, *519–520, 522–525*
Determinant, *455*
Differential calculus, *558*
Differentiation, *509–512*
Direction vector, *434*
Directrix, *171*
Direct variation, *119*
Discriminant, *34*
Distance formula, *49–51*
  in 3-space, *440–441*
Distributive axiom, *5*
Divergent sequence, *484–485, 487*
Divergent series, *492*
Division, *11*
Domain, *18, 95, 99*
Doppler effect, *359*
Dot product, *447–449*
Double-argument identities, *405–407*
Dummy variable, *469*

e, 295, 490, 554
Eccentricity (conic sections), *202–204*
Element (of set), *2–3*
Ellipse, *170, 186–189*
  axes of, *187*
  center of, *186*
  circle as, *197, 202–203*
  equation of, *187–188*
  focus of, *186*
  latus rectum of, *190*
  translation of axes, *197–200*
  vertex of, *187*
Elliptic cylinder, *441*
Elliptic paraboloid, *441*

Equal functions, *113*
Equal vectors, *425*
Equality axioms, *2*
Equation(s)
  of circle, *63–64*
  circular function, *412–414*
  of conic sections, *202–204*
  of ellipse, *187–188*
  exponential, *278–279, 280, 294–296*
  of hyperbola, *191–192*
  linear, *53–56, 66–70*
  logarithmic, *287, 291–292, 294–296*
  in matrix form, *454–455*
  of parabola, *171–175*
  parametric, *178, 360–362, 434, 444*
  of periodic functions, *394–396, 404*
  of a plane, *451–452*
  point-slope form of, *54–55*
  polar, *381–382, 383–385, 386*
  quadratic, *33–37, 63–64, 67–70*
  slope-intercept form of, *56*
  of sphere, *441*
  systems of, *66–70*
  in two variables, *53–56, 63–64, 66–70, see also* **Functions; Relations**
  vector, *434–435, 444–445*
Equivalent inequalities, *19*
Even function, *117*
Exponential function(s), *277–280, 282–285, 294–296*
  applications, *298–302, 306*
  graph of, *277–278, 284–285*
Exponential-growth principle, *300–301*
Exponents
  irrational, *282–283*
  laws of, *30–31, 283*

Factorial notation, *490*
Factoring, *13*
  meaning of, *7*
  quadratic equations and, *33–34*
Factor Theorem, *216*
Feasible region, *80, 82–84*
Fibonacci sequence, *459–460*
Field axioms, *5–7, 42*
Focus
  of ellipse, *186*
  of parabola, *171*

Formulas
  area under parabola, *539, 545*
  area under power function, *551*
  Celsius–Fahrenheit conversion, *122, 247*
  Celsius–Kelvin conversion, *122*
  distance, *49–51*
  distance formula in 3-space, *440–441*
  Doppler effect, *359*
  Heron's formula, *378*
  Law of Cosines, *373–375*
  Law of Sines, *368–371*
  midpoint, *50–51*
  point-to-line distance, *72–74*
  quadratic, *34–35, 43*
  Snell's Law, *356*
Fractional inequality, *236–237*
Fractions, operations on, *12*
Function(s), *99–102*
  absolute-value, *108, 515*
  addition of, *110–113*
  additive inverse of, *111–112*
  area, *551–554, 556*
  composite, *122–124, 269–271*
  constant, *107, 509*
  continuous, *228*
  cosecant, *327–329*
  cosine, *313–314, 316–318, 320–322, 324–325*
  cotangent, *330–333*
  derivative, *504–507, 509–512*
  equal, *113*
  even, *117, 141, 319*
  exponential, *277–280, 282–285, 294–296*
  $g(x) = f(ax)$, *335–337*
  $g(x) = f(ax + b)$, *340–341*
  graph of, *99–101, 107–109, 110–111, 115–117, 119, 127–131, 514–517*
  greatest integer, *108, 126*
  hyperbolic, *343*
  identity, *107*
  increasing and decreasing, *256–259, 515–517, 551*
  inverse, *247–250, 251–254, 259, 261–263, 265–268, 269–271, 344–347*
  limit of, *238–239, 503, 509*
  linear, *119*
  logarithmic (L), *286–288, 290–292, 294–296, 552–554*
  maximum and minimum values of, *153–155, 230–232*
  multiplication of, *114–116*

  notation for, *100–101*
  odd, *117, 319*
  one-to-one, *251–254*
  periodic, *317, 335–337, 340–341*
  polynomial, *209–213, 215–218*
  power, *265–268, 545–549*
  quadratic, *137–139, 141–144, 153–155, 171–175*
  rational, *238–241*
  reciprocal, *107, 327–328*
  secant, *327–328*
  sequence, *101–102, 459–462, 464–467*
  signum, *108*
  sine, *313–314, 316–317, 320–322, 324–325*
  slope, *499–502, 504–505*
  squaring, *116*
  subtraction of, *112*
  tangent, *330–333*
  temperature conversions, *122*
  trigonometric, *356–358,* see also **Cosine function; Sine function; Tangent function**
  vertical-line test, *100*
  wrapping, *309–311*
  zeros of, *104–105, 220–223, 225–227, 228–229*
Fundamental Theorem of Algebra, *213, 479*
Fundamental Theorem of Calculus, *556–558*

Galois, Evariste, *272*
Generator (line), *170*
Geometric mean, *149*
Geometric sequence, *464–467.* See also **Sequence**
Golden Mean, *488*
Graph(s)
  axis of symmetry, *141–144*
  of cosecant function, *327–329*
  of cosine function, *316–317, 320–322, 324–325*
  of cotangent function, *322*
  critical points on, *515*
  of exponential function, *277–278, 284–285*
  of function, *99–101, 107–109, 110–111, 115–117, 119, 127–131*
  of $g(x) = f(ax)$, *335–337*
  of $g(x) = f(ax + b)$, *340–341*
  of increasing and decreasing functions, *256–259*

  of inverse function, *250, 253–254, 261–263, 266–267*
  line, *53–56, 59–60, 66–70*
  of logarithmic function, *288, 291*
  of polar equation, *381–382*
  of polynomial function, *212, 220–221, 228*
  of quadratic function, *141–144, 153–155, 171–175*
  of quadratic inequalities, *150–151*
  of rational function, *238–241*
  of relations, *95–97, 99–101*
  relative maximum point on, *514*
  relative minimum point on, *514*
  of secant function, *327–328*
  of sine function, *316–317, 320, 322, 324–325*
  of systems of equations, *66–70*
  of systems of inequalities, *77–80, 81–84*
  of tangent function, *331–332*
  of tangents to curves, *499–502, 504–507*
  vertex on, *141–144*
"Greater than," *15*
Greatest integer function, *108, 126*
Group, *272–273*

Half-argument identities, *409–410*
Half-life (of isotope), *300–301*
Half-open interval, *22*
Half-plane, *77*
Harmonic series, *492*
Heron's Formula, *378*
Horizontal-line test, *253–254*
Hyperbola, *170, 191–195*
  asymptotes of, *193*
  auxiliary rectangle, *193*
  center of, *192*
  equation of, *191–192*
  translation of axes, *197–200*
  vertices of, *192*
Hyperbolic functions, *343*

i, *40–44*
Identities, *391–393*
  addition and subtraction, *398–401*

Index **595**

double-argument, *405–407*
half-argument, *409–410*
Pythagorean, *314*
Identity axiom, *5*
Identity function, *107*
Identity matrix, *454*
Imaginary axis, *416*
Increasing function, *256–259, 515–517, 551*
Induction hypothesis, *474*
Inequalities, *15–20*
  involving circular functions, *321–322, 414*
  compound, *22*
  exponential, *279, 295*
  fractional, *236–237*
  linear, *19*
  polynomial, *234–237*
  problem solving and, *17–18, 79–80, 86–88*
  quadratic, *146–148, 150–151*
  systems of, *77–80, 81–84*
  in two variables, *77–80, 81–84*
Infinite series, *490–494*
Integers, *1*
Integral calculus, *558*
Intermediate-value property, *229*
Interval, *22–23, 257*
Inverse
  additive, *6*
  multiplicative, *6*
Inverse axiom, *5*
Inverse cosine function, *346*
Inverse function(s), *247–250, 251–254, 259, 261–263, 265–268, 269–271, 344–347*
  graph of, *250, 253–254, 261–263, 266–267, 271*
Inverse sine function, *344–345*
Inverse tangent function, *347*
Inverse variation, *120*
Irrational exponents, *282–283*
Irrational numbers, *2, 295, 489, 490*
Irrational zeros, *228–229*

Latus rectum
  of ellipse, *190*
  of parabola, *176*
Law of Cosines, *373–375*
  Pythagorean Theorem and, *374*
Law of Sines, *368–371*
Leading coefficient, *209*
Least upper bound, *24*

Length (of vector), *425*
Leonardo of Pisa (Fibonacci), *459*
"Less than," *15*
***L* function,** *552–554*
Limit (of function), *238–239, 503, 509*
Limits, *483–487*
Line(s)
  of best fit, *159–162*
  parallel, *58–59*
  perpendicular, *59–60*
  of reflection, *141*
  slope of, *54–56*
  in 3-space, *442–445*
  x-intercept of, *56*
  y-intercept of, *56*
Linear equations, *53–56, 66–70*
  graph of, *53–56, 59–60, 66–70*
Linear functions, *119*
Linear inequality, *19*
Linear programming, *81–84*
  feasible region, *80, 82–84*
  lines of constant value, *83*
  problem solving and, *79–80, 86–88*
Linear speed, *354–355*
Locus, *167–168, 171–172, 186, 191, 362*
Logarithm(s), *see also* **Logarithmic functions**
  common, *287*
  natural, *295, 553*
  notation for, *287, 288, 295*
  properties of, *290–292*
Logarithmic equation(s), *287, 291–292, 294–296*
Logarithmic function(s), *286–288, 290–292, 294–296, 553–554*
  graph of, *288, 291*
Lower bound, *24*

Magnitude (of vector), *425*
Major axis (of ellipse), *187*
Mathematical induction, *473–477*
  induction hypothesis, *474*
  principle of, *475–477*
Matrix (matrices), *454–455*
  column, *454*
  determinant of, *455*
  dimensions of, *454*
  equations and, *454–455*
  identity, *454*
  row, *454*

square, *454*
Maximum value (of function), *153–155, 230–232, 519–520*
Measure (of angles), *353–355*
Method of least squares, *159–162*
Midpoint formula, *50–51*
Minimum value (of function), *153–155, 230–232, 519–520*
Minor axis (of ellipse), *187*
Modulus, *417*
Multiplication, *3*
  of functions, *114–116*
  scalar, *432–435*
  of sequences, *485–486*
Multiplication axiom, *15*
Multiplicative inverse, *6*

Natural logarithm, *295, 553*
Natural logarithm function, *295, 552–554*
Natural numbers, *1*
  as domain of sequence, *459*
  mathematical induction and, *473–477*
Negative numbers, *15*
Normal vector, *451*
***n*th root,** *25*
Numbers, *1–3*
  complex, *40–44*
  integers, *1*
  irrational, *2, 295, 489, 490*
  natural, *1*
  ordered pairs of, *95*
  pure imaginary, *41*
  radicals, *23–26*
  rational, *1*
  real, *2, 5–7, 9–13, 15–20*
  whole, *1*

Odd function, *117*
One-to-one function, *251–254*
  horizontal-line test, *253–254*
Open interval, *22*
Operations
  binary, *3*
  on fractions, *12*
  on sets of numbers, *3, 9–13*
Order axioms, *15–17, 42*
Ordered pair(s), *95*
  argument, *95*
  Cartesian product, *95*
  value, *95*
  vector as, *425*

Ordered triples, *439, 442*
Orthogonal curves, *185*
Orthogonal vectors, *447*

Parabola(s), *170, 171–175*
  applications, *177–179*
  area under, *533–534, 538–539, 545*
  directrix of, *171*
  equation of, *171–175*
  focus of, *171*
  latus rectum of, *176*
  tangents to, *181–184*
  translation of axes, *198–200*
  vertical, *171*
Paraboloid, *441*
Paragraph proof, *16*
Parallel lines, *58–59*
Parameter, *178, 362*
Parametric equations, *178, 360–362, 434, 444*
Pascal's Triangle, *39*
Perihelion, *188*
Periodic function(s), *317, 335–337, 340–341*
Periodic function(s)
  equations of, *394–396, 404*
Perpendicular lines, *59–60*
Pi (π), *489*
Plane(s), *450–452*
  equation of, *451–452*
  in 3-space, *439–441*
Point(s)
  distance between, *49–51*
  locus of, *167–168, 362*
  in plane, see **Coordinate plane**
  position of, *360*
  in 3-space, *439–441*
Point-slope form, *54–55*
Point-to-line distance formula, *72–74*
Polar axis, *380*
Polar coordinates, *380–382, 383–385, 386*
Polar equations, *381–382, 386*
Pole (point), *380*
Polynomial function(s), *209–213, 215–218*
  calculator evaluation of, *210–213*
  coefficients of, *209*
  constant term of, *209*
  degree of, *209*
  factor theorem, *216*
  graph of, *212, 220–221, 228*
  intermediate-value property, *229*
  irrational zeros of, *228–229*
  maximum and minimum values of, *230–232*
  rational zeros of, *220–223, 225–227*
  remainder theorem, *215–216*
  synthetic substitution, *211–212, 217–218*
Polynomial inequality, *234–237*
Positive numbers, *15*
Power function, *265–268*
  area under, *545–549*
Principal nth root, *25*
Principal square root, *25*
Principle of mathematical induction, *475–477*
Problem solving, see **Applications to real world**
Properties, see also **Axioms**
  cancellation, *6*
  reflexive, *2*
  symmetric, *2*
  substitution, *2*
  transitive, *2, 15*
  of vectors, *430, 433, 443*
Pythagorean identity, *314*

Quadratic equation(s), *33–37, 63–64, 67–70.* See also **Quadratic functions**
  discriminant, *34*
  quadratic formula, *34–35, 43, 137*
  radicals in, *36–37*
  solutions, number of, *34–35*
Quadratic formula, *34–35, 43, 137*
Quadratic function(s), *137–139, 141–144, 171–175*
  graph of, *141–144, 153–155, 171–175*
  maximum and minimum value of, *153–155*
Quadratic inequalities, *146–148, 150–151*
Quadric surfaces, *441*
Quartic polynomial, *220*

Radians
  angle measure and, *353–355*
  calculator and, *314*
Radicals, *23–26*
  principal roots, *25*
  in quadratic equations, *36–37*
Range, *95, 99*
Rational functions, *238–241*
Rationalizing denominator, *26*
Rational numbers, *1*
Rational zeros, *220–223, 225–227*
Rational-zero Theorem, *222–223*
Real axis, *416*
Real numbers, *2, 5–7, 9–13, 15–20*
  division of, *11*
  subtraction of, *9–10*
  system of, *5*
Reciprocal function, *107, 327–328*
Reduced polynomial, *217*
Reflection, line of, *141*
Reflexive property, *2*
Region (in coordinate plane), *96, 531–534.*
Relation(s), *95–97.*
  See also **Functions**
  domain of, *95, 99*
  graph of, *95–97, 99–101*
  range of, *95, 99*
Relative maximum point, *514*
Relative minimum point, *514*
Remainder Theorem, *215–216*
Repeating decimal, *1–2*
Right triangles, trigonometric ratios and, *364–366*
Rise (slope of line), *54*
Roots
  of complex numbers, *479–482*
  nth, *25*
  square, *23–25*
Row vector, *454*
Run (slope of line), *54*

Scalar, *432*
Scalar multiple, *432*
Scalar multiplication, *432–435*
Secant function, *327–328*
  graph of, *327–328*
Segment, slope of, *53–54*
Sequence(s), *101–102, 459–462, 464–467, 468–471*
  arithmetic, *460–462*
  common difference, *461*
  common ratio, *465*
  convergent, *484–487*
  divergent, *484–485, 487*
  Fibonacci, *459–460*
  limit of, *483–487*
  multiplication of, *485–486*
  notation for, *460, 468–471, 483*
  nth term of, *459–460*
  sigma notation, *468–471*

sum of, *485–486*
sum of first *n* terms, *461, 465–466*
terms of, *101–102, 459–460*
Series
 convergent, *492*
 divergent, *492*
 harmonic, *492*
 infinite, *490–494*
 partial sum, *490*
Set(s), *1–3*. See also **Numbers**
 bounded, *23–24*
 Cartesian product, *95*
 closed, *3*
 element of, *2–3*
 operations on, *3, 9–13*
 of ordered pairs, see also
  Functions; Relations
 subset, *2*
Sigma notation, *468–471*
 dummy variable, *469*
Signum function, *108*
Sine function, *313–314, 316–317, 320–322, 324–325*
 graph of, *316–317, 320, 322, 324–325*
Sines, Law of, *368–371*
Slope
 of horizontal segment, *53*
 of line, *54–56*
 of nonvertical segment, *53–54*
 of vertical segment, *53*
Slope function, *499–502, 504–505*
Slope-intercept form, *56*
Snell's Law, *356*
Solution, *18*
Solution set, *18*
Space, see **3-space**
Specials, *39, 76, 126, 158, 170, 243, 272–273, 306, 343, 386, 404, 454, 489, 503, 536–537*
 Analytic Geometry—The Conic Sections, *170*
 An Application, *76*
 An Application to Physics, *158*
 The Binomial Theorem, *39*
 Carbon Dating, *306*
 Computer Approximations, *536–537*
 Conic Sections in Polar Coordinates, *386*
 Groups, *272–273*
 Hyperbolic Functions, *343*
 int and the Computer, *126*
 Interval Bisection, *243*

The Limit of a Function, *503*
 Matrices, *454–455*
 The Number π, *489*
 Wave Addition, *404*
Sphere, *441*
Spiral, *382*
Square matrix, *454*
Square root, *23–25*
Square-root function, *266*
Squaring function, *116*
Subscripts, *102*
Subset, *2*
Substitution property, *2*
Subtraction, *9–10*
 of functions, *112*
 of vectors, *430–431*
Surfaces (in 3-space), *441*
Symmetric property, *2*
Symmetry, axis of, *141–144, 171, 186–187*
Synthetic division, *217–218*
Synthetic substitution, *211–212, 217–218*
Systems of equations, *66–70*
Systems of inequalities, *77–80, 81–84*

Tangent
 to circle, *499*
 to curve, *499–502, 504–507*
 to parabola, *181–184, 499*
Tangent function, *330–333*
 graph of, *331–332*
Terms (of sequence), *101–102*
Tests, see **Chapter Tests**
3-space, *439–441*
 lines in, *442–445*
 vectors in, *442–445*
Transitive axiom (order), *15*
Transitive property, *2, 15*
Transverse axis (hyperbola), *192*
Triangle(s)
 area of, *377–378*
 congruence of, *374*
 right, *364–366*
Trichotomy axiom, *15*
Trigonometric functions, *356–358*. See also **Cosine function; Sine function; Tangent function**
 right triangles and, *364–366*

Unit circle, *309, 330*
Unit vector, *433*
Upper bound, *23*

Value (in ordered pair), *95*
Variable, *18*
Variation, *119–120*
 constant of, *119*
 direct, *119*
 inverse, *120*
Vector(s), *425–427*
 addition of, *428–430*
 algebraic properties of, *430, 433, 443*
 angle between, *447–448*
 components of, *425*
 direction, *434*
 dot product, *447–449*
 equal, *425*
 equations involving, *434–435, 444–445*
 magnitude of, *425*
 normal, *451*
 notation for, *425*
 orthogonal, *447–449*
 in physics, *437–438*
 scalar multiple of, *432*
 scalar multiplication, *432–435*
 standard representative of, *426*
 subtraction of, *430–431*
 in 3-space, *442–445*
 unit, *433*
 zero, *430*
Vector sum, *428*
Velocity, *522–525*
Vertex (vertices)
 of cone, *170*
 of ellipse, *187*
 of graph of function, *141–144*
 of hyperbola, *192*
 of parabola, *171*
Vertical-line test, *100*

Whole numbers, *1*
Wrapping function, *309–311*

***x*-intercept**, *56*

***y*-intercept**, *56*

Zero(s)
 of a function, *104–105, 220–223, 225–227, 228–229*
 multiplication and, *9*
 multiplicity of, *220*
Zero vector, *430*